Arboriculture: Cultivation and Management of Trees, Shrubs and Vines

Arboriculture: Cultivation and Management of Trees, Shrubs and Vines

Edited by **Suede Crawford**

SYRAWOOD
PUBLISHING HOUSE
New York

Published by Syrawood Publishing House,
750 Third Avenue, 9th Floor,
New York, NY 10017, USA
www.syrawoodpublishinghouse.com

Arboriculture: Cultivation and Management of Trees, Shrubs and Vines
Edited by Suede Crawford

International Standard Book Number: 978-1-68286-289-6 (Hardback)

Printed in the United States of America.

Contents

Preface

Over the recent decade, advancements and applications have progressed exponentially. This has led to the increased interest in this field and projects are being conducted to enhance knowledge. The main objective of this book is to present some of the critical challenges and provide insights into possible solutions. This book will answer the varied questions that arise in the field and also provide an increased scope for furthering studies.

Arboriculture is a significant discipline in the current scenario. Cultivation and management of trees and shrubs is an essential part of planning of urban areas. This book on arboriculture extensively delves into concepts like selection and planting of plants, pest and pathogen control, etc. and provides a thorough understanding of the field to the readers. Different approaches, evaluations, methodologies and advanced studies on arboriculture have been included in this book. It is a vital tool for all researching or studying arboriculture or associated disciplines.

I hope that this book, with its visionary approach, will be a valuable addition and will promote interest among readers. Each of the authors has provided their extraordinary competence in their specific fields by providing different perspectives as they come from diverse nations and regions. I thank them for their contributions.

Editor

1

Genetic diversity and population structure of 151 *Cymbidium sinense* cultivars

Jiangjie Lu[1,2], Xu Hu[1], Junjun Liu[3] and Huizhong Wang[1]*

[1]Key Laboratory for Systems Biology of Medicinal and Ornamental Plants, Hangzhou Normal University, Hangzhou 310018, China.
[2]Institute of Soybean, Nanjing Agricultural University, Nanjing 210095, China.
[3]Pacific Forestry Centre, Canadian Forest Service, Natural Resources Canada, Victoria V8Z 1M5, Canada.

Cymbidium sinense cultivars exhibit an incredible range of diversity in the foliar morphology as well as the range of flower colors and shapes, which make them more popular among horticultural plants with great economic value. Understanding the genetic diversity and population structure in target populations will be of great importance for germplasm collection, breeding improvement and conversation of this species. In this study, Inter-simple sequence repeat (ISSR) markers were used to assess the genetic diversity and population structure of 151 _C. sinense_ cultivars collected from China and Japan. _C. sinense_ cultivars exhibited moderate levels of genetic diversity (H = 0.24, I = 0.38, Ppl = 100%) and genetic differentiation (_Gst_ = 0.17) with gene flow estimate _Nm_ of 2.4 among six geographical groups. With neighbor joining (NJ) analysis, 151 cultivars were clustered into seven main groups, and approximately related to their geographical distribution. Population structure analysis revealed six subpopulations, generally consistent with NJ-clustering. The results in our study suggest that different provenance collection and _in situ_ conversation are important for _C. sinense_ conservation and genetic improvement.

Key words: *Cymbidium sinense,* DNA polymorphism, genetic diversity, inter-simple sequence repeat, population structure.

INTRODUCTION

In the Orchidaceae family, many species are found with high morphological diversity. Many orchids are horticulture plants, including ancestral cultivars, adapted cultivars and natural hybrids. Orchid cultivation has become one of the favorites for the main leisure culture in China since Song Dynasty (960 to 1279 A. D.) and a Chinese book about orchid, the Pedigree of Orchid in Jin Zhang, was first published as early as 1233 A.D. (Liu et al., 2006).

Cymbidium sinense is a terrestrial orchid whose native habitat spreads from India through Thailand and into China as a lithophyte at an elevation up to 2000 m (David et al., 2007). *C. sinense* is also called Mo orchid in China ("Mo" means black, pitch-dark in Chinese), because its young leaves are light green and turn darker as they mature and the multi-flora are fragrant with dark color.

Cultivation of this species is relatively later than other orchids and the first record on *C. sinense* was about 200 years ago in China (Liu et al., 2006). *C. sinense* shows as erect plant-type with thick and beautiful flower stalks. Their blossoms are colored elegantly or gorgeously with highly ornamental value (Yan, 2001; Xu, 2002; Liu et al., 2006). These *C. sinense* cultivars exhibit an incredible range of diversity in the foliar morphology as well as the range of flower colors and shapes (Wu, 1993; Liu et al., 2006). Improvement of the cultivars with different horticulture traits in breeding needs us to understand the genetic diversity and population structure of this species. Related genetic diversity study would also help the conventional classification of *C. sinense* cultivars, which is mainly based on morphological traits of leaf and flower.

According to orchid horticultural classification (Jin and Yao, 2006), *C. sinense* cultivars are mainly grouped into

*Corresponding author. E-mail: whz62@163.com.

Figure 1. Morphological diversity of some *C. sinense* horticultural classes (Jin and Yao, 2006). A: Qihua class Heban type, B: Qihua class Meiban type, C: Qihua class Dieban type, D: Qihua class Shuixian type, E: Yeyi class, F: Sehua class Caihua type.

three classes in China (Figure 1). Qihua (flower with peculiar shape patterns), Sehua (whole flower with different brilliant colors), and Yeyi (leaves with various colors and shape patterns, Figure 1E). Based on sepal morphology, Qihua class is subdivided into Heban type (lotus-petal-like sepals, Figure 1A), Meiban type (plum-petal-like sepals, Figure 1B), Dieban type (butterfly-petal-like sepals or petals, Figure 1C), Shuixianban type (narcissus-petal-like sepals, Figure 1D). The Sehua class contains Suxin type (white labellum with a single color for other structures of the flower) and Caihua type (various brilliant colors for the whole flower, Figure 1F). The Yeyi class includes at least four types: Xianyi type (yellow or white lines/spots with inner green leaf), Aizhong type (dwarf leaf), Shuijing type (hyalo-white leaf tissue in apexes or margin, and even in the vein of leaf), and Duoyi type (the cultivars exhibiting combined morphological characteristics of multiple Yeyi types). Meanwhile, some cultivars share almost identical morphological characters with only a few differences, which usually results in confusion on the assessment of genetic resources at species level.

Characterization and assessment of plant genetic resources and diversity are essential for both germplasm collection and conversation. With the development of biotechnology, various molecular genotyping techniques have been applied in the studies of genetic diversity and (or) identification of *Cymbidium* plants. The first application of molecular polymorphism to measure *Cymbidium* genetic variation was enzyme polymorphism marker (Obara-Okeyo et al., 1998a). After that, various types of DNA markers, such as random amplified polymorphic DNA (RAPD) marker (Obara-Okeyo et al., 1998b; Wang et al., 2004; Choi et al., 2006), amplification fragment length polymorphism (AFLP) marker (Wang et al., 2004) have been explored to evaluate the genetic diversity and study the genetic relationship in this genus.

In this study, we explored the usefulness of the Inter-simple sequence repeat (ISSR) approach in study of genetic diversity and population structure of *C. sinense*. The ISSR technique, no need to define PCR primers for an individual species, easy to carry out, treated as dominant markers, provides genomic information for a range of applications including: population genetics, hybridizations and gene mapping (Wink, 2006), it has been applied widely in plants, notable in the conservation of rare species, used to estimate the genetic diversity at both plant species and population levels (Zietkiewicz et al., 1994; Treutlein et al., 2003; Gobert et al., 2006; Kothera et al., 2007). The ISSR technique was used for assessment of the genetic diversity and molecular identification of *Dendrobium* species (Wang et al., 2009a, 2009b). More recently SRR and EST-SSR markers were also reported to study diversity in the genus *Cymbidium*resources in horticulture, it is of interest to characterize genetic diversity of this species. Here, we collected 151 *C. sinense* cultivars with various horticultural types originated in China and Japan. Their genetic variation (Huang et al., 2010; Moe et al., 2010).

For sustainable management of *C. sinense* genetic was

investigated at both species and geographical group levels by ISSR markers for the research objectives:

1) To reveal ISSR fingerprinting profiles for molecular identification of *C. sinense* cultivars.
2) To assess genetic diversity and population structure of *C. sinense* cultivars using those identified ISSR markers.

MATERIALS AND METHODS

Plant materials

One hundred and fifty-one cultivars of *C. sinense* with various horticultural types were collected from 5 provinces of China and Japan (Table 1) as six geographical populations: Taiwan population (No. 1 to No. 97), Guangxi population (No. 98 to No. 104), Guangdong population (No. 105 to No. 138), Fujian population (No. 139 to No. 141), Yunnan population (No. 142 to No. 145), and Japan population (No. 146 to No. 151). These cultivars covered all the Chinese traditional horticultural classification types (Table 1).

DNA extraction and ISSR analysis

Genomic DNA was extracted from young leaves by using the cetyltrimethyl-ammonium bromide (CTAB) method (Wang et al., 2004). Concentration and purity of total genomic DNA were determined by UV spectrometer and DNA electrophoresis with 1% agarose gel respectively.

Two ISSR primer sets, UBC # 9 (University of British Columbia UBC, Biotechnology Laboratory, Vancouver Canada) and # I (Wang et al., 2009) were synthesized (Sangon Co., Ltd., China). Four genomic DNA samples were used for primer screen. Eventually, 18 ISSR primers with production of clear and reproducible bands were selected for amplification of all genomic DNA samples (Table 2).

The ISSR amplification was performed as described previously (Wang et al., 2009). PCR products were separated by electrophoresis on 1.5% agarose gel. The electrophoretic patterns of the PCR fragments were visualized with ethidium bromide staining and recorded digitally with a Gel-Doc 2000 image analysis system (Bio-Rad, Philadelphia, PA, USA). The presence of amplified bands with different intensities and locations were detected and analyzed with the software Quantity One 4.62 (BioRad, Hercules, CA, USA) ISSR is a dominant marker, amplified fragments were considered as di-allelic and bands were scored as the two alleles of a locus (1 for presence and 0 for absence) to produce a set of binary data for genetic similarity comparison among cultivars.

Data analysis

For genetic diversity analyses, POPGENE32 Version 1.32 (Yeh, 2000) was used to measure mean Nei's (1973) gene diversity (*H*), mean Shannon's (Lewontin, 1972) information index (*I*), the percentage of polymorphic loci (Ppl), observed number of alleles (Na), effective number of alleles (*Ne*) (Kimura and Crown, 1964), the total genetic diversity all over the populations (Ht), the average genetic diversity within populations (*Hs*), and differentiation (*Gst*) and gene flow estimate (*Nm*) among *C. sinense* cultivar of different geographical populations.
The UPGMA distance method was used to analysis the band data usually. However, this method does not fully take into account the evolutionary pattern and does not search for the optimal tree. Therefore, we used neighbor-joining (NJ) cluster analyses with Power Marker version 3.25 (Liu and Muse, 2005) and the tree was displayed using MEGA 4 (Tamura et al., 2007).

To infer the approximate number of genetic clusters present in the data set (*K*) and to assign individuals to these cluster, we used STRUCTURE version 2.2 (Pritchard et al., 2000; Falush et al., 2007). The optimum number of population (*K*) was selected after four independent runs of a burn-in of 500,000 iterations followed by 500,000 iteration for each value of *K* (testing from $K = 2$ to $K = 10$). A model-based clustering with distinctive allele frequencies placed individuals into *K* clusters, where *K* is chosen in advance but can be varied for independent runs of the algorithm. The most likely number of cluster (*K*) was selected by comparing the logarithmized probabilities of data [Pr(*X*|*K*] and α value for each value of *K* according to Pritchard et al. (2000).

RESULTS

High ISSR polymorphism in *C. sinense*

Eighteen ISSR primers of UBC # 9 and # I primer sets, which produce clear and reproducible bands with 4 DNA samples of *C. sinense* cultivars, were selected for the genomic DNA amplification of all 151 *C. sinense* cultivars. As showed in Table 2, 15 primers were for di-nucleotide repeats while 3 were for tri-nucleotide repeats. Of 15 di-nucleotide repeat primers, 10 were (AG or GA) repeat primers and 5 were (AC or CA) repeat primers. All of these primers were informative and they revealed polymorphic loci in *C. sinense.*

A total of 14, 478 DNA fragments with sizes ranging from 100 to 2500 bp were scored from 151 cultivars with 18 ISSR primers. An example ISSR profile is shown in Figure 2. DNA fragments with the same size from the same primer were considered as the same ISSR locus. Finally, 251 ISSR loci were revealed across 151 cultivars, with an average of 13.94 ISSR loci per primer, ranging from 9 (UBC840) to 19 loci (UBC825). Of all 18 primers used in the present study, each primer revealed 100% polymorphism of ISSR loci at the species level.

Moderate genetic differentiation among *C. sinense* geographical populations

One hundred and fifty-one cultivars originated from 6 geographical regions were found with 100% ISSR polymorphic oci at species level (Table 2). Genetic diversity was measured in six geographical populations (Table 3). Within them, the Guangdong population has the highest genetic diversity ($H = 0.2489$, $I = 0.3894$, and Ppl = 92.43%). In contrast, the Fujian population showed the lowest genetic diversity ($H = 0.1420$, $I = 0.2084$, and Ppl = 35.86%). The diversity was moderate ($H = 0.2380$, $I = 0.3780$, Ppl = 100%), for all 251 ISSR loci at species level. Coefficient of genetic differentiation (Gst) was 0.1715, with estimate of the gene flow *Nm* of 2.4149

Table 1. *Cymbidium sinence* cultivars materials used in this study.

Cultivar code	Cultivar name	Origin	Traditional classification	Simulation population
1	Shi Ba Jiao	Taiwan	Sehua	V
2	Wen Shan Xian Die	Taiwan	Qihua (Dieban)	II
3	Wen Shan Qi Die	Taiwan	Qihua (Dieban)	IV
4	Jin Ying	Taiwan	Qihua (Shuixian)	IV
5	Jin Niao Huang Zhua	Taiwan	Sehua	V
6	Liu Xian Nv	Taiwan	Qihua (Shuixian)	IV
7	Feng Huang Zhua	Taiwan	Sehua	V
8	Bi Lv Yi	Taiwan	Sehua	IV
9	Fu Ji Cui	Taiwan	Yeyi	IV
10	Yu Fei Guan	Taiwan	Yeyi	IV
11	Xin Pu Wang Yue	Taiwan	Qihua (Heban)	VI
12	Huang Yu	Taiwan	Sehua	II
13	Yu Shi Zi	Taiwan	Sehua	IV
14	Shi Men Zhua	Taiwan	Yeyi	V
15	Da Shi Men	Taiwan	Yeyi	IV
16	Bao Dao Qi Hua	Taiwan	Sehua	V
17	Wen Shan Long	Taiwan	Yeyi	IV
18	Hong Yu	Taiwan	Sehua	V
19	Wan Dai Fu Zhua	Taiwan	Yeyi	IV
20	Da Mo Shi Ren Gong Yi	Taiwan	Qihua (Heban)	V
21	Long Feng Guan	Taiwan	Yeyi	III
22	Tian Luo Tou	Taiwan	Qihua (Heban)	V
23	Bi Lv	Taiwan	Sehua	IV
24	Wen Shan Jia Long	Taiwan	Sehua (Suxin)	V
25	Yin Long	Taiwan	Yeyi	II
26	Huang Guan Jin Die	Taiwan	Qihua (Dieban)	V
27	Dai Dai Fu	Taiwan	Yeyi	IV
28	Guo Xiang Mu Dan	Taiwan	Qihua (Meiban)	V
29	Cai Long	Taiwan	Qihua (Meiban)	I
30	Xin Niang	Taiwan	Sehua	V
31	Bai Weng	Taiwan	Qihua (Dieban)	II
32	Da Mo	Taiwan	Yeyi	II
33	Hua Guang Die	Taiwan	Qihua (Dieban)	II
34	Wu Cai Long	Taiwan	Qihua (Meiban)	IV
35	Tai Yang	Taiwan	Yeyi	III
36	Shui Jing Long Liu	Taiwan	Sehua (Suxin)	II
37	Fu Lu Shou	Taiwan	Qihua (Shuixian)	I
38	Xue Bai Zhua	Taiwan	Yeyi	II
39	Jin Niao	Taiwan	Yeyi	IV
40	Tao Ji Zhua	Taiwan	Yeyi	IV
41	Tian Long	Taiwan	Qihua (Heban)	VI
42	Yu Qi Lin	Taiwan	Qihua (Meiban)	IV
43	Lv Yun	Taiwan	Qihua (Heban)	VI
44	Ri Xiang	Taiwan	Yeyi	IV
45	Feng Huang	Taiwan	Sehua	IV
46	Niao Jin Zhi Bao	Taiwan	Sehua	IV
47	Da Mo Guan	Taiwan	Yeyi	II
48	Jin Ru Yi	Taiwan	Qihua (Meiban)	V
49	Gui Fu Ren	Taiwan	Sehua	III
50	Fu Cui	Taiwan	Qihua (Heban)	II

Table 1. Contd.

51	Rui Bao	Taiwan	Sehua (Suxin)	V
52	Chang Le Shou	Taiwan	Yeyi	IV
53	Hua guang Die Zhua	Taiwan	Sehua (Suxin)	V
54	Da Tun Qi Lin Lv Mao	Taiwan	Yeyi	II
55	Rui Hua	Taiwan	Qihua (Meiban)	V
56	Wen Han Qi Die Lv Zhua	Taiwan	Sehua	II
57	Da Tun Qi Lin Lv Zhua	Taiwan	Sehua	V
58	Hao Guang Si She	Taiwan	Yeyi	II
59	Da Mo Liu He	Taiwan	Yeyi	II
60	Da Tun Qi Lin	Taiwan	Sehua	VI
61	Yu Guan Yin	Taiwan	Qihua (Dieban)	II
62	Wan Dai Fu Zhua	Taiwan	Yeyi	II
63	Da Mo Zhong Tou Yi	Taiwan	Yeyi	II
64	San Xiong Hong Die	Taiwan	Qihua (Dieban)	III
65	Hong Ju	Taiwan	Sehua	II
66	Yan Ji	Taiwan	Qihu (Heban)	VI
67	Bai Yu	Taiwan	Qihua (Shuixian)	IV
68	Hua Wang Jin	Taiwan	Yeyi	IV
69	Long Feng Zhua	Taiwan	Yeyi	IV
70	Nian Nian Fu	Taiwan	Yeyi	I
71	Yu Jin Jiao	Taiwan	Yeyi	II
72	Yu Fei	Taiwan	Qihua (Heban)	V
73	Yang Ming Jin	Taiwan	Yeyi	II
74	Lv Bao	Taiwan	Yeyi	V
75	Wang Dai Fu	Taiwan	Yeyi	II
76	Lan Yang Qi Die	Taiwan	Qihua (Dieban)	V
77	Ji Fu Long Mei	Taiwan	Qihua (Dieban)	IV
78	Jin Bao Bei	Taiwan	Sehua (Suxin)	II
79	Shuang Mei Ren	Taiwan	Yeyi	II
80	Tian Tang Niao	Taiwan	Qihua (Meiban)	V
81	Long Feng	Taiwan	Yeyi	IV
82	Feng Guan Shui Jin	Taiwan	Yeyi	I
83	Hong Bao Shi	Taiwan	Qihua (Meiban)	V
84	Liu Feng	Taiwan	Qihua (Meiban)	II
85	Xin Gao Shan	Taiwan	Yeyi	I
86	Huan Dao Zhua	Taiwan	Yeyi	II
87	Shuang Long Guan	Taiwan	Yeyi	I
88	Bao Shan Zhua	Taiwan	Yeyi	II
89	Fu Gui	Taiwan	Qihua (Meiban)	V
90	Tai Zhong Wang Yue	Taiwan	Yeyi	II
91	Ai Zhi Bao	Taiwan	Yeyi	V
92	Tao Ji	Taiwan	Sehua	IV
93	La Ba Ji	Taiwan	Sehua	I
94	Yu Zhou Wang	Taiwan	Yeyi	I
95	Shi Ba Jiao Mei Zhua	Taiwan	Yeyi	I
96	Xu Huang	Taiwan	Yeyi	II
97	Xu Huang Zhua	Taiwan	Yeyi	VI
98	Xue Yu	Guangxi	Yeyi	III
99	Wu Cai Fei Long	Guangxi	Qihua (Meiban)	III
100	Feng Lai Chao	Guangxi	Yeyi	III
101	Jin Ying	Guangxi	Qihua (Shuixian)	III
102	Gong Zhou Long	Guangxi	Sehua (Suxin)	III

Table 1. Contd.

103	Chao Yang He	Guangxi	Qihua (Heban)	I
104	Yi Die	Guangxi	Qihua (Dieban)	IV
105	Bai Yu Su Jin	Guangdong	Yeyi	I
106	Tian Fu	Guangdong	Qihua (Shuixian)	II
107	Ji Guan Shui Jing	Guangdong	Yeyi	V
108	Yuan Dong He	Guangdong	Qihua (Heban)	VI
109	Jin Feng Die	Guangdong	Qihua (Dieban)	I
110	Gui He	Guangdong	Qihua (Heban)	III
111	Shan She Qi Hua	Guangdong	Qihua (Meiban)	VI
112	Cai Xin Jin Si Ma Wei	Guangdong	Yeyi	I
113	Qing Yu	Guangdong	Sehua	I
114	Jin Hua Shan	Guangdong	Yeyi	III
115	Zhu Hai Yu Nv	Guangdong	Qihua (Dieban)	III
116	Nan Guo Shui Xian	Guangdong	Qihua (Shuixian)	VI
117	Dong Fang Ming Zhu	Guangdong	Qihua (Shuixian)	VI
118	Shi Zi Tou	Guangdong	Yeyi	V
119	Huang Zhong Wang	Guangdong	Sehua	I
120	Yi Pin He	Guangdong	Sehua	VI
121	Yin Feng Bai Hua	Guangdong	Sehua	VI
122	Tang Shan Hu	Guangdong	Yeyi	VI
123	Jin Yao Shi	Guangdong	Yeyi	IV
124	Shan Dian	Guangdong	Yeyi	V
125	Nan Hai Mei	Guangdong	Qihua (Meiban)	III
126	Shen Zhou Qi	Guangdong	Qihua (Shuixian)	III
127	Yin Feng	Guangdong	Yeyi	II
128	Fu Se Hua	Guangdong	Sehua (Suxin)	III
129	Hu Po Jin Long	Guangdong	Yeyi	III
130	Hu Bi Jin Long	Guangdong	Yeyi	III
131	Huang Jin Guan	Guangdong	Yeyi	II
132	Huang Fei	Guangdong	Sehua	III
133	Bai Tian E	Guangdong	Yeyi	III
134	Jin Bi	Guangdong	Sehua	I
135	Xin Pin Die	Guangdong	Qihua (Dieban)	II
136	Yin Hua	Guangdong	Yeyi	III
137	Tian Fu Shou	Guangdong	Sehua	III
138	Shui Jing Long	Guangdong	Yeyi	III
139	Tian Sheng	Fujian	Qihua (Shuixian)	III
140	He Zhi Hua	Fujian	Sehua	III
141	Ying Zui Shui Jing	Fujian	Yeyi	III
142	Qing Long Jian	Yunnan	Qihua (Shuixian)	VI
143	Huang Chun Shui Jing	Yunnan	Yeyi	III
144	Feng Die	Yunnan	Qihua (Dieban)	VI
145	Bai Mei	Yunnan	Qihua (Meiban)	VI
146	Ai Guo	Japan	Yeyi	V
147	Da Xun Zhua	Japan	Sehua (Suxin)	II
148	Sheng Ji Guang	Japan	Yeyi	VI
149	Jin Yu Man Tang	Japan	Qihua (Dieban)	VI
150	Huang Jin Yang Lao	Japan	Qihua (Shuixian)	IV
151	Ri Yu	Japan	Sehua	I

among six populations. This means that the main genetic diversity (82.8%) is within the population and only

Table 2. Polymorphism of ISSR markers in *Cymbidium sinense* cultivars.

Primer name	Primer sequence [a]	Total bands	Ratio of polymorphic ISSR loci [b] (100%)	Fragment size range (bp)
UBC807	$(AG)_8T$	753	14/14	150-1500
UBC808	(AG)8C	1131	16/16	100-1600
UBC809	(AG)8G	1040	16/16	200-1400
UBC811	(GA)8C	819	17/17	150-1750
UBC825	$(AC)_8T$	789	19/19	180-2000
UBC826	$(AC)_8C$	549	14/14	200-1400
UBC827	$(AC)_8G$	775	16/16	250-2100
UBC834	(AG)8Y*T	907	14/14	250-1100
UBC835	(AG)8Y*C	676	10/10	200-1300
UBC836	(AG)8Y*A	882	18/18	180-1700
UBC840	(GA)8Y*T	415	9/9	150-2000
UBC864	$(ATG)_6$	647	11/11	160-1350
UBC866	(CTC)6	905	9/9	100-1750
I2	A $(CA)_8T$	643	11/11	220-1900
I4	$A(CA)_8G$	807	14/14	350-2000
I34	(AG)8AA	1018	12/12	160-1550
I39	$(ACG)_6$	823	15/15	150-1100
I65	(AG)8CC	899	16/16	250-1300
Total		14 478	251/251	100-2100

[a)] Y=C + T, [b)] Number of polymorphic fragment in total number of fragments amplified by each primer

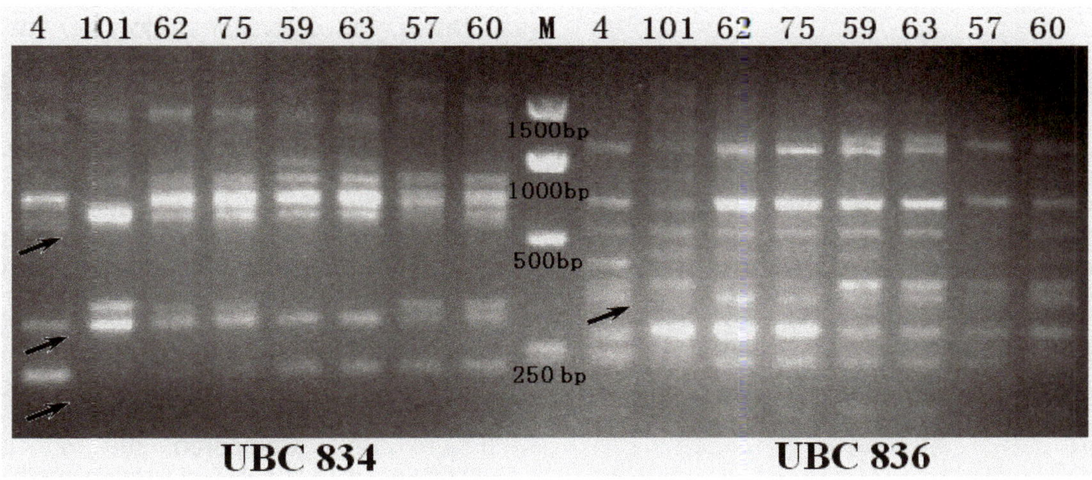

Figure 2. Example of the ISSR profiles in the *C. sinense* cultivars. Electrophoresis patterns from two ISSR primers (UBC 834 and UBC 836) are shown for identification of cultivars No. 4 and No. 101 with the same horticultural name "Jin Ying". Three pairs of cultivars: No. 57 and No. 60, No. 59 and No. 63 or No. 62 and No. 75, were very similar with each other respectively. Arrows indicate the PCR bands amplified differentially between cultivars No. 4 and No. 101. The same ISRR band patterns are observed in three sets of close cultivars.

moderate genetic differentiation (17.2%) occurs among these populations. *C. sinense* was revealed with the total genetic diversity all over the populations (Ht = 0.2336) and the average genetic diversity within populations (Hs = 0.1935).

NJ clustering analysis in *C. sinense*

The genetic similarity of 151 cultivars in this study was calculated by means of Dice's coefficient of total 14,478 DNA fragments from 251 ISSR loci. The lowest pairwise

Table 3. Genetic diversity within morphological populations of *Cymbidium sinense* cultivars based on ISSR data.

Horticultural population	Cultivar number	Na	Ne	*H*	*I*	Npl	Ppl (%)
Taiwan	97	1.97±0.17	1.36±0.32	0.22±0.17	0.36±0.23	244	97.21
Guangxi	7	1.64±0.48	1.32±0.34	0.20±0.18	0.30±0.26	161	64.14
Guangdong	34	1.92±0.27	1.40±0.33	0.25±0.16	0.39±0.22	232	92.43
Fujian	3	1.36±0.48	1.25±0.37	0.14±0.20	0.21±0.29	90	35.86
Yunnan	4	1.44±0.50	1.27±0.36	0.16±0.19	0.24±0.28	111	44.22
Japan	6	1.58±0.49	1.32±0.37	0.19±0.19	0.29±0.28	146	58.17
Species level	151	2.00±0.00	1.38±0.31	0.24±0.16	0.38±0.21	251	100.00
Group level	4-97		164	65.34			

* Na = Observed number of alleles; Ne = Effective number of alleles (Kimura and Crown1964); *H*= Nei's (1973) gene diversity; *I* = Shannon's Information index (Lewontin 1972); Npl = The number of polymorphic loci; Ppl = The percentage of polymorphic loci.

genetic similarity (39.3%) was between cultivars No. 29 and No. 135, and highest (86%) between cultivar No. 59 and No. 66 among the 151 cultivars. The dendrogram tree (Figure 3A), constructed by NJ analysis with the Nei's similarity among ISSR loci, showed that 151 *C. sinense* cultivars could be divided into two major clusters upon seven groups. Group 2, 4, 5 and 7 contained 80 of 97 cultivars originated from Taiwan in one cluster. Group 1, 3 and 6 contained 30 of 34 cultivars originated from Guangdong in the other cluster. The cultivars originated from Guangxi, Fujian, Yunnan and Japan were not clustered in an independent group, but scattering in others (Figure 3A).

Bayesian structure analysis in *C. sinense*

The model-based simulation of population structure using ISSRs showed that likelihood was maximized and minimized when the number of populations was set at six, suggesting that these cultivars can be grouped into six subpopulation, inferred from the model, showed as POP 1, POP 2, POP 3, POP 4, POP 5 and POP 6, respectively (Figure 4, Tables 1 and 4). The genetic structure of *C. sinense* is obviously complex.

DISCUSSION

This study is aimed to measure molecular biodiversity and investigate genetic structure of 151 *C. sinense* cultivars using ISSR markers. With advantages of low-cost and high efficiency, the PCR-based ISSR genome fingerprinting technique has gradually become a common molecular tool for genetic variation assessment among plant populations originated in different geographical sites since its development in1994 (Zietkiewicz et al., 1994; Fang and Roose, 1997; Raina, 2001; Pradeep et al., 2002). In view of the large and increasing numbers of both cultivars and wild exotic germplasm available for collection and *in situ* conversation, it is necessary for us to evaluate genetic diversity and population structure among various populations from different geographical origins for sustainable management of the orchid species *C. sinense*. Our investigation findings on genetic diversity, population structure and gene-flow among 151 cultivars would provide valuable information for horticultural practice in *C. sinense*. The present study indicates that ISSR technique is useful in *C. sinense*, as 18 primers revealed 251 ISSR loci with an average of 13.94 loci per primer across 151 cultivars (Table 2). We found 35.86 to 97.21% ISSR polymorphisms inside geographical populations (Table 3), suggesting a considerably high level of genomic polymorphism within these populations.

Due to its cultivation history for many centuries in China, *C. sinense* has been divided into numerous horticultural classes and types. Largely for their horticultural value, some *C. sinense* cultivars were transplanted from place to place in the history, which result in the confusion and difficulty to trace their evolutionary origins. In our study, NJ clustering and structure analysis both showed some information of this, as cultivars originated from Taiwan (in Groups 2, 4, 5 and 7) and Guangdong (in Groups 1, 3 and 6) separated each other. Because of sampling difficulty, only a few cultivars were collected from Guangxi, Fujian, Yunnan and Japan, their NJ-clustering may not reveal the real relationship of these four geographical populations with other two (Figure 3A). Compared the structure population simulation with the NJ clustering, both showed similar results in general (Figures 3B, 4 and Table 4), 12 of 18 POP1 cultivars were found in Group 1; 25 of 31 POP 2 cultivars were found in Group 2; 16 of 25 POP 3 cultivars were in found Group 3; 12 of 18 POP 6 cultivars were found in Group 6; 19 of 29 cultivars in POP 5 came from Groups 5 and 7; while the POP 4 contained 6 of 8 Group 4 cultivars and others scattering in all the seven NJ groups.

Some confusion of horticultural classification also came from documentation with different cultivars given the same name or the same cultivar named differently by flower breeders and growers. Taken two cultivars in our current study for example the cultivars No. 4 (originated

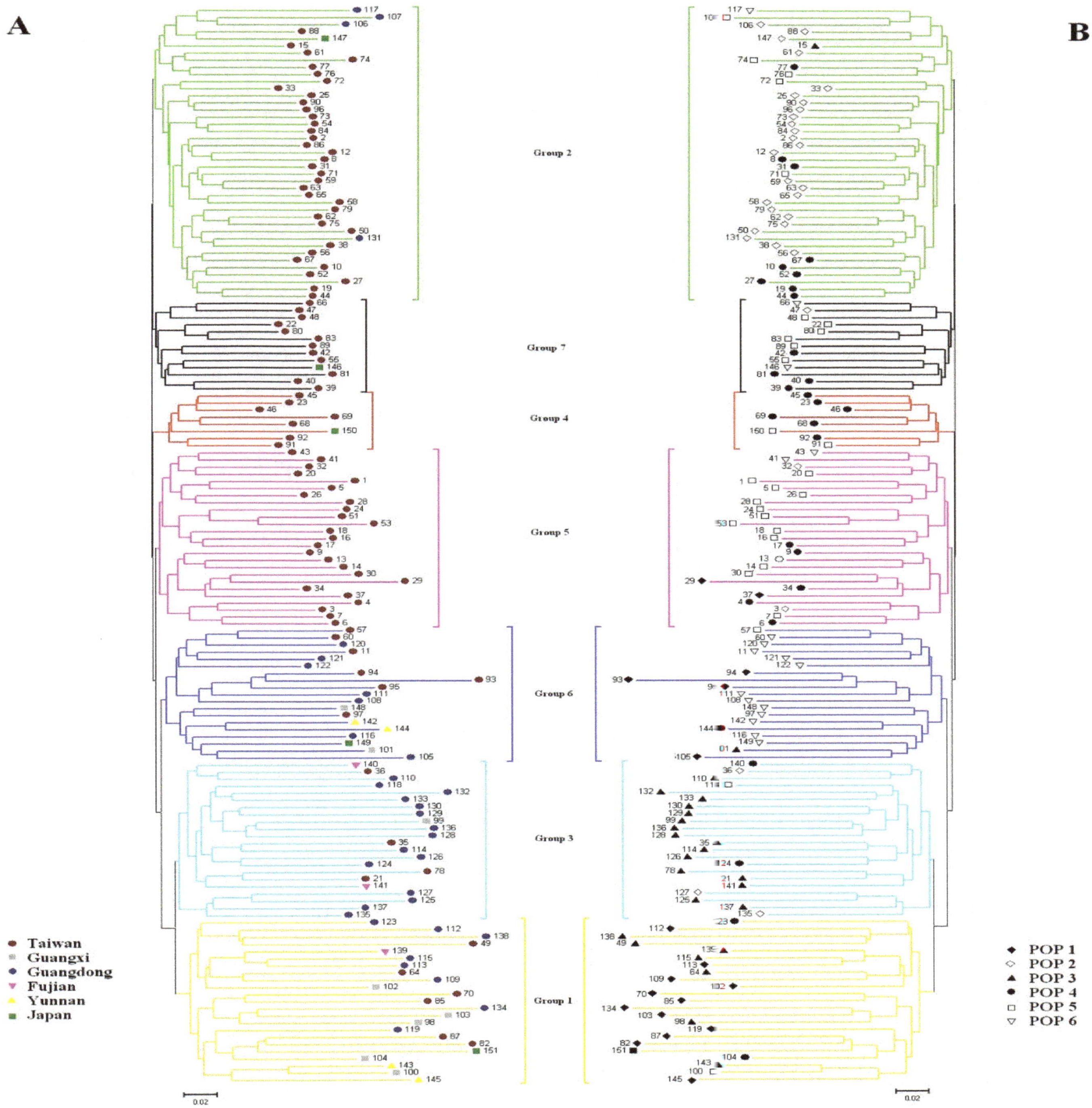

Figure 3. The comparison of simulation populations with geographical populations in a dendogram produced using NJ cluster analysis. 151 *C. sinense* cultivars were clustered into seven groups (Groups 1 to 7) by NJ analysis based on Nei's similarity of ISSR markers. Each individual cultivar was labeled according to their six geographical origins (A), or labeled according to their distribution among six simulation populations (B). Cultivars numbers were shown as in Table 1.

in Taiwan) and No. 101 (originated in Guangxi) have the same name "Jin Ying" (Table 1), but they were distinguished from each other on ISSR fingerprinting profiles (Figure 2). This suggests that these two cultivars might share certain common flower features despite their different genetic backgrounds resulting from geographical isolation.

The long-term horticultural selection of *C. sinense* has

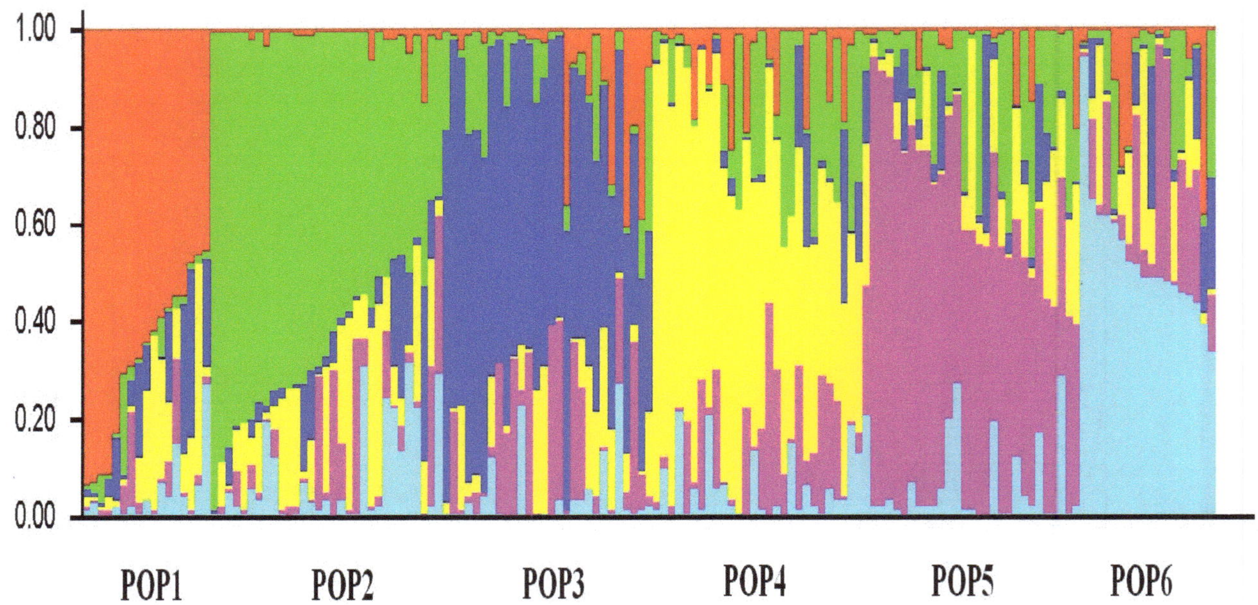

Figure 4. Population structure of 151 *C. sinense* cultivars based on 251 ISSR loci (K = 6). Each cultivar is represented by a thin vertical line, which can be partitioned into six colored segments that represent the estimated membership probabilities (Q) of the individual to the six clusters, and then all accessions were sorted by Q.

Table 4. The number of *C. sinense* cultivars in simulation structure populations (Pop 1-6) and NJ clusters (Groups 1-7).

Population	Cluster							
	Group 1	Group 2	Group 3	Group 4	Group 5	Group 6	Group 7	Total
Pop 1	**12**	0	0	0	2	4	0	18
Pop 2	0	**25**	3	0	2	0	1	31
Pop 3	7	1	**16**	0	0	1	0	25
Pop 4	2	9	2	**6**	6	1	4	30
Pop 5	1	5	1	2	**13**	1	**6**	29
Pop 6	1	1	0	0	2	**12**	2	18
Total	23	41	22	8	25	19	13	**151**

resulted in large numbers of cultivars within this orchid species. Nuclear mutations that occurred during the selection process probably lead to the polymorphisms of a set of functional genes, which may be responsible for the phenotypic variations of horticultural traits that enable us to select vigorous orchid plants with desired leaves and flower characteristics. In our collection, some cultivars were selected and bred from bud mutants of individual plant, so they shared close relationship with each other based ISSR profiling (Figure 2A). For example, our ISSR study found 86% genetic similarity between cultivars No. 59 (Da Mo Liu He) and No. 63 (Da Mo Zhong Tou Yi). Cultivars No. 59 was bred from a floral mutant of cultivar No. 63. Both No. 59 and No. 63 belonged to Yeyi class based on leaf characteristics while No. 59 could also be classified as a Heban-type cultivar when floral features are considered. Sehua-class cultivar No. 57 (Da Tun Qi Lin Lv Zhua), featured with special green petals, was cultured from a Yeyi-class cultivar No. 60 (Da Tun Qi Lin), both sharing 85.09% ISSR genetic similarity. Similarly, a mutant of the cultivar No. 75 (Wan Dai Fu) resulted in a novel cultivar No. 62 (Wan Dai Fu Zhua), both cultivars had 85.57% genetic similarity between each other. Some *C. sinense* cultivars may have closer phylogenetic relationship if they have originated directly or indirectly from mutants of the same ancestor cultivar though their morphological characters vary widely. Horticultural classification of *C. sinense* generally emphasizes only one or two characters, such as shape and color patterns of plant leaf and flower for economic purposes, and its taxonomic value is very limited. Those cultivars of the same horticultural class with similar horticultural characters were scattered throughout the NJ cluster (Figure 2A), suggesting that the ISSR marker-based genetic diversity is far beyond that underlying variations of morphological traits selected in horticultural practice. The ISSR-based phylogenetic relationship probably provides useful information on the origin and biology of *C. sinense* cultivars, but it may not reflect genetic variation of a small set of genes for the horticultural traits with cultivation interests. Future investigation with other genotyping technologies would provide further insight into evolutionary origin of the *C. sinense* cultivars and natural germplasm.

We applied the ISSR genotyping system for genetic characterization of *C. sinense* cultivars. Our results of ISSR molecular biodiversity and genetic structure among 151 cultivars are important for further discovery of genes responsible for commercially important traits such as flower colors and fragrance, and structure alterations in future research on *C. sinense*. An indirect gene flow estimate of Nm was 2.4 among 6 geographical regions, within Nm range from 1 to 4, indicating that gene flow among populations inhabiting different regions was sufficient to deter population differentiation if they were at equilibrium between migration and random genetic drift (Wright, 1931). High variation levels ($Gst = 0.1715$) within geographical groups suggest that *C. sinense* cultivars in each region are equally important for germplasm collection and genetic conservation.

ACKNOWLEDGEMENTS

This research was funded in part by the National Natural Science Foundation (No. 30670199, 30870180 and 31070298), the Natural Science Foundation of Zhejiang Province (No. Y3090426, X305692 and No. X301406), the Zhejiang Scientific and Technological Program (2008C12081, 20090233T15, and 20101032B25), the Hangzhou Scientific and Technological Program (20080432T06), and Qianjiang Scholar Program.

REFERENCES

Choi SH, Kim MJ, Lee JS, Ryu KH (2006). Genetic diversity and phylogenetic relationships among and within species of oriental *cymbidiums* based on RAPD analysis. Sci. Hort., 108(1): 79-85.

David DP, Cribb P (2007). The Genus *Cymbidium*. Royal Botanic Gardens, Kew.

Falush D, Stephens M, Pritchard JK (2007). Inference of population structure using multilocus genotype data: Dominant markers and null alleles. Mol. Ecol. Notes, 7: 574-578.

Fang DQ, Roose ML (1997). Identification of closely related citrus cultivars with inter-simple sequence repeat markers. Theor. Appl. Genet., 95(3): 408-417.

Gobert V, Moja S, Taberlet P, Wink M (2006). Heterogeneity of three molecular data partition phylogenies of mints related to *M. x piperita* (Mentha Lamiaceae). Plant Biol., 8(4): 470-485.

Huang Y, Li F, Chen K (2010). Analysis of diversity and relationships among Chinese orchid cultivars using EST-SSR markers. Biochem. Syst. Ecol., 38: 93-102.

Jin WT, Yao SP (2006). Cultivation and appreciation of noble Spring Orchid cultivars. Guangdong Science and Technology Press, Guangzhou, China.

Kimura M, Crow JF (1964). The number of alleles that can be maintained in a finite population. Genetics, 49: 725-738.

Kothera L, Richards CM, Carney SE (2007). Genetic diversity and structure in the rare Colorado endemic plant *Physaria bellii* Mulligan (Brassicaceae). Conserv. Genet., 8(5): 1043-1050.

Lewontin RC (1972). Apportionment of human diversity. Evol. Biol. 6: 381-398.

Liu ZJ, Chen XQ, Ru ZZ (2006). The genus *Cymbidium* in China. Science Press, Beijing.

Liu K, Muse SV (2007). PowerMark: Integrated analysis environment for genetic marker data. Bioinformatics, 21: 2128-2129.

Moe KT, Zhao W, Song H-S, Kim Y-H, Chung J-W, Cho Y-I, Park PH, Park H-S, Chae S-C, Park Y-J (2010). Development of SSR markers to study diversity in the genus Cymbidium. Biochem. Syst. Ecol.,. doi:10.1016/j.bse.2010.07.004.

Nei M (1973). Analysis of gene diversity in subdivided population. Proc. Natl. Acad. Sci. USA, 70: 321-323.

Obara-Okeyo P, Fujii K, Kako S (1998a). Isozyme variation in *Cymbidium* species (Orchidaceae). Sci. Hort., 33(1): 133-135.

Obara-Okeyo P, Kako S (1998b). Genetic diversity and identification of *cymbidium* cultivars as measured by random amplified polymorphic DNA (RAPD) markers. Euphytica, 99(2): 95-101.

Pradeep RM, Sarla N, Siddiq EA (2002). Inter simple sequence repeat (ISSR) polymorphism and its application in plant breeding. Euphytica 128(1): 9-17.

Pritchard JK, Stephens M, Donnelly P (2000). Inference of population structure using multilocus genotype data. Genetics, 155: 945-959.

Raina SN, Rani V, Kojima T, Ogihara Y, Singh KP, Devarumath RM (2001). RAPD and ISSR fingerprints as useful genetic markers for analysis of genetic diversity, varietal identification, and phylogenetic relationships in peanut (*Arachis hypogaea*) cultivars and wild species. Genome, 44(5): 763-772.

Tamura K, Dudley J, Nei M, Kumar S (2007). MEGA4: Molecular Evolutionary Genetics Analysis (MEGA) software version 4.0. Mol. Biol. Evol., 10.1093/molbev/msm092.

Treutlein J, SmithGF, van Wyk BE, Wink M (2003). Evidence for the polyphyly of Haworthia (Asphodelaceae Subfamily Alooideae; Asparagales) inferred from nucleotide sequences of rbcL, matK, ITS1 and genomic fingerprinting with ISSR-PCR. Plant Biol., 5(5): 513-521.

Wang HZ, Feng SG, Lu JJ, Shi NN, Liu JJ (2009a). Phylogenetic study and molecular identification of 31 *Dendrobium* species using inter-simple sequence repeat (ISSR) markers. Sci. Hort., 122(3): 440-447.

Wang H-Z, Wu Z-X, Lu J-J, Shi N-N, Zhao Y, Zhang Z-T, Liu J-J (2009b). Molecular diversity and relationships among *Cymbidium* goeringii cultivars based on inter-simple sequence repeat (ISSR) markers. Genetica, 136: 391-399.

Wang HZ, Wang YD, Zhou XY, Ying QC, Zheng KL (2004). Analysis of genetic diversity of 14 species of *Cymbidium* based on RAPDs and AFLPs. Acta Biol. Exp. Sin., 37(6): 482-486.

Wink M (2006). Use of DNA markers to study bird migration. J. Ornithol., 147: 234-244.

Wright S (1931). Evolution in Mendelian populations. Genetics, 16: 97-159.

Wu YS (1993). *Cymbidium* plants in China, 2nd edn. Chinese Forestry Press, Beijing.

Xu DS (2002). Superior *Cymbidium sinense* of China. Chinese Agriculture Press, Beijing.

Yan XL (2001). Chinese *Cymbidium sinense.* Guangdong Sci Tech Pre, Guangzhou.

Yeh FC, Yang RC, Boyle TBJ, Ye ZH, Mao JX (2000). POPGENE ver. 1.32: The user-friendly shareware for population genetic analysis. Molecular Biology and Biotechnology Centre, University of Alberta Canada.

Zietkiewicz E, Rafalski A, Labuda D (1994). Genome fingerprinting by simple sequence repeat (SSR)-anchored polymerase chain reaction amplification. Genomics, 20(2): 176-183.

2

Relation between catalase activity, salt stress and urban environments in *Citrus aurantium* L.

Khelifa, S.[1], M'Hamdi, M.[2], Rejeb, H.[3], Belbahri, L.[4]* and Souayeh, N.[5]

[1]Sylvopastoral Institute of Tabarka, B.P 345 - 8110 Tabarka, Tunisia.
[2]Department of Agronomic and Economic Sciences, Higher School of Agriculture of Kef, 7119, Kef Tunisia.
[3]Department of Biological Sciences and Plant Protection, Higher Institute of Agronomy of Chott Mariem, 4042, Sousse, Tunisia.
[4]Agronomy Department, HEPIA, School of Engineering of Lullier, 150 Route de Presinge, 1254 Jussy, Switzerland.
[5]University of Gabes, Cité Erriadh 6072, Zrig, Gabes, Tunisia.

In order to evaluate the degree of adaptation of urban *Citrus aurantium* L. trees to osmotic stress and to explore the effect of seed's origin on stress tolerance/adaptation, catalase activity levels were measured. Before carrying out the analysis, seedlings issued from high and low plant mother's vitality were grown on NaCl stressed media during 2 months. Catalase activities revealed different levels according to cities, plant mother vitality and salt dose. It increased in stressed medium for the same seedlings origin. Thus, tissues superoxide dismutase (SOD) activity seems to be initiated on plant mother trees and seedlings behaviour is correlated with seed formation environment.

Key words: Seedlings, origin, urban trees, salt stress, catalase.

INTRODUCTION

A remarkable diversity in trees' vitality is noticed between urban trees even at the level of the same row, thus demonstrating stress adaptation expressions with certain subjects rather than others (Rejeb et al., 1999). Osmotic stresses (draught, salinity and heat) are considered to be the most important abiotic factors responsible of urban trees heterogeneous vitality (Tomiczek, 2003; Ledoigt and Coudret, 1992). Stresses induce the production and the accumulation of toxic oxygen species (Foyer et al., 1994; Bowler and Fluhr, 2000) which can damage cellular components such as lipids, proteins, carbohydrates and nucleic acids (Monk et al., 1989). Cells have an enzymatic and non-enzymatic antioxidant system to neutralize these free radicals. In higher plants, three major enzymes are implicated in the AOS detoxification: Superoxide dismutase (SOD), ascorbate peroxidase (APX) and catalase (CAT) (Dodet, 1991; Willekens et al., 1995; Scandalios, 2002). Stress tolerance/adaptation seems to be correlated with stimulation of antioxidant enzymes and the enhanced ability to remove AOS and a higher concentration of CAT is noted (Bettaieb et al., 2007; Xu and Huang, 2004; Vieira Santos et al., 2001). CATs and SOD are principle enzymes which scavenge active oxygen species and avoid lipid peroxidation, cell membranes damage and chlorophyll degradation (Smirnoff, 1993; Foyer et al., 1994). CATs control of H_2O_2 level in plant cells, reduce the germination rate of seeds (Wadsworth and Scandalios, 1990) and participate to photosynthetic process (Willekens et al., 1997).

Nevertheless, plant antioxidant response is dependent on exogenous parameters such as plant development environment leading to resistance or sensitivity (Xu and Huang, 2004; Alonzo et al., 2001). According to this, and considering the fact that citrus response to stress is correlated with seed formation conditions (Moya et al., 1999; Bedhioufi et al., 2008), CAT activity was analyzed for different urban origins seedling plants and submitted to three levels of salinity. The choice of low salt doses is based on the fact that SOD activity increased if initial plant material is exposed to moderate salt stress (Viera Santos et al., 2001). The aim of this research was to learn about the degree of adaptation of seedlings collected in different urban regions in Tunisia and to search for some elements in relation with stress tolerance

*Corresponding author. E- mail: lassaad.belbahri@hesge.ch.

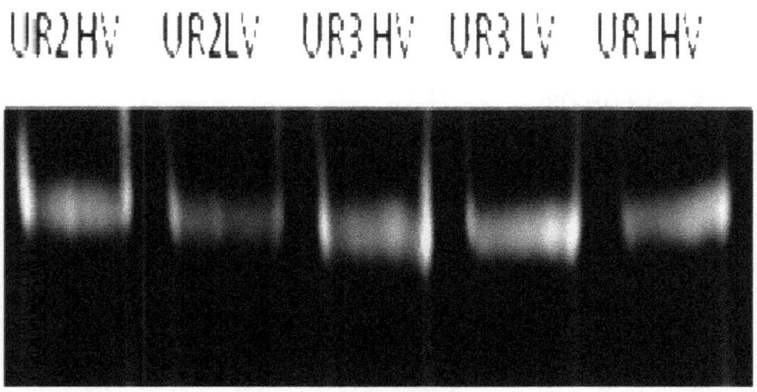

Figure 1. Zymogram analysis of catalase activity in different vitality *Citrus aurantium* urban trees (HV, high vitality ; LV, low vitality) in three urban regions (UR) of Tunisia.

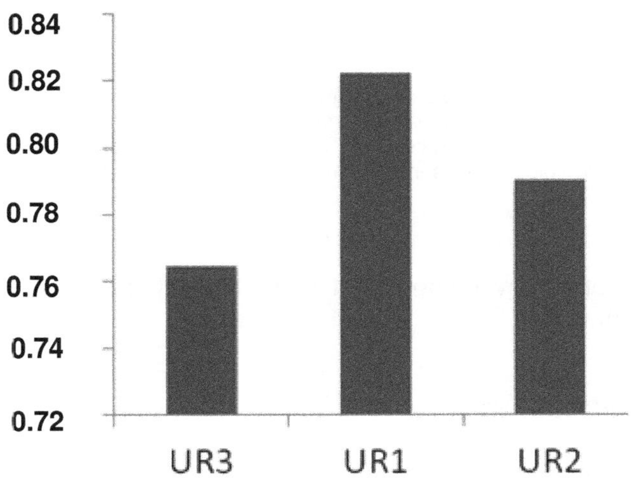

Figure 2. Fv/Fm ratio of urban Sour orange trees cultivated in three urban regions of Tunisia.

adaptation.

MATERIALS AND METHODS

Plant material and growth conditions

Seed *Citrus aurantium* L. trees showing heterogeneous vitality were selected one year in advance in different urban regions (UR) in Tunisia. Some ripe fruit were collected on hard high and low vitality *C. aurantium* L. Three cities in different bioclimatic stages were chosen for this study : UR1 a coastal humid area, UR2 a high bioclimatic semi-arid level and UR3 a typically arid region. Fruit were washed with some running water, left to soak for 10 min in bleach at 12°C, then for 15 min in a fungicide and thiophanate-methyl solution (250 g/hl). They were then left to dry in the shadow under natural temperature for 3 days and finally canned at 4°C in polyethylene bags. Seeds were grown under controlled environment with a photoperiod of 16°C and a darkness period of 8 h at 19°C. The nutrient solution was supplemented with increasing concentrations of NaCl (0 mM, 15 mM and 30 mM). Salt treatments were given for two months.

CAT analysis

Frozen fresh leaves were grounded to a fine powder in a pestle with liquid nitrogen and mixed with 1volume of extraction buffer (50 mM potassium phosphate pH 7.6, 10 mM sodium metabisulfite, 1mM ascorbic acid, 1 mM EDTA, 20% (w/v) sorbitol, 2% (w/v) polyvinylpolypyrrolidone and centrifuged at 12,000 g for 20 min at 4°C). The supernatant was collected and the protein concentration was determined using Bradford's method (1976). CAT activity was measured essentially as described by Clairbone (1985), with some modifications. The assay contained 15 mM H_2O_2 in 50 mM phosphate buffer (pH 7.0) and 50 mg of protein extract in total volume of 1 ml. CAT activity was estimated by a decrease of H_2O_2 absorbance at 240 nm and one unit of CAT was defined as the amount of enzyme dismounting 1 mM of H_2O_2 per min. Statistic analysis were carried out using statistical analysis system (SAS) program.

Catalase activity of mother trees showing high and low vitality was revealed on gel as follows ; the gel was washed three times (15 min each) with water, then incubated for 10 min in 0.88 mM H_2O_2 solution, which was rinsed again with water, and finally incubated with 1% (w/v) of ferric chloride and potassium ferricyanide solution until yellow bands appeared on a green background).

Chlorophyll fluorescence

Beside enzymatic measures, analysis of chlorophyll fluorescence was undertaken. The photosynthetic plant mother tree activity was estimated equal to the chlorophyll fluorescence of their leaves using a rotating system (FIM 1500) analytical development limited (ADC). The maximal measured fluorescence Fm, the initial fluorescence F0 and the variable fluorescence Fv (obtained by the difference between F0 and Fm of dark adapted leaves) were used to calculate the ratio Fv/Fm. These parameters were measured on a sample of 24 leaves for each urban region UR1, UR2 and UR3.

RESULTS

Staining gel showed superiority in the catalase activity of the adapted mother trees compared to sensitive ones even if it seems to be variable from one urban region to another (Figure 1). The higher CAT activity was observed for both, high and low *C. vitality*, cultivated in UR3. This tendency goes hand in hand with chlorophyll fluorescence values measured in each city. Infact, chlorophyll fluorescence values measured (Figure 2) showed that UR3 plant mother trees witnessed a low photosynthetic activity once compared to those of the UR1 and UR2 translated by a photochemical ratio (Fv, Fm) much lower. The weak values of Fv/Fm of urban sour orange photosynthetic system of the UR3 better reflect a lack of vitality as a result of a much higher level of stress. This superiority seems to be heritable and a highest CAT activity's intensity was observed for seedlings issued from trees showing high vitality compared to those issued from sensitive ones (Figure 3). The CAT activity averages for each NaCl level increased

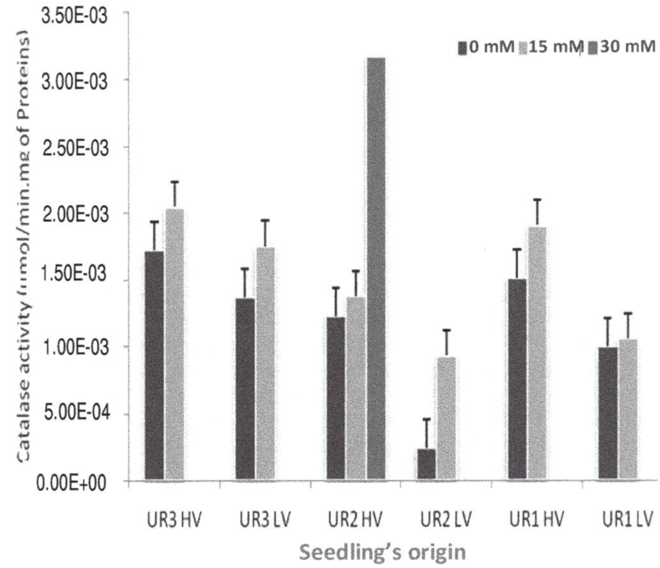

Figure 3. Catalase activity in leaves of urban *Citrus aurantium* seedlings collected in three cities of Tunisia (RU3, arid zone ; RU2, high bioclimatic semi-arid level; RU1, subhumid level) submitted to different NaCl doses (0, 15 mM and 30 mM).

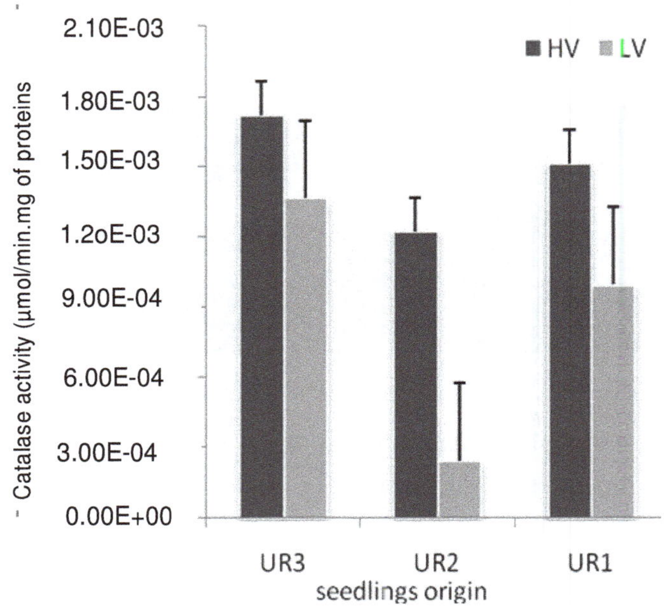

Figure 5. Catalase activity in leaves of *Citrus aurantium* control seedlings issued from high vitality plant mother trees collected in three cities of Tunisia (Gafsa : arid zone, Testour : high bioclimatic semi-arid level and Nefza : subhumid level).

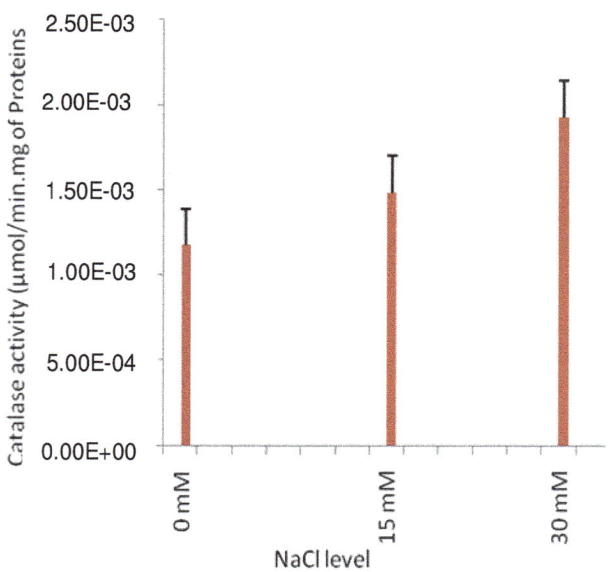

Figure 4. Averages of catalase activity in leaves of urban *Citrus aurantium* seedlings submitted to three salt doses.

significantly ($p = 0.0006$) with salt doses (Figure 4). A positive evolution of antioxidant capacity of urban seedlings for all tested NaCl doses was observed. This rule is verified for all origins but with different degrees (Figure 3) and the variable UR had hardly affect the CAT activity ($p = 0.0005$). Variability of catalase activity between cities could reveal different adaptation levels face to salt which increases from the less stressed plant mother environment UR1 to the highest one UR3 (Figure 5).

DISCUSSION

The results of this study experiments demonstrate that salt exposure induce an accumulation of CAT that play a role in the photosystem II (PSII) operation (Willekens et al., 1997). It is probable that seedlings increased CAT activity in order to neutralize H_2O_2 and thus avoid cellular damage caused by accumulation of the substrate (Willekens et al., 1994). The increase of CAT activity for all stress levels is predicable according to the fact that SOD activity increased if the initial plant material is exposed to moderate salt stress (Viera Santos et al., 2001). Despite the importance of the antioxidant systems in stress tolerance and their sensibility to exogenous conditions, no study explored the physiologic effect of the mother plant disturbance on seedling response to osmotic stress. It seems likely that seeds origin act hardly on tolerance/adaptation capacity of seedlings and involve stimulation of antioxidant enzymes. This result converge with previous researches where stress tolerance were correlated with a higher concentration of CAT and the ability to remove AOS (Bettaieb et al., 2007; Xu and Huang, 2004; Vieira Santos et al., 2001), but, it is in contradiction with others (Jiang and Huang, 2001). Effectively, plant antioxidant response is dependent on exogenous parameters, such as plant development conditions leading to resistance or sensitivity (Xu and Huang, 2004; Alonzo et al., 2001).

The increase of CAT activity for all salt doses deduce that urban environments are stressful and that seed formation conditions involved different tolerance levels to stress according to previous works (Moya et al., 1999; Bedhioufi et al., 2008). UR1 is a costal city located at a humid climatic stage whereas UR2 is located at a high bioclimatic semi-arid level and UR3 is situated in atypically arid region. The hypothesis of acclimation of seeds which confer *Citrus vitro* plants the capacity to tolerate salt stress can be deduced. In mandarin, cold acclimation induce physiological and biochemical changes that confer tolerance to photo-oxidative event (Pietrini et al., 2005). These data support the idea that salt tolerance/adaptation may be correlated with a stimulation of antioxidant enzymes (Viera et al., 2001). Effectively, seedlings collected in the UR3 (arid region) are the most tolerant to NaCl, which is shown by the higher specific activity of CAT for both high and low mother trees vitality.

Conclusion

The evaluation of the catalase activity measured in leaves of various *Cirus aurantium* urban seedlings showed different values of specific activity from a city to another. All young plants noted an increase of CAT activity when NaCl stress increase. The increase of antioxidant activity values could deduce a tolerance/adaptation embryo's cells stimulation, on plant mother, induced by urban environment. This hypothesis is supported by the fact that UR3 seedlings noted the highest antioxidant activity for both high and low mother tree vitalities. Thus, this study suppose that antioxidant enzyme system is more stimulated. Thus, these results concluded that seedlings behaviour is correlated with seed formation environment and have learnt to question if urban stressed trees could be used to produce rootstock for *Citrus orchards*.

REFERENCES

Alonso R, Elvira S, Castillo FJ, Gimeno BS (2001). Interactive effects of ozone and drought stress on pigments and activities of antioxidantive enzymes in *Pinus halepensis*. Plant, Cell Environ., 24 : 905-916.

Bedhioufi KS, Rejeb H, Souayah N (2008). Preliminary observations of *Citrus aurantium* submitted to urban stress : Behavior analysis of *vitro* plants face to salt stress. J. Food Agric. Environ., 6(2): 418-421.

Bettaieb T, Mhamdi M, Ruiz de GJI, Du JP (2007). Relation between the low temperature stress and catalase activity in gladiolus somaclones (*Gladiolus grandiflorus* Hort.). Sci. Hortic., 113: 49-51.

Bowler C, Fluhr R (2000). The role of calcium and actived oxygen as signals for controlling cross tolerance. Trends Plants Sci. Rev., 5: 241-246.

Bradford MM (1976). A rapid and sensitive method for the quantification of microgram quantities of protein utilizing the principle of protein-dye binding. Anal. Biochem., 72: 248-254.

Dodet B (1991). La chasse aux radicaux libres oxygene´s. Biofutur, May 23-34.

Foyer CH, Descourvieres P, Kunert KJ (1994). Photoxidative stress in plants. Physiol. Plant, 92: 696-717.

Jiang Y, Huang B (2001). Drought and Heat Stress Injury to Two Cool-Season Turfgrasses in Relation to Antioxidant Metabolism and Lipid Peroxidation. Crop Sci., 41: 436-442.

Ledoigt G, Coudret A (1992). Stress hydrique: étude des mécanismes moléculaires et des modification de l'expression du genome. Bulletins de la Société Botanique Française, 139 : 175-190.

Monk LS, Fagerstedt KV, Crawford RMM (1989). Oxygentoxicity and superoxide dismutase as an antioxidant in physiological stress. Physiol. Plant, 76: 456-459.

Moya JL, Primo ME, Talon M (1999). Morphological factors determining salt tolerance in Citrus seedlings the shoot to root ratio modulates passive root uptake of chloride ions and their accumulation in leaves. Plant Cell Environ., 22: 1425-1433.

Pietrini F, Chaudhuri D, Thapliyal AP, Massacci A (2005). Analysis of chlorophyll fluorescence transients in mandarin leaves during a photo-oxidative cold shock and recovery. Agric. Ecosyst. Environ., 106: 189-198.

Rejeb H, Bettaieb T, Krichen R (1999). Remarks on behaviour of trees in the main Tunisian cities. Acta Hortic., 496: 369-373.

Scandalios JG (2002). The rise of ROS. Trends Biochem. Sci., 27 : 483-486.

Smirnoff N (1993). The role of active oxygen in the response of plants to water deficit and desiccation. New Phytol., 125: 27-58.

Tomiczek C (2003). The phytomedical situation of plants under urban conditions. Second International Symposium of Plant Health in Urban Horticulture. Berlin, Germany, August 27-29.

Viera SZL, Campos A, Azevedo H, Caldeira G (2001). *In situ* and *in vitro* senescence induced by KCl stress: nutritional imbalance, lipid peroxidation, and antioxidant metabolism. J. Exp. Bot., 355(52): 351-360.

Wadsworth GL, Scandalios JG (1990). Molecular characterization of a catalase null allele at cat3 locus in maize. Genetics, 125: 867-872.

Willekens H, Chamnogpol S, Davey M, Schravdner M, Langebartels C, Van MC, Inze´ D, Van CW (1997). Catalase is a sink for H2O2 and is indispensable for stress in C3 plants. EMBO J., 16: 4806-4816.

Willekens H, Inze´ D, Van MM, Van CW (1995). Catalases in plants. Mol. Breed., 1: 207-228.

Willekens H, Langebartels C, Tire C, Van MM, Inze´ D, Van CW (1994). Differential expression of catalase genes in *Nicotiana plumbaginifolia*. Proc. Natl. Acad. Sci. U.S.A., 91: 10450-10454.

Xu Q, Huang B (2004). Antioxidant Metabolism Associated with Summer Leaf Senescence and Turf Quality: Decline for Creeping Bentgrass. Crop Sci., 44: 553-556.

3

Provenance and family variation in seed mass and fruit composition in *Sclerocarya birrea* sub-species caffra

Cliff S. Dlamini

Department of Forest and Wood Science, Faculty of Agrisciences, University of Stellenbosch, South Africa.
E-mail: csdlamini@uniswa.sz.

Sclerocarya birrea **(A. Rich.) Hochst. subsp. caffra (Sond.) Kokwaro (Marula) is one of the traditionally important indigenous multi-purpose fruits trees in Africa, which has been commercialized as its fruits and processed nuts/kennels have entered local, regional and international trade. This study was an analysis of the variation between and within provenances of S. birrea sub-species caffra with respect to seed mass of dried fresh seed, total soluble solids (total sugars) content of fresh fruits, and the vitamin C content of fresh fruits as well as family heritabilities of fruit parameters and seed mass. A nested sampling procedure was followed for where there were four areas (provenances), eight to ten families (parent trees) and fifty fruits per family. There were significant differences in provenance and family in seed mass (p=0.06 and p=0.0001) and vitamin C (p=0.001 and p=0.0001), total soluble solids lacked provenance variation (p=0.35) but had wide family variation (p=0.0001). There was no correlation between parameters except a significant correlation between rainfall intensity and total soluble solids content; lower the rainfall higher the sugar content and the higher the rainfall the lower the sugar content (p=0.06). The most variable parameter was vitamin C with the highest family heritability, however all traits had relatively high family heritabilities (above 0.5). It was concluded that the results of fruit composition and seed mass could be of remarkable importance in the identification of superior genotypes, making use of the wide family variation that exists. A study to capture a wider range of provenances and families including studying the environmental parameters of soils, rainfall and altitude for more precision.**

Key words: Traits, parameters, correlation, variance.

INTRODUCTION

Sclerocarya birrea subspecies caffra (Marula) is a widely distributed species found throughout the semi- rid, deciduous savannas of much of southern and East Africa (Peters, 1988; Dlamini, 1998; Dlamini, 2010). The fruit, seeds, leaves, wood and bark of this species have been used extensively by rural communities for many hundreds of years in most countries in which it is found. Probably one of the most important uses of this multi-purpose tree has been, and still is, the production of an alcoholic beverage (commonly known as wine or beer) from its fruit. Evidence indicates that this brew was made as long ago as 9000 BC after piles of marula nuts were found in caves in the Matobos Hills in Zimbabwe (Walker, 1989). Other important uses include the consumption of fresh fruit and kernels, the extraction of oil from the kernels for a variety of purposes from cooking to cosmetic (although the oil is not used in the study area - Bushbuckridge), the harvesting of leaves and bark for traditional medicine, and the use of wood for fuel, fencing and carving (Dlamini, 1998; Shackleton et al., 2002; Dlamini and Geldenhuys, 2009; Dlamini, 2010).

Variation in seed mass

A positive relationship between seed weight and amount of food stored in the seed storage tissue has been demonstrated in numerous plant species. A large seed normally indicates a large endorsperm, or a large embryo and cotyledons. In albuminous seed, the endosperm is the reservoir of energy necessary to facilitate the germination process and in exalbuminous seeds such as those of Fidherbia albida (Marunda, 1993; Dlamini, 1998; Dlamini, 2010). In most tree species need size has been found to be strongly correlated with germination rate and seedling size, however, the advantage could be short-lived depending on the species. Marunda (1993) reported that for Acacia nilotica, F. albida and Azandirachta indica, the ease of germination did not vary with seed weight, but early seedling growth did, more in particular height increment. Similarly, Shivkumar and Benerjee (1986) and Dlamini (1998) revealed the provenances of Acacia nilotica with bag seeds produced plants which were strong growers (tall and thick) while smaller seed germinated with difficulty and germination and early seedling growth were slower.

Seed mass is thought commonly to be an important focus of selection in life histories of plants (Harper et al., 1970; Janzen, 1977; Dlamini, 1998), because the likelihood of dispersal (Howe and Vande Kerckhove, 1980), germination (Putievsky, 1980), and survival (Black, 1957; Schaal, 1980) can all depend on seed mass or size. As a consequence, demographic patterns within plant populations may result partly from the distribution of seed masses in the seed pool. Nonetheless, little is known about variation in seed mass within and among individuals within natural plant population. Studies of seed mass have usually been based on variation between lots of a given number of seeds (often 100) rather than on individual seed masses (Cawers and Harper, 1966; Twamely, 1967; McWilliams at al., 1968; Maun and Cavers, 1971; Baker, 1972; Schimpf, 1977; Bentley et al., 1980; Thompson, 1981; Dlamini, 1998). Such analysis can be used in identifying differences in means between treatments in an experiment and between populations. The few studies of variation in individual seed mass based on large samples in natural population have shown variation up to 5.6 fold (Janzen, 1977; Janzen, 1978; Schaal, 1980; Dlamini, 1998). These values for variation provide an estimate of the range of plasticity in seed mass within a population, but are for plants with varied (uncontrolled) histories. There appears to be no estimated, however, of the magnitude and partitioning of variation in individual seed mass within and among total seed crops of uncultivated plants grown from seedlings under similar conditions, which would provide a baseline for evaluating variability in seed masses in natural plant population. A study by Thompson (1984) on the variation among individual seed masses in Lamatium grayi (Umbeliferae) under controlled conditions show much more variability in individual seed masses than indicated by the common ecological impression that seed masses tend to be fairly constant.

Chemical composition of fruit chemical composition

Information on the chemical composition of indigenous fruits in eastern and southern Africa has been scarce and limited until recently (Wehmeyer, 1966; FAO, 1983; Saka, 1994; Dlamini, 1998). A study of the chemical composition of edible indigenous fruits in Malawi was initiated by Saka (1994). The vitamin C of 28 fruits tree species was analyzed and some results were published (Saka, 1994). From these findings it was concluded that the pulp of the fruit and the seed kernel of some species contain vitamin C levels comparable to those of exotic and domesticated fruits (Williamson, 1975; Wills et al., 1983). Among the fruits analyzed Adansonia digitata, Diospyros usamberensis, Bauhinia thonningii and Vitex payos contain about 70 mg per 100 g fresh weight and are thus excel ent source of this vitamin. D usamberensis had the highest levels of vitamin C at 337 mg per 100 g. The concentration of ascorbic acid (Vitamin C) is higher when the fruit is harvested in June. Saka (1994) produces a summary of the vitamin C content of 28 different species where a wide variation in vitamin C content between species of the same genus was seen. The concentration of other important vitamins such as A and B has not been analyzed; however, a study was recently started at the University of Malawi by Saka (1994). The seed kernels or nuts of numerous fruits species in Southern Africa are edible and from an important role in the diet of moral people (Shone, 1979; A-Ogle, 1990; Taylor et al., 1995; Dlamini, 1998; Dlamini and Geldehuys, 2009; Dlamini, 2010). For example the edible nuts of such species as A. digitata, Telifania pedata, Terminalia catappa, Treculia Africana, Parkia filicoredea and Parinari curatellifolia are a good source of vegetable oil and are rich in protein. The seed kernel of A. digitata has been analysed for proximate composition by Saka (1994). Mwamba (1989, 1994) studied variation in fruit composition of Uapaca Kirkiana and effects of in situ silvicultural treatments. The parameters studies were fruit colour, fruit size and yield, skin %, seed %, pulp weight and total soluble solids concentration. The findings revealed that all other parameters are physiologically controlled and can be silviculturally altered except fruit colour.

Objectives

1. To study genetic variation in seed mass and fruit composition between and within provenances of *Sclerocarya birrea* sub-species caffra in four areas in Swaziland.
2. To study genetic variation, between and within provenances, in fruit composition;
3. To study genetic variation in seed mass between and within provenances;
4. To study correlation between traits (height and root collar diameter);
5. To study correlation between mean annual rainfall and the total soluble solids and vitamin C content;
6. To study correlation between seed mass and altitude;
7. To correlation between mass and vitamin C content;
8. To study the variance components for formulation of family heritabil ties in sugars, vitamin C and seed mass.

Table 1. ANOVA for Total Soluble Solids (TSS), Seed Mass (SM) and Vitamin C (VTC).

SOV	Mean Squares (MS)			
	Df	TSS	SM	VTC
Provenance	3	97.5 NS	93.1*	1200465.8**
Fam (Provenance)	33	86.2***	35.6***	184867.7***
Error	1812	1.39867	0.49100	7225.60
Total	1848			

NS=not significant, *=significant at 0.05 probability, **=significant at 0.001, ***=significant at 0.0001.

MATERIALS AND METHODS

Study area description

Fruits of four different provenances that were at least 100 km apart were harvested from Swaziland. Swaziland is one of the smallest countries in Africa (17 364 km^2), yet it has a remarkable geographic variability with six clearly defined physiographic zones, each with distinct vegetation and climatic features. In this study the fruit collection sampled a wide range of altitude and latitude. The altitude ranged from 340 to 500 m and the mean annual rainfall ranged from 500 to 750 mm (Dlamini, 1998).

Experimental design

The design of the experiment was a double nested sampling with trees used as repetitions for area, and fruit/seed used as repetitions for trees. This implies that parent trees were nested in provenances and individual fruits were nested in parent trees. There were four provenances from which eight to ten parent trees were identified, and in turn 49 to 51 fruits were selected per tree and analyzed for Total soluble solids and seed mass. In the case of vitamin C analysis instead of having 50 fruits analyzed there were only three composite fruits samples analyses per tree (Dlamini, 1998).

The field sampling procedure had two levels of randomization. The first level was random selection of trees in the provenances and the second level was random selection of fruits form trees. The same procedure was followed in the laboratory analysis (Dlamini, 1998).

Field and laboratory work

In each provenance eight to ten trees that were at least 150 m apart were identified and ripe fruits were picked from each. Specially made 4 m wattle rods were used to shake branches of the selected Marula trees to shed fruits on to the ground. Shaking was done all around the tree to ensure maximum coverage to obtain a relatively representative sample of the whole tree. A random sample of 360 fruits was picked from the ground out of which only a further random sample of 50 fruits was picked as a working sample (Dlamini, 1998).

Fruit collection was followed by cleaning of fruits before storage in the fridge for a least a week before analysis. Fruits were then analyzed for their composition specifically in total soluble solids (mainly sugars) and ascorbic acid (vitamin C).

Determination of total soluble solids was done through a hand refractometer (Type A.S.T. Japan 0-32).

Vitamin C was determined by the AOAC Method no. 967.21 from the Association of Analytical Chemists, as described by Kirk and Sauyer (1990). Seed mass was measured seed by seed using the "Mattler pm 300".

Statistical analysis

The analysis of variance was performed using SAS statistical software version 6.12 (SAS Institute INC, 1997). The Shapiro-Wilk statistic was performed to test for normality of the data (Shapiro and Wilk, 1965). The data was considered normally distributed as $W > 0.95$ and $p > 0.05$. All significance tests were computed at the 5% level of probability. Duncan's Multiple Range test was used to compare treatment means. Correlation analyses were undertaken following Pearson Correlation Coefficients to study the relationship between the parameters: seed mass, vitamin C content, total soluble solids, altitude and rainfall ($p=0.05$).

Variance components estimates

The variance components, following the analysis done for a two factor experiment (n observations per cell) with factor B nested in factor A, were used to estimate heritability for fruit parameters as described by Ott (1993)

Analysis of variance components for the differences between provenances and between families within provenances, assumed fixed effects for provenances and random effects for families within provenance, and used the mixed model.

RESULTS AND DISCUSSION

Total soluble solids

The analysis of variance indicated no significant differences in the total soluble solids content between provenances ($p=0.35$). Similar results were found by Mwamba (1989, 1994) in a variation in the fruits of Uapaka kirkiana. Families within provenances were significantly different ($p=0.0001$) (Table 1). Significant differences existed between individual families ignoring provenances ($p=0.0001$), in total soluble solids (Table 2). Some families within provenances were more variable in the total percentage sugar content compared to others, those of Dvokolwako showing the highest variance. Family D3 of Dvokolwako had the highest mean of 14.11% and family D6 of Dvokolwako had the lowest at 7.65%. This implies that the differences between families were not due to the localities. The average annual rainfall of the area of origin of the parent trees appeared to have had some influence on the percentage sugar content of the fruits: this may mean only that the water in the fruit may be more and the amount of sugar per fruit may be

Table 2. ANOVA for TSS, SM and VTC for 37 families ignoring provenances.

SOV	Df	MS TSS	MS SM	SOV	Df	MS VTC
Family	36	87.1***	40.4***	Family	36	269500.0***
Error	1812	1.3	0.4	Error	74	7225.6
Total	1848			Total	110	

***=significant at 0.0001 probability.

Table 3. Provenance mean Vitamin C and seed mass.

Provenance code	Mean SM (g) ±Std Error	Provenance code	Mean VTC (mg/100 ml juice) ±Std Error
Lungungu (4)	4.1±0.06 [a]	Kalanga (1)	694.2±48.33 [a]
Kalanga (1)	3.9±0.04 [ba]	Lungungu (4)	625.3±56.24 [a]
Dvokolwako (2)	3.9±0.05 [ba]	Dvokolwako (2)	539.2±53.60 [a]
eBuhleni (3)	3.1±0.04 [ba]	eBuhleni (3)	224.6±19.41 [b]

NB. Means with same letter are not significantly different according to the Duncan's new multiple range test.

less with less rain. The provenance with the lowest rainfall (Kalanga) yielded the highest total sugar content, while the area with the highest rainfall (Lungungu) yielded the lowest total sugar content. There was no clear trend between the altitude of origin of parent trees and the total sugar content, for example, in this study areas lying on the same altitude gave different total sugar contents; Kalanga and Dvokolwako are 340 m above sea level, yet they yielded different total sugar content.

Total soluble solids (sugars)

There were significant differences between provenances in the sugar content of the fruits sampled, with the highest with 11.6% and the lowest with 10.6%. However, there were significant differences (p=0.0001) between families. The highest family had 14.11%, which was highly significant from the lowest family which had 7.65%, and these coincidentally come from the same provenance (Dvokolwako). Similar results were found by Mwamba (1994) in Uapaca kirkiana. According to Barker and Taylor (1982) total sugar contents in a sample of Marula fruits ranged between 10.4% and 16.0%. Mwamba (1994) reported total sugar contents ranging from 8.8% to 21.1% in the fruits of Uapaca kirkiana. Wills, Schriven and Greenfield (1983) reported that rich vegetables and fruits should have between 8% and 18% total sugars, and this study falls within this range. As a result, *Sclerocarya birrea* may then be classified under rich fruits and vegetables. The wide variation in total sugars between families within provenances can be utilized to propagate more trees with high sugar levels within a short space of time, which would greatly benefit Marula jam producing companies.

The results show a very low coefficient of variation of 10.7% which implies that more provenances need to be sampled to increase provenances variation. Though results are not statistically different, the provenance with the highest mean was the hottest and driest while the provenance with the lowest mean was the coolest and wettest. This could mean available water and sunlight have an effect on the sugar content of the fruits. However, the "sugar levels" are a % and therefore it is a ratio and may not mean that the amount of sugar per fruit (g) is different. A future study to determine the g sugar per fruit can be used to check if it is not simply the water increases per fruit that reduces the % sugar content.

Vitamin C content (ascorbic acid)

The analysis of variance shows that there was a statistically significant differences between areas in the vitamin C content of fruits analyzed (p=0.001). Kalanga had a mean of 694.2 mg/100 ml juice, Lungungu had 625.3 mg/100 ml juice, Dvokolwako had 539.2 mg/100 ml juice all of which were significantly different from eBuhleni with 224.6 mg/100 ml juice (Table 3). Differences between families within areas were highly significant (p=0.0001) (Table 1). The vitamin C content for families within areas was the most variable parameter. Further analysis of variance to compare families on individual bases revealed highly significant statistical differences between families (p=0.0001) (Table 2). Family K8 of Kalanga had the highest vitamin C content while family E6 of eBuhleni had the lowest. The driest area had the highest vitamin C content, and it appears the solar radiation has an effect on the vitamin C content in Marula trees.

Shone (1979) and Dlamini (1998) reported an average Vitamin C of 237.6 mg/100 g and 195 mg/100 g in the Marula fruit skin and flesh, respectively.

Vitamin C content

There were significant differences between provenances in vitamin C contents of fruits (p=0.001). However, the Duncan's multiple comparisons indicated no significant differences between Kalanga, Lungungu and Dvokolwako, while eBuhleni recorded about half the amount of vitamin C content of the other three provenances. Research has shown that vitamin C content differ drastically even among fruits of the same species reported Saka (1994) on vitamin C contents of numerous indigenous tree species that were analyzed. Variation discovered in this study could be of great importance, particularly in the preliminary selection of superior trees for vitamin C. Progeny trials may be necessary to prove whether this superiority is indeed genetic or it is just due to conducive environmental parameters like sufficient rainfall, optimum temperatures and rich fertile soils. The co-efficient of variation is low at 14.09% between provenances and 16.1% between families. Though the sunlight intensity was not critically studied, the Kalanga provenance is known to be the hottest amongst the four. According to Fourie (1996) the important environmental factor in the determination of level of vitamin C is sunlight; the greater the amount of sunlight received during growth the higher the vitamin C content, and results of this study show that the highest vitamin C came from the provenance with highest sunlight intensity.

Seed size (seed mass)

The ANOVA for seed mass indicated statistically significant difference between provenances in seed mass (p=0.06, Table 3). A similar trend like in total soluble solids occurred here, Lungungu had a mean of 4.10 g, Kalanga had 3.93 g, Dvokolwako had 3.9 g, all of which were higher than eBuhleni with a mean of 3.10 g .Highly significant differences existed between families within provenances (p=0.0001). Likewise highly significant differences were found between families analyzed on individual basis (p=0.0). Family L6 of Kalanga had the highest mean of 6.50 g while family E1 of eBuhleni had the lowest of 2.35 g. This means that provenances had weak effect on the family variation. Mean annual rainfall of the provenances seems to have no influence on the seed mass, since provenances like eBuhleni and Dvokolwako with the same rainfall regime of 650 mm per annum yielded completely different seed masses. Although the provenance with the highest rainfall had the biggest seeds, amazingly the driest provenance came second.

Seed mass

Significant differences in seed mass between provenances were noted. Multiple comparison indicated that the provenances with the highest seed mass (4.10 g) was significantly higher than the provenance with the lowest seed mass (3.10 g). Kalanga and Dvokolwako had a mean of 3.9 apiece (not significantly different from neither the highest nor the lowest). Rainfall and altitude did not have much effect on seed mass, since the provenance with the highest altitude of 500 m and the highest annual rainfall of 750 mm produced the biggest and heaviest seeds, but the provenance with the lowest annual rainfall of 500 mm and the lowest altitude of 340 m did not produce the smallest but produced the second biggest/heaviest seeds. These results are different from those in Salazar (1986) in *Gliricidia sepium*, where driest provenances produced the biggest seeds and wet provenances produced smallest seeds. Variability may reflect differences in pollen that fertilized the ovules of the different flowers. What is clear at this point is that within provenance, among families, there is more variability in seed mass than would be expected from usual ecological impression that seed masses that seed masses are fairly constant (Obeid, 1967; Dlamini, 1998). The observation that the smallest seeds are almost half the size of the largest seeds in study are similar to findings made by Maun and Cavers (1971) with seeds of Silene alba, where seeds were divided into three classes using a soil sieve, and indicated that the masses of seeds with largest class are two times that of seeds in smallest class. Marunda (1993) reported that bigger seeds are the best for selection for early seedling growth and development. Consequently seeds for Lungungu provenance can be selected as root stocks and for other traits (vitamin C and sugars) then other provenances may be chosen and used as scions in the vegetative propagation of superior genotypes in tree improvement programmes.

Correlation between parameters

There was no correlation between fruit and seed parameters, between these parameters and rainfall and altitude in all cases, except for total soluble solids and rainfall, p=0.06, see Table 4. The driest area produced the highest total soluble solids levels and the wettest had the lowest total soluble solids.

Variance components estimates

The range of the family heritabilities of the traits was comparable to that found by Cotterill and Zed (1980), which ranged from 0.40 to 0.97, in family heritabilities of Pinus radiata progeny tests in South Australia. Vitamin C had the highest family heritabilities at 0.89 and total soluble solids and seed mass had 0.54 and 0.58, respectively (see Table 5). Generally one needs more than 100 families to get stable estimates of heritabilities, especially if the heritability is low (Dlamini, 1998). Analysis of variance components for total soluble solids %, seed mass and vitamin C is presented in Table 6.

Table 4. Pearson Correlation Coefficients for fruit and environmental parameters.

Variable	Vitamin C	Total Soluble Solids	Altitude	Rainfall
Seed mass	0.14035 0.4074 NS	-0.21153 0.2088 NS	0.13137 0.4383 NS	0.02805 0.8691 NS
Vitamin C	-	0.04706 0.7821 NS	0.06372 0.7079 NS	-0.14994 0.3758 NS
Total Soluble Solids	-	-	-0.24029 0.1520 NS	-0.30426 0.0671*
Altitude	-	-	-	0.80996 0.0001***
Rainfall	-	-	-	-

NS= not significant, *= significant at 0.05 probability and ***= significant at 0.0001

Table 5. Family heritabilities for TSS, VTC and SM.

Trait	Family heritability
TSS	0.54
VTC	0.89
SM	0.58

Table 6. Analysis of Variance Components Estimates (VCE) for Total Soluble Solids (TSS), Seed Mass (SM) and Vitamin C (VTC).

SOV	DF	VCE		SOV	DF	VCE
		TSS	SM			VTC
Areas	3			Areas	3	
Families (Areas)	33	1.66	0.71	Families (Areas)	33	59214.0
Error	1812	1.39	0.49	Error	74	7225.6
Total	1848			Total	110	

NB. VCE are only calculated for random effects not fixed effects, as a result there are none for areas.

Family heritabilities

Family heritabilities in the context of this paper can be used to determine expected responses from selection of superior trees according to the performance of their progeny. The range of family heritabilities of the traits (sugars, vitamin C and seed mass) was comparable to that found by Cotterill and Zed (1980) in family heritabilities of Pinus radiata progeny trials in South Australia and those of Foster (1986) on Pinus taeda. From family heritabilities found in sugar content, seed mass and vitamin C content it was clear that significant genetic gains can be expected from family trials for selection for any of these traits. The total sugars of 0.54 in Marula fruits in this study is almost the same with that of 0.50 in peaches reported by Janick and Moore (1996), Dlamini (1998).

CONCLUSION AND RECOMMENDATIONS

The results of the fruit composition studies showed little provenances variation (only one provenance was significantly different from the rest), but has wide family variation which suggests a family study could be of great value in identifying superior genotypes. Furthermore the family variation needs to be tested through the establishment of progeny trials of the superior phenotypes to see whether the superiority is genetic or due to conducive environmental conditions. Alternatively *in situ* silvicultural treatments may be established in the wild population to determine the response of these families to the treatments. Such treatments may be re-spacing, thinning, no thinning and so forth.

There were significant differences between

provenances in seed mass and the smallest were almost half the size of the biggest, but rainfall and altitude did not have much effect on this and biological basis for the variation in *Sclerocarya birrea* in not known at present. Variability may be due to differences in pollen sources that fertilized the ovules of the different flowers.

This current study considered general soil type, the average annual rainfall and the altitude. It thus recommended that a detailed research study in the same provenances is carried out to do in-depth soil analysis for macro and micro elements and minerals for each parent tree (family), the minimum and maximum daily temperatures, the previous season's rainfall distribution (and not the annual rainfall which is an estimate) for each provenance. This will seek to find out whether or not variation is environmental or genetic.

Family heritabilities for sugar content, vitamin C and seed mass were high, and it is concluded that genetic improvement of these traits can be easy and faster through establishment of family trials.

Eventually the study has shown that provenance means alone may often be insufficient in the understanding of variability in seed mass and fruit parameters in *Sclerocarya birrea*. Ultimately families within provenances provide a good basis for studying between populations of trees. In this case had it not be family variation the results would have indicated that there was no variation at all.

ACKNOWLEDGEMENTS

I would like to express my sincere gratitude and appreciation to the following persons and organizations: 1. All the rural communities from which *Sclerocarya birrea* fruits were collected in Swaziland; 2. Professor G. van Wyk, who was my supervisor throughout this study at the University of Stellenbosch; 3. Dr. J.M. Theron for his valuable comments during the write-up of this paper; The Home Economics Department of the University of Swaziland for allowing me to use their laboratory for seed mass and fruit analysis 4. The Canadian International Development Agency for funding this research study and lastly; The Agriculture Research Council for assistance during statistical data analysis.

REFERENCES

A-Ogle, B (1990). Dietary Use of Wild plant resources in rural Swaziland. Proceedings of the Twelfth Plenary Meeting of AETFAT, Symposium Viii, Hamburg. Mitt. Inst. Allg. Bot. Hamburg, 23b: 895-910.

Barker N, Taylor FW (1982). Report on indigenous fruit tree research, August-November. Veld Products Research, Botswana.

Bentley S, Whittaker JB, Malloch AJC (1980). Field experiments on the effects of grazing by a chrysomelid beetle (*Gastrophysa oridula*) on seed production and quality in *Rumex obtusifolius* and *Rumex crispus*. J. Ecol., 68: 671-674

Black JN (1957). The early vegetative growth of three stains of sub-tettainain clover (*Trifolium subterraneum* L.) in relation to size of seed. Australian J. Agric. Res., 8: 1-14.

Cotterill PP, Zed PG (1980). Estimates of Genetic Parameters for Growth and Form Traits in Four *Pinus radiata* D. Don. Progeny Tests in South Australia. Austr. J. Agric. Res., 10(2): 155-67.

Cawers PB, Harper JL (1966). Germination polymorphism in *Rumex crispus* and *Rumex obtusifolius*. J. Eco., 54: 367-382.

Dlamini CS (1998). Provenance and family variation in germination, early seedling growth, fruit composition and seed mass in *Sclerocarya birrea* (A. Rich) Hochst. sub-species caffra (Sond). Kokwaro (Anacardiaceae) - The Marula. M.Sc. Thesis, University of Stellenbosch, Stellenbosch.

Dlamini CS (2010). Provenance and family variation in germination and early seedling growth in *Sclerocarya birrea* sub-species caffra. J. Hortic. For., 2(9): 229-235. Available at: www.academicjournals.org/JHF/abstracts/abstracts2010/Nov/Dlamini.htm

Dlamini CS, Geldenhuys CJ (2009). The Socio-economic Status of the Non-timber Forest Product Sub-sector in Swaziland. South. For., 71(4): 311-318.

FAO (1983). Food and fruit-bearing forest species: examples from Eastern Africa. FAO Forestry Paper, 44(1): 1-172.

Foster GS (1986). Trends in genetic parameters with stand development and their influence on early selection for volume growth in Loblolly pine, For. Sci., 32(4): 944-959.

Fourie PC (1996). Chemical composition and nutritional value of fruit and vegetables. In: Integrated management of post-harvest quality (ed) J.C. Combrink, Agriculture Research Council, Infruitec, Stellenbosh.

Harper JL (1977). Population biology of plants. Academic Press, New York. USA.

Howe HF, Vande KGA (1980). Nutmeg dispersal by tropical birds. Science, 210: 925-927.

Janick J, Moore JN (1996). Tree and tropical fruits. Fruit Breed., 1: 392-392.

Janzen DH (1977a). Promising directions of study in tropical animal-plant interactins. Annals of the Missouri Botanical Gardens, 64: 706-736/1978. Inter and intra-crop variation in seed weight of Costa Rican Ateleia Herbert-smithii Pitt. (Leguminosae). Breuesia, 14-15: 311-323,

Kirk RS, Sauyer R (1990). Pearson/s composition and analysis of foods. 9th edition. Longman Scientific and Technical, Singapore, John Wiley & Sons Inc., 605, Third Avenue, New York.

Marunda CT (1993). Genetic variation and physiological studies of *Acacia albida*. MSc. Thesis, Australian National University, Canberra.

Maun MA, Cavers PB (1971). Seed production and dormancy in Rumex crispus. I. The effects of removal of cauline leaves at anthesis. Can. J. Bot., 49: 1123-1130.

McWilliams EL, Landers RQ, Mahlestede JP (1968). Variation in seed weight and germination in population of *Amaranthus retroflexus* L. Ecology, 49: 290-296

Mwamba (1989) Natural variation in fruits of Upaca kirkeana in Zambia. For. Ecol. Manage., 26: 299-303

Obeid M, Machim D, Harper JL (1967). influence of density on plant to plant variation in flore flax, *Linum usitatissimum* Crop Sci., 7: 477-473.

Ott RL (1993). An introduction to statistical methods and data analysis. 4th edition. Duxbury Press, Belmont, California.

Putievsky E (1980). Germination studies with seed of caraway, coriander and dill. Seed Sci. Technol., 8: 245-254.

Saka JDK (1994). The nutrition value of edible indigenous fruits: present research status and future directions. In: Improvement of indigenous fruit trees of the miombo woodlands of southern Africa, eds. J.A. Maghembe, Y. Ntupanyama & P. W. Chirwa, Proceedings of a conference held on 23-27 January at Club Makokola, Mangochi, Malawi.

Salazar R (1986). Genetic variation in seeds and seedlings of ten provenances of *Oliricadia sepium* (Jacq) otend. For. Ecol. Manage., 16: 391-400.

SAS Instit. Inc 1997. SAS/STAT. User's Guide, SAS Inst. Inc. Cary, NC. 2(4): 6-12

Schaal BA (1980). Reproductive capacity and seed size in Lupinus texensis. Am. J. Bot., 67: 703-709.

Shapiro SS, Wilk MB (1965). An analysis of variance test for normality (complete samples). Biometrika, 52(3-4): 591-611.

Shivkumar P, Benerjee AC (1986). Provenance of Acacia nilotica. J. Tree Sc., 5 (1): 53-56.

Taylor FW, Butterworth KJ, Mateke SM (1995). The importance of indigenous fruit trees in semi-arid areas of Southern and Eastern African. *Sclerocarya birrea* sub species caffra. A paper presented at the African Academy of Sciences' Second Roundtable Discussion on Non-wood/timber products, Pretoria.

Thompson PA (1981). Variations in seed size within population of *Silence dioica* (L) Clairv. In relation to habitat. Ann. Bot., 47: 623-634.

Twamley BE (1967). Seed size and seedling vigour in birds-foot trefoil. Canadian, J. Plant Sci., 47: 603-609.

Wehmeyer AS (1966). The nutrient composition of some edible wild fruits in the Transvaal. South Afr. Med. J., 40: 1102-1104.

Williamson J (1975). Useful plants of Malawi. Revised ed. Zomba. University of Malawi.

Wills RB, Schriven M, Greenfield H (1983). Nutrient composition of stone fruit (*Prunus* spp.) cultivars: cultivars apricots, cherry, nectarine, peach and plum. J. Sci. Food Agric., 34: 1383-1389.

Wills RBH, Lim SK, Greenfield H (1986). Composition of Australian Foods. 31. Tropical and subtropical foods.

Host location and ovipositional preference of *Elaeidobius kamerunicus* on four host palm species

Adaigbe V. C.[1]*, Odebiyi J. A.[2], Omoloye A. A.[2], Aisagbonhi C. I.[1] and Iyare O.[1]

[1]Entomology Division, NIFOR, P.M.B 1030, Benin City, Edo State, Nigeria.
[2]Crop Protection and Environmental Biology, University of Ibadan, Ibadan, Nigeria.

The response of the oil palm pollinating weevil *Elaeidobius kamerunicus* Faust (Coleoptera: Cucurlionidae) to the inflorescence of different palm species was tested in a choice, non- choice and a four arm- olfactometer bioassay. Weevils preferred significantly the inflorescence of oil palm (*E. guineensis* Jacq) for oviposition. In both choice and non-choice tests using mated adult females, significantly more eggs were laid on oil palm than coconut, livingstona and royal palms. *E. kamerunicus* is able to locate its host plant by olfactory stimuli. These experiments support the hypothesis that pollen–seeking insects use pollen odour cues to locate its oviposition site.

Key words: *Elaeidobius kamerunicus*, *Elaeis guineensis*, *host* preference, oviposition, olfactometer.

INTRODUCTION

Herbivore-host plant interactions can have an important influence on the dynamics and management of insect populations. The relationship between ovipositional preference and host suitability has been a central problem in theories on the evolution of host associations between phytophagous insects and plants (Thompson, 1988). Host preference is associated with higher oviposition rates, increased female longevity, shorter developmental time and higher survival in all life stages (Lenteren and Noldus, 1990). The oil palm, *Elaeis guineensis* Jacq, belonging to the family Arecaceae is an entomophilous crop. In West Africa, approximately ten species of insects were identified as being responsible for oil palm pollination (Syed, 1981). These species included *Elaeidobius kamerunicus* F., *E. singularis* F., *E. subvittatus* F. (Coleoptera: cucurlionidae), *Mystrops costaricensis* Gillogly (Coleoptera; Nitidulidae), *Thrips lawaiilensis* Morgan (Thysanoptera: Thripidae) and bees (*Apis* spp.) (Corrado, 1985; Genty et al., 1986; Mariau and Genty, 1988). Among these, *E. kamerunicus* is the most efficient pollinator (Syed, 1979, 1981). *E. kamerunicus* feed on the anther filaments and oviposit on them during feeding. The hatching larvae feed on the anther of the post-anthesized male inflorescence until pupation in 9 to 11 days and the total developmental period from egg hatching to adult emergence range from 10 to 13 days. The emerged adult upon visiting the anthesizing male inflorescence for feeding and oviposition, carries some pollen load which are deposited on the female inflorescence when visited, thereby achieving pollination.

The host association status of *E. kamerunicus* population is still unknown. The objectives of the present study are:

1) To determine ovipositional preference of adult female *E. kamerunicus* when provided a simultaneous choice of four palm species;
2) To determine the ability of *E. kamerunicus* to locate suitable host by olfactory response.

MATERIALS AND METHODS

Oviposition preference

This study was conducted using ‘choice’ and ‘no choice’ experiments to determine the oviposition preference on the test palm species.

Choice preference test for oviposition

Four economically important palm species were selected including *E. guineensis* (oil palm), *Cocos nucifera* (coconut), *Livingstona*

*Corresponding author. E-mail: Vicadaigbe@yahoo.com.

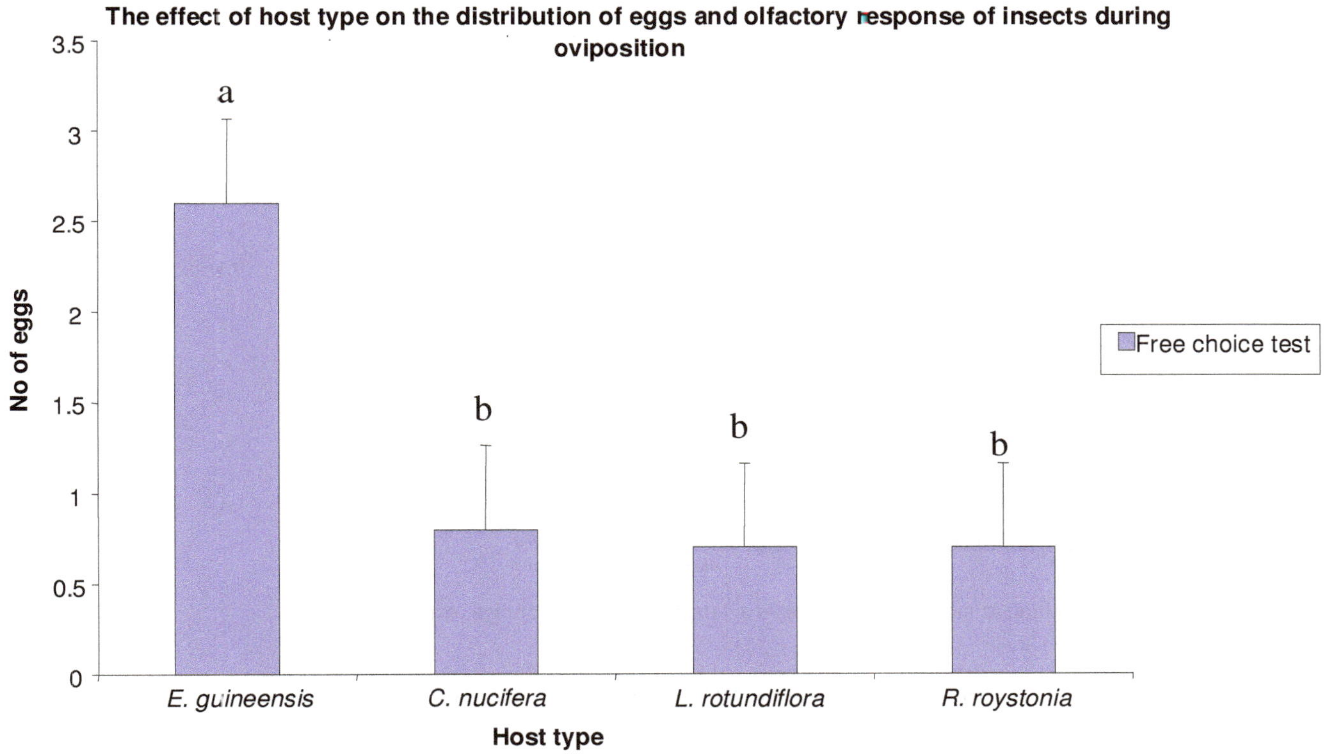

Figure 1. Mean number of eggs on different palm species (free choice test).

rotundiflora (livingstona) and *Roystonia olerecea* (royal palm). Excised inflorescence at anthesis from all host plants were kept in a refrigerator for 48 h to ensure plants were free of both pollinating weevil and other insects prior to experiment. Each inflorescence, spaced 5 cm apart, was placed at the base of a plastic bowl measuring 44 cm in diameter and 15 cm deep. Ten mated *E. kamerunicus* females were added to the bowl and the opening covered with a Nylon mesh held in place with an elastic band. The exposed females were left to oviposit for 24 h only and thereafter the inflorescence was removed from the oviposition chamber and the number of eggs laid on each inflorescence was counted under stereomicroscope.

No choice preference test for oviposition

Each palm inflorescence used earlier was placed in an individual aerated polythene bag in ten replicates. Ten mated *E. kamerunicus* females were then released into each polythene bag. The females were allowed to oviposit for 24 h, thereafter the inflorescence was removed from the oviposition chamber and the numbers of eggs laid on each inflorescence was counted under the microscope.

Olfactory response of *E. kamerunicus* to the odour of different palm species

The culture of *E. kamerunicus* was initiated from individuals collected from the field. The weevils were reared at laboratory temperature between 26.5 to 28.9°C, 69 to 92% R.H, and L12:D12 photoperiod on male "oil palm" inflorescence. Adult females were collected and used within 48 h of adult emergences. Weevils were starved for 24 h before experiments to increase their motivation to search for food. Four anthesizing inflorescences of different palm comprising oil palm, coconut, livingstona and royal palms were placed in transparent polythene bag (17.5 x 12.6 cm). The olfactometer consisted of plastic pipes, each measuring 12.4 cm long and 0.4 cm in diameter. The pipes were fixed to a transparent plastic container of 11 cm in diameter and 4.5 cm deep and were vertically secured 6.5 cm from each other. The free ends of the plastic pipes were now inserted through the opening of the polythene and tied with twine. Cotton wool was placed round the tube before tying to ensure no weevil escaped. Ten adult *E. kamerunicus* were introduced into the plastic container. Weevil left the container climbing through the pipes to meet the host of their choice.

The number of females trapped within each plant species through each plastic pipe was recorded. The experimental design was a "completely randomized design" with six replicates.

RESULTS AND DISCUSSION

Ovipositional preference

In the free-choice test, weevils laid significantly higher eggs ($P < 0.001$) on oil palm than on other palms (Figure 1).

No – choice

There were significant differences in eggs oviposited on

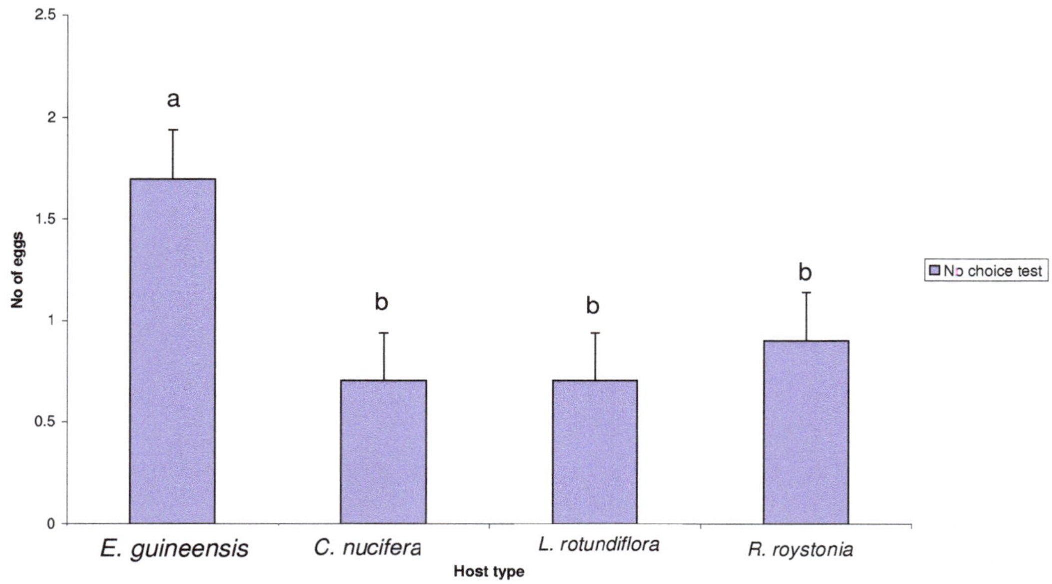

Figure 2. Mean number of eggs on different palm species (no-choice test).

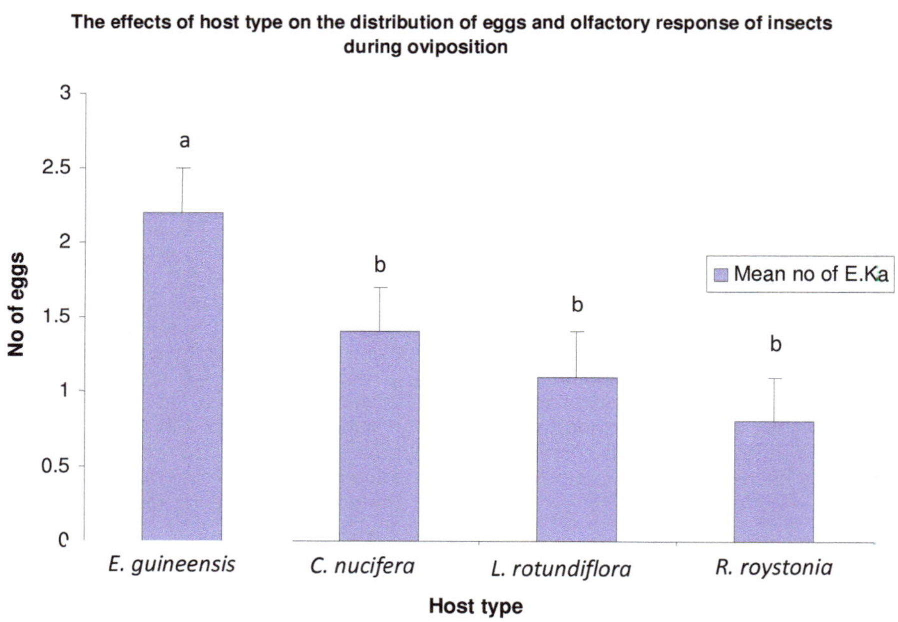

Figure 3. Response of *E. kamerunicus* to odour emitted from inflorescence of different palm species in 4-arm bioassay.

the different host plant species ($P < 0.001$) (Figure 2), when the palm species were tested in the no-choice laboratory test. These results corroborate those of the free-choice test.

Olfactory response of *E. kamerunicus* to inflorescence of different palm species

The weevils preferred the plant odour emitted by their main host plant significantly to the odours of the three non-host plants; however, the odour from the non-hosts palms also attracted the beetles (Figure 3). Data obtained in this study shows *E. kamerunicus* reared in the laboratory preferred oil palm inflorescence for ovisposition compared to other palm species. This result confirms the report that *E. guineensis* is a more suitable host for *E. kamerunicus* (Syed et al., 1982). It also suggests that *E. kamerunicus* were able to distinguish host and non-host species through olfactory response.

ACKNOWLEDGEMENT

The authors are grateful to the Executive Director, NIFOR for giving us a conducive environment to carry out this work.

REFERENCES

Corrado F (1985). Confirmation of oil palm bunches (*Elaeis guineensis* Jacq) in certain Colombian plantations. Oleagineux, 40: 173-187.

Genty P, Garzon A, Lucchini F, Delvare G (1986). Oil palm insect pollination in Tropical America. Oleagineux 41: 99-112.

Lenteren JCV, Noldus PJJ (1990). Whitefly-plant relationships: Behavioral ecological aspects. In: D. Gerling (ed), Whiteflies: their bionomics, pest status and management. Intercept Ltd., Andover, Hants, pp. 47-89.

Mariau D, Genty P (1988). Contribution de l'IRHO á l'átude de insect pollinisateurs du palmier á huile en Afrique, amerique du sud et Indonesis. Oleagineux, 43: 233-240.

Syed RA (1979). Studies on oil palm pollination by insects. Bull. Entomol. Res., 69: 213-224.

Syed RA (1981). Insect pollination of oil palm: Feasibility of introducing *Elaeidobius* sp. into Malaysia. Oil palm News, 25: 2-16.

Syed RA, Law IH, Corey TH (1982). Insect pollination of oil palm: Introduction, establishment and pollinating efficiency of *Elaeidobius Kamerunicus* faust in Malaysia. Planter, Kuala Lumpur, 58: 561–574.

Thompson JN (1988). Evolution and ecology of the relationship between oviposition preference and performance of offspring in phytophagous insects. Entomol. Exp. Appl., 47: 3-14.

5

Plant performance of *Anthurium andreanum* as affected by shade conditions and different conventional nutrient sources

Nuchhungi Khawlhring[1]*, James L. T. Thanga[2] and F. Lalnunmawia[3]

[1]Department of Horticulture, Aromatic and Medicinal Plants, Mizoram University, Tanhril-796009, Aizawl, Mizoram, India.
[2]Department of Economics, Mizoram University, Tanhril-796009, Aizawl, Mizoram, India.
[3]Department of Forestry, Mizoram University, Tanhril-796009, Aizawl, Mizoram, India.

Investigation was done on the performance of *Anthurium andreanum* plant as affected by shade conditions as well as the effects of different conventional nutrient sources. Anthuriums were planted on beds containing a media of cocopeat, charcoal and vermicompost. The beds were constructed under two shade conditions, that is, shade house (netlon with 75% shade) and natural tree shade. Different sources of nutrients such as NPK, biofertilizers, cattle manure, pig and poultry manure were supplemented. Better performance in terms of consistent growth as well as taller plant heights were obtained from those grown under shade house. Under shade house, NPK and biofertilizer application resulted in best performance of plant height whereas cattle manure treatment resulted in tallest average plant height under tree shade.

Key words: Anthurium, shade effect, nutrient sources, plant height.

INTRODUCTION

Anthurium andreanum is a member of the Araceae family, which includes more than 100 genera and about 1599 species, chiefly from the tropics (Higaki et al., 1994). In the global market, anthurium is second, only to the orchid, among tropical cut flowers (Galinsky and Laws, 1996). Anthurium plantlet growth begins with a monopodial vegetative period during which no flowers are produced (Dufour and Gue´rin, 2003) and the duration of this period was strongly influenced by fertilization produced (Dufour and Gue´rin, 2005).

The mild climatic condition of Mizoram state, India, permits anthurium cultivation without the use of temperature control facilities. It has been cultivated since the year 2002, with approximately 300 families actively engaged in the cultivation. Cultivation is done usually on shade house, with soilless cultivation. Soilless cultivation systems require expensive equipment, so the yield and quality of the flowers have to be very high (Dufour and Gue´rin, 2003). The present experiment is conducted to study the performance of plants under natural tree shade,as compared to the conventional shade house being used in Mizoram, so as to understand the plant perfor-mance as affected by shading conditions, and also to investigate the effect of different nutrient sources on the plant performance.

*Corresponding author. E-mail: puii.kh@gmail.com.

MATERIALS AND METHODS

The experiment was conducted at Mizoram University campus, Tanhril, Mizoram, which is situated at 21°58' and 24°35'N latitude and 92°15' and 93°29'E longitude and at an altitude of 750 MSL. The environment is a warm, humid sub-tropical climate, with an annual average temperature of 11 to 21°C in winter and 20 to 30°C in summer, and 250 cm per annum rainfall. The experiment was conducted during 2008 to 2009, and plant height reading was taken from the third month after planting, that is, during September 2008 to August 2009.

Plant materials and cultivation conditions

Tissue cultured plantlets of anthurium cv. Evita (AVO *Anthurium Vogels*, Holland) were planted on the flower beds of 1.2 m^2, on a media consisting of cocopeat, charcoal, and vermicompost in the ratio 6.5:1:1. The plantlets were each measuring 20 to 25 cm (pots) and nine (9) plants were planted on each plot at a distance of 40 cm. Two plots were selected, one under a forest where terrace was constructed, leaving the trees to provide shade to the terraces. The flower beds were made on these terraces with a brick lining. For the other plot, terraces were made on barren land, where shade house was constructed with a netlon that provides 75% shade (Rakshak Agro-shading, White x Black, 75%).

Treatments

The first treatment was that of the cultivation condition, that is, on shade house and under forest trees. Different sources of nutrient were applied as:

i. Control –no additional nutrient source
ii. NPK(19:19:19) at 220 kg/ha/year
iii. Biofertilizers (Azospirillum and Phosphotika) at 500 gm/acr/year.
iv. Cattle manure at 10 tonnes/ha/year.
v. Pig manure at 10 tonnes/ha/year.
vi. Poultry manure at 5 tonnes/ha/year.

Each treatment having three replications under two different conditions, as T_1R_1, T_1R_2, T_1R_3, T_2R_1,...$T_{12}R_3$. Among these, T_1 to T_6 represents the plants cultivated under shade house, while T_7 to T_{12} are the plants grown under natural tree shade. Nine plants were planted within each of these treatments.

Plant growth measurements

Three representative plants, out of nine plants, were randomly selected for measurement from each plot. The height of each plant was observed by measuring the plant from base to petiole end of the tallest leaf.

Statistical analysis

To analyze if there was a systematic difference between different treatments being applied in the experiment on the growth performances of plant height during the twelve month period, the method of 2-way ANOVA with m (that is, 3) observations per cell was adopted to each of the shade house and tree shade separately. Using this method, three possible effects, namely, treatment, durations (month) and interaction were examined. In addition, an attempt had been made to examine if there is significant difference between the two types of experimental plots, that is, shade house and tree shade, using 3-way ANOVA with m (that is, 3) observations per cell.

RESULTS AND DISCUSSION

As it is presented in Tables 6 and 7, all the effects of treatment and duration are found to be significant under shade house and tree shade; while interaction is not significant. That is, the growth performances of plant height have shown significantly different behaviour from one treatment to another and duration (months) has also played a significant role in the growth performance of plant height. Meanwhile, Table 8 revealed that, whereas the shade effect is significant, its interaction with treatment is found to be indeterminate. However, the result being observed in the overall performances is that there are significant differences in the overall treatment averages between shade house and tree shade, with shade house showing a better performance throughout.

Among the different treatments, T_2 recorded the most number of tallest plant height (that is, five times) followed by T_3 (four times) during the twelve months recorded period. Also, plant heights of both T_2 and T_3 are significantly at par with the tallest recorded plant height in most of the readings (ten times and eleven times respectively). We can thus conclude that treatment with NPK (T_2) and biofertilizer (T_3) resulted in the best performance in terms of plant height, under shade house condition (Table 1).

From the observations taken from the plants grown under natural tree shade (Table 2) among the different treatments, T_{10} recorded the most number of tallest plant height (seven times) during the twelve month recorded period. Also, the record of T_{10} on third month was the tallest plant height record of all the treatments. Besides, plant height record of T_{10} was significantly at par with the tallest record in all the months except for the second month. Thus, we can conclude that treatment with Cattle manure (T_{10}) recorded best performance in terms of plant height during the twelve month recorded period under natural Tree shade conditions. Also the performance of plant height was fairly good with treatments of Pig manure (T_{11}) and NPK (T_8).

The average plant height under shade house increases up to the third month (November, 2008), which lowers down in the following months, rises again after the ninth month up to the tenth month (May to June, 2009) and then falls down. However, the plant height under natural tree shade rises initially up to the third month, which falls from the fifth month and continues to lower. This fluctuation in the pattern of plant height is attributed to the adverse climatic condition especially water stress during the winter seasons (December to February, that is, fourth to sixth months), and also pruning of the oldest leaves during the peak growth season to avoid overcrowding of the plantation. However, Agasimani et al. (2011) revealed that different varieties of anthurium performed well in winter from September to December compared to rainy season, that is, May to August with respect to foliage and floral characters. The plants under shade house recovers from the adverse climatic conditions once favourable climatic conditions prevails in the ninth and tenth month, whereas the plants exposed to natural conditions, even under the shade of trees, do not recover as well as those grown under shade house.

Under shade house (Table 3), among the different treatments, T_2 (NPK) have the tallest average plant height during twelve month observation, followed by T_3 (biofertilizer treatment) whereas T_5 (pig manure treatment) shows lowest average plant height. Again, under natural tree shade, treatment with cattle manure (T_{10}) recorded tallest average plant height, whereas poultry manure treatment (T_{12}) shows lowest average plant height (Table 4

Table 1. Growth performance of plant height with different nutrient sources during 12 months (shade house).

Month																							
1		2		3		4		5		6		7		8		9		10		11		12	
T_3	38.60	T_6	37.33	T_4	44.00	T_2	42.83	T_3	43.67	T_4	42.33	T_2	40.33	T_3	38.50	T_3	41.33	T_2	44.67	T_2	38.67	T_2	41.67
T_1	35.73	T_2	37.00	T_5	42.67	T_6	42.50	T_6	43.33	T_2	41.67	T_3	39.03	T_1	37.67	T_6	40.67	T_3	44.33	T_4	36.33	T_3	39.67
T_5	34.80	T_4	37.00	T_6	42.67	T_3	41.50	T_2	42.33	T_3	39.33	T_6	37.50	T_2	37.67	T_2	40.17	T_1	43.67	T_1	36.00	T_4	39.00
T_4	34.27	T_5	36.33	T_3	42.33	T_1	40.00	T_5	41.33	T_5	39.33	T_1	36.67	T_5	37.67	T_5	38.50	T_5	41.33	T_6	36.00	T_6	38.00
T_6	33.67	T_3	35.67	T_1	41.67	T_5	39.00	T_1	41.00	T_6	39.33	T_5	36.17	T_6	37.60	T_1	38.00	T_4	41.00	T_3	35.33	T_5	36.67
T_2	33.50	T_1	35.00	T_2	39.67	T_4	38.83	T_4	39.33	T_1	37.33	T_4	35.67	T_4	34.33	T_4	37.50	T_6	39.67	T_5	34.67	T_1	36.00
Average	35.09		36.39		42.17		40.78		41.83		39.89		37.56		37.24		39.36		42.44		36.17		38.50

5% level critical difference for treatment means = 3.33.

Table 2. Growth performance of plant height with different nutrient sources during 12 months (tree shade).

Month																							
1		2		3		4		5		6		7		8		9		10		11		12	
T_7	37.13	T_{11}	41.67	T10	45.67	T_{10}	42.83	T_{12}	42.33	T_8	38.00	T_8	39.67	T_{10}	40.83	T_{10}	41.37	T_{10}	38.00	T_{10}	39.67	T_{10}	38.67
T_8	36.83	T_8	40.00	T_7	42.67	T_8	41.67	T_{10}	42.00	T_{11}	38.00	T_{11}	39.67	T_{11}	40.43	T_8	38.80	T_{11}	36.33	T_{11}	38.67	T_{11}	37.67
T_{10}	36.53	T_{12}	39.00	T_8	42.33	T_{11}	41.03	T_{11}	42.00	T_{10}	36.67	T_7	38.33	T_8	39.33	T_{11}	37.77	T_9	35.67	T_8	35.67	**T7**	33.67
T_9	36.47	T_7	38.67	T_9	42.00	T_7	40.67	T_8	41.00	T_{12}	36.00	T_{10}	38.00	T_9	37.33	T_9	34.67	T_8	35.00	T_9	35.00	T_9	33.67
T_{11}	35.83	T_9	38.00	T_{11}	42.00	T_9	40.00	T_9	40.00	T_9	35.50	T_9	33.07	T_7	36.00	T_7	34.43	T_7	33.00	T_7	34.67	T_8	32.00
T_{12}	35.40	T_{10}	37.50	T_{12}	39.67	T_{12}	38.83	T_7	39.67	T_7	35.00	T_{12}	33.00	T_{12}	31.67	T_{12}	33.43	T_{12}	32.00	T_{12}	30.00	T_{12}	27.67
Average	36.37		39.14		42.39		40.84		41.17		36.53		36.96		37.60		37.10		35.00		35.61		33.89

5% level critical difference for treatment means = 4.20.

Table 3. Growth of plant height (cm) during 12 months (shade house).

Treatment											
Months	**T_1**	**Months**	**T_2**	**Months**	**T_3**	**Months**	**T_4**	**Months**	**T_5**	**Months**	**T_6**
10th	43.67	10th	44.67	10th	44.33	3rd	44.00	3rd	42.67	5th	43.33
3rd	41.67	4th	42.83	5th	43.67	6th	42.33	5th	41.33	3rd	42.67
5th	41.00	5th	42.33	3rd	42.33	10th	41.00	10th	41.33	4th	42.50
4th	40.00	6th	41.67	4th	41.50	5th	39.33	6th	39.33	9th	40.67
9th	38.00	12th	41.67	9th	41.33	12th	39.00	4th	39.00	10th	39.67
8th	37.67	7th	40.33	12th	39.67	4th	38.83	9th	38.50	6th	39.33
6th	37.33	9th	40.17	6th	39.33	9th	37.50	8th	37.67	12th	38.00

Table 3. Contd.

7th	36.67	3rd	39.67	7th	39.03	2nd	37.00	12th	36.67	8th	37.60
11th	36.00	11th	38.67	1st	38.60	11th	36.33	2nd	36.33	7th	37.50
12th	36.00	8th	37.67	8th	38.50	7th	35.67	7th	36.17	2nd	37.33
1st	35.73	2nd	37.00	2nd	35.67	8th	34.33	1st	34.80	11th	36.00
2nd	35.00	1st	33.50	11th	35.33	1st	34.27	11th	34.67	1st	33.67
Average	38.23		40.01		39.94		38.30		38.21		39.02

5% level critical difference for duration means = 2.35

Table 4. Growth of plant height (cm) during 12 months (tree shade).

					Treatment						
Months	**T_7**	**Months**	**T_8**	**Months**	**T_9**	**Months**	**T_{10}**	**Months**	**T_{11}**	**Months**	**T_{12}**
3rd	42.67	3rd	42.33	3rd	42.00	3rd	45.67	3rd	42.00	5th	42.33
4th	40.67	4th	41.67	4th	40.00	4th	42.83	5th	42.00	3rd	39.67
5th	39.67	5th	41.00	5th	40.00	5th	42.00	2nd	41.67	2nd	39.00
2nd	38.67	2nd	40.00	2nd	38.00	9th	41.37	4th	41.03	4th	38.83
7th	38.33	7th	39.67	8th	37.33	8th	40.83	8th	40.43	6th	36.00
1st	37.13	8th	39.33	1st	36.47	11th	39.67	7th	39.67	1st	35.40
8th	36.00	9th	38.80	10th	35.67	12th	38.67	11th	38.67	9th	33.43
6th	35.00	6th	38.00	6th	35.50	7th	38.00	6th	38.00	7th	33.00
11th	34.67	1st	36.83	11th	35.00	10th	38.00	9th	37.77	10th	32.00
9th	34.43	11th	35.67	9th	34.67	2nd	37.50	12th	37.67	8th	31.67
12th	33.67	10th	35.00	12th	33.67	6th	36.67	10th	36.33	11th	30.00
10th	33.00	12th	32.00	7th	33.07	1st	36.53	1st	35.83	12th	27.67
Average	36.99		38.36		36.78		39.81		39.26		34.92

5% level Critical difference for duration means = 2.97.

Table 5. Five percent critical limit of plant height w.r.t treatment during 12 months.

Treatment	Shade house	Tree shade
	Mean ± $t_{(.025)}$. SE (mean)	**Mean ± $t_{(.025)}$. SE (mean)**
Control	38.23 ± 1.26	36.99 ± 1.64
NPK	40.01 ± 1.30	38.36 ± 1.40
Biofertilizer	39.94 ± 1.19	36.78 ± 1.33
Cattle manure	38.30 ± 1.28	39.81 ± 1.49
Pig manure	38.21 ± 1.26	39.26 ± 1.23
Poultry manure	39.02 ± 1.32	34.92 ± 1.72

The average plant height was taller under shade house as compared to tree shade with different treatments except with cattle manure and pig manure treatments (Table 5). Also, the plant height readings show better consistency under shade house in all the cases except for pig manure treatment, which have slight difference in the SE values with that under tree shade. This shows that plants grown under shade house showed good consistency in their growth during twelve month observation, whereas those under tree shade shows highly variable growth rate.

Thus, we can conclude that anthuriums grown under shade house results in better consistent growth as compared to those grown under natural tree shade in Mizoram. Also, the average plant heights under different nutrient sources were relatively taller under shade house as compared to natural tree shade. Under shade house, NPK and biofertilizer application results in best performance in terms of plant height, while under natural tree shade, cattle manure treatment show tallest average

Table 6. Two-way ANOVA for plant height during 12 month period (shade house).

Source	SS	df	MS	F	Prob > F
Treatment	130.28	5	26.06	3.00	0.01
Month	1250.89	11	113.72	13.11	0.00
Interaction	375.09	55	6.82	0.79	0.84
Residual	1249.00	144	8.67		
Total	3005.26	215	13.98		

Table 7. Two-way ANOVA for plant height during 12 months period (tree shade).

Source	Partial SS	df	MS	F	Prob > F
Treatment	590.50	5	118.10	8.54	0.00
Month	1381.16	11	125.56	9.08	0.00
Interaction	528.81	55	9.61	0.7	0.94
Residual	1990.39	144	13.82		
Total	4490.87	215	20.89		

Table 8. Three-way ANOVA with 3 observation per cell for plant height.

Source	Partial SS	df	MS	F	Prob > F
Shade effect	303.40	1	303.40	26.97	0.00
treatment	720.78	10	72.08	6.41	0.00
month	1857.41	11	168.86	15.01	0.00
Shade effect*treatment	0.00	0			
Shade effect*month	221.10	11	20.10	1.79	0.06
Treatment*month	903.90	110	8.22	0.73	0.97
Residual	3239.39	288	11.25		
Total	7669.28	431	17.79		

plant height.

REFERENCES

Agasimani AD, Harish DK, Imamsaheb SJ, Patil VS (2011). Anthurium Varieties Performance in Rainy and Winter Seasons under Greenhouse. Res. J. Agric. Sci., 2(2): 337-339

Dufour L, Gue´rin V (2003). Growth, development features and flower production of *Anthurium andreanum* Lind. in tropical conditions. Sci. Hortic., 98: 25–35.

Dufour L, Gue´rin V (2005). Nutrient solution effects on the development and yield of *Anthurium andreanum* Lind. in tropical soilless conditions. Sci. Hortic., 105: 269–282.

Galinsky R, Laws N (1996). Anthurium market. RAP Mark. Inform. Bull., No. 11.

Higaki T, Lichty JS, Moniz D, (1994). Anthurium Culture in Hawaii. University of Hawaii, HITAHR Res. Ext. Ser. 152, 22p.

6

Scion and stock diameter size effect on growth and fruit production of *Sclerocarya birrea* (marula) trees

Simon A. Mng'omba[1]*, Gudeta W. Sileshi[1], Ramni Jamnadass[2], Festus K. Akinnifesi[1] and Jarret Mhango[3]

[1]World Agroforestry Centre (ICRAF), P. O. Box 30798, Lilongwe 3, Malawi.
[2]World Agroforestry Centre (ICRAF), P. O. Box 30677-00100, Nairobi, Kenya.
[3]Mzuzu University, Private Bag 201, Luwinga, Mzuzu 2, Malawi.

***Sclerocarya birrea* (marula) is a highly valued fruit tree in southern Africa because of its products which have gained both regional and international markets. However, its fruits have been harvested from the wild with a few provenances being managed on farmland. Grafted marula trees have shown wide variations in scion and rootstock growth (for example, stem diameters), but effects of these variations on tree survival, growth and fruit production have not been evaluated. The objectives of this study were to (1) assess variations in growth of scions and their stocks and (2) evaluate the effect of scion and stock growth differences on overall grafted tree growth, survival, flower and fruit production on farmland, six to eight years after planting. Over 70% of the grafted trees had similar scion and stock diameters, while 24% had thicker (i.e, slow growing) scions than their stocks. Trees with similar scion and stock growth rate had a high flowering percentage (> 40%), but there was also significant fruit abortion during the three years of fruit production. Grafted trees with thicker scions than their stocks produced significantly more ($P < 0.05$) fruits than the rest. Such trees were also significantly ($P < 0.0126$, $N = 50$) taller than the rest. There was a low survival for grafted trees with either thick or thin scions relative to the stock diameters. It can be concluded that unequal growth rate of scion and stock in grafted marula trees has a negative effect on early tree survival and dwarfing characteristics, but this may not reduce fruit yield in the early years. Therefore, scion and stock selection is critical to reduce growth variations which have negative effects on grafted tree growth and fruit productivity.**

Key words: Fruit abortion, graft compatibility, phenotypic variations, tree dwarfing.

INTRODUCTION

Sclerocarya birrea (A. Rich.) Hochst. subsp. *caffra* (Sond.) Kokwaro also known as Marula is an economically important indigenous fruit tree in southern Africa (Leakey et al., 2005). It belongs to the Anacardiaceae family just like mango and cashew. Its fruits and other products have gained regional and international markets (Shackleton, 2002; Wynberg et al., 2002). The fruits are processed into traditional beer, 'Amarula cream' liqueur, jams, wine, juice and many other products (Leakey et al., 2005). The Amarula cream liquor, produced by Distell Corporation in South Africa, is commercially sold in 63 countries world-wide (Ham, 2005; Ham et al., 2008). It is estimated that the total value of the commercial marula trade to the rural communities is worth US$ 160,000 a year in South Africa (Mander et al. 2002). The marula fruits and kernels are potential sources of high quality oil that is rich in tocopherol (Leakey, 2005) used in the cosmetic industry (Wynberg et al., 2002). Despite the significant contributions of the marula trees to the livelihoods of many communities of southern Africa, there have been limited studies undertaken on the field cultivation and

*Corresponding author. E-mail: smngomba@cgiarmw.org.

management of these trees. Currently, harvests of marula fruits and its other products depend entirely on wild populations, but this may not be sustainable as wild harvesting, especially on a large scale, can lead to overexploitation (Akinnifesi et al., 2008).

Marula is a deciduous tree with a dioecious reproduction system, but there are also claims that it is monoecious (Shackleton, 2002). It grows in a wide range of soils and preferably well-drained soils. The ideal altitude varies from 0 to 1800 m above sea level and the annual rainfall range from 200 to 1500 mm. However, it is sensitive to frost (Shackleton, 2002). The marula trees propagated from seeds have a long vegetative (juvenile) phase, but there are mixed reports regarding the time they take to first fruiting. Emanuel et al. (2005) reported that female trees take 19 years to first fruiting, while Moganedi et al. (2007) indicated about 8 to 10 years to first fruiting after planting. At Makoka in southern Malawi, marula trees fruited after 10 to 12 years, although the fruit loads were low (Akinnifesi et al., 2006). Differences in time to first fruiting could depend on both genetic and environmental factors as marula trees exhibit wide genetic variations (Kadu et al., 2006). According to Moganedi et al. (2007), grafted marula trees started producing fruits four years after planting in South Africa, but there were phenotypic variations in terms of flowering time and other traits within some lines. Morphological and molecular studies have shown growth variations among marula provenances (Rukuni et al., 2001; Chirwa et al., 2007; Kadu et al., 2006), but the effects of variations in scion and rootstock growth on the overall tree growth and fruit yield have remained unclear. Although, grafting is desirable to achieve fruiting precocity and dwarfing, graft failure has been reported in marula (Moganedi et al., 2007). The main cause of graft failure has remained unclear and complex as it is often linked to a number of factors including genetic constitution, graft incompatibility and the environment (Usenik and Štampar, 2000). Therefore, the objectives of this study were to (1) assess variations in growth of scions and their stocks and (2) evaluate the effect of scion and stock growth differences on overall grafted tree growth, survival, flower and fruit production on farmland six to eight years after planting.

MATERIALS AND METHODS

Study site

The study was carried out at ICRAF-Makoka in Zomba, Malawi with an altitude of 1029 m above sea level (masl), 15° 30' S latitude and 35° 15' E longitude. It has annual rainfall variations ranging from 560 to 1600 mm and temperatures vary between 16 and 32°C (Akinnifesi et al., 2008). The soils are Ferric Lixisol and the soil texture is 46% sand, 46% clay and 8% silt. The clay content increases with soil depth, but major chemical characteristics are relatively constant to >1 m depth. The soil chemical properties in the top 20 cm soil are: pH 5.62, 1.08% organic carbon, 38.31 mg kg^{-1} phosphorus, 1.30 cmol kg^{-1} potassium, 12.11 cmol kg^{-1} calcium and 0.86 cmol kg^{-1} magnesium. The soil organic carbon content decreased with depth. The soil exchangeable cations (Ca, Mg and K) were generally higher in the top 20 and 60 to 100 cm depths (Akinnifesi et al., 2008).

Plant materials

Marula fruits were collected from forest in Zomba and their seeds were used to raise rootstocks. Scions were collected from mature and bearing marula trees in Mangochi (14° 28′ S latitude, 35° 14′ E longitude and 469 masl with maximum temperatures of more than 32°C) in Malawi. Mangochi is more than 100 km from ICRAF-Makoka in Zomba. The scions were grafted onto one-year old marula seedling rootstocks at pencil thick size (10 mm in diameter) and 40 to 50 cm tall using a splice grafting method (Akinnifesi et al., 2008). At the time of grafting, all scions were grafted to rootstocks of almost similar sizes (that is, diameter size). After 6 months in the nursery, 96 grafted marula trees were planted at ICRAF-Makoka in December 2003 at the spacing of 2 m within rows and 5 m between rows (planting holes of 60 cm width and 60 cm deep) and in three blocks (arranged in a randomized complete block design). These grafted plants received different soil amendment (manure and fertilizer) and dry-season irrigation treatments in the first three years after planting, in order to improve their survival and growth. However, none of the different treatments either singly or in combination significantly influenced survival of the grafted marula trees (Akinnifesi et al., 2008). This trial has been managed by weeding only since then. The morphological growth similarities and differences between scions and their rootstocks (2009 to 2011) of 65 grafted trees were assessed visually and differences in scion and stock diameter were noted. As shown in Figure 1a and b, some grafted plants had similar scion and rootstock diameter, while others had thicker or thinner scion compared to their respective rootstocks. The second assessment was based on measurements to verify the observed growth variations between scions and their rootstocks and 50 grafted marula trees were measured. In this assessment, grafted trees were grouped into having either thinner, thicker or same (similar) scions compared to their respective rootstocks. "Thinner" represents faster growth in diameter of the rootstock relative to the scion, while "thicker" represents the opposite. "Similar" represents the same rate of growth in diameter of the scion and rootstock. The number of trees (out of a total of 50 trees) in each category (thinner, thicker or similar) was also calculated based on either ± 0.25, 0.5, 0.75, 1.0 or ±1.5 cm stem diameter size differences, and these measurement differences were selected to minimize subjectivity and match visual assessment and the actual measurements collected.

Statistical analysis

Tree height, flowering, fruit yield, scion and rootstock diameters and tree survival data were statistically analyzed using PROC MIXED procedure of the Statistical Analysis System (SAS) (SAS, 1999). Three different scion and rootstock growth variations identified (thick, thin or similar scion diameter sizes) formed the random treatment components in the statistical analysis. Means, standard errors and percentages were calculated from the counts based on the five (5) categories of scion and rootstock diameter variations (that is, ± 0.25, ± 0.5, ± 0.75, ± 1.0 or ± 1.5 cm) selected during the scion/rootstock diameter measurement assessment.

RESULTS

Variations in tree survival

There was a significant increase ($P = 0.0141$) in grafted

Figure 1. Grafted *Sclerocarya birrea* (marula) trees showing morphological growth differences between scion and rootstock (a) uniform scion and rootstock stem diameter (b) enlarged scion on a small rootstock at ICRAF-Makoka in Malawi.

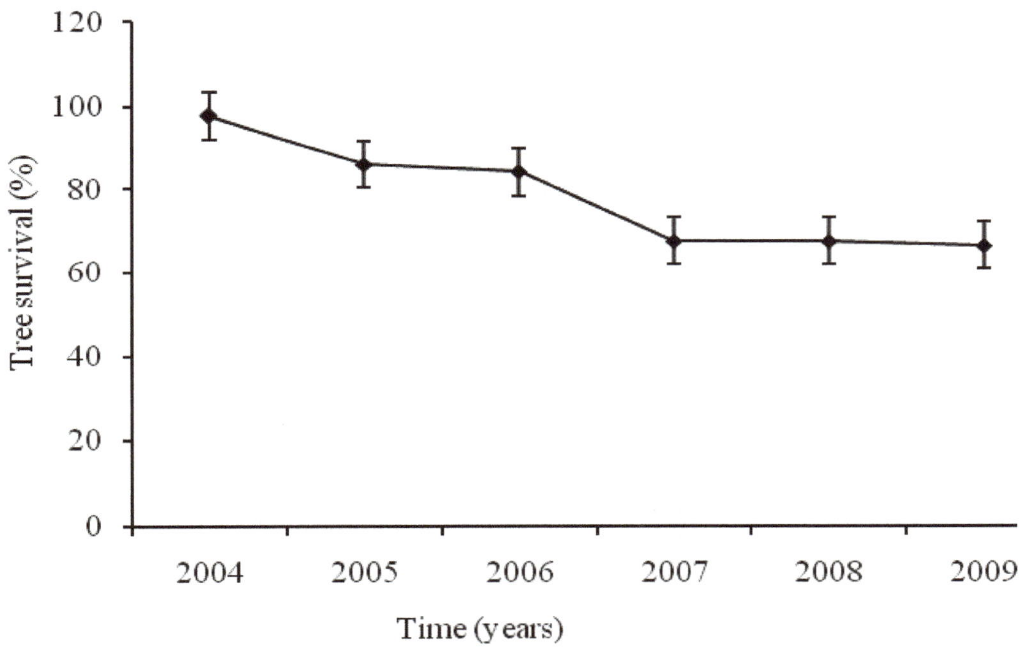

Figure 2. Survival (%) of *Sclerocarya birrea* (marula) trees seven years after planting at ICRAF-Makoka in Malawi.

tree mortality (67%) six years after planting (Figure 2). The highest tree mortality (16.5%) was from the third year to the fourth year after planting (2006 to 2007) and then tree survival remained almost stable until 2009 (Figure 2).

Scion and stock variations

Over 70% of grafted trees had similar scions and rootstocks (that is, diameter), while 6% had scions thinner than their rootstocks and 24% had scions thicker than the respective rootstocks (Table 1) based on visual assessment. Measured scion and rootstock diameters (Table 2) indicated an increase in number of trees whose scions and rootstocks were similar in stem diameter at any selected interval (with a mean of 67.6% ± 5.27). This was followed by those with thicker scions than their rootstocks (a mean of 21.6% ± 3.37). At ±1.5 cm stem diameter size differences between scions and their rootstocks, there were 84% of grafted trees whose scions

Table 1. Visual assessment of *Sclerocarya birrea* (marula) morphological growth patterns at the graft union interfaces at ICRAF- Makoka fruit orchard (N = 65).

Scion/stock stem diameter size	Number of trees (%)
Scions similar to rootstocks	70
Scions thicker than rootstocks	24
Scions thinner than rootstocks	6

Table 2. Counts based on stem diameter size differences of grafted *Sclerocarya birrea* (marula) trees (N = 50).

Stem diameter	0.25 cm	0.5 cm	0.75	1.0 cm	1.5 cm	Mean
Similar size	27	30	33	37	42	33.8±2.63
Thicker Scions	15	13	11	10	5	10.8±1.69
Thinner Scions	8	7	6	3	3	5.4±1.03
Percentages (%) of each stem diameter size						
Similar size	54	60	66	74	84	67.6±5.27
Thicker scions	30	26	22	20	10	21.6±3.37
Thinner scions	16	14	12	6	6	10.8±2.06

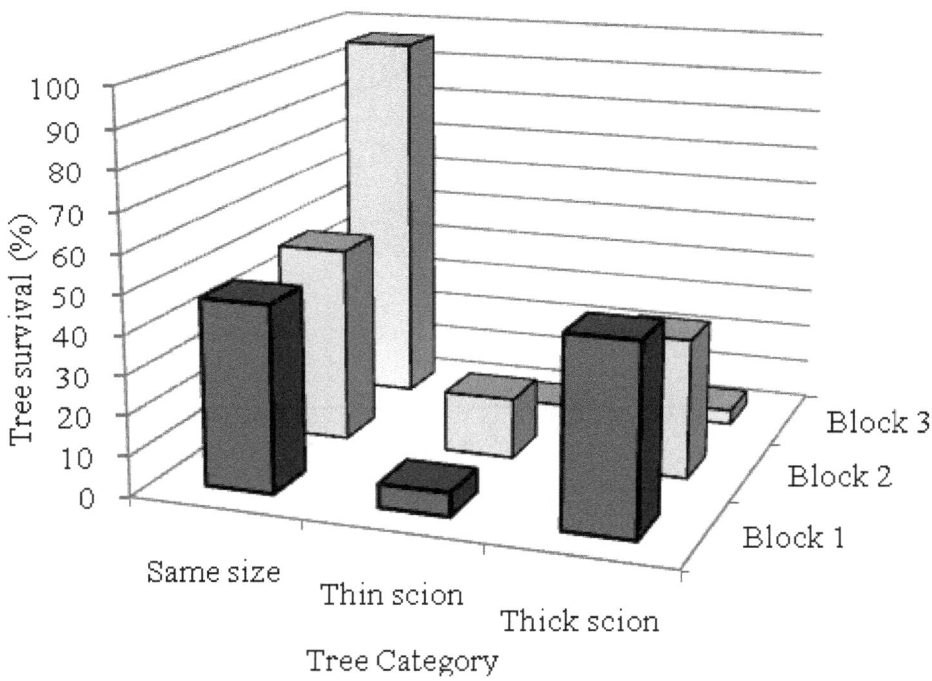

Figure 3. Survival (%) of *Sclerocarya birrea* (marula) trees with respect to the three different scion diameter sizes (same/similar, thinner and thicker scions) in three blocks.

and rootstocks were similar in diameter, while only 10% of them had thicker scions than their rootstocks (Table 2). The data in Figure 3 shows distribution of grafted marula trees per block with respect to survival (N = 65). A high mean percentage (64% ± 16) of trees whose scions and rootstocks were similar in stem diameters survived. This was followed by those trees whose scions were thicker than their rootstocks (29% ± 13). The remaining grafted trees (7% ± 4.4) had thinner scions relative to their respective rootstocks. Grafted trees with similar sized

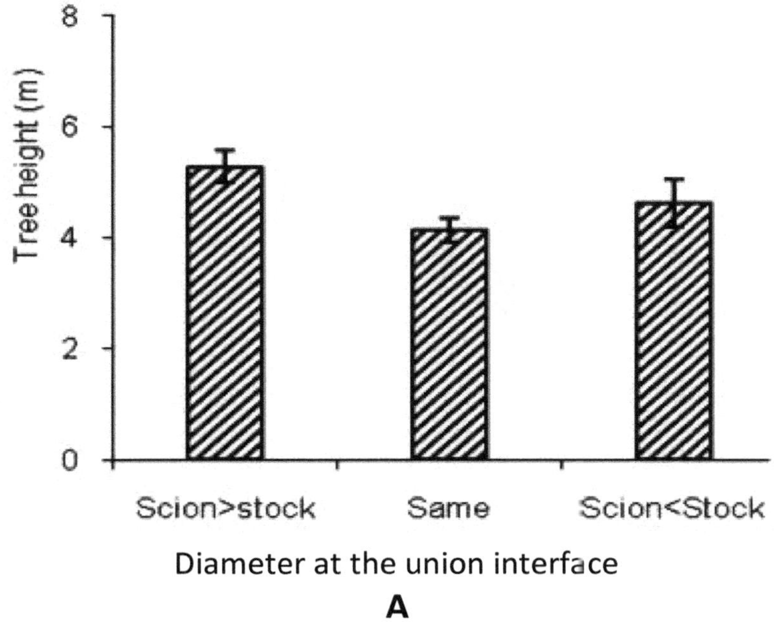

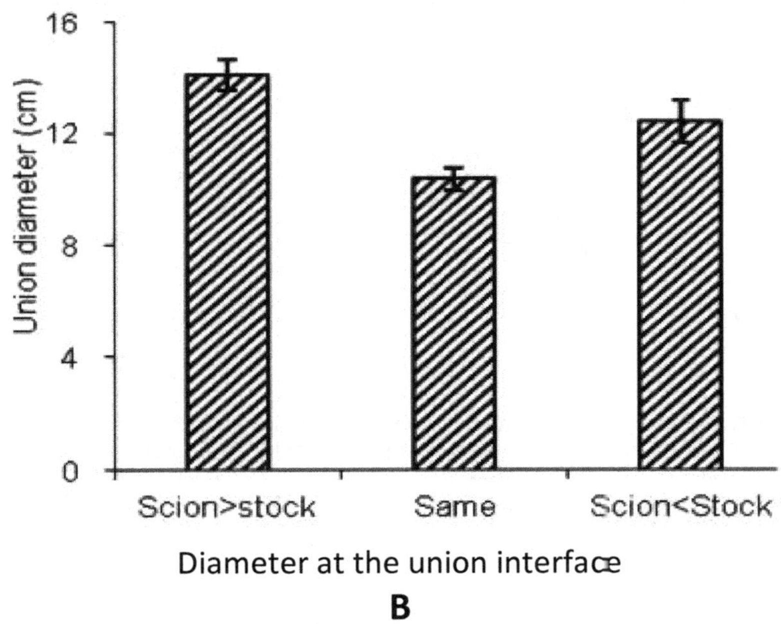

Figure 4. *Sclerocarya birrea* (marula) growth (a) tree height (m) in relation to their stem diameter sizes of the scions and rootstocks (b) stem diameter at the graft union.

scions and rootstocks were significantly ($P = 0.0126$, $N = 50$) shorter (4.14 ± 0.21 m) than the rest (Figure 4a). A similar trend is also shown with respect to stem diameter size at the union interface (Figure 4b).

Variations in flower and fruit production

The grafted trees with similar scion and rootstock diameter produced significantly ($P = 0.0401$) larger number of flowers six years after planting (Figure 5a). A similar trend was also obtained seven years after planting (Figure 5b). This was followed by those with thicker scions than their rootstocks.

Grafted trees with scions thicker than their respective rootstocks had a significantly ($P < 0.050$) higher fruit load than the rest of the tree categories, six to eight years (2009 to 2011) after planting (Figure 6). However, there

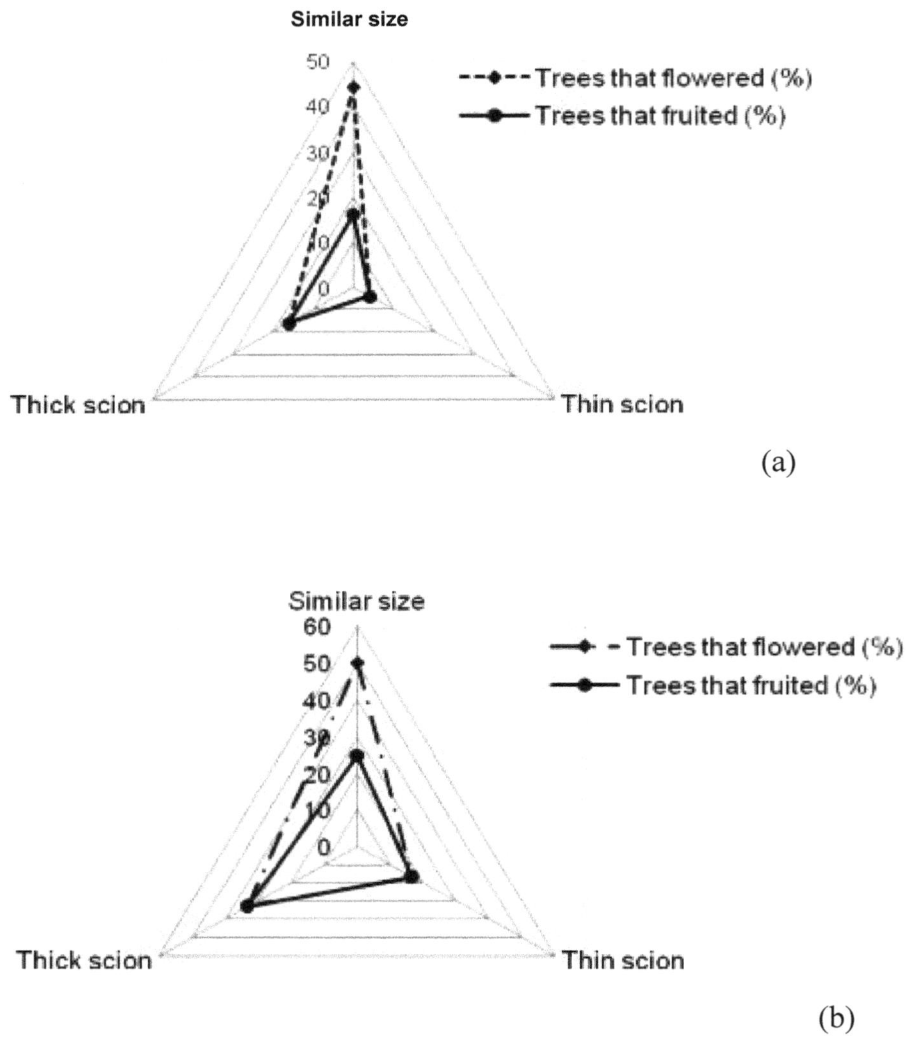

Figure 5. *Scerocarya birrea* (marula) flowering and fruiting patterns at ICRAF Makoka (a) six (2009) and (b) seven (2010) years after planting at ICRAF-Makoka in Malawi.

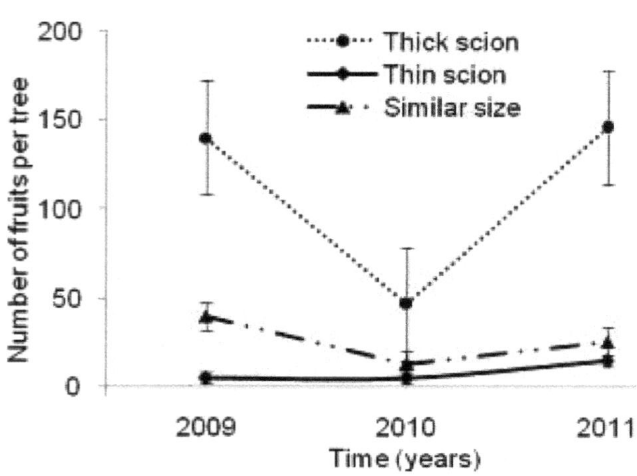

Figure 6. *Sclerocarya birrea* (marula) fruit yield after six to eight years after planting at ICRAF-Makoka in Malawi.

was a lower fruit production in the second year (2010) followed by an increase in fruit load in next fruit production season (2011).

DISCUSSION

An increase in tree mortality indicates that graft failure in marula can occur in an orchard after a successful graft-take in the nursery. The harsh field conditions, scion and rootstock growth variations and other factors could promote early mortality of the grafted trees. In this study, high tree mortality, three years after planting, could partially be attributed to no further application of dry-season irrigation and soil amendment (Akinnifesi et al., 2008). They found low growth of grafted marula trees at 33 months, despite soil amendment and dry-season irrigation. Mateke (2000) reported that survival of grafted marula trees ranged from 0 to 100% in Botswana, and this depended upon the location. Furthermore, Mhango and Akinnifesi (2001) reported low survival and growth of grafted marula trees in the field. These reports indicate that the low survival and growth of the grafted marula trees could not be attributed to poor soil and water stress alone. Other factors such as graft incompatibility, irregularities and/or failure could significantly contribute to poor survival of marula trees. Genetic and environmental differences between the source populations for scions (from Mangochi district) and rootstocks (from Zomba district) could have contributed to the observed graft problems. This study highlights the fact that morphological growth irregularities could occur if scions are collected at random from different mother trees with no traceable relation to the rootstocks. This can contribute to scion/rootstock morphological growth irregularities which could lead to graft compatibility problems. Leakey (2005) reported tree-tree phenotypic variations for marula provenances in South Africa and Namibia, and hence selection of compatible scions and rootstocks is critical to reduce growth irregularities.

In this study, both assessments (observations and measurements) indicated that trees whose scions and rootstocks were similar in stem diameters were abundant (>50%), then followed by those with thicker scions (Tables 1 and 2). Also, findings obtained through observation assessment matched with the actual measurements of the stem diameters, especially at ± 1.0 cm stem diameter difference. Assessment by observation was also reliable in classifying the three categories of scion and rootstock stem diameters. The presence of many trees with either thicker or thinner scions than their rootstocks contributed to high tree mortality. For instance, block 3 with a high number of trees (96%) whose scions and rootstocks were similar in stem diameters had a higher total tree survival than the other scion and rootstock categories. This reinforces the fact that morphological growth irregularities between scions and their rootstocks bring some graft problems. According to Rouphaela et al. (2010), tissue and structure growth differences can bring graft problems apart from other factors. In this study, grafted trees with similar scion and rootstock sizes exhibited significant dwarfing in growth, compared to those where scions and rootstocks grew at different rates. Tree dwarfing is a desirable trait, as it allows harvesting of fruits easily and efficient resource utilization, including close tree spacing (less land area) and less tree pruning. Grafted marula trees with morphological growth irregularities were tall and this is critical as marula fruits fall when ripe (Emanuel et al., 2005), which could increase incidences of fruit damage due to impact upon hitting the ground. Furthermore, marula trees with thick scions compared to their rootstocks are highly prone to graft failure due to a heavy load of the scion (huge mass of scions and high fruit load) being imposed on small rootstocks. This could be a threat to the survival of such trees. Grafted marula trees whose scions and rootstocks grew at the same rate produced more flowers than the rest (unlike trees with irregular scion and rootstock growth). Surprisingly, trees with irregular scion and stock growth rate had minimal fruit abortion. Marula trees with thick scions compared to their rootstocks had a high fruit load. Possibly, their prolific growth (tall trees with thick stem sizes) contributed to prolific fruit production in this study. The trees exhibited alternate bearing as evidenced by low fruit production in 2010 and high in both 2009 and 2011 fruiting season. Alternate bearing is common in mango trees, which belong to the same Anacardiaceae family. However, this may need further research for confirmation, as the present fruit yield data (three years) collected might not be sufficient to be conclusive. Marula trees are known to be prolific in fruit production. Mojeremane et al. (2010) reported production of 50,000 to 90,000 fruits per year from trees derived from seeds/seedlings. In this study, the trees were still in their early years of production and possibly the low temperatures at the study site (Zomba) might be too low for optimal marula fruit production, compared to their usual (natural) environment. In Mangochi, marula fruit production has been observed to be high and this could be attributed to fairly drier and high temperatures. Furthermore, marula trees in Mangochi have been observed to start bearing fruits six years after planting.

Conclusion

There have been significant morphological growth irregularities for both the marula scions and rootstocks and these scion and rootstock growth variations have negative effect on grafted tree survival and dwarfing characteristics in a marula fruit orchard. Although, these morphological growth variations might not have negative effect on fruit production, graft failure poses a great threat to the old, huge and fruit loaded grafted marula trees, especially those with thicker scions than their rootstocks. The findings indicate that scion and rootstock selection is critical in grafting marula trees to reduce scion and rootstock morphological growth irregularities which could bring graft failure in an orchard. This study warrants further studies to select marula genotypes or provenances with insignificant morphological growth irregularities between scions and their rootstocks when grafted. Furthermore, other practices such as tree pruning and training needs investigations to ascertain their effects on dwarfing, scion/rootstock growth patterns and promoting early fruit production for the grafted marula trees.

ACKNOWLEDGEMENTS

We thank BMZ and GTZ for providing the funds to

implement the project on "Domestication and Marketing of Indigenous Fruit Trees for Improved Nutrition and Income in Southern Africa", and CIDA Canada for funding the Zambezi Basin Agroforestry Project.

REFERENCES

Akinnifesi FK, Kwesiga F, Mhango J, Chilanga T, Mkonda A, Kadu CAC, Kadzere I, Mithöfer D, Saka JDK, Sileshi G, Ramadhani T, Dhliwayo P (2006). Towards the development of miombo fruit trees as commercial tree crops in southern Africa. For. Trees Livelihoods 16:103-121.

Akinnifesi FK, Mhango J, Sileshi G, Chilanga T (2008). Early growth and survival of three miombo woodland indigenous fruit tree species under fertilizer, manure and dry-season irrigation in southern Malawi. For. Ecol. Manage. 255:546-557.

Chirwa PW, Bwanali RJ, Meke G, Sagona W, Munthali CRY, Mwabumba L (2007). Growth and phenology of a three- to four-year-old *Sclerocarya birrea* international provenance trial in Malawi. Southern Hemisphere For. J. 69:49-54.

Emanuel PL, Shackleton CM, Baxter JS (2005). Modelling the sustainable harvest of *Sclerocarya birrea* subsp. caffra fruits in the South African lowveld. For. Ecol. Manage. 214:91-103.

Ham C (2005). Plant for food and drink. In: Mander M, Mckensie M (eds) Southern African trade directory of indigenous natural products. Commercial products from the wild group, South Africa pp. 17-22.

Ham C, Akinnifesi FK, Franzel S, Jordaan D du PS, Hansmann C, Ajayi OC, de Kock C (2008). Opportunities for commercialization and enterprise development of indigenous fruits in southern Africa. In: Akinnifesi FK, Leakey RRB, Ajayi OC, Sileshi G, Tchoundjeu Z, Matakala P, Kwesiga FR (eds) Indigenous fruit trees in the tropics: domestication, utilization and commercialization. CABI, UK p. 438.

Kadu CAC, Imbuga M, Jamnadass R, Dawson IK (2006). Genetic management of indigenous fruit trees in southern Africa: A case study of *Sclerocarya birrea* based on nuclear and chloroplast variation. S. Afr. J. Bot. 72:421-427.

Leakey R, Pate K, Lombard C (2005). Domestication potential of Marula (*Sclerocarya birrea* subsp caffra) in South Africa and Namibia: 2. Phenotypic variation in nut and kernel traits Agroforest. Syst. 64:37-49.

Leakey R (2005). Domestication potential of Marula (*Sclerocarya birrea* subsp. caffra) in South Africa and Namibia: 3. Multiple trait selection. Agroforest. Syst. 64:51-59.

Mateke S (2000). Cultivation of native fruit trees of Kalahari sandveld: studies on the commercial potential, interactions between soil and biota in Kalahari sands of southern Africa. Project Report. Velds Product, Botswana p. 49.

Mander M, Cribbins J, Shackleton SE, Lewis F (2002). The commercial marula industry in South Africa: A sub-sector analysis. Institute of Natural Resources, Scottsville, 3209, South Africa. Report for the DFID Forest Research Programme: Winners and losers in forest product commercialization, Project No. R7795. Centre for Ecology and Hydrology, Wallingford, OX10 8BB, UK.

Mhango J, Akinnifesi FK (2001). On-farm assessment of farmer managementand perception of priority indigenous fruit trees in southern Malawi. In:Proceedings of the 14th Southern Africa Regional Review and Planning Workshop. Harare, Zimbabwe.

Moganedi KLM, Colpaert N, Breyne P, Sibara MM, Goyvaerts EMA (2007). Determination of genetic stability of grafted marula trees using AFLP markers. Sci. Hort. 111:293-299.

Mojeremane W, Legwaila GM, Mmolotsi RM, Madisa ME, Rampart M (2010). Indigenous food plants in rural household food security in Botswana. A paper presented at the regional symposium on economically useful plants of Southern Africa, 21-23 June. Lusaka, Zambia.

Rouphaela Y, Schwarzb D, Krumbeinb A, Collac G (2010). Impact of grafting on product quality of fruit vegetables. Sci. Hort. 127:172- 179.

Rukuni D, Maruzane D, Zirobwa M, Shumba EM (2001). Provenance variation in a *Sclerocarya birrea* provenance trial ten weeks after planting at Enyandeni in Zimbabwe. The newsletter of the SADC tree seed centre network pp. 5-8.

SAS Institute Inc (1999). The SAS system for windows. SAS Institute Inc. SAS Campus drive. Cary, North Carolina, USA.

Shackleton SE (2002). The informal marula beer traders of Bushbuckridge, Limpopo Province, South Africa, Report to UK DFID Forestry Research Programme (Project No R7795). Rhodes University, Grahamstown, South Africa p. 25.

Usenik V, Štampar F (2000). Influence of various rootstocks for cherries on ρ-coumaric acid, genistein and pruning content and their involvement in the graft incompatibility process. Gartenbauwissenschaft 65(6):245-250.

Wynberg R, Cribbins J, Leakey RRB, Lombard C, Mander M, Shackleton SE, Sullivan CA (2002). A summary of knowledge on marula (*Sclerocarya birrea* subsp. caffra) with emphasis on its importance as a non-timber forest product in South and Southern Africa. 2. Commercial use, tenure and policy, domestication, intellectual property rights and benefit-sharing. Southern Afr. For. J. 196:67-77.

7

Inventory and distribution of the Annonaceae along elevation gradient on Mount Cameroon

Mekou Y. Bele[1]*, Derek A. Focho[2,3], Enow A. Egbe[3] and Bindeh G. Chuyong[3]

[1]Center for International Forestry Research (CIFOR), Central Africa Regional Office. P.O. Box 2008, Yaounde-Cameroon.
[2]Faculty of Science, University of Dschang, P. O. Box 222 Dschang, Cameroon.
[3]Faculty of Science, University of Buea, P. O. Box 63 Buea, SW Cameroon.

This work was carried out on Mount Cameroon, South West Region of Cameroon. It focused on producing an inventory and distribution of Annonaceae along the elevation gradient on the mountain. As a matter of fact, how Annonaceae flora change along an elevation gradient has never been studied for Mount Cameroon. The Complete Census Method was used. Twelve sites, located at three elevation categories: low (lowland: sea level to 700 m), mid (submontane: 800 to 1800 m), and high (montane: 2200 to 3500 m) were investigated. At each site, 8 transects of 500 m × 10 m were sampled and all the Annonaceae species enumerated. To supplement the information gathered on measured plots and to generate a complete list of target species, plotless sample method was used. This entailed the collection of materials along footpaths and within particular habitats such as ravines, seasonal streams, and disturbed areas. A total of 49 species in 21 genera including climbers, shrubs and trees were collected and identified from both plot and plotless samples. Low and mid elevations were rich in Annonaceae with 37 species in 17 genera and 31 species in 16 genera, respectively. The family was nearly absent from high elevation with only one species. Sixteen species in 10 genera were restricted to low elevation and 11 species in 8 genera confined to the mid elevation. Only *Uvariodendron connivens* was abundantly found at both low and mid elevations. Three species were frequent and confined to low elevation only. One and five species appeared occasionally and were restricted to low and mid elevations, respectively. 23, 18 and 1 species were rare at low, mid and high elevations respectively. Low elevation showed a higher species diversity (H' = 2.65) than the mid elevation (H' = 0.48). The high elevation with two individuals in a single species was the least diverse (H' = 0.15). The species were more even in their abundance distribution within the low elevation (J = 0.76) than in the mid elevation (J = 0.15).

Key words: Inventory, elevational distribution, Annonaceae, Mount Cameroon.

INTRODUCTION

Of the world's major ecosystems, the tropical forest is the most complex both architecturally and floristically and this complexity is least understood (Locatelli et al., 2008). Lying between 8° to 16°E and 2° to13°N (Amougou et al., 1985), Cameroon is a diversified territory with abundant natural resources. The sustainable exploitation and conservation of these resources is only possible if quailtative and quantitative estimations are made. Letouzey (1985)estimated the flora of Cameroon to be about 75,000 species distributed in about 1,800 genera belonging to 230 families. Annonaceae are among the most represented plant families in Cameroon particularly within Mount Cameroon (Cable and Cheek, 1998). It is one of the groups of primitive families of flowering plants sometimes termed the Annoniflorae, often with indefinite number of free floral parts and spirally arranged stamens.

*Corresponding author. E-mail: yoube_bele@yahoo.com.

It is a large family of the angiosperms with about 2,500 species in 130 genera worldwide. There are 900 species in 40 genera in Africa. 111 species in 24 genera are reported in the Flora of West Tropical Africa (Hutchinson and Dalziel, 1954). 121 species in 29 genera are reported in the Democratic Republic of Congo, 57 species in 16 genera in Central African Republic, 73 species in 24 genera in Angola, and 119 species in 29 genera in Gabon (Le Thomas, 1969). There are 4 species in 3 genera reported in Gambia (Steentoft, 1988) and 92 species in 27 genera recorded in Kenya (Beentje, 1994). In Cameroon, 22 species in 13 genera have been collected and identified in the 'Herbier National Camerounais' (YA). On Mount Cameroon, 37 species in 18 genera have been collected and identified in the Limbe Botanic Garden Herbarium (SCA).

In Africa, most of the Annonaceae genera are confined in the Guinean-Congolian region. A number are monotypic, such as, Meiocarpidium, Balonga, Pseudartabotrys and Exellia. Annonaceae are centred in the lowland rainforests and tend to change with elevation but are generally absent from elevations above 1,500 m (Dassanyayaka and Fosberg, 1985). To test the hypothesis that Annonaceae flora change with elevation, we focused our work on the distribution of the species of this family along three elevation categories on Mount Cameroon: low (lowland: sea level to 700 m), mid (submontane: 800 to 1800 m), and high (montane: 2200 to 3500 m). This work drew its origin on recommendations from previous works on the mountain by Tchouto (1995; 1996), Cable and Cheek (1998) and Ndam (1998) based on: (1) although plant species of Mount Cameroon are described in a comprehensive volume (Cable and Cheek, 1998), there was an urgent need for an account, not merely of the occurrence of species on the mountain, but also of their distribution especially along the elevation gradient for families whose flora tends to change along an elevation gradient such as the Annonaceae; (2) botanical collectors, naturally enough, have concentrated their collection of the lowland species of Annonaceae especially on accessible places generally near forest paths, villages, main roads and towns, and with accessible flowers for much of the year. In addition, climbers have often been omitted. Foresters get off the beaten track, but their main interest is on large commercial trees. Therefore, further inventory was needed to have a full view of an elevational distribution of the family on the mountain; (3) in addition, according to WCMC (1995) and IUCN (1998) some species within the family have been reported to be threatened in Cameroon and especially within Mount Cameroon (Cable and Cheek, 1998).

Therefore, this study aimed to assess the distribution of the Annonaceae species on the mountain by providing frequency of species occurrence, variation of species diversity, individual species abundances and the list of rarity along the three elevation categories.

The objectives were to undertake an inventory of Annonaceae on Mount Cameroon, to augment herbarium specimens and to determine the distribution patterns of the various species along the three elevation categories.

METHODS

Description of the study area

This work was carried out on Mount Cameroon in the South-West Region of Cameroon, on the coastal belt of the Gulf of Guinea (Biafra) (Figure 1). Mount Cameroon is the highest mountain in Central and West Africa, rising up steeply from sea level to 4,095 m at the summit (Cable and Cheek, 1998; Ndam, 1998). This huge volcanic mass has its long axis (about 45 km long and 30 km wide) running SW to NE between 3°57' to 4°27'N and 8°58' to 9°24'E. The main peak is at 4°7'N and 9°10'E (Tchouto, 1996; Fraser et al., 1998).

Mount Cameroon has a humid tropical climate and the climatic pattern is sharply modified by the influence of the topography (Courade, 1972). The main annual rainfall varies between 2,085 mm near Ekona to 9,086 mm at Debundscha (Fraser et al., 1998).

The mean annual temperature is about 25°C and this decreases by 0.6°C per 100 m of ascent, to 4°C at the summit (Boughey, 1955; Barry, 1992). The soil temperature, measured at 10 cm depth, varies from 25°C at 200 m through 20°C at 1,100 m to 15°C at 2,200 m (Payton, 1993). The air humidity remains at 75 to 80% throughout the year on the South Western side of the Mountain due to the marine influence and the incidence of mist and orographic cloud formation. The persistent cloud cover and mist make Mount Cameroon one of the areas that receive the lowest values of annual sunshine hours in West Africa. This starts with 900 to 1,200 h per annum at sea level and decreases with altitude. Between 1,200 to 2,000 m, the mountain is characterized by semi-permanent mists and cloudiness (Payton, 1993). Snow is rare on the summit (4,095 m) of Mount Cameroon, though frost is not uncommon.

At sea level, there is often a gentle breeze at the coast, but generally air movements are slow. Hurricanes are unknown. At the beginning of the wet season however, the opening frontal storms can be violent and this is the main season for natural tree fall.

Mount Cameroon is an active Hawaiian type volcano without a central crater, subject to continued fissure eruptions on the flank of the mountain that result in lava flows and small cinder cones. The mountain is built up by a successive series of eruptions of basaltic lava, tuff and ash and materials that surround and cover to a considerable depth, the base of the mountain. This results in a landscape of steeply sloping (15 to 35°C) rocky ridges conforming to individual lava flows separated by numerous ravines.

The soil types are mainly volcanic and relatively fertile. The lower slopes of the mountain have ancient ferralitic soils on highly weathered substrates. They are much deeper, nutrient-rich andosols, less stony and therefore intensively cultivated. Above 1,000 m, the volcanic lava gives rise to brownish or blackish soils. The surface of the mountain peak is covered by latest lava flows, cinder cones and other pyroclastic materials (Tchouto, 1996; CDC, 1997).

Human factors that affect the mountain come in the form of burning for hunting purposes and farming activities, although above 1,200 m the terrain becomes too rough for farming. At lower levels, the forest has greatly been replaced by shifting cultivation or organized commercial plantations, mainly oil palms, banana and rubber scattered over the Southern slopes of the mountain from sea level to 950 m (Payton, 1993).

Desktop preparation and reconnaissance surveys

The work involved review of the plant specimens found in both the

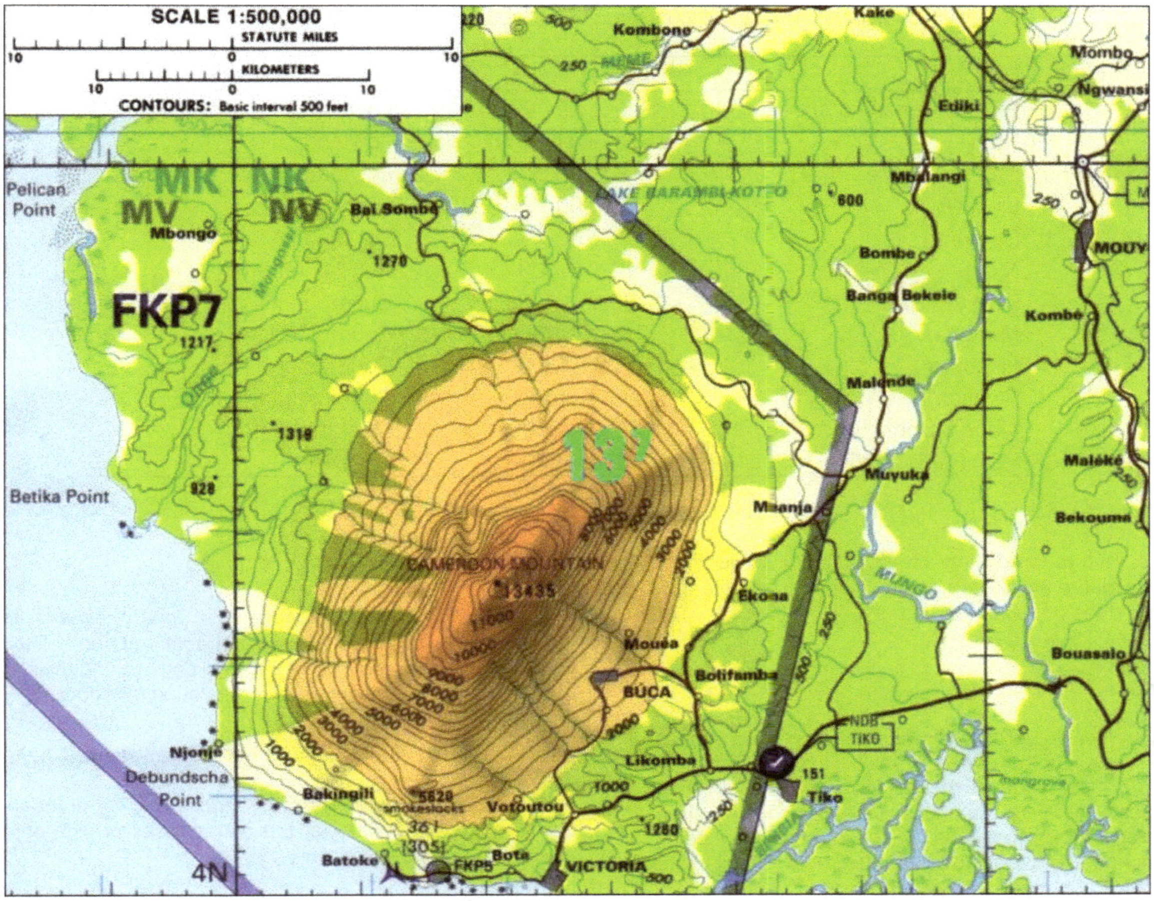

Figure 1. Map of the study area.

Herbier National Camerounais (YA)' and the Limbe Botanic Garden Herbarium (SCA). Reconnaissance surveys aimed to choose representative sites for plant collection on the mountain were done by taking into consideration the elevation category and locations on the mountain. The surveys entailed going through the accessible parts of the vegetation of the mountain. It was also a period to get acquainted with village communities prior to proper field surveys.

Field surveys and collection sites

Since Mount Cameroon is too large for this study in the time available, representative sites were selected for more detailed study after preliminary reconnaissance surveys. 12 sites were selected at three elevation categories (Table 1). The choice of the sites was based on the differences in structure of the vegetation, physiognomy and crown form. Other criteria such as geology, soils, climate and ecological factors (e.g. fire) were also taken into consideration. Field surveys and collections were also made at these sites at the three elevation categories.

Sampling strategy and techniques

Transects were used for a better option for estimation, detection, and distribution of Annonaceae species. They were laid along the altitudinal gradient in each of the three elevations (lowland, submontane and montane). Based on the considerations outlined in Peters (1994 and 1999) and Sunderland (1999), sampling consisted of a series of parallels of 10 m wide established at irregular intervals at a random starting point and were 500 m in length (Ideally, these would have been longer (1 to 2.5 km, Peters, 1994 and 1999- but time constraints dictate the need for more numerous, shorter transects)and were run out along a predetermined bearing using a compass. Distances were corrected for slopes in order to get horizontal distances using:

$$D= d \times \cos\alpha,$$

Where D= horizontal distance of the transect, d= slope distance and α = the angle of the slope.

Transects were covered by walking, counting and recording the sample species. Eight transects were established in each of the five sites at low and mid elevations. At high elevation, only two sites were selected and eight transects were established at each of the sites. However, given the fact that the measured plot samples method is inefficient for generating species list and that some species, particularly those with very restricted habitats or those which are scarce, may almost certainly be missed, plotless sample method was used. This method was used to supplement the information gathered on measured plots and to ensure that target species were collected or recorded. It entailed the collection of materials along footpaths and within particular habitats such as ravines, seasonal streams and disturbed areas. These distinctions were very important for evaluating range data in particular as the sample plot method has most statistical meaning for defining

Table 1. Field surveys and collection sites.

Site	Elevation (m)	Elevation category
Bimbia-Bonadikombo	Sea level - 200	Lowland
Njonji	Sea level - 600	Lowland
Bakingili	200 – 700	Lowland
Liwenyi	200	Lowland
Bomana	150	Lowland
Njonji	800 - 1800	Submontane
Bakingili	800 – 1000	Submontane
Mapanja	800 –1800	Submontane
Likombe	700 - 1800	Submontane
Bwassa	800 – 1800	Submontane
Mann's Spring	2200 – 3500	Montane
P&T Radio Station	2200 – 3500	Montane

associations (allows prediction of distribution) while the plotless method provides the least ambiguous distribution data (that is, it really does or does not occur in the area) (Sheil, 1997).

Data collection

Once transects were laid out, the enumeration team moved slowly along these transects and carefully searching or scanning a 10 m-band (5 m on either side of the central line) for individuals of all target species. The 5 m distances were periodically checked for borderline trees. Each individual of the appropriate species once spotted, checked and identified by the field botanist was recorded in the datasheet. Each individual tree, shrub and liana was recorded. Information on physical characteristics of the habitat, vegetation structure and elevation was also recorded.

Plant collection, treatment and identification

Herbarium specimens were made for species not well represented in the SCA herbarium. In addition, herbarium collections were automatically made in case of doubt in identification in the field and for suspected species for further identification in the herbarium. Adequate samples of all the organs available were collected whenever possible. Plant specimen preparation methods generally followed those of Foreman and Bridson (1989). On returning to the herbarium, specimens were inspected and interleaved with blotters and corrugated sheets before placing them in the dryer. The presses in the dryer were frequently inspected. During inspection, specimens that had dried up were removed from the dryer. Identification was done in the Limbe herbarium (SCA) using already existing herbarium specimens and illustrated flora. Some doubtful specimens were taken to the National Herbarium for identification. Undermined specimens were then piled for future identification at Kew by the family specialist.

Data analysis

The species encountered were compiled according to the three elevation categories. Excel and Statsgraphics software were then used for different analyses. Species diversity and similarity were measured to provide information about rarity and commonness of species at the three different elevation categories. The following equations were used to calculate indexes:

$D = 1\text{-}S(pi)$ or $H' = S(ni/N)(\log_n ni/N)$

Where D= Simpson's index; H'= Shannon-wiener index; ni=number of individuals of a species; N=total number of individuals. These index measures accounted for both abundance and evenness of the species present at each elevation category.

$J = S(ni/N)(\log_n ni/N)/\log_n S$

Where J= Equitability or evenness and S= number of species. This measures how similar the abundances of different species were at each elevation category.

$Cj = a/(a+b+c)$ or $Cn = 2a/(2a+b+c)$

Where Cj= Jaccard's Coefficient, Cn= Sorensen's coefficient, a= number of species in Samples A and B (joint occurrence), b= number of species in Sample B but not in Sample A, and c= number of species in Sample A but not in Sample B. These measures helped compare the similarity and diversity of sample sets at different elevation categories.

$t = \chi - \mu/S\chi$

Where t= student t-test, χ= deviation, μ= mean, and $S\chi$= standard deviation. This test was used to check the normality of the distribution.

Mann-Whitney test: $W = n1\ n2 + n1(n1+1)/2 - R1$ and $W' = n2\ n1 + n2(n2+1)/2 - R2$

Where n1= Number of observations in community 1, n2 = Number of observations in community 2, R1= Sum of rank in community 1, R2= Sum of rank in community 2. This test was used as an alternative to the independent group t-test in case the assumption of normality or equality of variance was not met.

RESULTS

Species encountered

41 species in 20 genera were collected and identified for a total of 96 transects of 10 m × 500 m surveyed from 12 sites at low, mid and high elevation (Table 2). In addition,

Table 2. List of Annonaceae species recorded at three elevation categories.

Species code no.	Species code	Scientific name	Elevation		
			Low	Mid	High
1	ANNICH	*Annickia chloranta* (Oliv.) Setten & P.J.Maas	X	X	
2	ANNOSE	*Annona senegalensis* Persoon			X
3	ANONFL	*Anonidium floribundum* Pellegr	X		
4	ARTAAR	*Artabotrys aurantiacus* Engl.et Diels	X		
5	ARTARH	*Artabotrys rhopalocarpus* Le Thomas	X	X	
6	ARTAVE	*Artabotrys velutinus* Scott-Elliot		X	
7	BOUTPL	*Boutiquea platypetala* (Engl.et Diels) Le Thomas	X	X	
8	FRIEEN	*Friesodelsia enghiana* (Diels) Verdc.	X	X	
9	FRIEGR	*Friesodelsia gracilipes* (Benth.) v.Steenis	X	X	
10	FRIEMO	*Friesodelsia montana* (Engl.et Diels) v. Steenis		X	
11	GREESU	*Greenwayodendron suaveolens* (Engl.et Diels) Verdc.	X	X	
12	ISOLZE	*Isolona zenkeri* Engl	X		
13	MEIOLE	*Meiocarpidum lepidotum* (Oliv.) Engl.et Diels		X	
14	MISCEL	*Mischogyne elliotianum* (Engl.et Diels) R.E.Fries		X	
15	MONAFO	*Monanthotaxis foliosa* (Engl.et Diels) Verdc.		X	
16	MONOBR	*Monodora brevipes* Benth	X		
17	MONOCR	*Monodora crispata* Engl.et Diels	X	X	
18	MONOMY	*Monodora myristica* (Gaertn.) Dunal	X	X	
19	MONOTE	*Monodora tenuifolia* Benth	X	X	
20	PACHST	*Pachypodanthium staudtii* (Engl.et Diels) Engl.et Diels	X		
21	PIPTFA	*Piptostigma fasciculata* (De Wild Boutique)		X	
22	PIPTMU	*Piptostigma multinervum* Engl.et Diels	X	X	
23	PIPTPI	*Piptostigma pilosum* Oliv	X	X	
24	POLYPA	*Polyceratocarpus parviflorus* (Bak.f.) Ghesq	X	X	
25	UVARAN	*Uvaria anonoides* Bak.f.	X	X	
26	UVARBA	*Uvaria baumannii* Engl.et Diels	X		
27	UVARIN	*Uvariastrum insculptum* (Engl.et Diels) Spr.et Hutch.	X		
28	UVARPY	*Uvariastrum pynaertii* De Wild	X		
29	UVARZE	*Uvariastrum zenkeri* Engl.et Diels	X		
30	UVARCON	*Uvariodendron connivens* (Benth.) R.E.Fries	X	X	
31	UVARFU	*Uvariodendron fuscum* (Benth.) R.E.Fr.		X	
32	UVARGI	*Uvariodendron giganteum* (Engl.) R.E.Fries	X		
33	UVARBA	*Uvariopsis barkeriana* (Hutch.et Dalz.) Robyns & Ghesq	X	X	
34	UVARDI	*Uvariopsis dioica* (Diels) Rob. & Ghesq	X	X	
35	UVARK0	*Uvariopsis korupensis* Gereau & Kenfack	X	X	
36	XYLOAC	*Xylopia acutiflora* (Dunal) A. Rich	X	X	
37	XYLOAF	*Xylopia africana* (Benth.) Oliv.		X	
38	XYLOHYP	*Xylopia hypolampra* Mildbr.	X	X	
39	XYLOQU	*Xylopia quintasii* Engl.et Diels	X		
40	XYLORU	*Xylopia rubescens* Oliv.	X		
41	*XYLOST*	*Xylopia staudtii* Engl.et Diels	X		

8 species were collected and identified from plotless samples within particular habitats out of transects such as foothpaths, ravines, seasonal streams, and disturbed areas (Table 3). *Annona muricata* was the only cultivated species collected. The ornamental species recorded were *Greenwayodendron peltata* and *Cananga odorata*.

Distribution

From a total of 41 species obtained from measured plots, 12 (29%), 9 (22%) and 1 (3%) were restricted to low, mid and high elevation respectively while 19 (46%) were shared by both low and mid elevations (Table 2). From

Table 3. List of species found out of transects.

Species code	Scientific name	Elevation		
		Low	Mid	High
CLEIST	*Cleistopholis staudtii* Engl. et Diels	X	X	
FRIEHI	*Friesodelsia hirsuta* (Benth.) v.Steenis		X	
MONAFI	*Monanthotaxis filamentosa* (Diels) Verdc.		X	
PIPTCA	*Piptostigma callophyllum* Mildbr. et Diels	X		
PIPTOY	*Piptostigma oyemensis* Pellegr		X	
UVARHE	*Uvaria heterotricha* Pellegr	X		
UVARMO	*Uvariodendron molundense* (Engl. et Diels) R.E.Fries	X		
XYLOPA	*Xylopia parviflora (A. Rich.)* Benth	X		

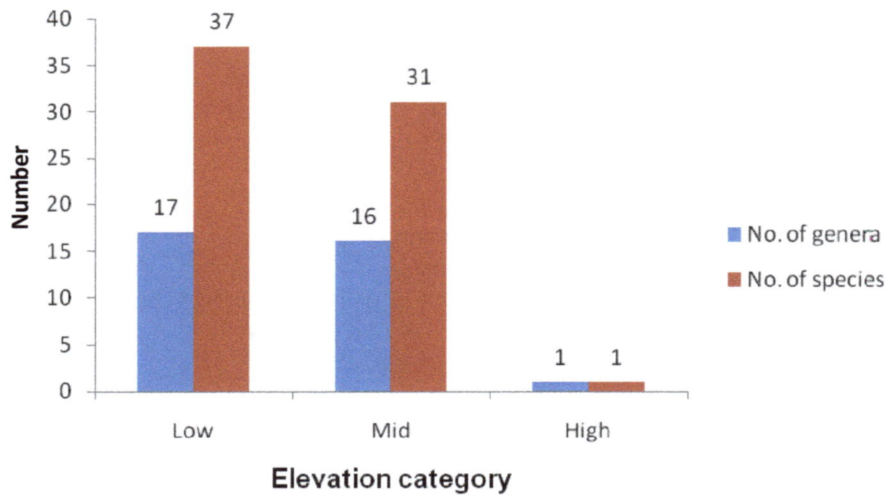

Figure 2. Number of genera and species at three elevation categories.

plotless samples, only *Cleistopholis staudtii* was shared by both low and mid elevation and 4 and 3 species restricted only to low and mid elevations respectively and nothing found at high elevation (Table 3).

The number of genera and species at each elevation category was calculated (Figure 2). Considering that species from plotless samples only served to generate a complete list of species and since their number of individuals was not the main interest here, the eight species from these samples were not more included in the further analysis. The distribution was therefore studied in terms of presence or absence, abundance as well as frequency from measured plot samples (Tables 4 and 5).

Frequency of occurrence of species within the three elevation categories

The frequency of each species was calculated at the three elevation categories using the formulae

$X/Y \times 100$

Where X= number of transects in which species is present, Y= total number of transects in each elevation category (Table 5).

The evaluation of each species in each elevation was made as follows:

1. Species present in 30 - 40 transects that is, 75-100%, meant that the species was abundant
2. Species present in 20 - 29 transects that is, 50-74%, meant that the species was frequent
3. Species present in 10-19 transects that is, 25-49%, meant that the species was occasional

4. Species present in less than 10 transects i.e. <25%, meant that the species was rare.

Code numbers used to represent these were as follows: Abundant= 4; Frequent= 3; Occasional = 2; and Rare = 1.

Table 4. Total number of individuals of each species encountered at each elevation category

Species code no.	Species name	Species code	Total		
			Low	Mid	High
1	*Annickia chloranta*	ANNICH	30	5	0
2	*Annona senegalensis*	ANNOSE	0	0	2
3	*Anonidium floribundum*	ANONFL	1	0	0
4	*Artabotrys aurantiacus*	ARTAAR	1	0	0
5	*Artabotrys rhopalocarpus*	ARTRH	5	4	0
6	*Artabotrys velutinus*	ARTAVE	0	4	0
7	*Boutiquea platypetala*	BOUTPL	1	1	0
8	*Friesodelsia enghiana*	FRIEEN	67	15	0
9	*Friesodelsia gracilipes*	FRIEGR	75	15	0
10	*Friesodelsia Montana*	FRIEMO	0	5	0
11	*Greenwayodendron suav.*	GREESU	3	3	0
12	*Isolona zenkeri*	ISOLZE	1	0	0
13	*Meiocarpidum lepidotum*	MEIOLE	0	2	0
14	*Mischogyne elliotianum*	MISCEL	0	2	0
15	*Monanthotaxis foliosa*	MONAFO	0	2	0
16	*Monodora brevipes*	MONOBR	1	0	0
17	*Monodora crispate*	MONOCR	2	2	0
18	*Monodora myristica*	MONOMY	15	11	0
19	*Monodora tenuifolia*	MONOTE	2	11	0
20	*Pachypodanthium staudtii*	PACHST	2	0	0
21	*Piptostigma fasciculate*	PIPTFA	0	3	0
22	*Piptostigma multinervum*	PIPTMU	1	3	0
23	*Piptostigma pilosum*	PIPTPI	21	5	0
24	*Polyceratocarpus parviflorus*	POLYPA	74	20	0
25	*Uvaria anonoides*	UVARAN	1	2	0
26	*Uvaria baumannii*	UVARBA	4	0	0
27	*Uvariastrum insculptum*	UVARIN	4	0	0
28	*Uvariastrum pynaertii*	UVARPY	2	0	0
29	*Uvariastrum zenkeri*	UVARZE	1	0	0
30	*Uvariodendron connivens*	UVARCON	130	1861	0
31	*Uvariodendron fuscum*	UVARFU	0	14	0
32	*Uvariodendron giganteum*	UVARGI	1	0	0
33	*Uvariopsis barkeriana*	UVARBA	13	5	0
34	*Uvariopsis dioica*	UVARDI	6	5	0
35	*Uvariopsis korupensis*	UVARK0	11	5	0
36	*Xylopia acutiflora*	XYLOAC	4	8	0
37	*Xylopia Africana*	XYLOAF	0	10	0
38	*Xylopia hypolampra*	XYLOHYP	3	5	0
39	*Xylopia quintasii*	XYLOQU	1	0	0
40	*Xylopia rubescens*	XYLORU	1	0	0
41	Xylopia staudtii	XYLOST	1	0	0

From Table 5, only one species (*Uvariodendron connivens*) was found abundant at both low and mid elevations. Three species were frequent and confined to low elevation only. 1 and 5 species appeared occasionally and were restricted to low and mid elevations respectively. 23, 18 and 1 species were rare at low, mid and high elevations respectively (Figure 4).

The diversity index (Table 6) was also assessed. Shannon-Wiener diversity index for the different elevation categories is given in Figure 5.

The similarity measures were also assessed (Table 7). The higher elevation with only 1 species with 2.

Table 5. Frequency of occurrence of each species at each elevation category.

		Low			Mid			High		
Species code	Species code	No. of transects in which species occur out of 40 (X)	Frequency (X/Y) × 100	Frequency code	No. of transects in which species occur out of 40 (X)	Frequency (X/Y) × 100	Frequency code	No. of transects in which species occur out of 16 (X)	Frequency (X/Y) × 100	Frequency code
1	ANNICH	14	35	2	4	10	1	0	0	0
2	ANNOSE	0	0	0	0	0	0	2	12.5	1
3	ANONFL	1	2.5	1	0	0	0	0	0	0
4	ARTAAR	1	2.5	1	0	0	0	0	0	0
5	ARTARH	3	7.5	1	5	12.5	1	0	0	0
6	ARTAVE	0	0	0	3	7.5	1	0	0	0
7	BOUTPL	3	7.5	1	1	2.5	1	0	0	0
8	FRIEEN	25	62.5	3	11	27.5	2	0	0	0
9	FRIEGR	22	55	3	11	27.5	2	0	0	0
10	FRIEMO	0	0	0	4	10	1	0	0	0
11	GREESU	3	7.5	1	5	12.5	1	0	0	0
12	ISOLZE	1	2.5	1	0	0	0	0	0	0
13	MEIOLE	0	0	0	3	7.5	1	0	0	0
14	MISCEL	0	0	0	2	5	1	0	0	0
15	MONAFO	0	0	0	3	7.5	1	0	0	0
16	MONOBR	1	2.5	1	0	0	0	0	0	0
17	MONOCR	2	5	1	2	5	1	0	0	0
18	MONOMY	9	22.5	1	7	17.1	1	0	0	0
19	MONOTE	1	2.5	1	7	7	1	0	0	0
20	PACHST	3	7.5	1	0	0	0	0	0	0
21	PIPTFA	0	0	0	1	1	1	0	0	0
22	PIPTMU	1	2.5	1	3	3	1	0	0	0
23	PIPTPI	9	22.5	1	5	5	1	0	0	0
24	POLYPA	26	65	3	16	16	2	0	0	0
25	UVARAN	1	2.5	1	1	1	1	0	0	0
26	UVARBA	3	7.5	1	0	0	0	0	0	0
27	UVARIN	1	2.5	1	0	0	0	0	0	0
28	UVARPY	1	2.5	1	0	0	0	0	0	0
29	UVARZE	1	2.5	1	0	0	0	0	0	0
30	UVARCON	30	75	4	39	39	4	0	0	0
31	UVARFU	0	0	0	12	12	2	0	0	0
32	UVARGI	1	2.5	0	1	1	0	0	0	0
33	UVARBA	8	20	1	7	7	1	0	0	0

Table 5. Contd.

37	XYLOAF	0	0	0	15	15	2	0	0	0
38	XYLOHYP	1	2.5	1	5	5	1	0	0	0
39	XYLOQU	1	2.5	1	0	0	0	0	0	0
40	XYLORU	1	2.5	1	0	0	0	0	0	0
41	XYLOST	7	17.5	1	0	0	0	0	0	0

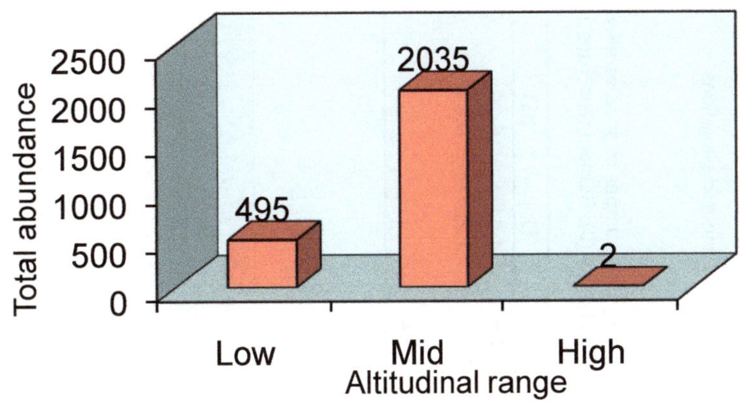

Figure 3. Total number of individuals at each elevation.

individuals was again isolated

To show an example of species richness, Bimbia Bonadikombo and Njonji were randomly selected for low and mid elevations respectively (Figure 6 and 7).

DISCUSSION

The family Annonaceae comprises climbers, shrubs, and trees. During this survey, 41 species in 20 genera were recorded in a total of 96 transects of 10 × 500 m surveyed from 12 sites at low, mid, and high elevations. Table 2 presents the list of all the species encountered in alphabetical order and their corresponding codes. There were 18 species of trees, 4 species of shrubs, 7 species of climbers, 12 species of shrubs/trees. In addition, 8 species were collected and identified from plotless samples within particular habitats out of transects such as foothpaths, ravines, seasonal streams and disturbed areas (Table 3). *A. muricata* was the only cultivated species collected. The ornamental species recorded were *G. peltata* and *C. odorata*. From a total of 41 species from measured plots, 12 (29%), 9 (22%), and 1 (3%) were restricted to low, mid and high elevation respectively while 19 (46%) were shared by both low and mid elevations only (Table 2). From plotless samples, only *Cleistopholis staudtii* was shared by both low and mid elevation and 4 and 3 species restricted only to low and mid elevations categories respectively and nothing found at high elevation (Table 3).

There were almost no significant differences in terms of the number of genera and species at low and mid elevation categories (Figure 2). However, the significant differences were quite clear high elevation category.

The differences in number of genera and species between low and mid elevations on one side and the high elevation on the other side could

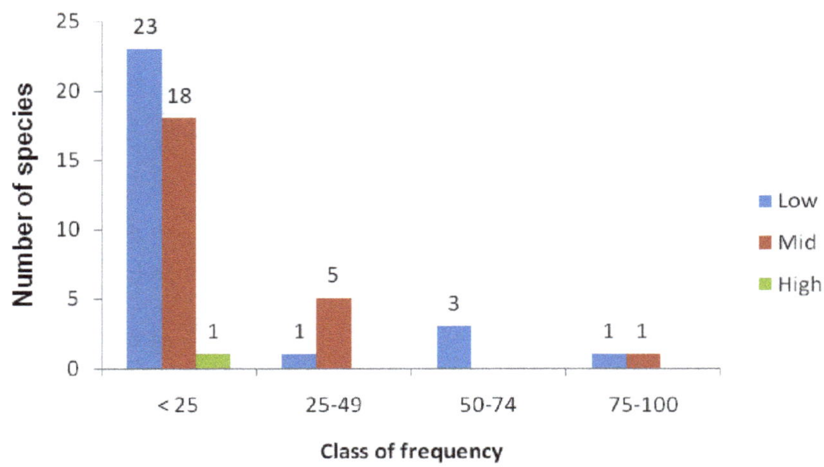

Figure 4. Classes of frequency of species at low and mid elevations.

Table 6. Number of plot (Np), number of individuals/elevation category (N), number of species/elevation category (S), Shannon-Wiener index (H'), Simpson's index (D), reverse of Simpson's index (1/D), Equitability (J), Student test (t-test), Mann-Whitney (w-test).

Elevation category	Np	N	S	H'	D	1/D	J	t-test	w-test
Low	40	494	32	2.65	4.61	0.22	0.76	t=27.85	w = 843.5
Mid	40	2035	27	0.48	1.23	0.81	0.15		p = 0.5094 α = 0.05
High	16	2	1	0.15					

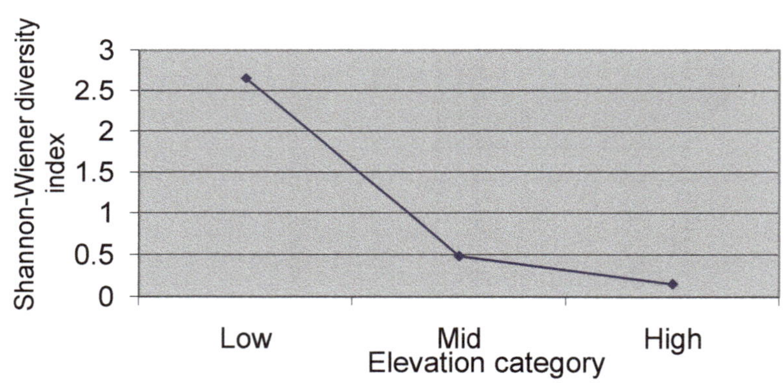

Figure 5. Graph of Shannon-Wiener diversity index for the three elevation categories.

be attributed to the fact that the differences in altitudes had a marked effect on temperature, relative humidity and soil which in turn affects the type of vegetation found on the mountain slope (Payton, 1993). Soils in the lowland forest are nutrient-rich andosols. Andosols are formed from volcanic cinders and weathered basalts from lava flows, which moved towards the lowland in the form of mud. This soil has relatively high natural fertility and good biological and physical properties (Cheek, 1992 and CDC, 1997). With increasing elevation, the humus content of the soil increases due to the slower oxidation rates and the soils are dark in colour. At higher elevations, these soils change rapidly to very young, shallow coarse-textured andosols containing very little water, interspersed with extensive areas of bare recent lava flows and cinder deposits. It has been suggested that recent volcanic activity and immature droughty soils are the major factors limiting tree growth at higher elevation (Hasselo, 1961; Payton, 1993).

Restriction of species distribution to one or two

Table 7. Similarity measures.

Measure	Value
Cj	0.488
Cs	0.656
Cm	0.187
Cmh	0.504

Cj= Jaccard measure (qualitative), Cs = Sorensen measure (qualitative), Cn = Sorensen measure (quantative data), Cmh = Morista-Horn measure (quantitative data).

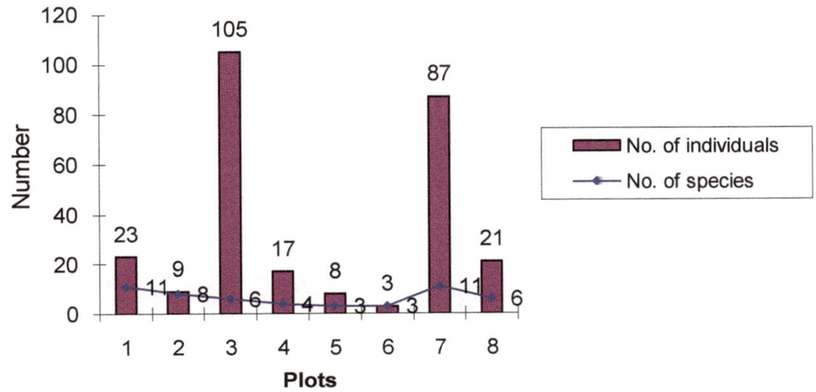

Figure 6. Number of species (S) against number of individuals (N) for all plots in Njonji.

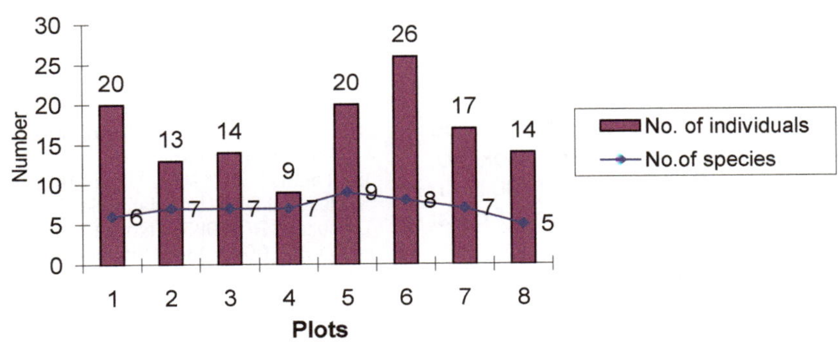

Figure 7. Number of species (S) against number of individuals/site (N) for all plots in Bimbia Bonadikombo.

elevation categories might be simply because dispersal has only just begun and they have not had enough time to expand their range into other habitats. On the other hand, they might be evolutionarily isolated and are particularly useful in indicating antiquity, isolation and diversification of habitats (Rexford, 1978; Tchouto, 1995). Restriction might also be attributed to the lack of dispersers that could carry seeds from one habitat to the other. It might also be because the species are habitat-characteristic or because the viability of their seeds is short. The poor species representation at the high elevation category can be attributed to low sampling size at that elevation category. In fact, only 16 transects were surveyed instead of 40 as in the other elevation categories. Species might also be killed by regular bush fires set by hunters during each dry season at the high elevation.

The abundance of certain species such as *Uvariodendron connivens, Friesodelsia gracilipes, Polyceratocarpus parviflora,* etc might be due to the fact that under climatic constraints, seeds have acquired the capacity to remain dormant during unfavourable periods.

Favourable conditions permit a continuous uninterrupted process of flowering, seed development and dissemination, germination and early seedling growth (Maury-Lechon, 1993). It might also be due to the fact that these species have long flowering and fruiting periods favouring pollination as compared to other species whose reproductive organs do not reach maturity probably because they are washed off by heavy rain at their early stage of development.

Only one species (*Uvariodendron connivens*) was found abundant at both low and mid elevations. Three species were frequent and confined to low elevation only. 1 and 5 species appeared occasionally and were restricted to low and mid elevations respectively. 23, 18 and 1 species were rare at low, mid and high elevations respectively (Figure 4). Rareness of species might be attributed to the fact that they do not flower yearly (intervals are usually irregular) and their seeds, generally fleshy and hydrated, lose their viability when dehydrated, so that they cannot be conserved ("recalcitrant" seeds) (Bras and Maury-Lechon, 1986; Corbineau and Côme, 1988; Tompsett, 1990).

There was an overall trend of decline of species diversity with elevation categories. The Shannon-Wiener index shows that low elevation (H' = 2.65) is more diverse than the mid elevation (H' = 0.48). Similar results were obtained for Simpson's index (Table 6). The higher elevation with two individuals of a single species was therefore the least diverse (H' = 0.15). Species were more even in their abundance distribution within at low elevation (J = 0.76) than at mid elevation (J =0.15). This shows that although the mid elevation contained more individuals; it has a less even species distribution than the low elevation category. Similar results have been obtained by Tchouto (1995) in the proposed Etinde Rainforest Reserve, Sunderland (2000) in Mokoko, Campo and Takamanda forests, and Mboh (2001) in Takamanda, Campo Ma'an, and Ejagham Forest Reserves.

A number of species were prominent in both the low and mid elevations. The Sorensen's Coefficient (Cs = 0.656) shows that low and mid elevations share similarity in species. A similar result was obtained for Jaccard's Coefficient (Cj = 0.488) and Morista-Horn's coefficient (Cmh = 0.504) of similarity measures. Mann-Whitney test also shows that there is no statistically significant difference at the 95% confidence interval between the low and the mid altitudinal elevation categories (Table 6). However, more studies are needed to better understand the pattern behind distribution of species at different elevation categories.

ACKNOWLEDGEMENTS

This project benefited enormously from the contributions of its research team especially Nkeng Philip Fonju; Maurice Betafor, Elias Ndive and Paul Ndumbe from the Limbe Botanic Garden. Sincere thanks to the Management of the Limbe Botanical and Zoological Gardens for their assistance, services, and facilities; and to all the reviewers of this work from the beginning to publication.

REFERENCES

Beentje HJ (1994). Kenya Trees, Shrubs, and Lianas, 722p.
Boughey AS (1955). The vegetation of the mountain of Biafra. Proc. Linn. Soc., 165: 144 -150.
Bras P, Maury-Lechon G (1986). Graines forestieres tropicales de type fortement hydraté: la conservation et ses effets, exemple du Symphonia globulifera L.f. de Guyane française. B.T.F. 212: 35-46.
Cable S, Cheek M (1998). The plants of Mount Cameroon: A conservation checklist. Royal Botanic Gardens, Kew, 112p
CDC (1997). Soil surveys and land evaluations for the second development of program of program of the Cameroon Development Corporation. FAO/UNDP/IRAF/ONAREST. Ekona, 7: 113
Cheek M (1992). A botanical inventory of the Mabeta-Moliwe forest. Report on Limbe Botanic Garen Conservation Project, 112p.
Corbineau F, Côme D (1988). Storage of recalcitrant seeds of four tropical species. Seed Sci. Technol., 16: 97-103
Courade G (1972). Regional Atlas. West 1, ORSTOM Cameroon.
Dassanyayaka MD, Fosberg FR (1985). A revised handbook to the Flora of Ceylan vol.5. Smithsonian institution and the Nation Science Foundation. Washington, D.C. pp. 475
Focho DA, Egbe EA, Chuyong GB, Fongod AGN, Fonge BA, Ndam N, Youssoufa BM (2010). An ethnobotanical investigation of the annonaceae on Mount Cameroon. J. Med. Plants Res., 4(20): 2148-2158.
Foreman L, Bridson D (1989). The herbarium handbook. Royal Botanic Garden, Kew, 333p.
Fraser PJ, Hall JB, Healey JR (1998). Climate of the Mount Cameroon region, long and medium term rainfall, temperature and sunshine data. SAFS, University of Wales Bangor, MCP-LBG, Limbe. pp. 56.
Hasselo HN (1961). The soils of the lower eastern slopes of Cameroon mountain and Their suitability for various perennial crops. H. Veemuneen, Wageningen, 67p.
Hutchinson J, Dalziel JM (1954). Flora of West Tropical Africa. 1: I-295.
IUCN (1998). Red date list of threatened species. World Conservation Monitoring Center
Le Thomas A (1969). Les Annonaceae. Flore du Gabon. Muséum National D'histoire Naturelle, 371p.
Letouzey R (1985). Notice of phytographic map of Cameroon at 1/1.500.000. International Institute of Cartography, Toulouse. 6 leaflets, 240p.
Maury-Lechon G (1993). Biological characters and plasticity of juvenile tree stages to restore degraded tropical forests: Programme proposition for genetic resource maintenance, pp. 37-47. In: Lieth H. and Lohmann M. (eds), Restoration of Tropical Forest Ecosystems. Kluwer Academic Publishers, Netherlands, 269p.
Mboh H (2001). Vegetation assessment of Takamanda forest reserve and comparison with Campo Ma'an and Ejagham forest reserves. Mémoire dissertation. Faculty of Agronomy and Agricultural Sciences, University of Dschang, Cameroon, 62p.
Ndam N (1998). Tree regeneration, vegetation dynamics and the maintenance of biodiversity on Mount Cameroon: the relative impact of natural and human disturbance. A Ph.D dissertation. School of Agriculture and Forest Sciences, University of Wales Bangor, U.K., 320p.
Payton RW (1993). Ecology, altitudinal zonation and conservation of tropical rainforest of Mount Cameroon Final Project Report R4600, ODA, London, Mount Cameroon Project, 70p.
Rexford D (1978). Plant geography. Academic Press. New York, San Francisco and London.
Steentoft M (1988). Flowering plants in West Tropical Africa. Cambridge University Press, 344p.
Sunderland TCH (2000). The Taxonomy, Ecology and Utilization of

African Rattan (Palmae & Calomoideae), PhD thesis, University College of London, UK. Pp. 308-320.

Tchouto P (1995). The vegetation of the proposed Etinde Forest Reserve, Mount Cameroon, and its conservation. M.Sc. dissertation, University of Edinburgh. Royal Botanic Garden, Edinburgh, 94p.

Tchouto P (1996). Forest inventory report of the proposed Etinde Rainforest reserve. Mount Cameroon Project, 68p.

Thompsett PB (1990). Predicting the storage life of orthodox tropical forest tree seeds; In: Proc. UIFRO Symp. Seed problems of multipurpose trees and other tropical and subtropical species, ACIAR, Canberra, pp. 93-98.

WCMC (1995). Cameroon-nationally threatened and endemic plant taxa: conservation Status Listing of Plants, 22p.

8

Screening of certain varieties of *Emblica officinalis* Gaertn. for resistance to insect pests and diseases in central India

P. B. Meshram[1]* and K. K. Soni[2]

[1]Forest Entomology Division, Tropical Forest Research Institute (ICFRE), P. O. RFRC, Mandla Road, Jabalpur- 482021, Madhya Pradesh, India.
[2]Forest Pathology Division, Tropical Forest Research Institute (ICFRE), P. O. RFRC, Mandla Road, Jabalpur- 482021, Madhya Pradesh, India.

Ten varieties of *Emblica officinalis* Gaertn. including Kanchan, Chakaiya, Francis, NA-7 (Narendra-7), NA-10 (Narendra 10), Anand-1, Anand-2, Krishna, Hatizola (Local) and Local- wild were screened against insect pests that is, gall forming insect (*Betousa stylophora* Swinhoe), leaf roller (*Garcillaria acidula* Forster), bark eating caterpillar (*Indarbela quadrinotata* Walker) and diseases that is vascular wilt (*Fusarium oxysporum f.* sp., *albedinis* Killian & Maire), fruit disease (*Alternaria* sp.). The results revealed that variety NA-10 (Narendra-10) followed by Kanchan was found to be least preferred by *B. stylophora*, *G. acidula*, *I. quadrinotata* and *Alternaria* sps. in clonal seed orchards (CSO). The other test parameters that is, fruit yields, weight/diameter of fruits and the plant morphological characters of the varieties were also studied. Variety of Hatizola (Local) followed by Francis showed less incidence caused by *F. oxysorum* in nursery stage.

Key words: Resistance, varieties, *Emblica officinalis,* insect pests, diseases.

INTRODUCTION

Indian Gooseberry or Aonla, *Emblica officinalis* (Gaertn. Syn. *Phyllanthus emblica*) is an important horticulture crop of India. It is a moderate sized deciduous tree. Fruit, bark and leaves are used in dyeing and tanning. It has aroused good deal of interest among the scientific workers because it is one of the richest natural sources of vitamin 'C' (Ascorbic acid). The fruits are major constituents of 'Chyavanprash' and 'Trifla'. Other uses of fruits are in pickle, marmalade, jam and sauce preparation (Shrivastava, 1990). This economically important tree species is damaged by various insect pests and diseases. Due to the repeated attack of various insect pests and diseases, the growth of plant is stunted (Beeson, 1941). Severe infestation of bark eating caterpillar *Indarbela quadrinotata* may results in the death of the damaged stem and branches but not in the main stem (Mann and Bindra, 1977). Studies on the varietal performance on guava have been reported by Sandhu et al. (1977, 1979), on ber cultivars Mann and Bindra (1977) and on citrus cultivars Sandhu et al. (1979). Variable pattern of susceptibility of *E. officinalis* plant to gall forming insect, *Betousa stylophora* Swinhoe has been observed by Patel et al. (1996). Perusal of literature revealed that, very few and scanty reports are available so far pertaining to the screening of varieties of *E. officinalis* against the insect pests and diseases. In the present study, ten different varieties of *E. officinalis* including Kanchan, Chakaiya, Francis, NA-7 (Narendra-7), NA-10 (Narendra-10), Anand-1, Anand-2, Krishna and Hatizola (Local) and Local- wild were screened against the key insect pests that is, gall forming insect *B. stylophora* Swinhoe, leaf roller *Garcillaria acidula* Forster, bark eating caterpillar *I. quadrinotata* Walker and

*Corresponding author. E-mail: pbmeshram@rediffmail.com.

Table 1. Screening of certain varieties of *E. officinalis* against key insect pests and diseases in clonal seed orchard.

Variety	Larval population of		No. of holes due to	No. of fruits damaged by disease,
	B. stylophora	*G. acidula*	*I. quadrinotata*	*Alternaria* sp.
Kanchan	14.00	10.00	2.0	12.0
Chakaiya	27.00	28.00	8.0	33.0
Francis	18.00	17.00	3.0	12.0
NA-7	18.00	12.00	3.0	13.0
NA-10	11.00	8.00	1.0	7.0
Wild	18.00	14.00	5.0	13.0
S.E. M. ±	7.31	3.05	3.84	3.38
C.D. at 5%	1.62	6.80	8.57	7.54

Figure 1. Gall forming insect, *B. stylophora.*

diseases that is, vascular wilt *Fusarium oxysporum* f.sp. *albedinis* (Killian & Maire), fruit disease *Alternaria* sps. in clonal seed orchards (CSO) and forest nursery which is one of the important component of integrated pest management (IPM).

MATERIALS AND METHODS

Studies were conducted in clonal Seed Orchard (CSO) At Hardi and Central Forest Nursery at Wamandehi (Research and Extension Circles, Reewa and Seoni, Madhya Pradesh, India) during 2006 to 2008. The clonal seed orchard of six varieties of *E. officinalis* including Kanchan, Chakaiya, Francis, NA-7, NA-10 and Local –wild were raised during 1997 to 1998 on 15 ha area following randomized block design (RBD) with spacing 6x6 m. Eight varieties of *E. officianlis* including Chakaiya, Francis, Kanchan, Anand-1, Anand-2, NA-7, Krishna and Hatizola (Local) were raised in polybags in central Forest Nursery at Wamandehi. The plants were found badly infested by insect pests and diseases. Ten numbers of trees in CSO and one bed (size 10 x 1 m) in nursery from each variety were considered for observations. The observations were recorded during peak period of infestation (August to September). The frass of *I. quadrinotata* was removed with the help of soft wire brush and active holes were counted. Some plant characters of varieties like tree height/girth, number of branches, color texture of shoot and yield parameters such as yield of fruits, weight diameter of fruits were also observed. Data were recorded and analyzed statistically.

RESULTS

Clonal seed orchard (CSO)

The analyzed presented data in Table 1 showed that, the larval population of insect pests and fruit damaged by diseases varied from 11.00 to 27.00; 8.00 to 28.00; 1.00 to 8.00 and 7.00 to 33.00. Out of six varieties of *E. officinalis* (viz. Kanchan, Chakaiya, Francis, NA-7, NA-10 and Local- wild), NA-10 (Narendra-10) followed by Kanchan were found to be least preferred by gall forming insect *B. stylophora* (Figure 1), defoliator *G. acidula*, bark eating caterpillar *I. quadrinotata* and fruit disease

Table 2. Screening of certain varieties of *E. officinalis* against test parameters in clonal seed orchard.

Variety	Fruit yield (kg/ tree)	Wt of per fruit (g)	Diameter of fruits (cm)	No. of branches	Height (m)	Girth (cm)
Kanchan	60	35.25	4.1	4.0	3.00	77.0
Chakaiya	45	31.22	4.0	4.0	2.70	63.0
Francis	50	19.65	3.6	4.0	2.40	73.0
NA-7	60	25.27	3.3	3.0	3.00	57.0
NA-10	65	38.37	4.2	4.0	3.30	77.6
Wild-Local	12	13.68	3.0	2.0	3.00	74.6
S.E.M. ±	2.36	1.97	1.45	5.77	3.72	2.82
C.D. at 5%	5.26	4.39	3.24	1.28	8.30	6.30

Table 3. Showing morphological characters of certain varieties of *E. officinalis* in clonal seed orchard.

Variety	Tree height	Tree form	Branch	Inter nodal length of shoot	Color of young intermediate shoot	Shoot surface	Prostrated hairs
Kanchan	Tall	Semi-spreading	Terete	0.39	Brownish red	Glabrous	Less-prominent
Chakaiya	Tall	Upright	Terete	0.35	Pinkish	Glabrous	Less-prominent
Francis	Semi-tall	Spreading	Angled	0.34	Brownish red	Glabrous	Prominent
NA-7	Tall	Semi-spreading	Angled	0.28	Pinkish	Glabrous	Prominent
NA-10	Semi-tall	Semi-spreading	Angled	0.40	Brownish	Non-glabrous	Prominent
Wild-local	Semi-tall	Semi-spreading	Terete	0.33	Brownish red	Non-glabrous	Less-prominent

Alternaria species when compared with other varieties.

The data based on other test parameters like fruit yield per tree, weight /diameter per fruit and height/girth of trees are also summarized in Table 2. The results revealed that, NA-10 followed by Kanchan recorded 65 and 60 kg fruits yield per tree, 38.37 and 35.25 g weight per fruit, 4.2 and 4.1 cm diameter of fruit, 4.00 and 4.00 numbers of branches, 3.30 and 3.0 m height and 77.6 and 77.0 cm girth in clonal seed orchard.

The data on other morphological characters of six different varieties are also presented in Table 3. The results indicated that, the difference might be due to non-preferences for diseases and oviposition or shelter to the insect pests. Further studies on plant morphology including volume of tree, bark character, branching habit (provide suitable shelter) and age of trees are needed to correlate the pest infestation.

The trends of results revealed that, NA-10 followed by Kanchan variety of *E. officinalis* showed more resistant in terms of larval population of *B. stylophora*, *G. acidula*, number of holes made by *I. quadrinotata*, number of fruits damaged by *Alternaria* sp., yield, weight / diameter of fruits and height/girth of trees.

Central forest nursery

The data in Table 4 showed that, the incidence of vascular wilt disease *F. oxysporum* (Figure 2) varied from 2.50 to 5.50%. Out of eight varieties of *E. officinalis* (Chakaiya, Francis, Kanchan, Anand-1, Anand-2, NA-7, Krishna, Hatizola-Local), the varieties including, Hatizola-Local (2.50%) followed Francis (2.75%) were least damaged by wilt disease in grafted seedlings in nursery stage. Varieties such as Kanchan (5.50%), Anand-2 (4.12%) and NA-7 (3.87%) showed more susceptible reaction to *F. oxysporum.*

DISCUSSION

Similar preference among fourteen varieties of *E. officinalis* to gall forming insect, *B. stylophora* attack in arid zone region of India was reported (Sandhu et al.,

Table 4. Screening of certain varieties of *E. officinalis* against vascular wilt disease, *F. oxysporum* in forest nursery.

Variety	Total seedlings per bed (size 10x1 m)	Dying of seedlings due to *F. oxysporum*	Incidence (%)
Chakaiya	1600	56	3.50
Francis	1600	44	2.75
Kanchan	1600	38	5.50
Anand-1	1600	28	3.50
Anand-2	1600	33	4.12
NA-7	1600	52	3.87
Krishna	1600	78	3.25
Hatizola (Local)	1600	40	2.50
S.E.M. ±		2.07	
C.D. at 5%		4.52	

Figure 2. Wilt, *F. oxysporum* on seedlings of *E. officinalis.*

1977) where *E. officinalis* varieties Faizabad and Kanchan were found less susceptible to *B. stylophora* than local varieties (Amla-I, Anand- I, III) (Patel et al., 1996). Vascular wilt disease, *F. oxysporum* also reported on the seedlings of *Buchanania lanzan* Spreng. in nursery stage in central India (Soni et al., 2005).

Conclusions

The present study revealed that, resistance and susceptibility against the insect pest and disease can be observed among the varieties of *E. officinalis* suggesting that, the resistant ones should be preferred for various afforestation and rural development programmes. It can be concluded that, varieties such as NA-10, Kanchan, Hatizola (Local) and Francis were least preferred by insect pests and diseases under tropical climate of central India. Therefore, it is suggested that these varieties may be used for further plantation activities to avoid the frequent damage and losses caused by the insect pests and diseases.

ACKNOWLEDGEMENTS

Authors are thankful to Dr. K. S. Negi, Director, Dr. N. Roychoudhuary, Head, Forest Entomology Division, Tropical Forest Research Institute, Jabalpur and Chief Conservators of Forest, Research and Extension Circles, Rewa and Seoni, Madhya Pradesh, India, for providing necessary field facilities.

REFERENCES

Beeson CFC (1941). The ecology and control of forest insects of India and neighbouring countries. Vasant Press, Dehradun, p. 767.

Mann GS, Bindra OS (1977). Field screening of different jujube cultivars against the bark eating caterpillar. Ind. J. Hortic., 34: 313-315.

Patel VB, Patel JR, Patel NC (1996). Varietal resistance in Aonla to gall forming black caterpillar, *Betousa stylophora*. Guj. Agric. Univ. Res. J., 22: 40-42.

Sandhu GS, Khangura JS, Deol IS (1977). Varietal susceptibility of Guava cultivars to bark eating caterpillar. Punj. Hortic. J., 17: 62-63.

Sandhu GS, Sohi AS, Batra RC (1979). Field performance of bark eating caterpillar *Indarbela quadrinotata* (Walker) on Citrus and Guava cultivars. Ind. J. Entomol., 41: 274-276.

Shrivastava SS (1990). Horticulture Science. Central Block House, Raipur, India, p. 230.

Soni KK, Asaiya, AJ, Nandeshwar DL, Jamaluddin (2005). *Fusarium* wilt of *Buchanania lanzan* Spreng- A new disease record from India. Ind. J. Trop. Biodiv., 13: 112-113.

9

Effects of shade on growth, production and quality of coffee (*Coffea arabica*) in Ethiopia

Adugna D. Bote[1,2] and Paul C. Struik[2]

[1]Jimma University College of Agriculture and Veterinary Medicine P. O. Box 307, Jimma, Ethiopia.
[2]Centre for Crop Systems Analysis, Wageningen University and Research Centre, Droevendaalsesteeg 1, 6708 PB Wageningen, The Netherlands.

The research work was conducted to evaluate the effect of shade on growth and production of coffee plants. To achieve this, growth and productivity of coffee plants growing under shade trees were compared with those of coffee plants growing under direct sun light. Different physiological, environmental and quality parameters were assessed for both treatments. Shade trees protected coffee plants against adverse environmental stresses such as high soil temperatures and low relative humidity. Shade, however, also triggered differences in physiological behaviour of the coffee plants, such as improved photosynthesis and increased leaf area index, resulting in better performance than possible in direct sun light. Consequently, coffee plants grown under shade trees produced larger and heavier fruits with better bean quality than those grown in direct sun light. Moreover, shaded plants had greater biochemical and physiological potential for high dry matter production which would help them to maintain high coffee yields in the long term. If growing coffee under shade trees would allow other sources of income such as fruits, fuel wood and timber to be produced, it could be socially more acceptable, economically more viable and environmentally more sustainable. Since in Ethiopia people are moving towards replacing coffee by 'chat' and/or growing coffee in an open sun, we support the recommendations of growing coffee in the shade and suggest that the future research should be directed toward deterring the development of fungal diseases and increase of coffee yields under shaded conditions.

Key words: Ethiopia, coffee, photosynthesis, shade, bean quality.

INTRODUCTION

Growing coffee under shade trees is one of the fundamental principles in traditional organic coffee growing systems (Beer et al., 1998). Shade trees reduce exces-sive light, mulch the soil with their litter, create hostile conditions for pests and diseases, and harbour a variety of predatory animals (Beer et al., 1998). Arabica coffee is a self-pollinated plant initiating heavy flowers that rapidly develop to fruits (Yunianto, 1986). During this period there is increasing carbohydrate absorption from both leaves and wood for flowers initiation and rapid fruits expansion. As a result roots are damaged, leaves are abscised and branches start dying from the tip and go back to the petiole. But, shade trees assist in maintaining coffee yields in the long term by reducing periodic over-bearing and subsequent die-back of coffee branches. In addition, shading delays the maturation of coffee berries resulting in a better bean filling and larger bean size resulting in better coffee quality (Muschler, 2001).

In Ethiopia, coffee was cultivated in this traditional way following the principles that Lammerts van Bueren and Struik (2004) called 'the concept of naturalness'. Soils were amended by applying compost, farm yard manure and green manure, while no chemical fertilizers, herbi-cides or fungicides were used. However, as demands for coffee production expanded, many coffee growers abandoned their traditional coffee growing system and started to grow coffee without shade trees. This new coffee production system was accompanied with inten-sive use of chemical fertilizers, insecticides, herbicides and fungicides resulting in blended and inferior coffee quality. Coffee plants in direct sunlight also showed a

*Corresponding author. E-mail: adugna_debela@yahoo.com.

higher incidence of premature death (Steiman, 2003). In addition, the genetic resources of *Coffea arabica* and its associated biodiversity are disappearing at an alarming rate and environmental degradation, including soil erosion and extreme river discharges, is becoming severe (Gole et al., 2002; Osman, 2001). People in Ethiopia became unable to nourish their families and frequently became dependent on food aid.

However, there is a growing worldwide movement to support and enhance organic coffee production systems under shade trees (Mark, 2005). This requires research into the effects of shade on growth, production and quality of coffee. We therefore carried out on-farm research to assess the effects of shade on physiology, growth, production and quality of coffee.

MATERIALS AND METHODS

Three different coffee farms were selected from Manna district, Jimma Zone, Ethiopia in the 2007 cropping season. The district is 22 km to the west of Jimma town and located between 7°54'N and 36°53'E (Garedew and Tsegaye, 2010) at an elevation of 1820 m.a.s.l with mean minimum and maximum temperatures of 13 and 25°C, respectively. The major soil types of the area are nitosol and combsols and the area receives an average annual rainfall of 1467 mm. *C. arabica* is the economically most important coffee of the area. Manna district is considered as one of the original *C. arabica* producing areas of the country. *C. arabica* produced in Manna has already been certified as organic coffee by the German BCS-Öko Garantie, GMBH (Schmitt and Grote, 2006).

Data on light intensity (lux), air temperature (°C), air relative humidity (%), and soil temperature at a depth of 10 cm (°C) were recorded using a light meter (EXTCH, Model EA30), thermo-hygrometer (HANNA, HI8564) and soil thermometer (Taylor Bi-Therm), respectively. Photosynthetic rate (A, µmol m^{-2} s^{-1} of CO_2), transpiration rate (E, mmol m^{-2} s^{-1} of water vapour), stomatal conductance (gs, mmol m^{-2} s^{-1} of water vapour or CO_2), Photosynthetically Active Radiation (PAR, µmol photons m^{-2} s^{-1}) and leaf temperature (lT, °C) were determined using a leaf chamber porometer (LCpro+). These parameters were assessed from five intact, young and fully expanded leaves located on the third and fourth pairs from the top of each plant in September and October 2007 between 10:00 and 12:00 a.m. The same leaves were subjected to a period of dark adaptation for 20 min. and then exposed to a light pulse of high (saturating) intensity (2000 µmol photons m^{-2} s^{-1}) using chlorophyll fluorometer (OS-30P) to assess chlorophyll fluorescence, indicating the electron transport efficiency of photosystem II (PSII). Chlorophyll fluorescence is expressed as the F_v/F_m ratio (no dimension) as a measure of relative quantum yield.

Leaf colour was assessed using a colour meter (AccuProbe, HH06) (Lab scale) and leaf area (m^2) was determined by scanning each sampled leaf with a leaf area meter (AM200). These leaves were dried in an oven at 75°C for 24 h to determine their dry weight; nitrogen content (mg per g dry matter) was assessed in the dried leaf material using micro - Kjeldahl Nitrogen apparatus (DK6) and data collected from leaves were averaged per plant per location (replication) for each leaf insertion.

Relative growth rate (RGR, cm cm^{-1} $month^{-1}$) of each plant was estimated from two selected primary plagiotropic branches whose length was measured at the beginning and at the end of the experiment. Fully ripened coffee berries were harvested from the selected individual trees and dried until a constant weight was reached (moisture content 9 to 12%). Coffee weight for each sample was registered on 1000 seed weight base (g per 1000 beans) and beans were made to pass through a series of sieves with round perforations of 0.635, 0.595 or 0.555 cm (hole's diameter) after which weight fractions retained on each sieve were converted to weight percentages of the total sample.

Raw quality (shape and make, colour and odour) was valued on a scale starting from 2, the minimum value (indicating small shape and make, brownish colour and strong/odd odour) to 15, the maximum value (indicating very good shape and make, bluish colour and clean odour). For each sample, 200 g of green coffee were roasted to a uniform light brown colour, at Jimma Agricultural Research Center laboratory roaster. Liquor quality (acidity, body and flavour) was tested by a panel of six experienced judges and values were given from 2, the minimum value (indicating lacking acidity, thin body and poor flavor) to 20, maximum value (indicating pointed acidity, full body and very good flavour (unit for both raw and liquor qualities: percentage of score).

The experimental design was a randomized complete block design with two treatments (shade and direct sun light) in which ten plants (five for each treatment) were randomly selected. This was replicated three times at different sites, resulting in a total of 30 plants (15 plants per treatment). At each measurement five leaves per plant were used for gas exchange and chlorophyll fluorescence. The collected data were subjected to analysis of variance and means were compared by a t-test at 5% probability.

RESULTS

Environmental variables

Soil temperature and light intensity of shaded coffee plants were significantly lower than those of coffee plants grown in direct sun light, whereas the relative humidity of the air of the shaded plants was significantly higher than of plants grown in full sun light. Air temperature did not differ significantly (Table 1).

Photosynthesis variables

Rate of photosynthesis (A) and F_v/F_m were higher for shaded plants than for plants in direct sun light. Treatments did not differ in transpiration rate or stomatal conductance. PAR at leaf level and leaf temperature were obviously lower for shaded plants than for plants in full and direct sun light (Table 2).

Plant variables

Coffee plants grown in the shade had higher values for SLA, LAI and leaf nitrogen content. Shaded leaves were also darker in leaf colour than leaves from plants grown in direct sun light (Table 3). The relative growth rate, however, did not differ significantly between the two treatments (Table 3).

Quality and yield

Bean weight and size assessment made on harvested coffee beans indicated that beans developed under shaded

Table 1. Averages and statistical analysis of various environmental variables for shaded coffee plants and coffee plants in direct sun light.

Variable	Treatment		Significance (two-tailed)
	Shaded plant	Plant in direct sun light	
Air temperature (°C)	25.5	26.7	ns
Soil temperature (°C)	19.7	20.8	0.00*
RH (%)	59.7	55.1	0.01*
Light intensity (lux)	557	1193	0.00**

ns indicates not significant (P > 0.05).* indicates significant difference at P < 0.05. ** indicates significant difference at P < 0.01.

Table 2. Averages and statistical analysis of various photosynthesis variables for shaded coffee plants and coffee plants in direct sun light.

Variable	Treatment		Significance (two-tailed)
	Shaded plant	Plant in direct sun light	
A (µmol CO_2 $m^{-2} s^{-1}$)	3.51	2.45	0.03*
F_v/F_m (-)	0.71	0.61	0.01*
Stomatal conductance (mmol $m^{-2} s^{-1}$)	100	60	ns
Transpiration rate (mmol $m^{-2} s^{-1}$)	1090	1140	ns
PAR (µmol photons m-2 s-1)	320	2260	0.00**
Leaf temperature (°C)	24.2	28.1	0.00**

ns indicates not significant (P > 0.05).* indicates significant difference at P < 0.05. ** indicates significant difference at P < 0.01.

Table 3. Averages and statistical analysis of various plant variables for shaded coffee plants and coffee plants in direct sun light.

Variable	Treatment		Significance (two-tailed)
	Shaded plant	Plant in direct sun light	
SLA (cm^2/g)	116	98	0.04*
LAI ($m^2 m^{-2}$)	3.8	2.8	0.01*
RGR(cm cm^{-1} $month^{-1}$)	12.3	9.7	ns
Leaf N content (mg g^{-1} leaf dry matter)	288	219	0.03*
Leaf colour (value on greenness scale)	-8.6	-7.6	0.00**

ns indicates not significant (P > 0.05). * indicates significant difference at P < 0.05.** indicates significant difference at P < 0.01.

Table 4. Averages and statistical analysis of various quality and yield variables for shaded coffee plants and coffee plants in direct sun light.

Variable	Treatment		Significance (two-tailed)
	Shaded plant	Plant in direct sun light	
Coffee weight (g/1000 beans)	148	134	ns
Raw quality (% of maximum score)	85	85	ns
Liquor quality (% of maximum score)	65	50	ns
Bean yield (Mg ha^{-1})	2.13	3.10	ns

ns indicates not significant (P > 0.05).

shaded condition were heavier and larger in size and had better liquor taste than those developed on plants grown in direct sun light (Table 4). No treatment difference was observed with respect to raw coffee quality. However, greater yields were obtained from coffee plants grown in the direct sun than from shaded plants (Table 4), although this difference was also not statistically significant.

DISCUSSION

Soil temperature plays a critical role in the survival of many organisms, but it varies in response to exchange processes that take place through the surface of the soil. The reduction in soil temperature, observed under shade, was mainly caused by the ability of shaded soil to stabilize the local thermal balances and also to reduce the heat flux caused by the accumulated plant based biomass (Morais et al., 2006). Siebert (2002) also reported that shading reduces and stabilizes the soil temperature by reducing the radiant flux reaching the soil and modifying the temperature amplitude at the soil surface.

The reduced air temperature registered for coffee grown under shade was in agreement with the result obtained by (Campanha et al., 2005). As they concluded the reduced air temperature was mainly due to the reduced direct incidence of solar radiation on the coffee canopy. Shading buffers the extreme temperature variations and provides a microclimate which attenuates extreme temperatures of air and soil and preserves surface soil humidity.

High rates of photosynthesis mean, according to Gulmon and Chu (1981), that there is high biochemical and physiological potential for a high carbon fixation capacity. To increase their carbon fixing potential, shaded plants undertake certain modifications such as developing thinner and larger leaves (Friend, 1984) with more thylakoids per granum and more grana per chloroplast (Fahl et al., 1994). These modifications allow them to efficiently capture and utilize the available light energy in order to increase their dry matter production. Plants having higher SLA exhibit higher productivity (Li et al., 2005) and have higher potential relative growth rate than those having lower SLA (Poorter and Werf, 1998).

Robakowski et al. (2003) concluded that SLA decreases as light intensity increases. In our experiment, lower SLA was obtained from coffee plants grown in an open sun condition which is in consistence with the findings of these authors. For shaded coffee plants, the increased SLA and the development of a darker green colour were mainly attributed to the higher nitrogen content found in their leaves. It is likely that the increased SLA and the development of dark green leaf colour under shaded coffee plants partly contributed for the higher rate of photosynthesis observed under this condition. Photosynthetic rate of sun grown coffee plants, on the other hand, was limited by stomata closure, high leaf temperature and low internal carbon dioxide concentration.

Since many of the physiological processes of plants are temperature dependent, under high temperature crops have great difficulty in maintaining photosynthetic activities and growth (Sethar et al., 2002). Coffee is exceptionally sensitive to fluctuations in leaf temperature, especially temperatures above 25°C. For the coffee plants grown in direct sun light, increased air temperature above this level resulted in subsequent lowering of stomatal conductance (data not shown) which in turn imposed a large limitation on the rate of CO_2 assimilation. Kasai (2008) has also found comparable results: stomatal conductance and photosynthetic rate were found significantly lowered by growing soybean plant under continuous light. High temperatures, according to Farquhar and Sharkey (1982), reduce the electron transport capacity and increase the rates of CO_2 evolution from photorespiration and other sources causing the photosynthetic rate to become lower.

As stomata are highly responsive to the factors that influence the rate of transpiration, their movements can also be affected by leaf-to–air vapour pressure difference (Kim et al., 2004). Under shade, however, reduced air temperature and light intensity increased the percentage relative humidity in the air around coffee plants and subsequently reduced the vapour pressure deficit (VPD) between the interior of the leaf and the atmosphere. This reduced VPD decreased rate of transpiration of the leaf resulting in increased leaf water potential. Under such small VPD, stomatal aperture increases providing better chance for CO_2 to be diffused into the leaf.

Due to the interception of solar radiation by shade trees, the incident solar radiation, PAR, was greatly reduced for coffee grown under shade. But, plants under shade had a higher photosynthetic rate. This paradox can be explained by the fact that plants under shade undertake certain morphological modifications and physiological adaptations, and their leaves are capable of absorbing more than 90% of the energy contained in the wavelengths between 400 and 700 nm (Lee, 1985). In addition, Bartlett and Remphrey (1998) indicated that there are no significant reductions in photosynthetic rate and growth of coffee plants grown under shade unless the level of shade exceeds 90%.

The darker green colour of coffee leaves developed in the shade was most likely associated with the larger amount of nitrogen accumulated in them (Titus and Pereira, 2005). Leaves having such dark green colour absorb more light, have chloroplasts with improved light-capturing capability and are cheap units of photosynthetic area (energetically) as they capture lower light intensity and utilize them efficiently to increase their photosynthetic rate.

The ecological behaviour of plants and their health status can be judged by considering certain parameters. For example, SLA reflect the growing conditions of the plants (Garnier et al., 2001), where as LAI and the F_v/F_m ratio are indicators of the health status of a given crop plant (Kitao et al., 2000; Malone et al., 2002). Higher LAI observed for coffee plants growing under shade indicated that these plants have higher potential for CO_2 assimilation and dry matter production than unshaded plants (McNaughton and Jarvis, 1983). Higher F_v/F_m ratio observed for shaded coffee plants illustrated the fact that these plants are less stressed by high light intensity than those grown in direct sun light. Sethar et al. (2002)

supported this idea indicating F_v/F_m ratio decreases significantly when plants are exposed to heat stress.

For plants grown under higher irradiance, the reduced value of F_v/F_m ratio is an indication of the damage of a proportion of PSII reaction centres, a phenomenon called photo-inhibition. Once a proportion of a PSII reaction centre is damaged, light energy utilization capacity of this centre decreases resulting in a reduced quantum yield of net photosynthesis (Rintamaki et al., 1995; Maxwell and Johnson, 2000). Light, as energy source for photosynthesis, is an essential prerequisite for plant life. Excess light, however, can inhibit photosynthesis and lead to photo-oxidative destruction of the photosynthetic apparatus, thereby decreasing the photosynthetic rate of the plant growing in direct sun light besides affecting its life span (Li et al., 2010). According to Fahl et al. (1994) and Ramalho et al. (2000), higher values of F_v/F_m obtained from shaded leaves can also be linked with higher leaf nitrogen content. Under stressed conditions the availability of more leaf nitrogen triggers photo-protective mechanisms against photo-oxidation by its ability in promoting the activation and backing up of the protective mechanisms (Fahl et al., 1994). Leaf nitrogen was also found to have a strong and positive correlation with carbon assimilation rates (data not shown) allowing shaded leaves to have better photosynthetic performance and greater vegetative growth rate than sun leaves. Therefore, coffee plants found in direct sun light were grown under environmental conditions that are more likely to lead to plants stress responses, compared with the environmental conditions under which shaded plants are grown.

The fact that shade resulted in heavier and larger coffee beans was mainly caused by its effect on temperature and the duration of the ripening period. Muschler (2001), who found comparable results, indicated that coffee bean size significantly and consistently increases even with increasing shade levels. Similarly, the shade effect on liquor taste was also the result of delayed fruit maturation and ripening. Morais et al. (2006) showed that shading enhances coffee quality, in terms of biochemical composition, including the contents of caffeine, oil and chlorogenic acid. The quality and size of coffee beans, and the taste of finished products, are therefore better under shade than in systems without shade trees.

Shade grown coffee gave lower yields than coffee in the direct sun, but this difference was not statistically significant. However, we surmise that a moderate yield will be more sustainable than a maximum yield and the former will be more important for traditional small farmers. Moreover, shade trees are alternative means of income via carbon sequestration, production of fuel wood, timber and fruits. Therefore growing coffee in an agroforestry system is socially more acceptable, economically more viable and environmentally more sustainable than growing coffee in the direct sun.

Conclusions

Coffee plants grown in the shade suffer less from environmental stresses and have higher biochemical and physiological potential for carbon fixation compared with coffee plants grown in the direct sun light. Shade grown coffee plants also produce larger and heavier beans with better cup taste than coffee plants grown in the direct sun. If growing coffee under shade trees would allow other sources of income such as fruits, fuel wood and timber to be produced, it could be socially more acceptable, economically more viable and environmentally more sustainable. We therefore recommended growing coffee in the shade.

In addition, shade trees serve as an alternative source of income for coffee producers. Nevertheless, highest yields per hectare were obtained from coffee plants in the direct sun light indicating the need for further research on determining proper plant density, compatibility between shade trees and coffee plants, as well as extent of competition between them for water and nutrients. Since in Ethiopia people are moving towards replacing coffee by 'chat' and/or growing coffee in an open sun, we support the recommendations of growing coffee in the shade and suggest that the future research should be directed toward deterring the development of fungal diseases and increase of coffee yields under shaded conditions.

ACKNOWLEDGEMENTS

The authors would like to thank all people who directly or indirectly were involved in this research activity and who brought it to this level. We also thank NUFFIC for financially supporting the senior author.

REFERENCES

Bartlett GA, Remphrey WR (1998). The effect of reduced quantities of photosynthetically active radiation on *Fraxinus pennsylvanica* growth and architecture. Can. J. Bot., 76: 1359–1365.

Beer J, Muschler R, Kass D, Somarriba E (1998). Shade management in coffee and cacao plantations. Agroforestry Systems, 38: 139-164.

Campanha MM, Silva RH, Freitas GB, Martinez HE, Gracia SL, Fing FL (2005). Growth and yield of coffee plants in agroforestry and monoculture systems in Minas Gerais, Brazil. Agroforestry Syst., 63(1): 75 – 82.

Fahl JI, Carelli MC, Vega J, Magalhaes AC (1994). Nitrogen and irradiance levels affecting net photosynthesis and growth of young coffee plants (*Coffea arabica* L.). J. Hort. Sci., 69: 161-169.

Farquhar GD, Sharkey TD (1982). Stomatal Conductance and Photosynthesis. Annu. Rev. Plant. Physiology, 33:317-345.

Friend DC (1984). Shade adaptation of photosynthesis in coffea arabica. J. Photosynthesis Res., 5(4): 325 – 334.

Garedew W, Tsegaye B (2010). Trends of Avocado (*Persea Americana* M) production and its constraints in Mana Woreda, Jimma Zone: A potential Crop for coffee diversification. Trends Hortic. Res., ISSN 1996 – 0735/ DOI: 10.3923/thr.2010.

Garnier E, Shipley B, Roumet C, Laurent G (2001). A standardized protocol for the determination of specific leaf area and leaf dry matter content. Funct Ecol., 15: 688-695.

Gole TW, Denich M, Demel T, Vlek PLG (2002). Human Impacts on

Coffea Arabica Genetic Pools in Ethiopia and the Need for its In situ Conservation. In: Managing Plant Genetic Diversity, Rao, R., A. Brown and M. Jackson (Eds.). CAB International and IPGRI, pp: 237 - 247.

Gulmon SL, Chu CC (1981). The effects of light and nitrogen on photosynthesis, leaf characteristics, and dry matter allocation in the chaparral shrub, *Diplacus aurantiacus.* Oecologia, 49:207-212.

Kasai M (2008). Effect of growing soybean plants under continuous light on leaf photosynthetic rate and other characteristics concerning biomass production. J. Agron., 7: 156-162.

Kim H, Gregory DG, Raymond MW, John CS (2004). Stomatal conductance of lettuce grown under or exposed to different light qualities. Ann. Bot., 94: 691 – 697.

Kitao M, Lie TT, Koike T, Tobita H, Maruyama Y (2000). Susceptibility to photoinhibition of three deciduous broad leaf tree species with different successional traits raised under various light regimes. Plant Cell Environ., 23(1): 81 – 89.

Lammerts van Bueren ET, Struik PC (2004). The consequences of the concept of naturalness for organic plant breeding and propagation. NJAS-Wageningen J. Life Sci., 52(1): 85 – 95.

Lee DW (1985). Duplicating foliage shade for research on plant development. Hort. Sci., 20: 116 – 118.

Li H, Yong Y, Li B, Jing R, Lu C, Li Z (2010). Genetic analysis of tolerance to photo-oxidative stress induced by high light in winter wheat (*Triticum aestivum* L.) J. Genet. Genom., 37(6):399-412.

Li Y, Douglas AJ, Yongzhong SU, Jianyuan CI, Tonghui Z (2005). Specific leaf area and leaf dry matter content of plants growing in sand dunes. Bot. Bull. Acad. Sinica., 46: 127 – 134.

Malone S, Herbert DA, Holshouser DL (2002). Relationship between Leaf Area Index and yield in Double-crop and full-season soybean systems. J. Ecol. Entomol., 95(5): 945–951.

Mark J (2005). Shade grown coffee and bird-friendly coffee. http://www.thenibble.com/REVIEWS/nutri/matter/organic-coffee4.asp [Accessed 24 January 2006].

Maxwell K, Johnson GN (2000). Chlorophyll fluorescence - a practical guide. J. Exp. Bot. 51: 659 – 668.

McNaughton KG, Jarvis PG (1983). Predicting effects of vegetation changes on transpiration and evaporation. In: Water Deficits and Plant Growth. Vol. 7, Kozlowski, TT (ed.), pp. 1–47 London Academic Press, New York.

Morais H, Caramori P, Ribeiro AM, Gomes JC, Koguishi MS (2006). Microclimatic characterization and productivity of coffee plants grown under shade of pigeon pea in Southern Brazil. Pesq. Agropec. Bras. 41: 5. 763-770.

Muschler RG (2001). Shade improves coffee quality in a sub-optimal coffee-zone of Costa Rica. Agroforestry Syst., 51(2): 131 – 139.

Osman M (2001). Rainfall and its erosivity in Ethiopia with special consideration of the central highlands. –Bonner Bodenkundl. Abh., 37, 249 S. Bonn.

Poorter H, Werf VA (1998). Is inherent variation in RGR determined by LAR at low irradiance and by NAR at high irradiance? A review of herbaceous species. In: Lambers, H, Poorter, H, Van Vuuren, MMI (eds.). Inherent variation in plant growth. Physiological mechanisms and ecological consequences. Leiden, Netherlands, Backhuys Publishers. pp. 309-336.

Ramalho JC, Pons TL, Groeneveld HW, Azinheira HG, Nunes MA (2000). Photosynthetic acclimation of high light conditions in mature leaves of Coffea arabica L.: Role of xanthophylls, quenching mechanisms and nitrogen nutrition. Aust. J. Plant Physiol., 27(1): 43 – 51.

Rintamaki E, Kettunen R, Tyystjärvi E, Aro EM (1995). Light dependent phosphorylation of D1 reaction-centre protein of Photosystem II: Hypothesis for the functional role *in vivo*. Physiol. Plant. 93: 191 – 193.

Robakowski P, Montpied P, Dreyer E (2003). Plasticity of morphological and physiological traits in response to different levels of irradiance in seedling of silver fir (*Abies alba* Mill.) Trees, (Berl.). 7: 431–441.

Schmitt C, Grote U (2006). Wild coffee production in Ethiopia: The role of coffee certification for forest conservation. Report project 'Conservation and use of wild populations of *Coffea arabica* in the montane rainforest of Ethiopia', Bonn, Germany.

Sethar MA, Pahoja VM, Chachar Q (2002). Photosynthetic acclimation of cotton genotypes at higher temperatures. Asian J. Plant Sci. 1: 261-263.

Siebert SF (2002). From shade- to sun-grown perennial crops in Sulawesi, Indonesia: Implications for biodiversity conservation and soil fertility. Kluwer Academic Publisher, the Netherlands. Biodivers. Conserv., 11(11): 1889 – 1902.

Steiman S (2003). Shade vs. Sun Coffee: A review PMicrosoft internet explorer. www.geocities.com/RainForest/Canopy/1290/basics.htm (accessed 24 January 2006).

Titus A, PereriaG (2005). Nitrogen Economy inside Coffee Plantations in Van Wijk MT Williams M, Laundre JA, Shaver GR (2003). Inter annual variability of plant phenology in tussock tundra: modelling interactions of plant productivity, plant phenology, snowmelt and soil thaw. Global Change Biol., 9:743–758.

Yunianto YD (1986). Overbearing dieback on Arabica coffee. J. Pel. Parkebu, 2(2):60-65.

Influence of spacing and organics on plant nutrient uptake of black nightshade (*Solanum nigrum)*

Sivakumar V.[1]* and Ponnusami V.[2]

[1]Agricultural research Station, Tamil Nadu Agricultural University, Bhavanisagar – 638451, Erode District, Tamil Nadu, India.
[2]Horticultural College and Research Institute, Tamil Nadu Agricultura University, Periyakulam - 625 604, Theni District, Tamil Nadu, India.

Investigations were carried out at the Horticultural College and Research Institute to study the effect of spacing and bio-stimulants on the plant nutrient uptake of *Solanum nigrum*. The experiments were laid out in split plot design with three replications. The treatments consisted of four levels of spacing (M_1 – 30 × 30 cm, M_2 – 45 × 45 cm, M_3 – 60 × 30 cm and M_4 – 60 × 45 cm) and nine bio-stimulants/organics (S_1– Recommended dose of fertilizer (RDF), S_2 – control, S_3 – panchagavya 2%, S_4 – panchagavya 3%, S_5 – moringa leaf extract 2%, S_6 – moringa leaf extract 4%, S_7 – humic acid, S_8 – cytozyme 1% and S_9 – vermicompost). Observations on nutrient uptake found that, the treatment closer spacing + RDF (M_1S_1) registered higher nitrogen uptake (223.83 kg ha^{-1}) and phosphorous uptake (22.02 kg ha^{-1}) followed by closer spacing + panchagavya 3% spray (M_1S_4) (202.64 kg ha^{-1}) (Nitrogen uptake - 202.64 kg ha^{-1} and Phosphorous uptake -20.03 kg ha^{-1}). Regarding the potassium uptake the treatment wider spacing + RDF (M_4S_1) combination recorded the highest K uptake (252.83 kg ha^{-1}) followed by the treatment wider spacing + humic acid combination (M_4S_7) (227.14 kg ha^{-1}).

Key words: *Solanum nigrum,* spacing, organics, plant nutrient uptake.

INTRODUCTION

Among various medicinal plant species, black nightshade, or (*Solanum nigrum* L. Mana thakkali, Family *Solananceae*) has great demand in the Indian system of medicine. This herb is recently gaining importance in pharmaceutical industry due to its amazing drug potential. The whole plant is recommended for use as cardiac tonic, alterative, diuretic, sedative, expectorant, diaphoretic, cathartic and anodyne. The active principle in this herb is solasodine; a glucosidal alkaloid, used as a starting material for the synthesis of hydrocortisones. It is in great demand in the pharmaceutical industry (Dhawan, 1986). Fertilizers and organic manures constitute important inputs in the production of all crop species. Plants require essential mineral nutrients to complete their life cycle and the quantities required for optimum growth and production vary with species. Increased fertilizers for the development of more efficient fertility management program. Organic formulations are chemical fertilizer cost and awareness of environmental pollution have necessitated the use of organic sources of apparently environment and farmer friendly renewable source of non-bulky, low cost organic agricultural inputs for improving soil quality status in consumable medicinal plants. Organic manures are fairly good source of nutrients which has direct influence on plant growth like other commercial fertilizers. Mukherjee et al. (1991) and Prasad and Singhania (1989) also reported that application of organic manures increased the leaf nutrient status. Jambhekar (1992) and Shivputra et al. (2004) also reported similar result. However, little information is available in this area; therefore this study was designed and conducted to identify the effect of spacing and organics on plant nutrient uptake of black nightshade.

MATERIALS AND METHODS

The Experiment was carried out at the Botanical Garden, Department of Floriculture and Medicinal plants, Horticultural College and Research Institute, Tamil Nadu Agricultural University, Coimbatore. The experiment was laid out in a split plot design with

*Corresponding author. E-mail: shivafruitscience@gmail.com.

13 treatment combinations replicated three times. The field lay out and randomizations of treatments to plots of size 2 × 2 m (4 square meters) were carried out as per the statistical methods suggested by Panse and Sukhatme (1985). The treatments consisted of four levels of spacing and nine bio-stimulants/organics.

Treatment details

Main plot treatments

Spacing: M_1 – 30 × 30 cm; M_2 – 45 × 45 cm; M_3 – 60 × 30 cm; M_4 – 60 × 45 cm.

Sub plot treatments

Bio-regulators/organics: S_1 – Recommended fertilizer dose; S_2 – Control (only FYM); S_3 – FYM + Panchagavya foliar spray – 2% at 15 days interval; S_4 – FYM + Panchagavya foliar spray – 3% at 15 days interval; S_5 –FYM + Moringa leaf extract foliar spray – 2% at 15 days interval; S_6 –FYM + Moringa leaf extract foliar spray – 4% at 15 days interval; S_7 – FYM + Humic acid 20 kg ha^{-1} as basal; S_8 – FYM + Cytozyme foliar spray – 1% at 15 days interval; S_9 – FYM + Vermicompost – 2.5 t ha^{-1} as basal.

The observations that were recorded are plant nitrogen uptake, plant phosphorous uptake and plant potassium uptake. The organics were sprayed at 15 days interval starting from 30 days after transplanting. The observations were recorded at 80 days after transplanting that is at harvest stage.

RESULTS AND DISCUSSION

Nitrogen uptake

The effect of M, S, M × S and S ×M exerted a significant influence on nitrogen uptake (Table 1). Closer spacing (M_1) recorded significantly higher nitrogen uptake (184.16 kg ha^{-1}) and wider spacing (M_4) registered the lowest nitrogen uptake (73.10 kg ha^{-1}). Of various organics and bio regulators tried, RDF (S_1) treated plants recorded more nitrogen uptake (139.68 kg ha^{-1}) followed by panchagavya 3% (S_4) sprayed plants (125.39 kg ha^{-1}). The uptake was low (105.84 kg ha^{-1}) in control (S_2). The interaction effect of closer spacing + RDF (M_1S_1) registered higher nitrogen uptake (223.83 kg ha^{-1}) which was followed by closer spacing + panchagavya 3% spray (M_1S_4) (202.64 kg ha^{-1}). Lesser nitrogen uptake of 67.80 kg ha^{-1} was recorded in wider spacing + control (M_4S_2) combination. The uptake of nutrients is primarily a function of total biomass production and nutrient content at cellular level. Increase in uptake of nutrients coincides with the rapid growth phase with a very low rate of uptake during initial vegetative phase. In the present study, nitrogen and phosphorus uptake was more under closer spacing + RDF followed by panchagavya and humic acid treatments. This may be attributed due to more population per plot. The increased uptake of nitrogen was supposed to be due to the better use efficiency of applied nitrogen fertilizer in the presence of humic acid coupled with the retarded nitrification process enabling the slow availability of applied nitrogen (Guminski, 1968). The principal physiological function of humic acid may be that they reduced the oxygen deficiency in plants which resulted in uptake of larger amount of nitrogen (Khristeva and Luk'Yanenko, 1962). According to Yadav (1989), the reactions of ammonium ion with humic acid yielding covalent compounds maintain the mobility of ammonium in solution and aids in the increase uptake of the same by plants. Another possible reason for enhanced nitrogen uptake in the presence of humic acid was suggested by Piccolo et al. (1992) who stated that the acid functionality of humic acid could simulate the nitrate uptake by plants. Nitrate reductase is the key enzyme for nitrogen assimilation by most of the plants (Campbell, 1999). Humic substances induced the activity of enzymes such as invertase and nitrate reductase and there by helped in increased assimilation of nitrogen by plants (Ferretti et al., 1991). Higher uptake of nitrogen by shoot could be due to the fact that nitrogen is the mobile element and it is utilized more in shoot than in root.

Phosphorous uptake

The effect of spacing, organics and bio regulators and interactions M × S and S × M exhibited a significant influence on phosphorous uptake (Table 1). Higher plant density (M_1) recorded more phosphorous uptake (17.99 kg ha^{-1}) and lesser plant density (M_4) registered lower phosphorous uptake (6.68 kg ha^{-1}). Among the different organics and bio regulators used, RDF (S_1) applied plants recorded greater phosphorous uptake (13.57 kg ha^{-1}) followed by panchagavya (S_4) sprayed plants (11.88 kg ha^{-1}) and humic acid (S_8) applied plants (10.72 kg ha^{-1}). The phosphorous uptake was lower (9.56 kg ha^{-1}) in control (S_2) treatment. Closer spacing + RDF (M_1S_1) combination recorded significantly higher phosphorous uptake (22.02 kg ha^{-1}) followed by M_1S_4 (20.03 kg ha^{-1}). The less phosphorous uptake (5.76 kg ha^{-1}) was observed in wider spacing + control (M_4S_2) interaction. Dell 'Agnola and Nardi (1987) concluded that humic substances extracted from soil stimulated ion uptake by roots. Humic acid is known to be surface active and increases the permeability of root membranes and thereby enhances the nutrient uptake as stated by Rauthan and Schnitzer (1991). Application of humic acid increased the root growth by 58% and this could help in better assimilation of phosphorus by plants (Vaughan and Malcolm, 1979). The increase in phosphorus uptake may be due to the prevention of phosphorus fixation in the soil and also by forming humophospho complexes which are easily assimilated by plants (Raina and Goswami, 1988). The increase in phosphorus accumulation in tomato shoot suggested that humic acid may enhance the uptake of phosphorus indirectly by complexing the iron (David et al., 1994). Addition of humic acid maintained high levels of acid phosphatase activity which increased phosphatase activity and holds key for increased

Table 1. Effect of spacing, organics and bio-regulators on N and P uptake (kg ha^{-1}) in black nightshade.

Treatment	N					P				
	M_1	M_2	M_3	M_4	Mean	M_1	M_2	M_3	M_4	Mean
S_1	223.83	116.304	124.67	93.90	139.68	22.02	11.50	11.19	9.59	13.57
S_2	166.40	89.86	99.28	67.80	105.84	16.50	7.42	8.59	5.76	9.56
S_3	170.57	93.37	103.10	68.81	108.96	16.54	8.00	9.15	6.12	9.95
S_4	202.64	105.83	116.93	76.18	125.39	20.03	9.20	11.06	7.24	11.88
S_5	169.19	94.12	101.8	68.06	110.25	16.62	7.55	8.13	6.09	9.87
S_6	173.93	95.44	102.38	70.31	110.52	16.64	8.03	8.61	6.39	9.91
S_7	191.00	102.07	113.53	72.82	119.86	18.69	8.66	9.04	6.49	10.72
S_8	180.99	97.33	106.36	69.88	113.64	17.23	8.04	8.33	6.13	9.93
S_9	178.92	97.03	105.80	70.16	112.98	17.61	8.38	8.62	6.30	10.22
Mean	184.16	99.91	108.21	73.10	116.34	17.98	8.65	9.19	6.67	10.62

	M	S	MS	SM	M	S	MS	SM
SEd	0.82819	0.97562	1.95123	2.01747	0.27667	0.21132	0.48510	0.42264
CD (0.05)	2.02659	1.94903	3.89806	4.18312	0.67702	0.42216	1.04000	0.84433

phosphorus uptake by plants (Malcolm and Vaughan, 1979). Organic phosphorus content in soil comprises of 30-70 percentage of the total phosphorus. Hydrolysis of these organic phosphorus compounds is essential for uptake by plants (Bhattacharyya and Ghosh, 2001) and humic acid brought out this hydrolysis (Adani et al., 1998).

Potassium uptake

The individual effect of spacing, organics and bio regulators and the interaction effect of M × S and S × M on potassium uptake were significant (Table 2). Among the spacing levels, potassium uptake was significantly higher (204.92 kg ha^{-1}) under closer spacing (M_1) followed by M_3 (117.35 kg ha^{-1}). Lesser score (71.45 kg ha^{-1}) was recorded in M_4 (wider spacing) level. Of different organics and bio regulators used, RDF (S_1) applied plants expressed more potassium uptake (150.32 kg ha^{-1}) followed by panchagavya 3% (S_4) given plants (134.11 kg ha^{-1}). The potassium uptake was less (113.03 kg ha^{-1}) in control (S_2). The effect of interaction on potassium uptake indicated that closer spacing + RDF (M_1S_1) combination recorded greater potassium uptake (252.83 kg ha^{-1}), which was followed by closer spacing + humic acid (M_1S_7) combination (227.14 kg ha^{-1}). The potassium uptake was less (64.78 kg ha^{-1}) at wider spacing + without spray (M_4S_2). From the present results it is evident that the potassium uptake was more in closer spacing + humic acid next to RDF, followed by panchagavya. David et al. (1994) suggested that the factors other than direct nutritional effect of humic acid may also operate in enhancing the potassium uptake by plants. According to Samson and Visser (1989), humic acid induced increase in the permeability of bio membranes for electrolytes accounted for increased uptake of potassium. Humic substances modify membrane bound ATPase activity and the relation between membrane ATPase activity, H^+ extrusion and the ion uptake suggested that humic substance influence active uptake of potassium by interfering with specific ion carrier. In the present study, the findings exhibited that more of nutrient uptake (N, P and K) was found in closer spacing than wider spacing, but the leaf nutrient content (N, P and K) was more in wider spacing than the closer spacing. This may be due to the fact that, population per unit area in closer spacing was more than under wider spacing. Leaf nutrient content namely nitrogen and phosphorous, recorded significantly higher values on wider spacing + panchagavya 3% spray next to RDF than humic acid but potassium content registered higher value in wider spacing + humic acid next to RDF than panchagavya. The reason may be that nitrogen and phosphorous content was more in panchagavya spray and more potassium content in humic acid. The same was observed in case of nutrient uptake, but closer spacing recorded significantly higher uptake than wider spacing.

Conclusion

In the present study, wider spacing + RDF (M_4S_1) treatment combination recorded significantly higher values for leaf nitrogen and phosphorous consent followed by the treatment Wider spacing + panchagavya (M_4S_4). Regarding lead potassium content and plant potassium uptake, the treatment wider spacing + RDF (M_4S_1) combination recorded the highest leaf K content

Table 2. Effect of spacing, organics and bio-regulators on K uptake (kg ha^{-1}) in black nightshade.

Treatment	K				
	M_1	M_2	M_3	M_4	Mean
S_1	252.83	123.48	135.02	89.94	150.32
S_2	183.97	95.97	107.42	64.78	113.03
S_3	186.08	102.26	112.73	67.98	117.26
S_4	219.38	109.36	122.58	71.12	134.11
S_5	186.84	100.78	109.94	66.96	123.43
S_6	186.36	103.92	112.91	68.92	118.03
S_7	227.14	111.35	123.51	74.44	130.61
S_8	196.96	105.36	117.14	69.88	122.34
S_9	204.7	104.97	114.90	69.06	117.89
Mean	204.92	107.17	117.35	71.45	125.22
	M	**S**	**MS**	**SM**	
SEd	1.13756	1.44489	2.95245	2.88977	
CD (0.05)	2.78364	2.88651	5.77302	6.09515	

(2.72%) followed by the treatment wider spacing + humic acid combination (M_4S_7).

REFERENCES

Adani F, Genevini P, Zaccheo, Zocchi G (1998). The effect of commercial humic acid on tomato plant growth and mineral nutrition. J. Plant Nutr., 21(3): 561-575.

Bhattacharyya P, Ghosh G (2001). Phosphorus use efficiency in Brinjal with FYM and sulphur. J. Indian Soc. Soil Sci., 49(3): 456-462.

Campbell WH (1999). Nitrate reductase activity, structure function and regulation: Bridging the gap between biochemistry and physiology. Plant Physiol. Plant Mol. Biol., 50: 277-303.

David PP, Nelson PV, Sanders DC (1994). Humic acid improves growth of Tomato seeding in solution culture. J.Plant Nutr., 17(1): 173-184.

Dell' Agnola G, Nardi S (1987). Hormone-like effect and enhanced nitrate uptake induced by depolycondenced humic fractions obtained from *Allolobophora rosea* and *A. caliginosa* faeces. Biol. Fertility Soils. 4: 115-118.

Dhawan BN (1986). Current research on medicinal plants in India, INSA, New Delhi. pp: 1-86.

Ferretti M, Ghisi R, Nardi S, Passera C (1991). Effect of humic substances on photosynthetic sulphate assimilation in maize seedlings. J. Soil Sci., 71: 249-242.

Guminski S (1968). Present day views on physiological effects induced in plant organism by humic compounds. Soviet Soil Sci., 19: 1250-1256.

Jambhekar HA (1992). Use of earthworms as potential source of decomposed organic wastes. In: Proceedings, National Seminar on Organic Farming. Mahatma Phule Krishi Vidyapeeth, Pune, pp.52-53.

Khristeva LA, Luk'Yanenko NV (1962). Role of physiologically active substances in the soil -humic acids, bitumens and vitamins B,C, P-P, A and D in the life of plants and their replenishment. Soviet Soil Sci., 3: 1137-1141.

Malcolm RE, Vaughan D (1979). Humic substances and phosphatase activities in plant tissues. Soil Biol. Biochem., 11: 253-259.

Mukherjee DS, Mitra, Das AC (1991). Effect of oilcake on changes in carbon and microbial population in soil. J. Indian Soc. Soil Sci., 39: 457 - 462.

Panse VG, Sukhatme PV (1985). Statistical methods for agricultural workers. ICAR, New Delhi.

Piccolo A, Nardi S, Concheri G (1992). Structural characteristics of humic substances as related to nitrate uptake and growth regulation in plant systems. Soil Biol. Biochem., 24(4): 373-380.

Prasad RA, Sighania RA (1989). Effect of different types of enriched manures and time of incubation on soil properties. J. Indian Soc. Soil Sci., 37: 329 - 322.

Raina JN, Goswami KP (1988). Effect of fulvic acid and fulvates on the growth and nutrient uptake by maize plant. J. Indian Soc. Soil Sci., 36: 264 - 268.

Rauthan BS, Schnitzer M (1981). Effects of a soil fulvic acid on the growth and nutrient content of cucumber (*Cucumis sativus*) plants. Plant Soil, 63: 491-495.

Samson G, Visser SA (1989). Surface active effects of humic acids on Potato cell membrane properties. Soil Biol. Biochem., 21: 343-347.

Shivputra SS, Patil CP, Swamy GSK, Patil PB (2004). Cumulative effect of VAM fungi and vermicompost on nitrogen, phosphorus, potassium and chlorophyll content of papaya leaf. Mycorrhiza News, 16(2): 15-16.

Vaughan D, Malcolm RE (1979). Effect of humic acid on invertase synthesis in roots of higher plants. Soil Biol. Biochem., 11: 247-252.

Yadav K (1989). Influence of different levels of humic acids on nutrients content and growth of maize (*Zea mays* L.). Proc. Natl. Seminar on "Humus Acids in Agriculture", Annamalai University. Tamil Nadu, India, pp. 125 - 130.

11

Effect of photoperiod on propagation of strawberry (*Fragaria × ananassa* Duch.)

Sayed M. Zain Hasan[1], Isam Al-Madhagi[1]*, Aziz Ahmad[2] and Wan Abdullah Bin Yusoff[3]

[1]Faculty of Agrotechnology and Food Science, University Malaysia Terengganu (UMT), 21030 Kuala Terengganu, Malaysia.
[2]Faculty of Science and Technology, (UMT) University Malaysia Terengganu (UMT), 21030 Kuala Terengganu, Malaysia.
[3]Malaysian Agricultural Research and Development Institute (MARDI), Cameron Highlands, Pahang, Malaysia.

The effect of photoperiod on multiplication of commercial strawberry cultivars was investigated at the Malaysian Agricultural Research and Development Institute (MARDI) Station in Cameron Highlands. Two strawberry cultivars, Camarosa and Camaroga, were grown under a plastic rain-shelter supplemented with three levels of photoperiod, 12 (normal day length of Cameron Highland), 15 and 17 h. The result showed that photoperiod, cultivars, photoperiod × cultivars had significantly affected on a number of runners per plant and increased the length of runner and number of plantlets per runner and per plant. 15 h long-day were found to be very efficient in multiplying the strawberry plant under which the number of plantlets per plant had significantly increased to about 19.79 as compared to about 7.4 under the 12 h photoperiod. The result also showed that cultivar Camaroga was capable to produce a significant number of about 17.31 plantlets per plant compared to about 13.06 in Camarosa. The analysis of photoperiod × cultivar interaction indicated that the highest number of about 22.06 plantlets per plant was obtained from Camaroga, grown under the 17 h photoperiod. This study indicated that propagating performance in Camaroga is better than Camarosa under Cameron Highland climatic condition.

Key words: *Fragaria × ananassa*, photoperiod, propagation.

INTRODUCTION

Under the normal practice, strawberry plants are comercially grown from the plantlets which are proliferated at the runner nodes of the adult plants. Therefore, the number of the runners produced by the adult plants is one of the major parameter that must be taken into consideration when propagating the strawberry. The growth of runner is a result of cell division and elongation in the internodes (Nishizawa and Hori, 1993).

The environmental condition is recognized as one of the major factors effecting runner production in strawberry. It was demonstrated that under the short day photoperiod, strawberry produced more crown than the runner. The runners began to proliferate when the day length was 12 h or longer with the temperature of above 10°C, and the runner production was very effective under the 15 h day-length photoperiod when the temperatures was not less than 22.7°C (Darrow, 1966).

It was also found that the number and length of strawberry runner were not only dependent on the photoperiod but also dependent on the variety factor and the interaction of the two factors (Pipattanawong et al., 1996). Serce and Hancock (2005) showed that *Senga Sengana* cv and *Tioga* cv gave the highest number and length of runner per plant under the long day photoperiod but the production was dependent on the genotype.

Cameron Highland is located in the tropics where the long day photoperiod is absent, and with the frequent cloudy weather condition, making the long day photo-period in Cameron Highland fewer. Therefore, a study was carried out with the aims to i) assessing the effects of photoperiod, cultivar and their interaction on the runner and plantlets production of strawberry, and ii) determining the runner and plantlets production in two commercial cultivars of strawberry grown under three different

*Corresponding author. E-mail: isam.madhagi@gmail.com.

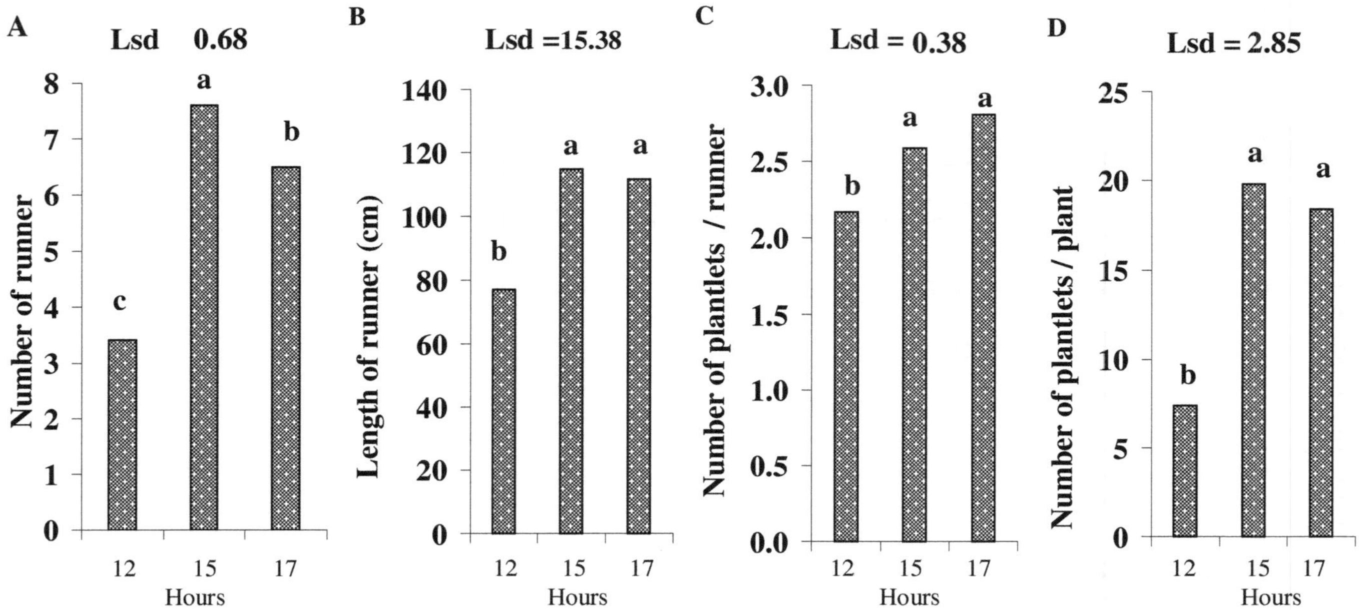

Figure 1. The effect of photoperiod on number of runner (A), length of runner (B), number of daughter plant per runner (C) and number of plantlets per plant (D). Bars with different letters indicate significant difference (P < 0.05) in the same chart.

photoperiods in Cameron Highland, Malaysia.

MATERIALS AND METHODS

Experimental design and plant material

The experiment was conducted at the Malaysian Agricultural Research and Development Institute (MARDI) Station in Cameron Highland, Malaysia, located at an altitude of 10,000 m above sea level, latitude 4°28'6.75"N and longitude 10°23'6.83"E. Two strawberry cultivars, Camarosa and Camaroga, were used. The uniform runner cuttings obtained from the adult plants were grown in nursery for plantlets propagation. A total of 594 plantlets of 3.5 leaves stage were transplanted in 15 cm diameter plastic pots containing 2 kg mixed growing media of coco peat and perlite at 9:1 ratio. All plants were placed under 400 m^2 plastic rain-shelter, watered and fertilized through the fertigation system. Pest and disease control were made using the standard practice applied in Cameron Highland.

The plants were exposed to three photoperiod levels, 12 (normal day length of Cameron Highland), 15 and 17 h by using 100 W fluorescent bulbs with an overall light density of approximately 50 lux. A total of 18 plants per experimental unit were arranged in a split plot factorial randomized completed block design (RCBD). Each treatment consists of three replications. The numbers and length of the runner, the number of the plantlets per runner and number of the plantlets per plant were measured at three and half months after treatment.

Statistical analysis

Analysis of variance (ANOVA) was used to determine the significant difference among the treatment and the multiple range test of Duncan (p > 0.05) for the significant separation of means among the treatments was calculated by using the MSTAT-C program.

RESULTS

The effect of photoperiod on the number of runner per plant is shown in Figure 1A. The number of runner had increased by 222.54 and 190.04% under 15 and 17 h long-day respectively, compared to those under the normal photoperiod. The highest number of runner was obtained from the plant grown under the 15 h photoperiod with an average of 7.6 runners per plant (Figure 1).

Overall, the result showed that photoperiod also increased the average runner length (cm) where the longest runner with nearly 114.65 cm was seen in plant grown under the 15 h photoperiod (Figure 1B). However, the runner length under the 15 h long-day was not significantly different with the length under the 17 h long-day.

The 15 and 17 h photoperiods had significantly increased the number of plantlets per runner to an average of 2.58 and 2.81 respectively comparing to about 2.17 by the 12 h photoperiod (Figure 1C). However, the number of plantlet per runner exposed to 15 h photoperiod was not significantly different with the number under 17 h photoperiods.

The number of plantlets per plant had also significantly increased to an average of 19.79 and 18.38 under the 15 and 17 h photoperiods respectively as compared to about 7.4 under the normal (12 h) photoperiod (Figure 1D). But, the number of plantlets per plant was not significantly different between the 15 and 17 h photoperiods treatments.

Figure 2 shows the effect of different cultivars on the multiplication of runner and plantlets in strawberry. It demonstrated that the number of runner per plant was

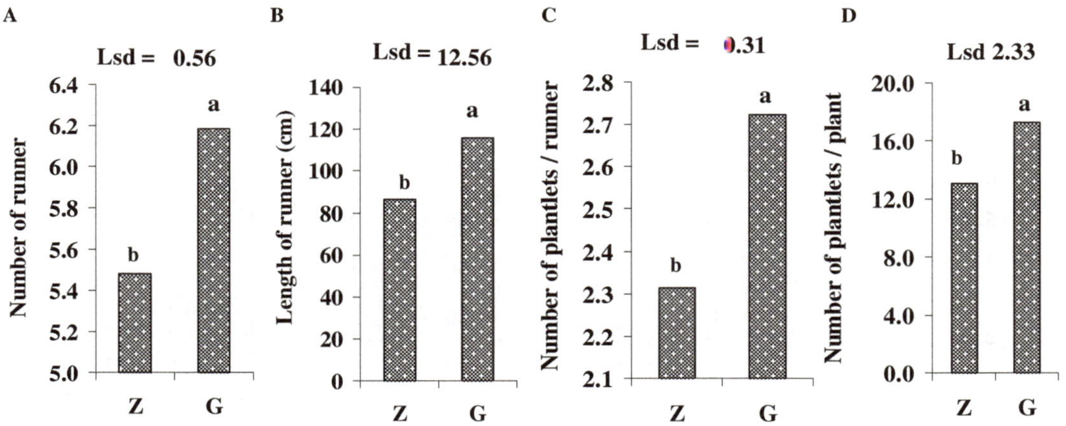

Figure 2. Shown the different between cultivars on number of runner (A), length of runner (B), number of daughter plants per runner and number of daughter plants per intact plant (D). Bars with different letters indicate significant difference (P < 0.05) in the same chart. Z = Camarosa cultivar, G = Camaroga cultivar.

significantly higher with an average of 6.19 in cultivar Camaroga compared to about 5.48 in cultivar Camarosa Figure 2A). Cultivar Camaroga also surpassed the Camarosa in terms of the runner length, where the (average length of about 115.8 cm was found in Camaroga as compared to about 86.4 cm found in Camarosa (Figure 2B).

The numbers of plantlets per runner and plantlets per plant were also significantly higher in Camaroga than those in Camarosa (Figures 2C and D). The average number of plantlets per runner in Camaroga and Camarosa were about 2.72 and 2.31 respectively, meanwhile, the number of plantlets per plant in the two cultivars were about 17.31 and 13.06 respectively.

Overall, the interaction of photoperiod × cultivar had significantly increased the runner and plantlets multiplication in strawberry (Figure 3). A significant highest average number of about 8.09 runners per plant was obtained from Camaroga grown under the 15 h photoperiod (Figure 3A). However, the number of runner in Camaroga was not significantly different between those grown under the 15 and 17 h photoperiods (Figure 3A).

The interaction effect of photoperiod × cultivar has also showed that the runner length of Camaroga was superb than Camarosa. Under the 15 and 17 h photoperiod, the average runner length of Camaroga reached nearly 130.90 cm per plant as compared to those in Camarosa of about 98.40 cm under the same photoperiods (Figure 3B).

The effect of photoperiod × cultivar interaction on the plantlets number per runner is shown in Figure 3C. It displayed that the most significant high number of about 3.17 plantlets per runner was produced by Camaroga when grown under the 17 h photoperiod. The number of plantlets per plant had also increased significantly under the 15 and 17 h photoperiods (Figure 4). The highest number of about 22.60 plantlets per plant was obtained from Camaroga, grown under the 17 h photoperiod. Meanwhile, the significant low number of plantlet per plant with about 7.03 and 7.76 has been shown by Camarosa and Camaroga respectively when grown under the 12 h photoperiod (Figure 3D).

DISCUSSION

The result of this study clearly indicated that the photoperiod had increased the number and length of runner per plant, the number of plantlets per runner and the number of plantlets per plant of strawberry. However, such effects are highly dependent on the cultivar of strawberry. Photoperiod in fact plays an important role in initiating runner bud proliferation (Darrow, 1966) and in differentiating the runner axillary buds of strawberry (Hytönen et al., 2009). The long day photoperiod led to increase in the level of endogenous gibberellins which encouraged bud proliferation in the plant (Taylor et al., 1994); this difference in peduncle length was attributed to the greater number of epidermal cells which indicated that cell division and length of the internodes is prolonged under LD conditions (Nishizawa and Hori, 1993; Nishizawa, 1994). This data is in agreement with the previous report by Snsteby et al. (2006) Nishiyama et al. (2003) and Konsin et al. (2001).

A little difference in the runner and plantlets productions among the 15 and 17 h photoperiod treatments in this study might be due to the reason that the 15 and 17 h above the normal (12 h) photoperiod in Cameron Highland gave the same effect on the runner and plantlet production, disregarding the difference of the extra hour length added to the normal photoperiod. The result of this study found that the runner and plantlet productions of

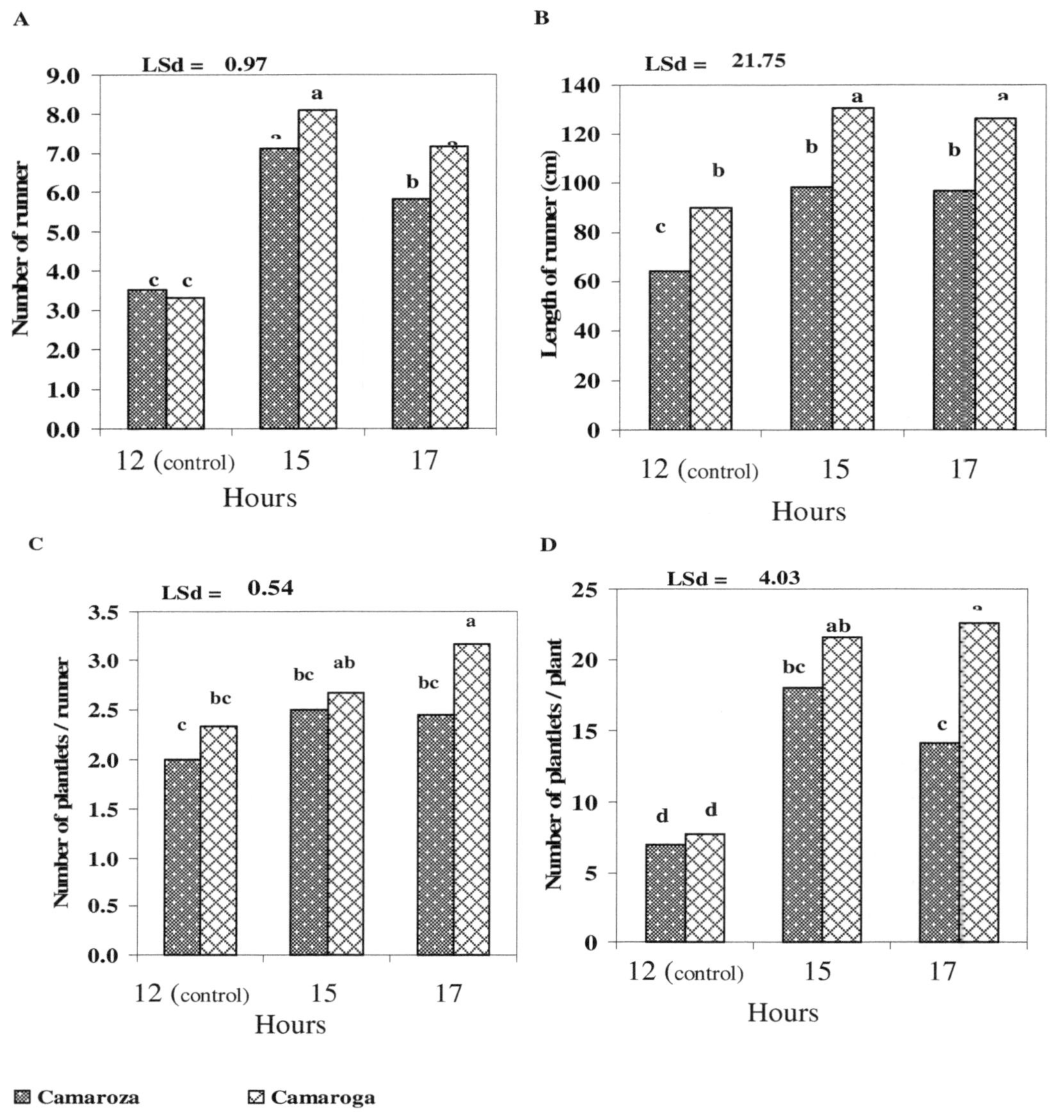

Figure 3. The interaction between photoperiod and cultivars on number of runner (A), length of runner per plant (B), number of daughter plant per runner (C) and number of daughter plant per intact plant (D). Bars with different letters indicate significant differences ($P < 0.05$) in same chart.

Camaroga are better than Camarosa, which confirms that the two varieties consist of two different genetic resources. Since the present cultivated strawberry varieties are believed to be derived from a diverse genetic resources of the wild parents of Fragaria chiloeniss and Fragaria virginiana (Serce and Hancock, 2005), therefore, it can be assumed that the different response to the photoperiod by the two cultivars is probably due to the different of their origin and stock.

Conclusion

It can be concluded that the 15 and 17 h photoperiods can induce the increase in number and length of runner per plant and the number of plantlet per runner and plantlet per plant of strawberry in Cameron Highland. However, cultivar Camaroga of strawberry showed a better performance than cultivar Camarosa for runner and plantlet production under Cameron Highland climatic condition.

REFERENCES

Darrow GM (1966). The Strawberry: History, Breeding and Physiology. New York: Holt, Rinehart and Winston.

Hytönen T, Elomaa P, Moritz T, Junttila O (2009). Gibberellin mediates daylength-controlled differentiation of vegetative meristems in strawberry (Fragaria × ananassa Duch). BMC Plant Biol., 9: 18

Konsin M, Voipio I, Palonen P (2001). Influence of photoperiod and duration of short-day treatment on vegetative growth and flowering of strawberry (Fragaria × ananassa Duch.). J. Hortic. Sci. Biotechnol., 76(1): 77-82.

Nishiyama M, Ohkawa W, Kanahama K (2003). Effect of photoperiod on the development of inflorescences in everbearing strawberry 'Summerberry' plants grown at high temperature. Tohoku J. Agric. Res., 53(3/4): 43-52.

Nishizawa T (1994). Effects of photoperiods on the length and number of epidermal cells in runners of strawberry plants. J. Jpn. Soc. Hortic. Sci., 63(2): 347-352.

Nishizawa T, Hori Y (1993). Elongation of strawberry runners in relation to length and number of cells. Tohoku J. Agric. Res., 43(3-4): 87-93.

Pipattanawong N, Fujishige N, Yamane K, Ijiro Y, Ogata R (1996). Effects of growth regulators and fertilizer on runner production, flowering, and growth in day-neutral strawberries. Jpn. J. Trop. Agric., 40(3): 101-105.

Serçe S, Hancock JF (2005). The temperature and photoperiod regulation of flowering and runnering in the strawberries, *Fragaria chiloensis*, F. virginiana, and *F. × ananassa.* Sci. Hortic. 103(2): 167-177.

Snsteby A, Heide OM, Grimsby S, Grimsby I (2006). Out-of-season strawberry production in Norway: yield responses of cv. Korona to photoperiod preconditioning treatment. Acta Hortic. (708): 371-374.

Taylor DR, Blake PS, Browning G (1994). Identification of gibberellins in leaf tissues of strawberry (Fragaria × ananassa Duch.) grown under different photoperiods. Plant Growth Regul., 15(3): 235-240.

12

Woody plant species used in urban forestry in West Africa: Case study in Lomé, capital town of Togo

Radji Raoufou*, Kokou Kouami and Akpagana Koffi

Laboratory of Plant Biology and Ecology, BP 1515 Lomé - Togo.

Many studies have been conducted on the flora of Togo. However, none of them is devoted to the ornamental flora horticulture. This survey aims to establish an inventory of the woody plant species in urban forests of Lomé, the capital town of Togo. It covers the trees planted along the avenues, in the gardens, courtyards, shady trees and trees used as fences for houses or trees at the seaside. In total, 297 plant species belong to 141 genera and 48 families were recorded. They are dominated by 79% of dicotyledonous, 13% of monocotyledonous and 8% of gymnosperms. Families that are best represented in terms of species are those of the Euphorbiaceae, Arecaceae and Acanthaceae. Alien species represent 69% and African species represent 31% out of which 6% are from Togo. According to the current threatening of the natural habitat by human activities, African native plant species could be more useful for ornamental purposes than exotic plants.

Key words: Ornamental horticulture, plant flora, green areas, valorisation, native flora.

INTRODUCTION

Urban forestry refers to trees and forests located in cities, including ornamental and grown trees, street and parkland trees, protected forests and green areas (Kuchelmeister, 2000). In European countries, urban forestry has mainly been developed for both aesthetic purposes and environmental benefits (Miller, 1997; Nilsson and Randrup, 1997). It therefore appears as a modern approach to conserve plants and forest stands with ornamental values in urban areas taking into account the long-term planning, as well as design and management (Nilsson and Randrup, 1996). All these functions led to the concept of "urban greening" or urban forestry (Miller, 1997).

Trees planted along the avenues in African capital cities, contribute to the plant diversity (IUCN, 1994). This also applies to the parks and green areas that provide a decent environment. The indirect benefits of an urban landscape covered with trees for the physical and mental health has been documented (Ulrich, 1984). It is demonstrated that plants have the capacity to absorb contaminants from the cities. They are thus called the lungs of cities (McPherson et al., 2005; Yang et al., 2005). They represent places of pastime, leisure, recreation and decoration necessary for human life. They contribute to the improvement of air quality, thus having a positive impact on health, with effects as obvious as the decrease in cases of respiratory diseases (McPherson, 2005).

In developed countries, several studies have been devoted to the study of trees in urban areas (Kuchelmeister and Braatz, 1993; Kuchelmeister, 2000). However, very little literature addresses the presence of trees in urban areas in Sub-Saharan Africa (SSA). Meanwhile, this knowledge of the urban flora is necessary to optimize the planning methods and the management of trees and urban green areas. It can help to promote several native species and bring under control those currently endangered in their natural habitats. It was therefore proposed in the framework of this study to make an inventory of the woody plant flora in Lomé, the

*Corresponding author. E-mail: pradji@hotmail.com.

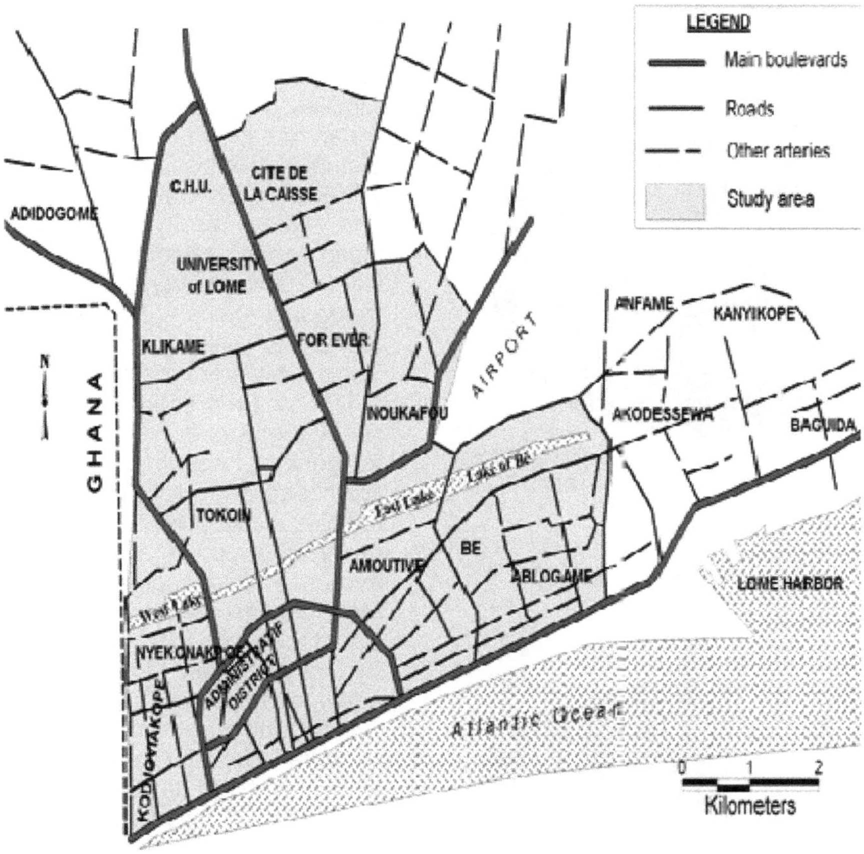

Figure 1. Study area.

capital city of Togo. A specific focus was done on the identification of trees planted in the city and the description of their current management methods. The challenge is that many programs and development projects are underway or planned. This stage of scientific knowledge is urgent to optimize the development plans in urban forestry and to create methods of managing trees and urban green areas. The main target is to promote consideration of diversity and importance of urban trees in an integrated and adapted planning framework. It finally allows emphasizing the importance of the indigenous flora in urban areas, facing the effect of climate change and natural habitat deterioration.

MATERIALS AND METHODS

Lomé, capital town of Togo was the study area (Figure 1). It is located in the southern coastal plain of Togo in sub-equatorial climate with two rainy seasons (April to July, September to October) and two dry seasons (November to March, August). Lomé was the driest station in the coastal plain with an annual average rainfall of 800 mm. The average temperature during the period of study was 27°C and the annual average humidity constantly high, varied between 80 and 90%. The annual average temperature is flattened (3 to 4°C) and kept within 25 to 29°C. The potential evapotranspiration (ETP Penman) is about 1610 mm per year.

The flora of the coastal plain is a mosaic of crops and fallow lands, thickets, shrubs, derived savannah, coastal grasslands and savannah with bushy termite mounds coexisting with forest patches (Ern, 1979 ; Brunel et al., 1984, Kokou, 1998). Floristic surveys were carried out, using a geographic map (IGN, 1/200 000). The botanical inventories were carried following a conventional sampling based on transect. In total, 87 transects of 2.5 Km by 5 m with a total surface of 103.5 km² (10350 ha) and over a coastal length of 10 Km, helped to record all the trees and shrubs of dbh (diameter at breast height) ≥ 10 cm. The species were identified scientifically on the field using the flora of Brunel et al. (1984), Grisvard et al. (1990) and Lebrun and Stork (2003, 2006, 2008). These identifications were later confirmed at the National Herbarium of Togo. The names of authors as well as the natural origins of the species were obtained using data from the International Code of Botanical Nomenclature (2007) and databases from Index Nominum Genericorum (ING), the International Plant Names Index (IPNI) and flowering plants of Africa by Lebrun and Stork (2003, 2006, 2008). For each species, the phytogeographical origin (Ake, 1984) and the life form

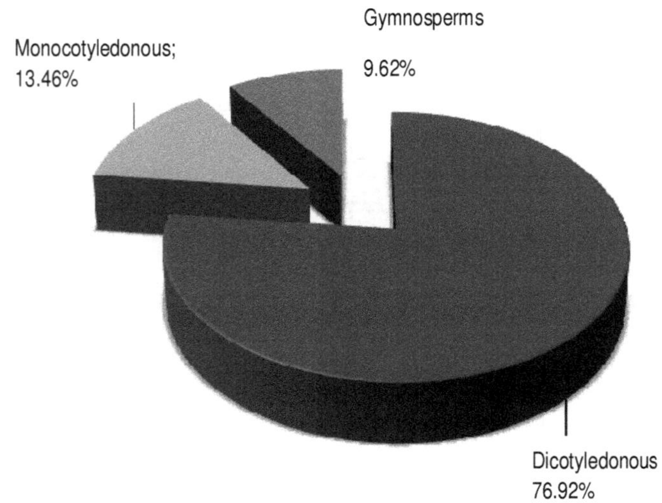

Figure 2. Taxonomic distribution of trees in Lomé, 2009

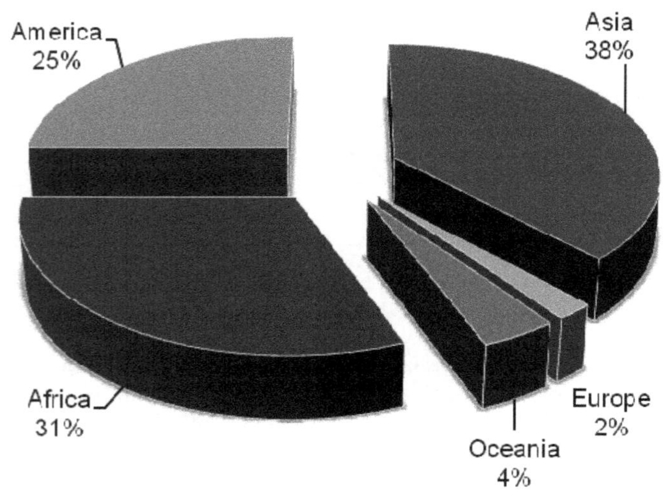

Figure 3. Distribution of trees in Lomé according to their native continents.

(Raunkier, 1934; Kalanda, 1981 in Cikuru, 1982) were determined.

On the basis of the findings of this inventory, a floristic list was established for the whole intakes. The floristic structure was analyzed by the determination of the abundance, the dominance and the taxon's frequency.

RESULTS

The following classification was identified: 1- trees planted at the roadsides; 2- trees in gardens of public services, inside houses, in front of the apartments and shady trees; 3- trees used as fences for houses; 4- trees planted inside uninhabited lands; 5- trees at the seaside. 297 plant species belonging to 141 genera and 52 families (Appendix 1: List of woody species present in urban forestry in Lomé) were recorded. The species were distributed among the major taxonomic groups with 76.92% of Dicotyledonous, 13.46% Monocotyledonous and 9.62% of Gymnosperms (Figure 2). Some large sized monocotyledons were considered as urban trees like *Cocos nucifera*, *Elaeis guineensis* and *Ravenala madagascariensis*. The inventory revealed that more than half of the listed species were not reported in the flora of Togo. Only 10 of them were reported as cash crops.

The best represented families were Euphorbiaceae (43 species grouped in 8 genera), Arecaceae (24 species in 20 genera), Acanthaceae (17 species from 11 genera), Moraceae and Araliaceae (15 species per family). These species accounted for 44% of all those listed. Families represented by a single species were Anacardiaceae, Bixaceae, Caricaceae, Flacourtiaceae Lecythidaceae, Moringaceae, Oxalidaceae, Polygonaceae, Sapindaceae, Sapotaceae, Sterculiaceae, Turneraceae, Zygophyllaceae, Musaceae and Strelitziaceae. Some edible fruit species including 7 species, (*Artocarpus altilis*, *Citrus sinensis*, *Citrus lemon*; *Coccoloba uvifera*, *C. nucifera*, *Mangifera indica* and *Terminalia cattapa*) were also considered as ornamental trees (Appendix 1).

Figure 3 showed that only 31% of the plant species encountered in Lomé are native to Africa while 69% were found to be from other continents including 38%, as native species from Asia and 25% as native species from American continents. Among dicotyledonous plants that accounted for 79% of the taxa, 54% originated from America; and monocotyledonous African species represented 15%. This percentage represents 20% of dicotyledons. More than 53% of the gymnosperms recorded on the field are native to America while only 10% are African native species. Trees planted at the roadsides, usually on both sides of the roadway are *Delonix regia*, *Khaya senegalensis* and *Tectona griandis*. Trees found in public services' gardens, houses and in front of apartments, as well as shady trees were characterized by species like *Guaiacum officinale*, *Polyalthia longifolia*, *Roystonea regia* and many Ficus. The trees used as fence of houses were *Ixora macrothyrsa*, *Pithecellobium dulce*, *Thevetia neriifolia* and *T. peruviana*. These species constituted strong hedges bordering private properties. *Cassia siamea*, *Eucalyptus* spp., *Millettia thonningii*, *T. cattapa*, *T. mantaly* were found to be tree species frequently planted in the uninhabited lands, courtyards or planted to fill the open spaces adjacent to buildings. The last type of urban forest includes the trees lining the sandy beach and characterized mainly by *C. nucifera* and *Casuarina*

Polyalthia longifolia in alignment in a garden

Mixed forest dominated by *Senna siamea* and *Azadirachta indica*

Monospecific forest made of *Senna siamea* (syn. *Cassia siamea*)

Khaya senegalensis along boulevard

Tectona grandis planted on both sides of an avenue

Forest of *Azadirachta indica* in the courtyard of a public institute

Figure 4. Some types of urban forests in Lomé.

equisetifolia.

DISCUSSION

The results of this study showed that only 31% of trees in Lomé are African native species and 69% are exotic, that is, originated from other continents. This percentage was found to be slightly low when compared to that pointed out by Aké (2002) in Côte d'Ivoire where exotic species account for 77%. Out of the 31% African native plant species found, less than 6% were from the local flora. These indigenous plant species were less spread, although they contained several useful ornamental species likely to be improved namely for the production of bright colours or large flowers. The ornamental plant market is booming (Jaffee and Masakure, 2005; Roy and

Thorat, 2008; Neven et al., 2009) while woody forest species are certainly meant for a promising future in the horticultural market. As shown in this study, wild plants produced in nurseries are scarce in this market. It is much better to use an indigenous plant, which comes from an environment similar to that in which it will be planted than buying exotic plants.

When the climate of a given continent gets closer to that of tropical Africa, the number of species from that continent gets high. This can be applied to America and Asia continents which represent 25 and 38% of the horticultural species in Africa, respectively. Plants from the Mediterranean and Australian climate were imported and they grow slightly well in Togo like in most African countries. Currently, 25% of the flora in Quebec is exotic and in Canada, out of a total number of 141 species purposely introduced, 73 or nearly 52% were considered as decorative or ornamental plants (Baudin, 2007). These results recall for reflection on the role of native species in the horticultural flora and the importance of their production.

Out of the classification of the urban forest proposed in the present study, many species belong to several types of trees are multipurpose (Kuchelmeister and Braatz, 1993). Furthermore, the trees found by the seaside and those grown along the avenues (Figure 4), currently with larger height do not receive silvicultural maintenance. They grow in an uncontrolled manner and whenever they receive any given treatment, the branches are cut bluntly without aesthetic or the trees are merely or simply felled. Courtyard or reforestation trees are usually planted in alleys and are subject to little maintenance. Trees used for fencing rarely exceed the human height (2 m): they are regularly subject to ornamental pruning. Trees found inside parks, as well as those serving as radiant hedges are subject to regular pruning maintenance. Their heights vary as slightly as the shapes given to the hedges. The pruning is made following the principles of "French park" characterized by polished, neat and well straight structures, with symmetries and panoramic views, axes easy to understand at a glance (Walpole, 2002; Baridon, 2004). However, the outer hedges are taller than those used as fences, demarcating areas of the same garden. In this case, the pruning is designed to have the aerial part of the tree meeting the needs of the owner. The latter is concerned with the need for security, preventing the inconveniences to nearby residents or aesthetics needs (Latchoumy, 2007). The products of the pruning are used as fuel wood for households (N'zala, 2002).

The number of identified species showed that, the plant flora of Lomé is fairly diversified. Development projects in the city of Lomé, especially repairing of the public roads include tree planting program. However, foresters and professional planners still play a minor role in the greening initiatives (N'zala, 2002) since they are rarely consulted. The ongoing management of plant heritage, carried out by the department of green area of the municipality should include the overall supervision, the maintenance and the plant heritage preservation. Most of the services could be assigned to such a business related provider on the basis of competitive bid for tender launched every three years to select the services provider. A pricelist would allow the financial management of provider's services.

Conclusion

The identification and knowledge of trees in Lomé are a vital factor in the development of green areas. This study yields a list of species that contribute to the beautification of the streets and avenues of the city, the production of shade and protection of the environment service. It has come out with a classification of forest types and pointed out the importance of local species in this plant diversity. Research is fundamental for the development of the native flora, the broadening of the presented list range and the streamlining of this heritage management.

This work is a prerequisite for any urban forestry work and urban plants management and for such a management in Lomé, the design of a database becomes a necessity. The forestry work carried out on trees in Lomé requires the training of skilled technical personnel for a rigorous and professional maintenance towards a sustainable urban forestry. Producers of ornamental plants should preferably be educated to develop appropriate methods for the growing market demand while respecting the resource and sustainable development standards.

REFERENCES

Aké AL (1984). Flora of Côte d'Ivoire: Descriptive study and biogeography, with some ethnobotanical notes. Th. Doct. Univ. Abidjan, p. 1206.

Aké AAE (2002). Contribution to the study of ornamental plants grown in areas of Abidjan and San Pedro, Cote d'Ivoire. Volume 1 - Text and iconography. Volume 2 - Illustrated Catalogue of 241 species and horticultural varieties. Th. Doct. 3e cycle, Univ. Cocody (RCI), p. 242.

Baridon M (2004). Gardens and landscapes in ancient times. Our History, 222: 18-20.

Beaudin E (2007). Invasive Alien Species. Green living, Reportage, Issue of November 14: 2.

Brunel JF, Hiepko P, Scholz H (1984). Analytical flora of Togo. GTZ ed. Eschborn, p. 750.

Cikuru B (1982). Flora and vegetation of ruderal Benqamisa (Haut-Zaire). Memory, unpublished UNIKIS, p. 111.

Dutrève B, Pinatel M, Besse F (1998). Urban forestry. Place the tree in two urban cities in West Africa: Nouakchott and Ouagadougou. Le Flamboyant, 47: 18–25.

Ern H (1979). The vegetation Togo. Gliederrung, endangerment, Erhat. Willdenowia, 9: 295-312.

FAO (2001). The urban forestry. Case study on developing countries. Rome. In Grey, G.W.et Deneke, F.J. 1978. Urban Forestry. 2nd edition. John Wiley and Sons, New York, p. 279.
Grisvard P Chaudun V Chouard P, Guillaumin A, Schneiter P (1990). The Good Gardener. Horticultural Encyclopedia, La Maison Rustique, Paris, 152nd ed. 2: 783.
International Code of Botanical Nomenclature (online). http://www.bgbm.fu-berlin.de/iapt/nomenclature/code/SaintLouis/0000St.Luistitle.htm (consulting on May, 19, 2010).
Jaffee S, Masakure O (2005). Strategic use of private standards to enhance international competitiveness: Vegetable exports from Kenya and elsewhere. Food Policy, 30: 316–333.
Kokou K (1998). Forest mosaics in southern Togo: Biodiversity, dynamics and human activities. Th doct. Univ. Montpellier II, p. 140.
Kuchelmeister G (2000). Trees for the urban millennium: An update on urban forestry, Unasylva, 200: 49-55.
Kuchelmeister G, Braatz G (1993). New Perspectives on Urban Forestry. Unasylva, 173: 3-12.
Latchoumy L (2007). The computerized management of heritage trees in the city of Avignon: a contribution to the development of GIS tree. Memory graduation Agro Paris Tech, Avignon ENGREF, p. 90.
Lebrun JP, Stork A (2003). Tropical African Flowering Plants. Ecology and distribution. Balanitaceae Annonaceae. Conservatory and Botanical Gardens of the City of Geneva. Occasional, 1(9): 793.
Lebrun JP, Stork A (2006). Tropical African Flowering Plants. Ecology and distribution. Dichapetalaceae Euphorbiaceae. Conservatory and Botanical Gardens of the City of Geneva. Occasional 2(9a): 306.
Lebrun JP, Stork A (2008). Tropical African Flowering Plants. Ecology and distribution. Mimosaceae Fabaceae. Conservatory and Botanical Gardens of the City of Geneva. Occasional, 3: 325.
Lebrun JP, Stork A (2008). Tropical African Flowering Plants. Ecology and distribution. Fabaceae (*Desmodium Zornia*). Conservatory and Botanical Gardens of the City of Geneva. Occasional, 4: 291.
Mailliet L (1989). Methodological approach to tree management in the city: Some elements of the heritage inventory. French For. Rev., Special Issue: The tree in town: XLI: 119-124.
McPherson EG (2005). Trees With Benefits. Am. Nurseryman, pp. 34-40.
McPherson EG, Simpson JR, Peper, Paula J, Maco SE, Xiao Q (2005). Municipal Forest Benefits and Costs in Five US Cities. J. For., 103(8): 411-416(6).
Miller R (1997). Urban forestry: Planning and management of green space. Prentice Hall, Upper Saddle River, New Jersey, pp. 22-67.
Neven D, Odera MM, Reardon T, Wang H (2009). Kenyan Supermarkets, Emerging Middle-Class Horticultural Farmers, and Employment Impacts on the Rural Poor. World Dev., 37(11): 1802–1811.
Nilsson K, Randrup TB (1996). Urban forestry in the Nordic Countries. Proceedings of a Nordic workshop on urban forestry, Reykjavik, Iceland, from 21 to 24 September 1996. Danish For. Landscape Res. Inst., pp. 15-41.
N'zala D (2002). Trees and green spaces in Brazzaville (Congo). Woods For. Tropics, 272(2): 88-92.
Raunkaier C (1934). Life forms of plants and statistical plant geography, Clarendon, Oxford, UK, pp. 51-78.
Roy D, Thorat A (2008). Success in High Value Horticultural Export Markets for the Small Farmers: The Case of Mahagrapes in India. World Dev., 36(10): 1874–1890.
UICN (1994). Putting plans into action. Report of Metropolitan Open Space Systems (MOSS) International Conference, Durban, Afrique du Sud, February, 9-11.
Ulrich R (1989). The Role of Trees in Human Well-Being and Health. Proceedings of the fourth conference on the urban forest. St. Louis, Missouri.
Véron J (2007). Half the world population lives in cities. Population Soc. Paris, 435: 4.
Walpole H (2002). Essay on the art of modern gardens. Mercure de France, pp. 12-18.
Yang J, McBride J, Zhou J, Sun J (2005). The urban forest in Beijing and its role in air pollution reduction. Urban For. Urban Greening, 3(2): 65.

Appendix 1. List of woody plant species present in urban forest areas in Lomé (Togo's capital).

Araucariaceae
Araucaria cunninghamii Ait.
Araucaria excelsa R. Br.

Cupressaceae
Callitris intratropica R.T. Baker & H.G. Sim.
Cupressus sempervirens L.
Thuja occidentalis L.

Cycadaceae
Cycas circinalis L.
Cycas revoluta Thunb.

Pinaceae
Pinus canariensis Sm.
Pinus caribaea Morelet

Zamiaceae
Encephalartos ferox L.
Encephalartos gratus L.
Zamia dudgesiana L.
Zamia dominican Hort.
Zamia fisheri L.
Zamia floridana L.
Zamia furfuracea L.
Zamia fusca Hort.Paull. ex Regel
Zamia pumila L.
Zamia standley Hort.

Acanthaceae
Acanthus mollis L.
Acanthus pubescens Engl.
Aphelandra squarrosa Nees.
Barleria lupulina Lindl.
Crossandra infundibuliformis (L.) Nees.
Crossandra nilotica Oliv.
Graptophyllum pictum Nees
Jacobinia pohliana Benth. & Hook. f.
Pachystachys lutea Nees
Pseuderanthemum atropurpureum Hort. var.tonga
Pseuderanthemum reticulatum Hort.
Sanchezia nobilis Hook.
Strobilanthes dyerianus Mast.
Strobilanthes maculata Nees.
Thunbergia erecta (Benth.) T. Anders.
Thunbergia fragrans Roxb.
Thunbergia grandiflora (Roxb. ex Rottl.) Roxb.

Anacardiaceae
Anacardium occidentale L.

Annonaceae
Cananga odorata (Lam.) Hook. f. & Thoms.
Polyalthia longifolia L.
Polyalthia longifolia var. pendula L.

Apocynaceae
Adenium obesum (Forssk.) Roem. & Schult.
Allamanda cathartica L. var. hendersonii Hort.
Allamanda neriifolia Hook.
Allamanda violacea Gardn.
Nerium oleander L.
Nerium oleander L. var. variegatum plenum
Plumeria rubra L.
Plumeria rubra f. tricolor (Roem. & Pav.) Woodson
Thevetia neriifolia Juss. ex Steud.
Thevetia peruviana (Pers.) Merr.
Strophanthus gratus (Hook.) Franch.

Asclepiadaceae
Asclepias curassavica L.
Calotropis procera (Ait.) Ait. f.
Cryptostegia grandiflora Br. R. ex Lindl.
Cryptosegia madagascariensis Boj.
Vernonia colorata (Willd.) Drake

Bignoniaceae
Crescentia cujete L.
Jacaranda mimosaefolia Don D.
Millingtonia hortensis L. f.
Newbouldia laevis (Beauv.) Seem. ex Bureau.
Parmentiera edulis DC.
Spathodea campanulata P. Beauv.
Tecoma stans Griseb.

Bixaceae
Bixa orellana L.

Bombacaceae
Adansonia digitata L.
Ceiba pentandra (L.) Gaertn.

Annonaceae
Cananga odorata (Lam.) Hook. f. & Thoms.
Polyalthia longifolia L.
Polyalthia longifolia var. pendula L.

Appendix 1. Contd.

Caesalpiniaceae
Bauhinia acuminana L.
Bauhinia monandra Kurz.
Bauhinia rufescens Lam.
Bauhinia tomentosa L.
Caesalpinia pulcherrima Sw.
Cassia laevigata Willd.
Cassia obtusifolia L.
Cassia occidentalis L.
Cassia sieberiana DC.
Cassia spectabilis DC.
Cynometra megalophylla Harms
Parkinsonia aculeata L.
Senna alata (L.) Roxb.
Senna siamea (Lam.) Irwin et Barneby

Fabaceae
Delonix regia (Bojer) Raf.
Dialium guineense Willd.
Peltophorum pterocarpum (Decne.) Benth.

Flacourtiaceae
Flacourtia flavescens Willd.

Lauraceae
Cinnamomum zeylanicum L.
Persea americana Mill.

Lecythidaceae
Napoleonaea vogelii Hook. & Planch.

Lythraceae
Lagerstroemia indica L.
Lawsonia inermis L.

Malvaceae
Hibiscus arnottianus A. Gray
Hibiscus rosa-sinensis L.
Hibiscus rosa-sinensis L. var. california gold
Hibiscus rosa-sinensis L. var. cooperi
Hibiscus rosa-sinensis L. var. crown of Bohemia
Hibiscus rosa-sinensis L. var. mist
Hibiscus rosa-sinensis L. var. scarlet'
Hibiscus rosa-sinensis L. var. albus
Hibiscus rosa-sinensis L. var. natal
Hibiscus rosa-sinensis L. var. plenus
Hibiscus schizopetalus (Mast.) Hook. f.
Thespesia populnea Soland. ex Corrêa

Caricaceae
Carica papaya L.

Euphorbiaceae
Acalypha hispida Burm.
Acalypha wilkesiana Müll. Arg.
Acalypha wilkesiana Müll. Arg. var. hoffmanii
Acalypha wilkesiana Müll. Arg. var. moorea
Acalypha wilkesiana Müll. Arg. var. 'java white'
Acalypha wilkesiana Müll. Arg. var. macrophylla
Acalypha wilkesiana Müll. Arg. var. tricolor
Breynia disticha J.R & G. Forst.
Breynia nivosa Small.
Codiaeum variegatum (L.) Bl. 'Clipper'
Codiaeum gloriosum superbum'
Codiaeum frutescens var.'Wellpark-Beauty'
Codiaeum variegatum (L.) Bl.
Codiaeum variegatum (L.) Bl. var.' america'
Codiaeum variegatum (L.) Bl. var. aucubifolium
Codiaeum variegatum (L.) Bl. var. cornutum
Codiaeum variegatum (L.) Bl. var.craigii
Codiaeum variegatum (L.) Bl. var. 'delaruye'
Codiaeum variegatum (L.) Bl. var. 'elaine'
Codiaeum variegatum (L.) Bl. var. exotica
Codiaeum variegatum (L.) Bl. var.'General Paget'
Codiaeum variegatum (L.) Bl.var. gloriosum
Codiaeum variegatum (L.) Bl. var. imperialis
Codiaeum variegatum (L.) Bl. var. lineatum
Codiaeum variegatum (L.) Bl. var.Mortimer
Codiaeum variegatum (L.) Bl. var. punctum aureum
Codiaeum variegatum (L.) Bl. var.rubrum
Croton zambesicus Müll. Arg.
Euphorbia lactea Haw.
Euphorbia millii Des Moul. var. breonii
Euphorbia oncoclada Drake.
Euphorbia tirucalli L.
Euphorbia deightoni Croizat
Hura crepitans L.
Jatropha curcas L.
Jatropha gossypifolia L.
Jatropha multifida L.
Jatropha pandurifolia Andr.
Jatropha podagrica Hook.
Pedilanthus tithymaloides Poit.
Pedilanthus tithymaloides Poit. var. variegata
Phyllanthus angustifolius Sw.
Ricinus communis L.

Appendix 1. Contd.

Meliaceae
Azadirachta indica A. Juss.
Melia azedarach L.
Khaya senegalensis A.Juss.

Mimosaceae
Acacia auriculaeformis A.Cunn. ex Benth.
Acacia biflorus R.Br.
Acacia farnesiana (L.) Willd.
Acacia nilotica (L.) Willd. Ex Del.
Adenanthera pavonina L.
Calliandra surinamensis Benth.
Pithecellobium dulce(Roxb.) Benth.
Samanea saman (Jacquin) Merrill

Moringaceae
Moringa oleifera L.

Myrtaceae
Eugenia uniflora L.
Psidium guayava L.
Eucalyptus citriodora Hook.
Eucalyptus deglupta Blume
Eucalyptus tereticornis Hook.
Eucalyptus torrelliana F. Muell.

Nyctaginaceae
Bougainvillea glabra Choisy
Bougainvillea spectabilis Willd.
Bougainvillea spectabilis Willd. var.alba plena
Bougainvillea spectabilis Willd. var. 'Mary Palmer
Bougainvillea spectabilis Willd. var. rubra plena
Bougainvillea spectabilis Willd. var. variegata
Bougainvillea x buttiana Holttum & Standl.
Bougainvillea x *buttiana* Holtt. var. Mrs. McLean
Bougainvillea x *buttiana* Holtt. var. praetoria
Bougainvillea harrisii

Punicaceae
Punica granatum L.
Punica granatum L. florepleno

Dracaenaceae
Dracaena arborea (Willd.) Link
Draceana fragans massangeana Ker-gawl.
Dracaena godseffiana Bak.
Dracaena godseffiana Bak. var. friedmanii
Draceana marginata Link.
Dracaena sanderiana Hort.

Casuarinaceae
Casuarina equisetifolia L.

Combretaceae
Quisqualis indica L.
Terminalia cattapa L.
Terminalia mantaly H.Perrier

Moraceae
Artocarpus nobilis J.R. & G. Forst.
Ficus bengamina L.
Ficus bengamina L. 'Variegata'
Ficus benjamina L. 'Starlight'
Ficus benghalensis L.
Ficus craterostoma Mildbr. & Burret
Ficus elastica Roxb.
Ficus elastica Roxb.'Decora'
Ficus elastica Roxb. 'Doescheri'
Ficus elastica Roxb.'Schryveriana'
Ficus lyrata L.
Ficus polita Vahl.
Ficus retusa L.
Ficus rubiginosa Desf. ex Vent. var. variegata Guilf.
Ficus triangularis Warb.

Oleaceae
Jasminum nitidum Skan
Jasminum officinale L. affine (Royle ex Lindl.) Rehder

Oxalidaceae
Averrhoa caramibola L.

Papilionaceae
Cordyla pinnata (Lepr. Ex A.Rich.) Milne-Redhead
Erythrina indica Lamk. picta L.
Millettia thonningii (Schum. & Thonn.) Baker
Sesbania grandiflora (L.) Poir. Cv. Alba
Sesbania rostrata Brem.

Polygonaceae
Coccoloba uivifera L.

Rutaceae
Citrus aurantifolia (Christm.) Swingle
Citrus grandis (L.) Osbeck
Citrus limon Burm. f.
Citrus reticulata Blanco
Citrus sinensis Osbeck
Murraya paniculata (L.) Jacq.

Appendix 1. Contd.

Musaceae
Musa zebrina Van Houtte ex Planch.

Strelitziaceae
Ravenala madagascariensis Gmel. J.F.

Strelitziaceae
Ravenala madagascariensis Gmel. J.F.

Verbenaceae
Clerodendrum fragrans Vent.
Clerodendrum inerme L.
Clerodendrum speciosissimum Drapiez
Clerodendrum thomsonae Balf.
Clerodendrum x *speciosum* Gürke
Gmelina arborea Roxb.
Gmelina asiatica L.
Lantana camara L.
Stachytarpheta cayenensis (L.C. Rich.) Schau.
Stachytarpheta speciosa Hort.
Tectona grandis L. f.

Araliaceae
Aralia elata Seem.
Aralia elata Seem. Variegata
Dizygotheca elegantissima (Veitch ex Mast.) R. Vig. & Guillaumin
Dizygotheca reginae (Linden ex W. Richards) Hemsl.
Nothopanax filicifolia Bailey
Polyscias palapala Hort.
Polyscias balfouriana Bailey
Polyscias guilfoylei L.H. Bailey laciniata
Polyscias balfouriana Bailey pennockii
Polyscias filicifolia Bailey
Polyscias guilfoylei Bailey
Polyscias guilgoylei Bailey victorae Bailey
Polyscias scutellaria (Burm.f.) Fosberg
Schefflera arboricola (Hayata) Merr.
Schefflera farinosa (Blume) Merr.

Agavaceae
Cordyline fruticosa (L.) A.Chev.
Cordyline terminalis (L.) Kunth
Cordyline terminalis (L.) Kunth var. amabilis
Cordyline terminalis (L.) Kunth var. calypso queen
Cordyline terminalis (L.) Kunth var. firebrand
Cordyline terminalis (L.) Kunth var. hawaii bonsai
Cordyline terminalis (L.) Kunth var. in habitat
Cordyline terminalis (L.) Kunth var. liliput

Sterculiaceae
Sterculia foetida L.

Turneraceae
Turnera ulmifolia L.

Zygophyllaceae
Guaiacum officinale L.

Rubiaceae
Gardenia jasminoides fortuniana Ellis
Ixora chinensis Lamk.
Ixora coccinea Hort.
Ixora finlaysoniana L.
Ixora javanica DC.
Ixora macrothyrsa Hort
Ixora macrothyrsa 'Super King'
Ixora odorata Hook.
Pentas lanceolata (Forsk.) Deflers pallida
Mussaenda erythrophylla Schumach. & Thonn.
Mussaenda philippica A.Rich.

Arecaceae
Areca triandra Roxb.
Bismarckborassusia nobilis Hildebr. & H.Wendl.
Borassus aethiopum Mart.
Brahea armata Wats.
Caryota mitis Lour.
Chamaedorea elegans Mart.
Chamaerops humilis L.
Chrysalidocarpus lutescens Wendl.
Cocos nucifera L.
Coccothrinax elegans O. Muñiz & Borhidi 'Bella'
Coccothrinax fragrans Burret
Corypha umbraculifera L.
Elaeis guineensis Jacq.
Howeia forsteriana Becc.
Licuala grandis H. Wendl.
Licuala rumphii Blume
Licuala spinosa Wurmb.
Linospadix monostachya H. Wendl.
Livistona australis Mart.
Livistona chinensis Mart.
Pritchardia beccariana Jacq.
Rhapis excelsa Henry
Roystonea regia O.F. Hook.
Washingtonia filifera Wendl.

Appendix 1. Contd.

Poaceae	**Sapindaceae**
Bambusa nana Roxb.	*Blighia sapida* C. Koenig
Bambusa multiplex (Lour.) Raeusch. ex Schult. & Schult. f.	
Bambusa vulgaris Schrader ex Wendl.	

13

Effect of GA_3 and kinetin on growth, yield and quality of sprouting broccoli (*Brassica oleracea* var. italica)

Manjit Singh, D. K. Rana*, J. M. S. Rawat and S. S. Rawat

Department of Horticulture, Chauras Campus, HNB Garhwal University, Srinagar (Garhwal) 246174, Uttarakhand, India.

A field experiment was carried out during the winter season of 2009 to 2010 on sprouting broccoli cultivar Palam Samridhi at Horticultural Research Centre and Department of Horticulture, H.N.B Garhwal University, Srinagar (Garhwal) Uttarakhand, India. 4 weeks old seedlings were treated before transplanting by dipping their roots for 24 h in different concentration of GA_3 (gibberellic acid), kinetin and their combinations solutions. The GA_3, kinetin and their combination significantly influenced the growth performance, yield and quality characters of sprouting broccoli. GA_3 30 mg L^{-1} + kinetin 30 mg L^{-1} treatment gave maximum growth and yield of sprouting broccoli whereas, highest vitamin A content found with 40 mg L^{-1} GA_3 and vitamin C was found maximum in GA_3 20 mg L^{-1} + kinetin 20 mg L^{-1} dipping.

Key words: Gibberellic acid, kinetin, *Brassica oleracea var. italica*, quality, yield.

INTRODUCTION

Sprouting Broccoli (*Brassica oleracea* var. italica) is an important winter season vegetable crop, which resembles cauliflower. Broccoli is a member of the *Brassicaceae* family as a wild form of this family, which is found along the Mediterranean region (Decoteau, 2000). Broccoli is an Italian vegetable, native to the Mediterranean region, cultivated in Italy in ancient roman times and about 1720 in England. On the other hand, in the USA, it first appeared in 1806, but its commercial cultivation was started around 1923 (Decoteau, 2000). However, due to increase in its popularity, there is a trend to increase cultivation by farmers, as well as consumption by consumers. Broccoli is an important vegetable crop and has high nutritional and good commercial value (Yoldas et al., 2008). It is low in sodium food, fat free and calories, high in vitamin C and good source of vitamin A, vitamin B_2 and calcium (Decoteau, 2000). According to Brown and Hutchison (1949), it contains 137 mg vitamin C and 9000 IU vitamin A. Sprouting broccoli has about 130 times more vitamin A contents then cauliflower and 22 times more than cabbage (Singh, 2007). Nowadays, broccoli attracted more attention due to its multifarious use and great nutritional value (Salunkhe and Kadam, 1998; Talalay and Fahey, 2001; Rangkadilok et al., 2002, 2004). The plants form a kind of head consisting of green buds and thick fleshy flower stalk. The terminal head rather loose, green in color and the flower stalks are longer than cauliflower (Bose et al., 2002). The sprouts in the axils develop strongly specially after removal of terminal head. Both the terminal head and sprouts with buds are consumed by human as food. Broccoli is usually boiled or steamed, but may be eaten raw and has become popular as a raw vegetable in trays. It is also a rich source of sulphoraphane, a compound associated with reducing the risk of cancer (Singh, 2007).

In recent years a great deal of research work has been reported on the uses of plant growth regulators in vegetable crops. Plant growth regulators modify the physiological processes within the plant, which ultimately affect the yield and quality of the crop. Among, plant growth regulators, GA_3 and kinetin exhibited beneficial effect in several cole crops (Chhonkar and Singh, 1963; Badawi and Sahhar, 1978). So, to evaluate the response of GA_3, kinetin and their combinations on yield and quality improvement of sprouting broccoli is taken and conducted.

*Corresponding author. E-mail: drdkrana@gmail.com.

Table 1. Effect of GA_3 and kinetin on growth characters of sprouting broccoli.

Treatment	Plant height (cm)	No. of leaves	Days to central head formation	Days to secondary head formation	Plant frame (cm^2)	No. of secondary head	Head size (cm^3)	Harvest duration (days)
20 mg L^{-1}	60.73	13.73	61.40	70.93	75.85	10.40	1306.65	7.53
GA_3 40 mg L^{-1}	63.07	13.47	61.53	71.33	72.66	10.00	1369.48	7.90
GA_3 60 mg L^{-1}	60.93	14.53	57.86	66.20	79.63	8.53	1496.47	7.66
GA_3 80 mg L^{-1}	60.27	14.87	61.53	70.60	79.61	9.80	1199.60	8.56
GA_3 100 mg L^{-1}	58.37	14.80	60.40	70.00	72.70	8.80	1179.66	8.33
kinetin 20 mg L^{-1}	55.40	14.80	59.86	69.60	73.53	8.73	1356.81	8.33
kinetin 40 mg L^{-1}	61.47	15.67	59.87	69.93	76.72	10.13	1412.67	8.47
kinetin 60 mg L^{-1}	60.80	14.53	60.40	71.86	76.13	10.20	1509.20	7.54
kinetin 80 mg L^{-1}	57.93	15.40	61.66	71.90	77.43	9.67	1001.55	8.33
kinetin 100 mg L^{-1}	59.67	14.07	61.20	71.60	75.15	8.27	991.70	8.03
GA_3 10 mg L^{-1}+ kinetin 10mg L^{-1}	59.67	15.33	58.86	69.60	75.46	9.20	1272.16	8.30
GA_3 20 mg L^{-1}+ kinetin 20mg L^{-1}	57.83	15.20	61.13	71.53	75.47	7.93	1518.83	8.33
GA_3 30 mg L^{-1} + kinetin 30 mg L^{-1}	63.77	16.00	60.73	70.06	78.67	9.07	1542.66	8.43
GA_3 40 mg L^{-1} + kinetin 40 mg L^{-1}	58.47	13.60	61.86	71.26	78.66	9.00	1109.53	8.30
GA_3 50 mg L^{-1} + kinetin 50 mg L^{-1}	56.23	15.20	61.53	70.86	71.23	9.20	1119.25	7.60
control	58.27	13.33	63.60	74.53	67.23	8.60	884.64	8.01
S.Em.+	1.76	0.69	0.90	1.01	4.14	0.46	166.79	0.60
C.D at 5%	5.09	2.01	2.62	2.94	11.96	1.35	481.70	1.75

MATERIALS AND METHODS

The experiment was conducted during the winter season of 2009 to 2010 at the Horticultural Research Centre of H N B Garhwal University, Srinagar (Garhwal), Uttarakhand, India situated in the Alaknanda valley (78° 47' 30" E longitude and 30° 13' 0" N latitude and at an elevation of 540 m above msl), a semi-arid, sub-tropical climate with dry summer and rigorous winters with occasional dense fog in the morning hours from mid December to mid February. The experiment was laid out in randomized block design (RBD) with three replications and 16 treatments. The seedling treatments comprised of five different concentrations of GA_3 that is, 20, 40, 60, 80 and 100 mg L^{-1}, kinetin that is, 20, 40, 60, 80 and 100 mg L^{-1} and their combination at 10, 20, 30, 40 and 50 mg L^{-1} by dipping the seedling roots for 24 h in solutions of the aforementioned growth regulators before transplanting in field. For the control seedlings were dipped in distilled water. The treated healthy seedlings were planted in beds supplied with FYM 20 tones, nitrogen 80 kg, phosphorus and potash each of 100 kg/ha at a distance of 60 × 45 cm. During whole period of experiment, uniform cultural practices were adopted. Average number of leaves per plant was counted by taken plants from each treatment after two months of transplantation. The number of days taken from the transplanting to the day when first central head appeared is used as days to central head formation. Heads with stems were cut to 15 cm long (Decoteau, 2000) and the yield and its components were calculated and expressed on per hectare basis using electronic top loading balance.

The vitamin C content was determined by titration of sample against 2,6-dichlorophenol-indophenol and vitamin A content was estimated by the method given by Ranganna (2004). Data were subjected to analysis of variance (ANOVA) given by Cochran and Cox (1992).

RESULTS AND DISCUSSION

Growth characters

Growth is function of various vegetative characters put together namely, height of plant, number of leaves, plant spread, no of secondary head, days to central head formation, days to secondary head formation, head size and harvest duration. The different growth parameters were significantly affected by dipping the roots of seedlings in different growth regulators solution before transplanting (Table 1). GA_3 30 mg L^{-1} + kinetin 30 mg L^{-1} treatment had maximum influence over other treatments with respect to plant height (63.77 cm), number of leaves (16.00) and head

Table 2. Effect of GA_3 and kinetin on yield and quality characters of sprouting broccoli.

Treatment	Central head yield plant^{-1} (gm)	Secondary head yield plant^{-1} (gm)	Central head yield plot^{-1} (Kg)	Secondary head yield plot^{-1} (Kg)	Total yield hectare^{-1} (q)	Vit. A content (I.U/100 g)	Vit. C content (mg/100 g)
GA_3 20 mg L^{-1}	397.00	426.67	4.58	5.62	169.80	5460.71	78.80
GA_3 40 mg L^{-1}	461.67	376.67	5.09	5.43	175.29	5995.75	65.89
GA_3 60 mg L^{-1}	482.33	428.33	5.28	5.29	176.11	3214.62	71.14
GA_3 80 mg L^{-1}	430.67	410.00	4.52	5.52	167.33	3289.08	72.23
GA_3 100 mg L^{-1}	398.33	421.66	4.61	5.43	167.38	4660.91	70.75
kinetin 20 mg L^{-1}	396.33	460.00	4.07	5.76	163.88	5386.22	81.02
kinetin 40 mg L^{-1}	480.33	453.33	4.60	5.89	174.94	4461.94	82.48
kinetin 60 mg L^{-1}	433.33	493.33	4.94	6.16	185.16	4655.21	86.43
kinetin 80 mg L^{-1}	367.33	441.67	4.67	5.97	177.48	4553.35	77.13
kinetin 100 mg L^{-1}	322.33	285.00	3.45	5.31	145.74	3698.49	79.40
GA_3 10 mg L^{-1}+ kinetin 10mg L^{-1}	330.00	451.67	4.86	5.37	170.58	4914.64	114.33
GA_3 20 mg L^{-1}+ kinetin 20mg L^{-1}	471.67	390.00	5.19	6.03	187.05	5278.69	117.59
GA_3 30 mg L^{-1} + kinetin 30 mg L^{-1}	511.67	415.00	5.35	6.02	189.67	3695.63	103.63
GA_3 40 mg L^{-1} + kinetin 40 mg L^{-1}	318.33	381.67	4.34	5.64	166.37	4346.51	112.01
GA_3 50 mg L^{-1} + kinetin 50 mg L^{-1}	340.00	295.00	3.56	5.97	158.89	5110.45	103.87
Control	278.67	270.00	4.13	4.58	145.15	2289.08	60.94
S.E.m.+	53.07	39.22	0.60	0.46	14.52	77.2196	1.08
C.D at 5%	153.29	113.28	1.73	1.34	42.24	223.017	3.12

GA_3 – gibberellic acid mg L^{-1} – milligram per litre, cm-centimeter, cm^2- square centimeter, cm^3-centimeter cube, gm-gram, Kg-kilogram, q-quintal, Vit. A- vitamin A, Vit. C- vitamin C, I.U.- international unit, mg-milligram, S.E.m-standard error of mean, C.D.- critical difference.

size (1542.66 cm^3). Similar result were found by Gonzalez et al. (2007), Lone et al. (2005), Khan et al. (2002) and Kumar and Ray (2000) in both GA_3 and kinetin. So might be due to combination of both regulators. GA_3 60 mg L^{-1} treatment proves to be most effective among all the treatments and required minimum days to central head formation (57.86 days) and secondary head formation (66.20 days). This is in agreement with the result obtained by Patil et al. (1987) and Chhonkar and Singh (1963). Further, GA_3 60 mg L^{-1} concentration was found effective in increasing the plant frame significantly then other treatments. Similar results have been reported by Kumar et al. (1996). The plant growth regulators increased the plant height and number of leaves per plant. As a result of this there was an increase in plant spread. Number of secondary heads per plant found maximum in GA_3 20 mg L^{-1} treatment. Singh and Lal (2001) found same findings. GA_3 80 mg L^{-1} treatment taken longest duration of harvest between central head to secondary heads harvesting. The possible reason for increase in plant frame, number of secondary heads per plant and harvest duration of sprouting broccoli plants may be due the gibberllic acid, which promotes vegetative growth by way of cell elongation and cell division.

Yield and yield attributes

It is evident from Table 2 that GA_3, Kinetin and their combination increased the central and secondary heads and yield as compared to control. Combination treatment GA_3 30 mg L^{-1} + kinetin 30 mg L^{-1} produced central head of 511.67 gm and kinetin 60 mg L^{-1} produced secondary heads of 493.33 gm per plant, which was statically maximum to all the treatments. The treatment of GA_3 30 mg L^{-1} + kinetin 30 mg L^{-1} produced the maximum yield of 189.67 q ha^{-1}, while the lowest yield of 145.15 q ha^{-1} was found in absolute control. The effects of gibberellic acid on broccoli was studied by Wang and Yang (2008) and reported that inflorescence differentiation and curd yield was increased, but the vitamin C of curd was decreased or varied disobviously. Similarly, Wang et al. (2009) also studied effect of different concentration of cytokinin on broccoli and found that it increases ball-flower yield and plants economic coefficient, but weakened ball-flower quality. In present investigation, the combination of GA_3 and kinetin gives the best performance in terms of yield and quality. Similar results were reported by Chauhan and Bordia (1971), Arora et al. (1989), Chhonkar and Singh (1963) and Chauhan and Tandel (2009). The combination of both regulators gives best result in yield. The increasing central head weight and secondary heads weight by GA_3 and kinetin treatments might be due to the rapid and better nutrient translocation from roots to aerial parts of the plant. Both the regulators play important role such as gibberllic acid help in cell elongation, fruit growth, tissue growth and development and kinetin mainly done incitation of cell division, delay of senesce and cell enlargement etc. So, that the combinations showing the best result.

Quality attributes

The data presented in Table 2 indicate that the plant growth regulator treatments significantly increased the vitamin A and C content of sprouting broccoli. Significant increase in beta carotien content of the head was recorded with GA_3 40 mg L^{-1} soaking treatment which is significantly higher among all the treatments and followed by GA_3 20 mg L^{-1}. Whereas, ascorbic content significantly increased by soaking in combination GA_3 20 mg L^{-1}+ kinetin 20 mg L^{-1}, then other treatments. Chauhan and Tandel (2009) reported that application of GA_3 increased the ascorbic content in cabbage. Khalili et al. (2008) found that kinetin also affect vitamin C content in broccoli. The increased vegetative growth influence other characters might have promoted the metabolic activities of the plants due to application of plant growth regulators such as GA_3, kinetin and their combinations and increased the vitamin A and C content of the broccoli heads, which is ultimately improvement in quality of sprouting broccoli heads.

REFERENCES

Arora SK, Vashastha RN, Partap PS (1989). Effect of cytozyme and miraculan on yield of cauliflower, cabbage and knol-khol. Har. J. Hort. Sci., 18(1-2): 94-98.

Badawi MA, Sahhar EL (1978). Influence of some growth substances on different characters of cabbage. Egypt J. Hort., 6(2): 221-285.

Bose TK, Kabir J, Maity TK, Parthasarathy VA, Som MG (2002). Vegetable Crops in India. Naya Prokash. Calcatta.

Brown HD, Hutchison CS (1949). Vegetable Science. J.B. Lippincot Company. New York, pp. 21-30.

Chauhan KS, Bordia NS (1971). Effect of gibberellic acid, beta naphthoxyacetic acid and 2,4. Dichloropjhenoxyacetic acid as pre sowing seed treatment on growth and yield of cabbage. Ind. J. Hort., 28(1): 57-59.

Chauhan UM, Tandel YN (2009). Effect of plant growth regulators on growth, yield and quality of cabbage (*Brassica oleracea* var. capitata L.) cv. Golden Acre. Asian J. Hort., 4(2): 512-514.

Chhonkar VS, Singh R (1963). Effect of gibberellic acid on growth, yield and quality of cabbage (*Brassica oleracea* var. capitata L.). Ind. J. Hort., 21: 57-63.

Cochran WG, Cox GM (1992). Experimental Designs. John Wiley and Sons. Inc. New York.

Decoteau DR (2000). Vegetable Crops. Upper Rever Company. New Jersey. U.S.A.

Dhengel RP, Bhosale AM (2007). Effect of NAA and GA along with urea on certain quality attributes of cabbage (*Brassica oleracea var.* capitata). Asian J. Hort., 2(2): 30-32.

Gonzalez ML, Caycedo C, Velasquez MF, Florez V, Garzon MR (2007). Effects of gibberllic acid applicaton on growth of cauliflower (Var *botrystis*.). Agron. Col., 25(1): 54-61.

Khalili F, Shekarchi M, Mostofi Y, Pirali-Hamedani M, Adib N (2008). Cytokinins affect Fermentation product accumulation, vitamin C reservation and quality of stored broccoli in modified atmosphere packages. Iran J. Med. Plants, 7(26): 53-56.

Khan NA, Ansari HR, Khan M, Mir RS (2002). Effect of phytohormones on growth and yield of Indian mustard. Ind. J. Plant Physiol., 7(1): 75-78.

Kumar D, Gujar KD, Singh K (1996). Effect of GA_3 and NAA on growth, chlorophyll content and maturity of cabbage. Crop Res. 12(1): 116-119.

Kumar V, Ray N (2000). Effect of plant growth regulators on cauliflower cv. Pant Subhra. Orissa J. Hort. 28(1): 65-67.

Lone MT, Haripriya K, Maheswari TU (2005). Influence of plant growth regulators on growth and yield of Chilli (*Capsicum annum* L.) cv. K2. Crop Res. 29(1): 111-113.

Patil AA, Maniur SM, Nalwadi UG (1987). Effect of GA and NAA on growth and yield of cabbage. South Ind. Hort., 35(5): 393-394.

Ranganna S (2004). Handbook of Analysis and Quality Control for Fruit and Vegetable Products (IInd ed.). Tata McGraw-Hill Publishing Company Ltd. New Delhi.

Rangkadilok N, Nicolas ME, Bennett RN, Eagling DR, Premier RR, Taylor WJ (2004). The effect of sulfur fertilizer on glucoraphanin levels in broccoli (*B. oleracea* L. var. italica) at different growth stages. J. Agric. Food Chem., 52: 2632-2639.

Rangkadilok N, Nicolas ME, Bennett RN, Premier RR, Eagling DRP, Taylor WJ (2002). Determination of sinigrin and glucoraphanin in *Brassica* species using a simple extraction method combined with ion-pair HPLC analysis. Scientia Horticulturae. 96: 27-41.

Salunkhe DK, Kadam SS (1998). Handbook of Vegetable Science and Technology: Production, Storage and Processing, 1st edition. Marcel Dekker, Inc. Madison Avenue. New York, USA, pp. 256-280.

Singh DK, Lal G (2001). Effect of plant bioregulators on the growth and yield of tomato (*Lycopersicon esculantum* mill.). Prog. Hort., 33(1): 61-64.

Singh NP (2007). Basic Concept of Vegetable Science. International Book Distributing Co. Lucknow, p. 444.

Talalay P, Fahey JW (2001). Phytochemicals from Cruciferous plants protect against cancer by modulating carcinogen metabolism. Am. Soc. Nutr. Sci., 23: 3027-3033.

Wang T, Yang X (2008). The Effects Gibberellin of on Inflorescence Differentiation, Photosythesis Characteristics and Quality in Broccoli. Northern Hort, 01: 10-12.

Wang T, Yang X, Yang X (2009). Effects of Cytokinin Application on Chlorophyll Content, Quality and Yield of Broccoli. Acta Agric. Jiangxi, 5: 65-66.

Yoldas F, Ceylan S, Yagmur B, Mordogan N (2008). Effect of nitrogen fertilizer on yield quality and nutrient content in broccoli. J. Plant Nutr., 31: 1333-1343.

14

New data about *Hésperis pýcnotricha* in Romania

Iulian Costache

University of Craiova, Faculty of Horticulture, Department of Biology, 15 Libertăţii Street, 200583 Craiova, Romania.
E-mail: iuliuscostache@yahoo.com, icostache33@gmail.com.

There are presented new data about *Hésperis pýcnotricha* in Romania. In 2004, the species was first identified in Gura Motrului Locality in Romania. Since then, the species has been pursued and so far, we have come up with new data on its origin and chronology. To our joy, the plant (ornamentally planted) has been identified in the same locality in a backyard adjacent to the first resort. Subsequently, the plant has been identified in other locations such as at the edge of abandoned farms and their premises in Cerna Valcea District; in some local households in Ponoarele Mehedinti District. Travelling in Turkey, at Marmara University in Erasmus, the author had the opportunity to see ornamental plants, inside the yard of Dolmabahçe Palace. Comparing the material collected from Turkey with that from Romania, he realized that both samples have the same morphologically characteristics. From bibliography sources, at the European level (like in Moldova, Slovakia, Turkey in Europe and Ukraine), it is known as a native plant or entered in Krym, Czech Republic, Lithuania and Russian Federation. The species is also known in Sweden while in Lithuania, it is mentioned in the list of invasive plants. The presence of the species in Romania, although was pointed out from 2004, is not consigned at the European level.

Key words: New chronology, *Hésperis pýcnotricha*, Romanian, flora.

INTRODUCTION

From the genus *Hésperis*, there are 13 species mentioned in Europe (Ball, 1993), 9 in the Romanian flora (Nyárády, 1955) and lately, 5 (Ciocârlan, 2000) and 6 species (Ciocârlan, 2009) after including the species *Hésperis pýcnotricha* (Ciocârlan and Costache, 2004). At the European level (Moldova, Slovakia, Turcia in Europe and Ukraine), *Hésperis pýcnotricha* is known as a native plant or entered in Krym, Czech Republic, Lithuania and Russian Federation. This is got from the information given by Euro+Med Preliminary Checklist with Flora Europaea, Med-Checklist and Flora of Macaronesia (http://ww2.bgbm.org/_EuroPlusMed/). IFlora Europaea (Vol. 2), Ball (1964) mentions *H. pýcnotricha* Borbás and Degen, Magyar Bot. Lapok (1902), within the Matronalis Group (Vol 1:pg 269), considers it a possible subspecies to *Hésperis matronalis* L. He gives U.R.S.S. as the spreading area between longitude 30°E and latitude 56°N, Rs (C, W, K, E). The same author in Flora Europaea, the 2nd edition Vol. 1 (1993) gives up the idea of existing subspecies and comes up with new data on the spreading area of the species; thus: C. and S. Russia and S. Ukraina, Tracia, E. Czechoslovakia, Bu, Rs (C, W, Krim., E) Tu. Besides these chronological data given by Flora

Europaea, other data about the species come from: Russia (Czerepanov, 1995) and other digital sources (http://www.sevin.ru/natreserves/index); Turkey (Davis et al., 1988), in plant images being posted by Nasip (2007); Marhold and Hindok (1998) published a list with non-vascular and vascular plants from Slovakia in which *Hésperis pýcnotricha* Borbás and Degen [*Hésperis steveniana* Auct., non-DC.] can be found; Herloff (1999) pointed it out in two Swedish localities: 1. a grassy field near an allotment-garden in Goteborg (Gothenburg) in the South-west; 2. in the overgrown garden of an abandoned croft near Linkoping in Ostergotland in the South-east; advancing the invasive character of the species and in 2006 new information about the species appeared online (http://www.bfig.se/foredrag/20060330.html). Ebel (2002) mentioned it in Tomsk Province (located in the Southeastern part of West Siberian Plain); Gudzinskas (2004) included the species in the data base of the invasive terrestrial plants from Lithuania; invasive, being considered also by Pyšek (2009). From Karadag Natural Rezervation, Crimeea (2009) posted species images.

Figure 1. Detail of middle leaves (Costache 2005).

Figure 2. Detail of inflorescence (Costache 2005).

In 2004, from Romania, Hésperis pýcnotricha species was first identified in Gura Motrului locality, Mehedinti District, Oltenia Region (Costache, 2005; Ciocârlan and Costache, 2006). Subsequently, the plant has been identified in other locations: Cerna (Valcea District) in the edge of abandoned farms and its premises; Ponoarele (Mehedinti District) in some local households.

In May 2010, traveling in Turkey, Istanbul, Marmara University, in Erasmus, I had the pleasant surprise to find

Figure 3. Detail of inflorescence axis with bifurcated hairs (Costache 2005).

plants, inside the yard of Dolmabahçe Palace, being ornamental cultivated.

MATERIALS AND METHODS

The method used for this research was the observation method. During the research, the identified species were collected along with the collection date, the locality as well as the conditions of the station. Subsequently, the material was examined in details, both rough and preserved, using a binocular magnifying glass. The determinations were established according to the specialized literature found in the bibliography. The detailed images were taken by a digital camera provided with a magnifying device. Coordinates of the localities (FQ 93 - Gura Motrului; FQ 38/48 – Ponoarele; GQ 29/39 - Pojogi-Cerna) was established after Lehrer and Lehrer (1990).

RESULTS AND DISCUSSIONS

In the last 5 years, *H. pýcnotricha* species has been searched in Turkey, and other resorts in Oltenia are also looking for the way to come into the country. The plant has been identified in different locations.

The plant identified in Istanbul, in Dolmabahçe Palace, is ornamentally cultivated (Figures 5, 6, 7 and 8). On this occasion, we could compare the material identified in Istanbul with ours collected from Gura Motrului Locality and other states, noting that it is the same species (Figures 1, 2, 3, and 4). Thus, it requires a representation of the species, knowledge of its morphological characters to identify and distinguish it from the taxon *Hesperis matronalis*.

Data about the species, *Hésperis pýcnotricha* Borbás and Degen Magyar Bot come from: Lapok (1902) in the Matronalis Group (Vol.1:269); Ball (1964) in Fl. Eur. (Vol. 2: 276), Fl. Eur. 2nd ed. (Vol 1: 337); Marhold and Hindok (1998) in Check list non-vascular and vascular plants

Figure 4. Detail of calyx presenting the bifurcated hairs (Costache 2005).

Figure 5. Habitus and detail basal leaves (Original, inside the yard of Dolmabahçe Palace, 24. 05. 2010).

Figure 6. Detail of middle leaves (Original, inside the yard of Dolmabahçe Palace, 24. 05. 2010).

(*Hésperis steveniana* auct., non DC.) from Slovakia.

Figure 7. Picture with *Hésperis pýcnotricha* (Original, inside the yard of Dolmabahçe Palace, 24. 05. 2010).

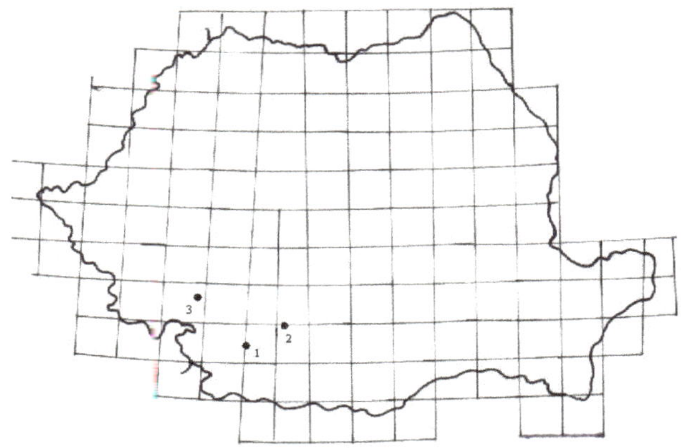

Figure 8. The chorology of the *Hesperis pycnotricha* in Romania: ● Localities: 1- Gura Motrului; 2- Pojogi-Cerna; 3- Ponoarele (original).

It is a perennial plant (H., V-VII, 2n=14, 26). It has a simple stems, sometimes ramified in the upper part of 40-90 (100) cm. The basal leaves (Figure 5) have an oblong-elliptic shape, the middle and upper leaves are ovate ± sessile (Figures 1 and 6), with gently toothed brims. The plant, in the whole, presents indumenta made from bifurcated hairs (the simple and the glandular hairs are missing). It has inflorescence axis and pubescent pedicels (Figure 3), with calyx of 6-8 (9) mm and pubescent (fig. 4). Its petals are 16-18 (19) mm (lamina of almost 10 mm and angle of 6-9 mm), of red colour In Romania, dichotomous key for determining the species is given by Ciocârlan (2009) as:

Ecology: It grows on dark, eu-mesobasical soils; it has heliophyle, mesoxerophyle, subtermophyle-termophyle, poorly neutral acid soil reaction (pH = 6-7); Station: It grows in the margins of agricultural places, Gura Motrului locality, Mehedinti District (Costache, 2005); New localities: It is found in Cerna (Valcea District), Ponoarele

(Mehedinti District), Oltenia Region; Geoelement: Pontic-Balkan.

For Station: It graws in the margins of agricultural places, FQ 93 - Gura Motrului locality, Mehedinti District (Costache, 2005).

For New localities: It is found in Pojogi-Cerna - GQ 29/39 - (Valcea District); Ponoarele - FQ 38/48 (Mehedinti District), Oltenia Region (Figure 8).

Conclusion

At this point, the species *Hésperis pýcnotricha* data chronology identified in Oltenia, Romania implicitly reveals three stations. The fact that the plant was identified in households and also in uncultivated places entitles us to conclude that the plant was introduced with the culture, becoming naturalized and sub-spontaneous. Perhaps, the plant is found also in other localities in Romania, but not recorded. This is because it is confused with *Hesperis matronalis*, being that they resemble.

ACKNOWLEDGMENT

This could not have been done without the help of Professor Vasile Ciocârlan, Ph.D. University of Bucharest.

REFERENCES

Ball PW (1964). Hesperis - In: Tutin TG, & al. (eds). *Flora Europaea.* Lycopodiaceae to Platanaceae. Cambridge: Cambridge University Press, 1: 275-277.

Ball PW (1993). Hesperis - In: Tutin TG, & al. (eds). *Flora Europaea.* 2nd ed. Psilotaceae to Platanaceae. Cambridge: Cambridge University Press, 1: 336-337.

Ciocârlan V (2000). *The Romanian Illustrated Flora. Pteridophyta* et *Spermatophyta.* Ceres Publishing House, Bucharest (in Romanian), 1141p.

Ciocârlan V, Costache I (2004). Completion to the Romanian Flora. *Acta Horti Bot. Bucurest.* Bucureşti 2006, 33: 93-98.

Costache I (2005). The flora and vegetation of the Lower hydrographic Basin of the River Motru. *PhD Thesis.* University of Bucharest. Romania (in Romanian).

Czerepanov SK (1995) Vascular Plants of Russia and Adjacent States (the Former USSR). Cambridge University Press, 5?9p. http://books.google.com/books.

Davis PH, Cullen JM, Coode JE (1988). *Flora of Turkey and the East Aegean Islands.* Edinburgh University Press. Edinburgh, 10: 562p.

Ebel A (2002). New Data on distribution of Brassicaceae species in South Siberia and East Kazakhstan. Turczaninowia, 5(2): 60-68.

Herloff B (1999). Hesperis pycnotricha found in Sweden. Source: Svensk-Botanisk-Tidskrift. Southern Appalachian Information Node Invasive Plant Pests Resource Collection for: Dame's Rocket - http://sain.sunsite.utk.edu/invasives/common27.shtml, 93(4): 213-214.

Lehrer A, Lehrer M (1990). The Cartography of the Romanian Flora and Fauna, Ceres Publ. House, Bucharest, (in Romanian), 291p.

Marhold K, Hindok F (1998). Cheklist of non-vascular and vascular plants of Slovakia. Veda-Bratislava Publishing House, 492p.

Nyárády EI (1955). Hesperis - In: Săvulescu Tr. & al. (eds). *Fl. Reipubl. Popularis Romanicae.* Editio Acad. Reipubl. Popularis Romanicae, Bucharest (in Romanian), 3: 183-197.

Pyšek P, Lambdon PW, Arianoutsou M, Künhan I, Pino J, Winter M (2009). *Handbook of alien species in Europe.* By DAISIE (Delivering Alien Invasive Species Inventories for Europe) Invading Nature. Springer Series in Invasion Ecology, Vol. 3 http://books.google.com/books.

15

Estimation of genetic variability in *Phyllanthus emblica* L. - Towards a contribution in sustainable rural development

Bhupendra Singh[1]*, A. K. Uniyal[1], J. S. M. Rawat[2] and D. K. Rana[2]

[1]Department of Forestry, Post Box-59, H.N.B. Garhwal University, Srinagar Garhwal, 246 174, Uttarakhand, India.
[2]Department of Horticulture, Post Box-59, H.N.B. Garhwal University, Srinagar Garhwal, 246 174, Uttarakhand, India.

The fruits of *aonla* (*Phyllanthus embilica* L.) were collected from four natural sites and were subjected to various morpho-chemical components for estimation. The observed variation was partitioned into heritable and non-heritable components in order to identify the characters for which selection will be most effective. This study revealed that morpho-chemical characters of *P. embilica* exhibited considerable genetic variability in fruit morphology, especially those based on fruit weight, which has high genotypic variation, higher heritability and greater potential for genetic gain. Vitamin C had moderate genetic variance, moderate heritability but greater genetic gain. Selection can therefore be effective based on these two characters and their phenotypic expression should be an indication of their high genotypic potential.

Key words: *Phyllanthus emblica,* fruit length, fruit weight, moisture percent, genetic variation, heritability.

INTRODUCTION

Phyllanthus emblica L. (V. Anwla or Aonla) is a deciduous tree of the family Euphorbeaceae, a native of India (Gaur, 2003), it has also been reported from Sri Lanka, Thailand, China, Japan and Vietnam. It is probably the only edible fruit to fill the gap of astringent food recommended by the Ayurvedic system of medicine for a balanced diet and sound health. The fruits are rich in vitamin C (Kumar and Singh, 2002) and are consumed fresh, pickled and processed into a variety of sweet and tasty food products. Aonla has great potential as a cash-generating tree for the farming communities in the Uttarakhand hills of India. The tree has been promoted as a multi-purpose crop suitable for cultivation in the mountain farms of Uttaranchal by the local government which has emphasized the importance of producing and processing high medicinal value agroforestry tree crops. This investigation was conducted to identify and select superior germplasm of Anola from natural populations in the Uttarakhand hills of Uttaranchal before too much genetic erosion takes place.

MATERIALS AND METHODS

Fruits of *P. emblica* were collected from four sites (Table 1) ranging from 30° 12'N to 30° 51' N latitude, to 78° 37' 4' E to 78 56' E longitude and at elevations ranging from 1050 to 1550 m asl. The average rainfall is 112 to 165 mm per year. The maximum average temperature ranges between 26 to 29°C and the minimum from 10 to 12°C (Table 1). During April 2005, ripe fruits of Aonla were collected from ten healthy trees at each of the four sites. Damaged fruits were rejected. The bulk sample of all the parent trees for each site was then put into cotton bags and allotted an accession number (seed source) before bringing them to laboratory. Each sample was stored in a cool incubator (5°C) until further analysis. The moisture content (%) of each fruit sample was determined on a fresh weight basis by drying the fruits at 103±3°C as per ISTA 1999 rules. Fruit length, diameter, weight and pulp: stone ratio was evaluated on five replicates samples, each consisting of 20 fruits randomly drawn from each bulked fruit lot. Fruit length and diameter was measured using a micrometer (Besto). The fruit weight of five more replicates samples (each of 100 fruits) was recorded using electronic top pan balance (ISTA, 1999). Total soluble solids (TSS) were determined at 20.0°C using a hand held refractrometer and values correlated. Acidity was determined by titrating the fruit pulp against 0.5 N/NaOH and expressed as percent malic acid. Vitamin C was analysed according to A.O.A.C (1975). Similarly, soluble sugar was determined by the method of Mc Cready et al. (1950).

The results were statistically analysed using ANOVA, Tukey test (Bartz, 1988), phenotypic coefficient of variation (PCV), genotypic

*Corresponding author: E-mail: butola_bs@yahoo.co.in.

Table 1. Morphological-chemical variation in fruit characters of *P. emblica.*

Source	Altitude	Fruit length (cm)	Fruit diameter (cm)	Fruit weight/100 fruits (g)	Moisture (%)	Pulp/stone ratio	Total soluble solids (%)	Sugar (%)	Vitamin C (mg/100 g)	Malic acid (%)
Gajeli	1050	$1.78±0.03^{b}$	$1.99±0.10^{ab}$	$498.48±19.76^{b}$	$75.13±0.63^{ab}$	$11.90±0.28^{bc}$	$14.32±0.11^{a}$	$4.25±0.06^{a}$	$508.00±2.71^{b}$	$2.64±0.34^{a}$
Thadiyar	1150	$1.67±0.06^{c}$	$1.88±0.05^{b}$	$415.03±7.24^{c}$	$77.37±0.75^{a}$	$11.31±0.36^{c}$	$12.960.26^{c}$	$2.90±0.16^{b}$	$529.53±2.28^{a}$	$2.01±0.34^{b}$
Dagar	1300	$1.83±0.06^{b}$	$2.13±0.17^{a}$	$527.49±38.08^{b}$	$76.14±0.66^{a}$	$12.68±0.90^{b}$	$13.200.14^{b}$	$4.39±0.05^{a}$	$491.73±7.23^{c}$	$2.30±0.08^{ab}$
Kandikhal	1550	$1.95±0.05^{a}$	$2.15±0.10^{a}$	$651.22±17.12^{a}$	$73.39±0.87^{b}$	$14.29±0.11^{a}$	$13.920.11^{a}$	$2.59±0.03^{c}$	$506.43±3.36^{b}$	$1.85±0.17^{b}$
	Mean	1.81	2.13	523.06	75.51	12.55	13.35	4.03	508.92	2.20
	C.V	6.63	12.68	18.70	2.22	10.28	7.19	8.93	3.06	15.91
	'r'	0.77**	0.84**	0.69**	-0.35^{NS}	0.64**	-0.09^{NS}	-0.90**	0.17^{NS}	0.77**

** Significant at P = 0.01, NS = non significant. Means followed by same later within each column are not significantly (P< 0.05) different, (± S.D.)

Table 2. Estimation of genetic variable for various fruit and biochemical characteristic of *P. embilica.*

Estimates	FL	F D	FW	M %	P/S ratio	TSS	Sugar	VC	MA
Phenotypic variance	0.016	0.026	9375.77	3.981	1.88	0.429	0.850	344.58	0.174
Genotypic variance	0.013	0.011	8773.12	2.080	1.57	0.371	0.845	191.38	0.091
PCV	6.99	7.57	18.51	2.64	10.92	4.91	22.88	3.15	18.96
GCV	6.29	4.92	17.91	1.01	9.98	4.56	22.81	2.35	13.71
Heritability (h^2)	0.81	0.42	0.94	0.52	0.83	0.86	0.99	0.55	0.523
Genetic advance	21.07	13.92	187850.16	213.16	234.24	115.15	187.62	2102.28	44.67
Genetic gain (%)	1164.08	653.52	3584.70	282.29	1866.15	862.54	4655.58	356.97	2030.45

Abbreviation: PCV - Phenotypic coefficient of variation, GCV - genotypic coefficient of variation, Heritability (h^2) - heritability (broad-sense), FL = fruit length, FD = fruit diameter, FW = fruit weight, M% = moisture percent, P/S ration = pulp/stone ratio, TSS = total soluble solid, VC = vitamin c, MA = malic acid.

coefficient of variation (GCV) were calculated as done by Barton and Devena (1953) while heritability (H), genetic advance (GA) and genetic gain were workout following Johnson et al. (1955) and simple correlation coefficients (Gupta, 1981).

RESULTS

Morphological characters

Fruit dimensions varied significantly (p = 0.05) among the four accessions. Fruit length ranged from 1.67 to 1.95 cm; diameter from 1.88 to 2.15 cm; fruit weight from 415.0 to 651.2 g/100 fruits; moisture contents from 73 to 77% and pulp:stone ratio ranged from 11.31 to 14.29. Values for all these characters were greatest at Kandikhal (1550 m asl) and lowest at Thadiyar (1150 m asl). The moisture content was highest at Thadiyar (1150 m asl) and lowest at Kandikhal (1550 m asl). Fruit length, diameter, weight and pulp:stone ratio were positively correlated (p = 0.01) with the mean elevation of the source (Table 2).

Biochemical variation

The total soluble solids ranged from 12.96 (Thadiyar) to 14.32% (Gajali); sugar from 2.59 (Kandikhal) to 4.39% (Dagar); Vitamin C from 491.73 (Dagar) to 529.53 mg/100 g (Thadiyar) and malic acid from 1.85 (Kandikhal) to 2.64% (Gajali). Total soluble solids and vitamin C were not correlated with elevation, while sugar content was negatively correlated with elevation (P=0.01) and malic acid was positively correlated (P=0.05) with elevation of the seed source (Table 2).

Genetic variability

Highest phenotypic (9375.77) and genotypic (8773.12) variance was found in fruit weight. However, the lowest values (0.016 and 0.011) were found in fruit length and fruit diameter for genotypic and phenotypic variance, respectively. The heritability (broad sense) values were highest (0.99) for sugar content whereas, and lowest (0.42) for fruit diameter. Genetic advance was found to be maximum (187850.27) for fruit weight and minimum (13.92) for fruit diameter. On the other hand, the highest genetic gain was for sugar (4655.58%) and least (282.29%) for moisture content.

Site variation

Analysis of variance found that the different physicochemical characteristics (except for fruit diameter, moisture content and malic acid) were highly significant (p = 0.01) between sites, but that all traits were not significantly different among the families (replicate sample) within the different sources.

DISCUSSION AND CONCLUSION

Morphological variation in fruit characters of *P. emblica* between populations could have been a result of differences in the environmental conditions, for example, nutrient, light or water to which the mother trees were subjected during growing season (Gutterman, 1992; Murali, 1997). Significant variations have also been observed in the different physio-chemical constituents of selected cultivars of *Emblica officinalis* by Sanjeev and Singh (2002). Similar variation in morphological characters of fruits have been previously reported for *Arachis* species (Chandran and Pandya, 2000), *Cola nitida*, (Adebola et al., 2002); *Dioscorea cayenensis* (Dansi et al., 1998); *Irvingia gabonensis* (Leakey et al., 2000); *Dacryodes edulis* (Waruhiu et al., 2004); *Sclerocarya birrea* (Leakey et al., 2005a, b); *Barringtonia procera* (Pauku et al., 2010) and for different chemical contents in *Choerospondies axillaries* (Paudel et al., 2002) and *Canarium indicum* (Leakey et al., 2008). Dubey (2000) has also reported variations in physicochemical properties in different cultivars of *Citrus sinensis*. The heritability provides measures of genetic variation upon which all the possibilities of improving the genetic composition of the species depend. While genetic gain express the degree of genetic improvement achieved within a selection process during the domestication of a species. Heritability estimates along with genetic gain are more useful than the heritability alone in predicting the resultant effect for selecting the best genotypes for a series of given traits (Volker et al., 1990; Singh and Chaudhary, 1993). The highest PCV was observed for sugar content followed by fruit weight and malic acid. Greater improvement could therefore be expected in the selection for these characters. The GCV values of present investigation were lower than the PCV values. A similar trend was found in *Prunus persica* (Khajuria et al., 1986); in *Sphenostylis stenocarpa* (Okoye and Ene-obong, 1992); in *Anacardium occidentale* (Vijulan et al., 1994), in *Prunus dulcis* (Sharma and Chandrababu, 1997), in *Cola nitida* (Adebola et al., 2002); and in *Musa* spp. genotypes (Kulkarni et al., 2002).

In this study, high heritability computed with high genetic advance was observed in fruit weight, Vitamin C, pulp:stone ratio, moisture content percent and sugar content, showing that the variability in these characters might be attributed to additive action. On the other hand, moderate to high heritability but low genetic advance was observed in fruit length, fruit diameter, total soluble solid and malic acid, suggesting non additive gene action and limited scope for improvement in respect of these characters. The genetic variability characters estimated for *P. emblica* reveals that fruit weight have high genotypic variation, higher heritability and greater genetic gain while Vitamin C have a moderate genetic variance, moderate heritability and greater genetic gain. Thus, selection can be effectively based on these two characters and their phenotypic expression should be an indication of their genotypic potentiality. The remaining characters had lower genetic variance, and genetic advance but moderate to higher heritability can be offered less scope for selection as they are much more under the influence of the environmental condition and accounted for non additive gene effects, as observed in tomato (Shan and Mishra, 1995). *P. emblica* is an important medicinal tree growing in traditional agroforestry systems, woodlots and forest near villages of Garhwal Himalayan region. The increasing pressure of human population and abrupt demand of plant materials consumed as fresh, pickled, variety of sweet and tasty food products those highly medicinal values has depleting at an alarming rate. Besides this, young trees in natural forests are lopped for fodder, fuel and timber, thus their over- exploitation is threatening the species in its natural habitat.

Future collection and evaluation of *P. emblica* should include new areas of Garhwal Himalaya. New parameters such as seed characters, germination potential, seedling growth should be assessed, and field trials should be considered for subsistence farming to enhance opportunities for poor farmers to improve their livelihoods and to maintain the sustainable status of this native species. This approach to agroforestry is seen as a means of developing multifunctional agriculture (Leakey, 2010).

REFERENCES

AOAC (1975). Official methods of analysis. Association of Official Analytical Chemists, Washington, D.C. Pp. 36.

Bartz AE (1988). One–way Analysis of Variance. In: *Basic Statistical concepts*, 3rd edition, MacMilan Publishing Company, New York, pp. 278-303.

Chandran K, Pandya SM (2000). Morphological characterization of *Arachis* species of section *Arachis*. Plant Genet. Resour. Newsl., 121: 38-41.

Dansi A, Mignouna HD, Zoundhihekpon J, Sangare A, Asiedu R, Quin FM (1998). Varietal identification key of the cultivated yams (*Dioscorea cayenensis*/ *Dioscorea rotundata* complex) of Benin Republic. Plant Genet. Resour. Newsl., 116: 18-25.

Dubey AK (2000). Comparative performance of certain sweet orange (*Citrus sinensis* Osbeck) cultivars at foothills situation of Arunanchal Pradesh. Progressive Hortic., 32 (2): 153-158.

Gaur RD (2003). Flora of the District Garhwal Northwest Himalaya with Ethnobotanical Notes. Transmedia Publication Center, Srinagar Garhwal, Uttaranchal, India, pp. 811.

Gupta SP (1981). Statistical Methods. 15th edition, Sultan Chand & Sons, New Delhi. p. 208.

ISTA (1999). International Rules for Seed Testing. Seed Science and Technology, 27, Supplement, Zurich, Switzerland. p. 333.

Khajuria HN, Nijjar GS, Chanana YR (1986). Variability studies in Peach (*Prunus persica* Batsch). In: Chadha, T.R., Bhutani, V.P., Kaul, J.L. editors. Advances in research on temperate fruits. Proceeding on the national symposium on temperate fruits; 1984 March 15-18, Dr. Y.S. Parmar University of Horticulture and Forestry, Solan, India, pp. 30-35.

Kulkarni VM, Srinivas L, Satdive RK, Bapat VA, Rao PS (2002). Dissection of the genetic variability in elite Indian banana genotypes. Plant Genet. Resour. Newsl., 132: 48-52.

Kumar S, Singh IS (2002). Physico-chemical studies of various cultivars of Aonla fruits. Progressive Hortic., 34(1): 102-104.

Leakey RRB (2010). Agroforestry: a delivery mechanism for Multi-functional Agriculture. In: *Handbook on* Agroforestry: Management Practices and Environmental Impact, pp. 461-471, Ed. Lawrence R. Kellimore, Nova Science Publishers. Environmental Science, Engineering and Technology Series, New York, USA.

Leakey RRB, Fondoun JM, Atangana A, Tchoundjeu Z (2000). Quantitative descriptors of variation in the fruits and seeds of *Irvingia gabonensis*. Agrofor. Syst., 50:47-58.

Leakey RRB, Fuller S, Treloar T, Stevenson L, Hunter D, Nevenimo T, Binifa J, Moxon J (2008). Characterization of tree-to-tree variation in morphological, nutritional and chemical properties of *Canarium indicum* nuts. Agrofor. Syst., 73: 77-87.

Leakey RRB, Pate K, Lombard C (2005b). Domestication potential of Marula (*Sclerocarya birrea* subsp *caffra*) in South Africa and Namibia: 2. Phenotypic variation in nut and kernel traits. Agrofor. Syst., 64: 37-49.

Leakey RRB, Shackleton S, du Plessis P (2005a). Domestication potential of Marula (*Sclerocarya birrea* subsp *caffra*) in South Africa and Namibia: 1. Phenotypic variation in fruit traits, Agrofor. Syst., 64: 25-35.

MC Cready RM, Guggot J, Silvaera V, Owen HS (1950). Determination of starch and amylase in vegetables. Ana. Chem., 22: 1156-1158.

Murali KS (1997). Patterns of seed size, germination and seed viability of tropical tree species in Southern India. Biotropica, *29*: 271-279.

Okoye FI, Ene-Obong EE (1992). Genetic variability and correlation studies in the African Yam bean (*Sphenostylis stenocarpa*). Nig. J. Bot., 5:75-83.

Paudel KC, Pieber K, Klump P, Laimer da Camara R, Machado M (2002). Collection and evaluation of germplasm of lapsi (*Choerospondias axillaries* (Roxb.) B. L. Burtt and A. W. Hill) an indigenous fruit tree of Nepal. Plant Genet. Resour. Newsl., 130: 36-46.

Pauku RL, Lowe A, Leakey RRB (2010). Domestication of indigenous fruit and nut trees for agroforestry in the Solomon Islands. Forests, Trees and Livelihoods, 19: 269-287.

Shan GS, Mishra RS (1995). Genetic divergence in tomato. Mysore J. Agric. Sci., 29:5-8.

Sharma RK, Chandrababu RJ (1997). Genetic variability studies in almund. Plant Genet. Resour. Newsl., 112:93-94.

Singh NB, Chaudhary VK (1993). Variability, heritability and genetic gain in cone and nut characters of Chilgoza pine (*Pinus gerardiana* Wall.). Silvae Genetica, 42:2-3.

Vijulan HCR, Suthanthira PIR, Tangavelu S (1994). Variability studies in metric traits in cashew (*Anarcadium occidentale* L.). South Indian Horti., 42: 13-140.

Volker PW, Dean CA, Tibbits WN, Ravenwood IC (1990). Genetic parameters and gain expected from selection in *Eukalyptus globules* in Tamilnadu, Silvae Genetica, 39 (1):18-21.

Waruhiu AN, Kengue J, Atangana AR, Tchoundjeu Z, Leakey RRB (2004). Domestication of *Dacryodes edulis*: 2. Phenotypic variation of fruit traits in 200 tress from four populations in the humid lowlands of Cameroon. Food Agric. Environ., 2: 340-346.

16

Ploidy estimation in West Indian gherkin, horned melon and wild *Cucumis* spp. by flow cytometry

Yuichi Matsumoto[1]* and Makoto Miyagi[2]

[1]United Graduate School of Agricultural Science, Tokyo University of Agriculture and Technology, Fuchu, Tokyo 183-8509, Japan.
[2]Plant Biotechnology Institute, Ibaraki Agricultural Center, Kasama, Ibaraki 319-0292, Japan.

The ploidy level of wild *Cucumis* spp. differs among species and accessions. Therefore, estimations of ploidy levels are important in the breeding of species and accessions of this plant. Flow cytometry is a powerful technique for estimating ploidy level. This study examined the applicability of flow cytometry analysis for estimating ploidy levels in wild *Cucumis* spp. West Indian gherkin and horned melon. The relative DNA content of these accessions was highly correlated with their ploidy levels. The results suggest that flow cytometry analysis can be used for estimating ploidy levels in these species.

Key words: West Indian gherkin, horned melon, *Cucumis* spp., ploidy level, flow cytometry.

INTRODUCTION

Plant breeding programs consistently require effective tools that can confer timesaving benefits and increased efficiency in improving cultivars; for example, generating doubled haploids is of great interest for the rapid production of pure lines (Griffing, 1975). Because the *Cucumis* plant is cross-pollinated, it has a long breeding cycle. Therefore, to produce pure lines in the shortest possible time, a collaborative study was set up to obtain irradiated pollen-induced parthenogenetic haploid embryos that were raised into complete haploid plants (Gursoz et al., 1991). The classic method of counting chromosomes is considered the most accurate way of determining ploidy level. However, other indirect methods can also be used satisfactorily. For example, flow cytometry (FCM) is a powerful technique for estimating plant nuclear DNA content and ploidy level because it permits sensitive measurements of the florescence intensity of large numbers of stained nuclei within seconds (Arumuganathan and Earle, 1991; Ushio et al., 2002). FCM analysis has been successfully used in melon (Ogawa et al., 2002) and cucumber (Kubaláková et al., 1996), but not yet in other *Cucumis* spp., for which microscopic counting of chromosomes in root-tip cells is conventionally done to determine the ploidy level (Dane and Tsuchiya, 1976). The genus *Cucumis* consists of about 30 species and is further classified into 2 subgenera representing different base chromosome numbers: *Cucumis* subgen. *Cucumis* (x = 7) and *Cucumis* subgen. *Melo* (x = 12) (Krkbride, 1993). The ploidy level of this genus differs among species or accessions, unlike that of melon and cucumber.

Variations in ploidy levels have been reported, especially in accessions of wild species of the subgenus *Melo*. Most accessions of wild species have been found to be 2n = 24 (*Cucumis africanus*, *Cucumis dipsaceus*, *Cucumis ficifolius*, *Cucumis hirsutus*, *Cucumis humifructus*, *Cucumis myriocarpus*, *Cucumis prophetarum*, *Cucumis pustulatus*, *Cucumis sagittatus*, *Cucumis sacleuxii* and *Cucumis zeyheri*). However, among these species, the accessions of 3 species have also been reported to be polyploid: *C. ficifolius*, *C. africanus*, *C. prophetarum*, *Cucumis heptadactylus* and *C. zeyheri*, 2n = 48; *C. pustulatus*, 2n = 48 or 72 (Dane and Tsuchiya 1976; Krkbride, 1993). Therefore, ploidy level estimations are necessary in breeding of these species and accessions. In this study, we examined the applicability of the FCM analysis for estimating ploidy levels in West Indian gherkin, horned melon and wild

*Corresponding author. E-mail: yu-matsumoto@agri.pref.ibaraki.jp.

Table 1. Accessions and ploidy level of *Cucumis* species used in this study.

Species	Accession No.	Ploidy level*
C. africanus	PI 275571	2x
	PI 299569	2x
C. anguria	PI 147065	2x
	PI 196477	2x
	PI 233646	2x
	PI 320052	2x
	PI 364475	2x
C. dipsaceus	PI 236468	2x
C. metuliferus	PI 292190	2x
C. myriocarpus	PI 203977	2x
	PI 282449	2x
	PI 282450	2x
	PI 299568	2x
	PI 374153	2x
C. sagittatus	PI 282441	2x
C. africanus	PI 299571	4x
C. ficifolius	PI 273648	4x
	PI 299570	4x
	PI 299572	4x
C. heptadactylus	PI 282446	4x
C. prophetarum	PI 193967	4x
C. subsericeus	PI 273650	4x
C. pustulatus	PI 343699	6x
	PI 343700	6x
	PI 343701	6x

* Ploidy level was according to Dane and Tsuchiya (1976).

Cucumis spp.

MATERIALS AND METHODS

Plant materials

Fresh leaf material of 10 wild *Cucumis* species, West Indian gherkin, and horned melon were collected, and a total of 25 accessions were used (Table 1). The seeds of these accessions were obtained from Germplasm Resources Laboratory (Agricultural Research Service, U.S. Department of Agriculture).

Furthermore, leaves of *Oryza sativa* L. cv. Nipponbare were used to internally calibrate the flow cytometer. The ploidy levels of each accession were taken from Dane and Tsuchiya (1976).

FCM analysis

FCM analysis was performed with a PA flow cytometer (Partec, Münster, Germany). Samples were prepared according to Ushio et al. (2002). Leaf segments of 0.5 cm^2 were chopped with a razor blade in the presence of a nucleus isolation solution (high-resolution DNA kit type P solution A; Partec). The suspension was filtered through a 30-μm nylon mesh, and nuclei were stained with 4',6-diamidino-2-phenylindole solution (high-resolution DNA kit type P solution B; Partec). In each sample, 5000 nuclei were analyzed, and the analysis was replicated 3 times. The gain of the instrument was calibrated so that the G_1 peak of 'Nipponbare' was set at channel 100. The relative DNA content of individual plants was calculated according to the following formula:

$$\text{Relative DNA content} = \frac{\text{Mean of the peak channel of the } G_1 \text{ of the sample}}{\text{Mean of the peak channel of the } G_1 \text{ of 'Nipponbare' nuclei}}$$

RESULTS AND DISCUSSION

The relative DNA contents of the 25 accessions analyzed were clearly classified into 3 groups: 0.7 to1.0, 1.6 to 2.0 and 2.4 to 2.8. These relative DNA contents were highly correlated with ploidy levels (R^2 =0.97)

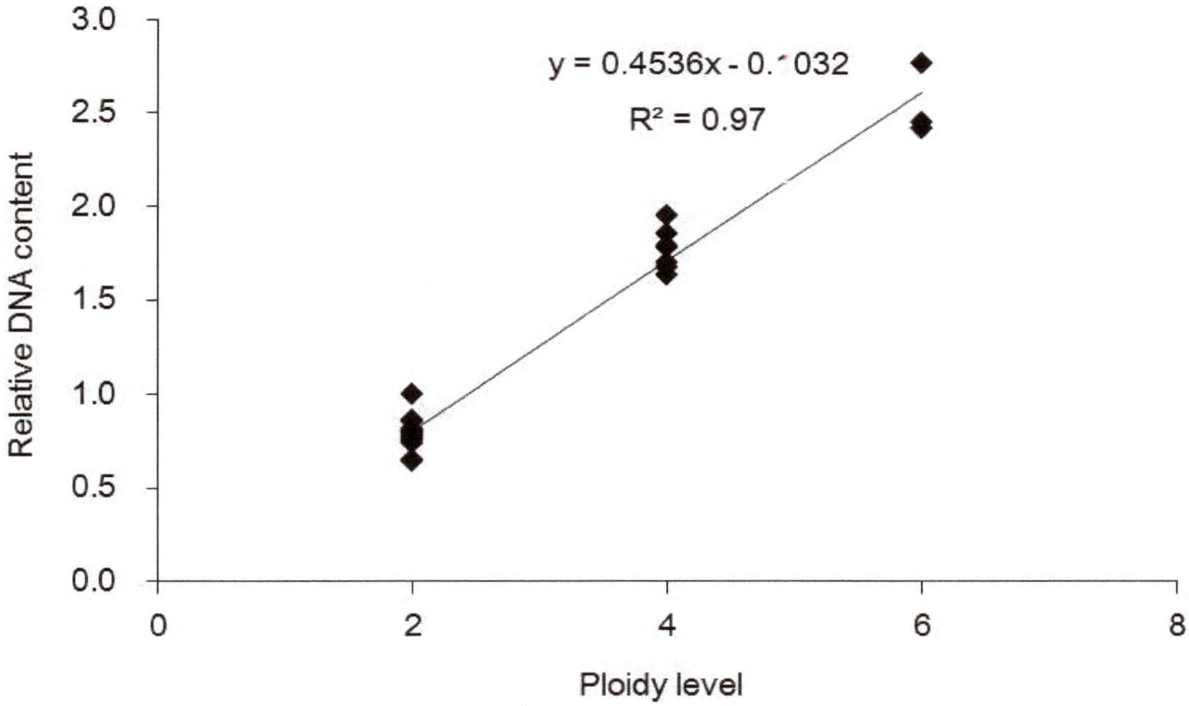

Figure 1. Relationship between the relative DNA content and ploidy level.

(Figure 1). The relative DNA contents of diploid, tetraploid and hexaploid accessions were 0.7 to 0.9, 1.6 to 2.0 and 2.4 to 2.8, respectively. Therefore, the results suggest that FCM analysis can be used for estimating the ploidy level of West Indian gherkin, horned melon and wild *Cucumis* spp. Furthermore, we used rice cultivar 'Nipponbare' as a calibration. 'Nipponbare' is one of the model plants, and widely used for genomic research (Ohmido et al., 2000). Therefore, these results could be used as comparison between other species. FCM analysis can be used as a simple and rapid method for estimating the ploidy level of a large number of accessions. Although, this technique has been used successfully in melon (Ogawa et al., 2002) and cucumber (Kubaláková et al., 1996), it has not been applied to West Indian gherkin, horned melon and wild *Cucumis* spp. The results of this study may have applications in the breeding of West Indian gherkin, horned melon and some *Cucumis* species, especially as an early and easy marker for ploidy manipulations (for example, polyploidization, haploidization and somatic hybridization).

Furthermore, FCM analysis might be suitable for detecting mixoploidy as well as for determining the ploidy level in progenies of conventional intra- and interspecific crosses of *Cucumis* spp.

REFERENCES

Arumuganathan K, Earle E (1991). Estimation of nuclear DNA content of plants by flow cytometry. Plant Mol. Biol. Rep., 9: 221–231.

Dane F, Tsuchiya T (1976). Chromosome studies in the genus *Cucumis*. Euphytica, 25: 367–374.

Griffing B (1975). Efficiency changes due to use of doubled-haploids in recurrent selection methods. Theor. Appl. Genet., 46: 367–386.

Gursoz N, Abak K, Pitrat M, Rode JC, Dumas de Vaulx R (1991). Obtention of haploid plants induced by irradiated pollen in watermelon (*Citrullus lanatus*). Cucurbit Genet. Coop. Rep., 14: 109–110.

Krkbride JH (1993) Biosystematic monograph of the genus Cucumis (Cucurbitaceae). Parkway Publishers Boone North Carolina. pp. 13-14.

Kubaláková M, DoleŽel J, Lebeda A (1996). Ploidy instability of embryogenic cucumber (*Cucumis sativus* L.) callus culture. Biologia Plantarum 38: 475–480.

Ogawa R, Oyabu T, Aoyama K (2002). Method of production of haploid plants by pseudofertilized ovule culture in melon. Res. Bull. Aichi Agric. Res. Ctr, 34: 61–66.

Ohmido N, Kijima K, Akiyama Y, de Jong JH, Fukui K (2000). Quantification of total genomic DNA and selected repetitive sequences reveals concurrent achnges in different DNA families in indica and japonica rice. Mol. Genet. Genom., 263: 388–394.

Ushio A, Onozaki T, Shibara M (2002). Estimation of polyploidy levels in Dianthus Germplasms by ytometry. Bull. Nat. Inst. Flor. Sci., 2: 21–26.

17

Studies on morphological and yield parameters of three varieties of Nigerian okra [*Abelmoschus esculentus* (L.) Moench]

O. A. Falusi[1]*, M. C. Dangana[1], O. Y. Daudu[1] and Jaime A. Teixeira da Silva[2]

[1]Department of Biological Sciences, Federal University of Technology, Minna, Nigeria.
[2]Faculty of Agriculture and Graduate School of Agriculture, Kagawa University, Miki-cho, Ikenobe, Kagawa-ken, Japan.

This study was carried out to assess the variation in morphological and yield parameters among three varieties (V_{35}, Jokoso and NH4e47.4) of okra [*Abelmoschus esculentus* (L.) Moench] from Nigeria. Okra seeds were collected and seedlings were raised to maturity at which stage they were examined for morphological and yield parameters. The distribution pattern in days to bud emergence and days to sexual maturity were not significantly different among the three varieties. On the other hand, the number of leaves/plant and plant height at maturity were significantly highest in V_{35}. Unlike the number of pods/plant, which was not significantly different ($P > 0.05$) among the three okra varieties, the number of seeds/pod was significantly higher in V_{35} and Jokoso than in NH4e47.4. The implications of these aspects with respect to future genetic improvement of okra are discussed.

Key words: *Abelmoschus esculentus*, morphological traits, yield traits.

INTRODUCTION

Okra [*Abelmoschus esculentus* (L.) Moench] belongs to the Malvaceae family. It is a native to tropical Africa where it serves as a staple vegetable crop (Indian Government, 2008; Kochhar, 1986; Hammon and Van Stolen, 1989). Hammon and Van Stolen (1989) also reported that there are 2,283 accessions out of which 2,029 were collected from the African continent of which 1,769 are from West Africa. The crop is therefore far more heavily represented in West Africa than in any other part of the world (Hammon and Van Stolen, 1989). It is an annual crop with a life-span ranging from 1 to 5 months depending on the species. The seeds are round, greyish and relatively large (about 1.5 to 3 mm). The importance of the crop as a vegetable lies in its wide acceptability due mainly to its flavour and viscosity which aids easy consumption of bulky and relatively hard foods like cassava pastes and pounded yam (NIHORT, 1987), in addition to being rich in vitamins A, B and C. The leaves also contain higher quantities of protein (2.7 to 3.0%) than the pods, and a significant amount of riboflavin and folic acid (Martin et al., 1979; Adelusi et al., 2006; Hegazi and Hamideldin, 2010). These popular uses of okra have fuelled an increasing demand for the crop, and a search for simple but viable ways of increasing supply of the product, independent of man-power and adequacy of farming conditions. Thus, attention has gradually shifted towards improving the genetic quality of the species through plant breeding and selection.

According to Ariyo (1992), vegetative traits of okra plants are positively correlated with yield and thus traits such as pod size must be given prime attention in the selection of okra species for breeding programmes to improve yield. In addition, the weight and number of pods/plant have been consistently identified as critical determinants of pod yield (Kault et al., 1978; Ariyo, 1989; Saifullah and Rabbani, 2009). On the other hand, Singh and Singh (1977) cited the number of pods/plant, days to flowering and plant height as being the most important parameters determining yield in okra. Akinyele and Osekita (2006) used Pearson's correlation coefficient to prove that morphological characters strongly influence

*Corresponding author. E-mail: falusiolamide@gmail.com.

Table 1. Mean values of morphological and yield parameters of 3 varieties of *A. esculentus* grown during the dry season in Minna.

Characters	Varieties		
	Jokoso	NH4e47.4	V_{35}
Bud emergence (days)	19.00 ± 0.82^a	19.90 ± 0.99^a	21.00 ± 0.00^a
Sexual maturity (days)	32.70 ± 2.79^a	33.10 ± 1.52^a	34.10 ± 0.52^a
Number of leaves/plant	9.20 ± 1.40^a	10.80 ± 1.93^a	12.40 ± 1.10^b
Plant height (cm)	19.10 ± 0.99^a	21.40 ± 1.17^a	25.10 ± 1.10^b
Number of pods/plant	6.50 ± 0.41^a	5.60 ± 0.77^a	6.10 ± 1.13^a
Number of seeds/pod	48.70 ± 5.52^a	54.70 ± 8.53^a	65.30 ± 8.18^b
Width of pod (cm)	5.08 ± 0.92^a	5.15 ± 0.63^a	4.95 ± 0.83^a
Length of pod (cm)	8.43 ± 0.37^b	5.38 ± 1.27^a	7.96 ± 1.79^b

*Values followed by the same letters within a row are not significantly different at $P > 0.05$ tested with DMRT.

okra yield. Variation in traits is a necessary requirement for a successful selection programme aimed at improving desirable traits. To better understand the relationship between variation in morphological and yield parameters in okra [*Abelmoschus esculentus*, L. Moench], the present study assessed such variation among three Nigerian varieties.

MATERIALS AND METHODS

Dried seeds of the three okra varieties (Jokoso, V_{35} and NH4e47.4) were obtained from the Institute of Agricultural Research, Ahmadu Bello University, Zaria, Nigeria. The viability of seeds was tested using the floatation method. They were planted in 5-L plastic buckets filled with sandy loam soil (4 seeds/bucket) at a depth of about 3 to 4 cm below the soil surface; these were later thinned to 2 seedlings/bucket, and 10 buckets/variety. Buckets were arranged in a completely randomized design (CRD). The planted seeds were watered once daily between 5.00 and 6.30 pm using bore well water from the Department of Biological Sciences, Federal University of Technology, Minna. The experiment was conducted between October and December, 2010 and the plants were monitored to avoid human and animal interference up to maturity. The morphological parameters investigated were assessed using standard procedures (Akinyele and Osekita, 2006). Specifically, bud emergence (BE), sexual maturity (SM), number of leaves/plant (NLP) were determined at maturity and plant height (PH) was determined at the flowering stage while yield was determined from the number of pods/plant (NPP), the number of seeds/pod (NSP), pod width (PW) and pod length (PL). Data were recorded regularly for 3 months using average data/plant. Data for each parameter and for each variety was pooled and used for statistical analysis. Differences between any 2 parameters were determine using a student's *t*-test and for more than 2 parameters using ANOVA.

Duncan's multiple range test (DMRT) was used to test for significant differences between means ($P < 0.05$).

RESULTS AND DISCUSSION

The days to BE among the three varieties ranged between 19.00 ± 0.82 days in Jokoso and 21.00 ± 0.00 days in V_{35} (Table 1). This indicates that the time taken for BE in V_{35} was slightly higher than in the other two varieties. The days to SM in the three varieties followed the same pattern as BE, suggesting that days to BE may not affect yield significantly in these varieties. The mean NLP was highest in V_{35} followed by NH4e47.4 and least in Jokoso (Table 1). V_{35} had consistently better vegetative growth, and may lead to increased yield performance. Mishra and Chhonkar (1979) and Nwangburuka et al. (2011) also reported considerable variation in okra vegetative and yield traits. PH was significantly higher in V_{35} than in Jokoso and NH4e47.4. According to Singh and Singh (1977) and Akinyele and Osekita (2006), days to BE and PH at maturity, among other morphological traits, are some of the most variable traits of okra that are necessary for selection programmes aimed at improving desirable traits. Since the trend of NSP for all three varieties was similar to that of other morphological parameters (Table 1), this might indicate that these factors are controlled by the same genetic variables. The highest NPP was recorded in Jokoso, the lowest PW was recorded for V_{35} in contrast to NSP and other morphological parameters that were consistently higher in V_{35} than in the other two varieties (Table 1). PL was highest in Jokoso even though this variety had consistently poor vegetative growth as measured by the other morphological parameters particularly NLP, PH and PW. Yield in okra might thus not be influenced exclusively by morphological variations. This agrees with the reports of Nwangburuka et al. (2011) but in contrast with that of Kaul et al. (1978) and Ariyo (1989) in which morphological variations (particularly NLP, PH and PW) are very important components of pod yield in okra.

The differences in yield parameters might be due to environmental influences while non-significant differences indicate that genetic components of varieties are still intact, as reported by Akinyele and Osekita (2006), Aladele (2009) and Udengwu (2009). This study indicates that vegetative productivity may not necessarily translate into yield. Therefore, any improvement programme intended for these okra varieties must target yield parameters, especially in Jokoso, which has proven to be very productive under the prevailing environmental conditions in Minna, Nigeria. Further studies need to be carried out on these varieties in other to determine certain genetic factors that may have triggered their performance.

REFERENCES

Adelusi AA, Makinde AM, and Folorunso AE (2006). Comparative studies of physic-biochemical parameters in *Abelmoschus esculentus* (L.) and *A. Mmoschatus* Res. J. Bot., 1(2): 104-109.

Akinyele BO, Osekita OS (2006). Correlation and path coefficient analysis of seed yield attributes in okra *Abelmoschus esculentus* Afr. J. Biotechnol., 5(14): 1330-1336.

Aladele SE (2009). Morphological distinctiveness and metroglyph analysis of fifty accessions of West African okra (*Abelmoschus caillei*) (*A.* Chev.) Stevels. J. Plant Breeding Crop Sci., 1(7): 273-280.

Ariyo OJ (1992). Factor analysis of component of yield and vegetative trait in okra. Indian J. Agric, Sci., 62: 83-84.

Hammon S, Van Stolen DH (1989). Characterization and evaluation of Okra. The use of plant genetic Res., pp. 173-174. http://horizon.documentation.ird.fr/exl-doc/pleins

Hegazi AZ, Hamildeldin N (2010). The effect of gamma irradiation on enhancement of growth and seed yield of okra [*Abelmoschus esculentus* (L.) Moench)] and associated molecular changes. J. Hortic. For., 2(3): 38-51.

Indian Government (2008). Biology of Okra (www.dbtbiosafety.nic.in/guidelines/okra). [An on-line article that describes taxonomy, morphology, growth and growing conditions of Okra (*Abelmoschus esculentus* (L.) Moench).

Kault T, Lal G, Peter KV (1978). Correlation and path coefficient analysis of components of earliness, pod yield and seed yield in okra. Indian J. Agric. Sci.,. 48(8): 459-463.

Kochhar SL (1986). Tropical Crop. A text book of economic botany. New Delhi Macmillan Co. Publishers, Indian Ltd, pp. 263-264. http://www.ajol.info/index.php/ajb/article/view/59202/47504.

Martin FW, Telek K, Ruberte R, and Samtiago AG (1979). Protein oil and gossypol content of vegetable curd made from okra seeds. J. food Sci., 44: 1517-1519.

Mishra RS, Chhonkar VS (1979). Genetic divergence in Okra. Indian J. Agric. Sci.,. 49: 244-246.

NIHORT (1987). Advances in fruit and vegetable Research at NIHORT (1976-1986) published by National Hortic. Res. Institution (NIHORT) Ibadan. pp. 31-34.

Nwangburuka CC, Kehinde OB, Ojo DK, Denton OA, Popoola AR (2011). Morphological classification of genetic diversity in cultivated okra, *Abelmoschus esculentus* (L.) Moench using principal component analysis (PCA) and single linkage cluster analysis (SLCA) Afr. J. Biotechnol., 10(54): 11165-11172.

Saifullah M, Rabbani MG (2009). Evaluation and characterization of Okra *(Abelmoschus esculentus* L. Moench.) genotypes SAARC J. Agri., 7(1): 92-99.

Singh KB, Singh HN (1977). Path coefficient analysis for yield in okra. Indian J. Agric. Sci., . 49: 244-246.

Udengwu OS (2009). Studies on Heterosis in *Abelmoschus esculentus* (L.) Moench and *A. callei* (A. Chev) Stevels Cultivars During Shorter Day Photoperiods in South Eastern Nigeria. Pakistan J. Biol. Sci., 12: 1388-1398.

Dormancy breaking of gladiolus cv. Jester for the mid hills of Nepal

Sharmila Piya*, Ajaya Shree Ratna Bajracharya, Jawahar Lal Mandal and Banhu Prasad Choudhary

Nepal Agricultural Research Council, Agriculture Research Station, Pakhribas, P. O. Box 29, Sunsari, Nepal.

The gladiolus is a very common and famous cut flower of Nepal. It is an economically very important and commercially grown cut flower for Nepal. Dormancy plays a vital role in flower production and its market price. In order to produce flower earlier than rainy season starts and to get good market price, different dormancy breaking treatments were tested at mid hill condition of Agriculture Research Station, Pakhribas (1740 masl). The treatments were gibberellins, water treatment, smoking, scales removal, chilling in moist sand and control (without any treatment). All the treatments were replicated five times and the design was complete randomized design (CRD) for lab experiment and randomized complete block (RCB) for field experiment. After more than fifty percent shoot initiation, the corms were planted in the field for field experiment. The rooting was found earlier in corms treated with moist sand (started from 16th day). At the 47th day of lab experiment, highest root initiation was from chilling treatment. Same result was also obtained for shoot initiation. At 62nd day, maximum shoot was found from chilling treatment and lowest from control treatment. Similarly, from field experiment, maximum flower size was recorded from chilling treatment followed by scales removal. In case of corm yield, highest corm yield was recorded from scales removal treatment. Thus, with these results, we can conclude that chilling corms in moist sand is the best and easy method to get early root and shoot.

Key words: Gladiolus, dormancy, rest period.

INTRODUCTION

Gladiolus, commonly known as sword lily (due to its sword like leaves) is a very popular cut flower among the world cut flowers. It is specially valued for floral arrangements. Its popularity is due to its wide color range, its nature to open in sequence (even in room condition) and its long lasting florets. It ranks as the number one cut flower in Nepal. Gladiolus is an economically important flower for Nepalese condition because floriculture business has been flourishing in Nepal since the early nineties (Yanai et al., 2007).

The main production season of gladiolus in Nepal is from April to August in the mid hills condition. In this season, flower quality deteriorates due to high rainfall, fusarium rot, etc. In addition to these problems, flower growers could not fetch good price due to over production. In this condition, farmers can get high price if they produce spike earlier than normal production. Gladiolus corms and cormels dormancy is reported as the major impediment in the commercial cultivation of gladiolus (Kumar and Raju, 2007). Gladiolus corms do not sprout immediately after harvesting. This is because of rest period and corm dormancy is attributed to the accumulation of growth inhibitory substances like abscisic acid (ABA). Ginzburg (1972.) also reported that dormant corm and cormels contain 5 to 10 times ABA than non dormant corms. According to Tsukamoto (1960), soaking the gladiolus corms in benzyladenine solution at 20 ppm for 24 h, followed three days later by re soaking in gibberellins solution at 100 ppm resulted in good sprouting and good growth of root and shoot. Denny (1930) reported that gladiolus rest period varies with varieties but usually, it is two to three months. In our condition, we observed that the corms of Jester variety

*Corresponding author. E-mail: sharmilapiya@yahoo.com.

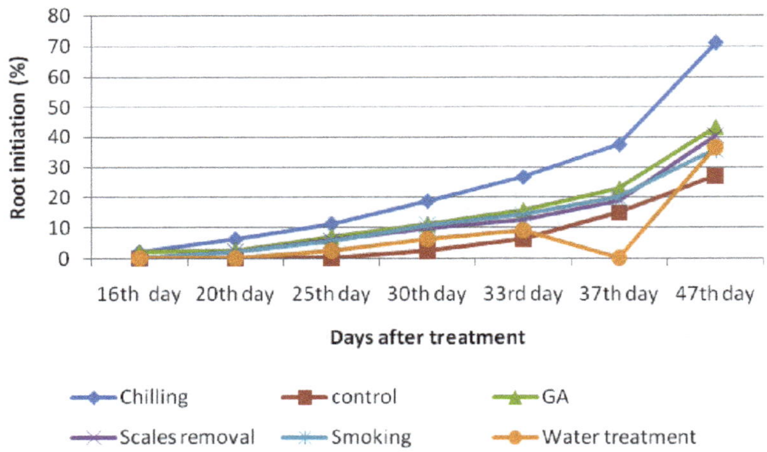

Figure 1. Root initation trend the first year.

normally rest from three to four months. Though, different chemicals treatments were recommended to reduce the rest period of gladiolus corms, these chemicals are expensive and easily not available in our market. Thus, we tried to identify the easy, effective and easily available materials to shorten the dormancy period of gladiolus corms cv. Jester.

MATERIALS AND METHODS

The experiments were set up two months after harvesting the corms. Later, the harvested corms were kept in shade condition. The design used for this experiment was completely randomized design. A total of six treatments were tested for dormancy breaking, which were as follows:

i. Gibberellins: Corms were dipped in 100 ppm giberellic acid for one hour.
ii. Water treatment: Corms were dipped in the solution of *carbendazim* (2 g/l) for one month.
iii. Smoking: Corms were smoked for one hour continue for three days
iv. Removal of scales: All the outer scales were removed
v. Chilling in moist sand: Corms were kept in layers of moist sand in trays. The soil temperature from chilling treatment was 0.9°C.
vi. Control: Corms without any treatment

There were a total of five replications. After the treatment, a total of 48 corms were kept in trays and kept in dark rook. The rooting and shoot initiation observations were taken in one week interval until root initiation and then every alternate day. The percentage data on root initiation and shoot initiation were transformed to arcsine (angular) and analyzed by Genstat Discovery Edition 3.

RESULTS FOR LAB EXPERIMENT

Root initiation in 16 days

There was significantly highest root initiation from corms treated with GA (16.51%) and chilling (15.03%) in the second year (Figure 1). In other treatments, root initiation was not started in the 16th day.

Root initiation in 30 days

40.63% root initiation was recorded from chilling treatment in second year which was the highest among all treatments. Least root initiation was found from control treatment (8.03%) from first year experiment. All treatments except water treatment were found superior over control treatment for root initiation of corms.

Root initiation in 47th days

At the 47th day of experiment, root initiation from the treatments GA and smoking were found for initiated root at par, from 58.57 and 58.54% respectively. Chilling treatment was superior to all treatments while control and water treatments were at par with 42.30 and 43.90% respectively. Figures 1 and 2 show the root initiation trend in the first and second year.

Shoot initiation

Observation on shoot initiation on the 25th and 30th day were not significantly different. However, shoot initation was started in corms treated with GA and chilling.

Shoot initiation at day 33

Only treatments were found significant and its interaction with year was not significantly different. Though, highest percentage of shoot initiation was found from treatment chilling (5.19%), percentage of shoot initation at 33 days

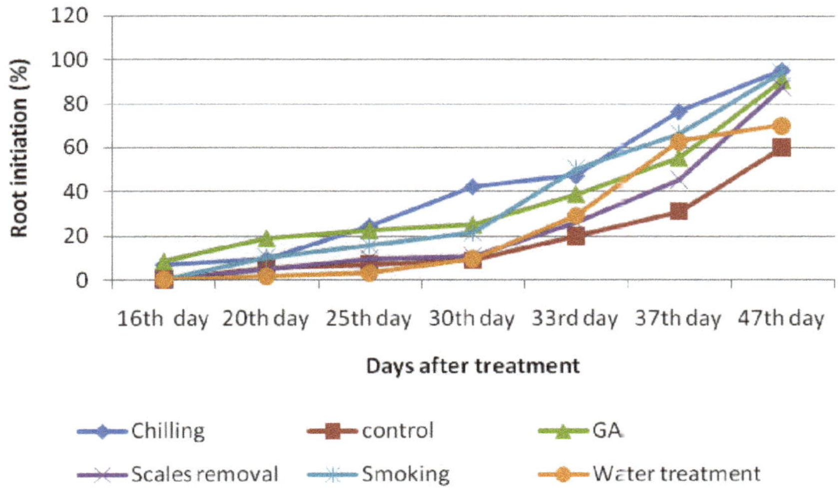

Figure 2. Root initiation trend in the second year.

from chilling treatment, scales removal (3.67%) and corms treated with GA (3.67%) were at par.

Shoot initation at day 37

Treatments for dormancy breaking was found highly significant but its interaction with year did not vary significantly for shoot initation. Highest shoot was initiated in chilling treatment, while lowest was in control and water treatment. Corms treated with GA, scales removal and smoking were at par with 5.56, 6.02 and 5.19% respectively.

Shoot initiation at day 55

Obverbations on percentage of shoot initiated at 55th day was found highly significant due to different treatments, year and their interaction. Among the different treatments, the chilling treatment (37.86%) was recorded as the highest percentage. Treatments scales removal and smoking were found also better with results of 26.97 and 18.95 %. While GA with 15.19% and water treatment with 12.96% were found at par. Second year of experimentation was better over first year. While we consider the interaction of treatments and year, chilling treatment of second year was the highest for shoot initiation (52.31%).

Shoot initation at day 62

Higly significant effect was observed in dormancy breaking due to different treatments, year and also its interaction. Chilled corms in sand layer produced highest percentage of shoot in second year. While lowest percentage was observed in control treatment in second year. Among the treatments, chilling treatment was the best followed by scales removal. After 55 days, the treatment of scales removal produced drastically highest shoot as compared to other treatments. Corms dipped in water with fungicide solution and control treatments did not give satisfactory shoots. Except water treatments, all the treatments were found superior as compared to control treatment. Figures 3 and 4 show the percentage shoot initiation in the first and second year.

Effect of different dormancy breaking treatments on flower parameters

Spike initiation was found highly significant and plants from control plots showed spike earlier than other treatments in both years. This might be due to control plots corms that were planted later due to late root and shoot initiation.

Other flower parameters like flower size, number of florets and spike length were observed not significantly different. But if we consider its mean value, flowers from scales removal treatment showed maximum flower size, florets number and spike length. Flower size was significant in second year; flowers from scales removal and chilling treatment had maximum flower diameter. Smallest flower was recorded from water treatment in second year (Tables 1 and 2).

Effect of dormancy breaking treatments on corm and cormel parameters

Highest corm weight per plant and total corm yield (t/ha) was recorded from treatment scales removal followed by chilling treatment. Scales removal, chilling, smoking and water treatment were found at par for total corm yield

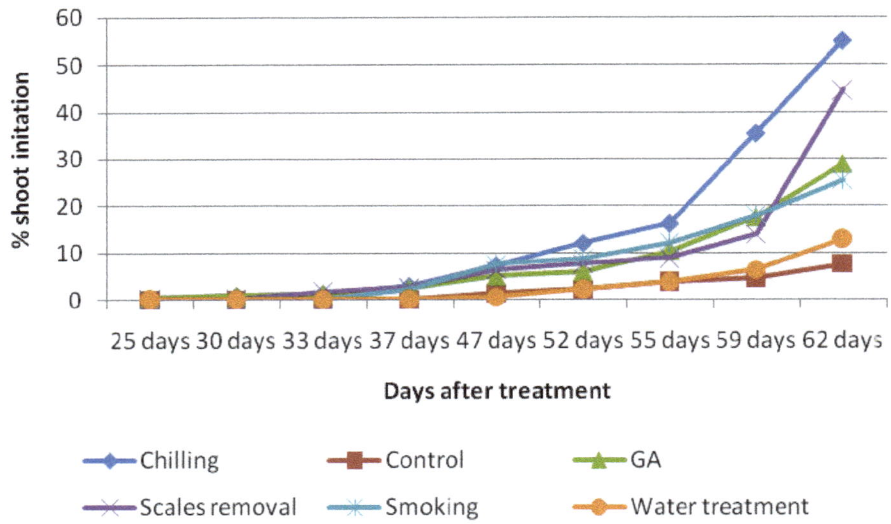

Figure 3. Percentage shoot initiation in first year.

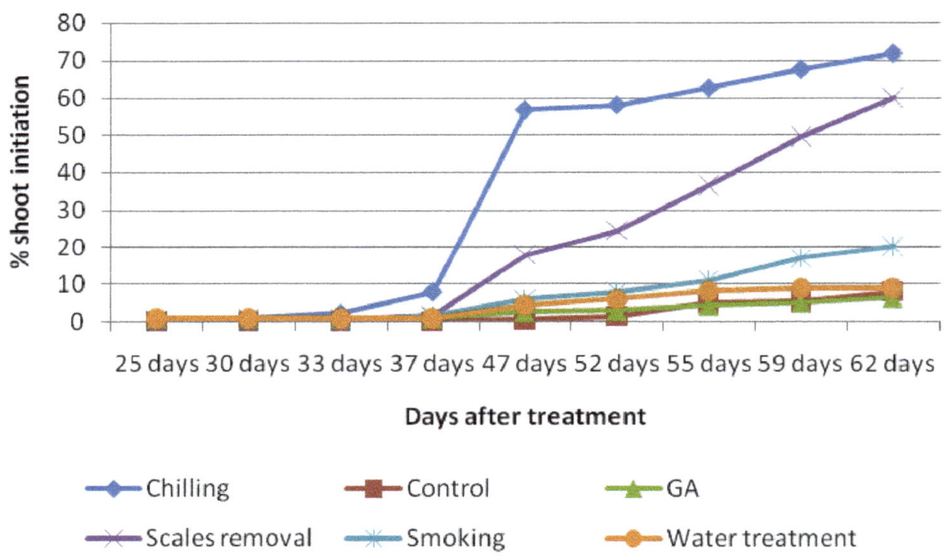

Figure 4. Percentage shoot initiation in second year.

Table 1. Flower parameters of first year.

Treatments	Spike initiation (days)	Harvesting (days)	Flower size (cm)	No. of florets	Spike length (cm)
Chilling	80.84b±5.2^{b}	92.5±7.7^{bc}	9.50	17.20	79.20
GA	69.46a±3.8^{a}	79.9±5.2^{ab}	9.19	15.15	76.40
Scales removal	79.92b±7.8^{b}	97.3±12.0^{c}	9.72	18.36	87.50
Smoking	68.08a±2.9^{a}	86.9±1.4^{abc}	9.18	15.99	76.60
Water treatment	70.28a±0.5^{a}	81.9±3.8^{abc}	9.25	17.08	79.50
Control	67.68±6.8^{a}	72.8±12.1^{a}	8.08	13.13	67.82
F test	***	*	NS	NS	NS
CV%	6.7	13.2			
lsd	6.420	14.86			

Table 2. Flower parameters of second year.

Treatments	Spike initiation (days)	Harvesting (days)	Flower size (cm)	No. of florets	Spike length (cm)
Chilling	84.28±5.4^{c}	100.97±6.6^{c}	10.31±0.77^{a}	19.20	94.10
GA	70.37±4.2^{ab}	90.03±1.1^{ab}	8.47±1.13^{b}	17.60	93.80
Scales removal	75.41±3.0^{c}	92.45±2.9^{b}	10.21±1.12^{a}	20.80	100.30
Smoking	67.88±2.3^{a}	85.32±2.6^{a}	9.75±0.81^{a}	19.00	93.30
Water treatment	70.61±2.6^{ab}	88.68±2.4^{ab}	8.00±0.73^{b}	17.40	95.40
Control	72.20±4.5^{bc}	84.45±8.3^{a}	8.23±1.13^{b}	18.00	95.00
F test	***	***	***	NS	NS
CV%	3.4	4.9	8.5		
Lsd	3.367	5.806	1.043		

Table 3. Corm and its parameters in first year.

Treatments	Average corm weight (g)	Average cormel weight/plant (g)	Corm diameter (cm)	Total corm yield (t/ha)	Total cormel no./plot
Chilling	98.20±7.0^{ab}	12.56 ±3.0^{a}	9.19	13.78±1.61^{ab}	37.20±6.0^{a}
GA	83.90±2.6^{bc}	7.26±2.3^{b}	6.36	9.64±2.53^{b}	22.20±4.6^{bc}
Scales removal	106.20 ±22.1^{a}	9.56±1.4^{ab}	6.90	14.43±2.26^{a}	33.10±8.5^{ab}
Smoking	67.00 ±14.8^{c}	8.68±1.0^{ab}	5.77	12.80±0.63^{ab}	24.20±1.7^{bc}
Water treatment	81.80±2.4^{bc}	7.16±0.1^{b}	5.44	12.36±0.19^{ab}	17.40±2.9^{c}
Control	74.80±14.1^{c}	6.23±2.9^{b}	5.73	10.02±2.15^{b}	20.20±8.8^{c}
F test	**	*	NS	*	**
CV%	17.7	33.9		24.3	35.0
lsd	19.93	3.837		3.899	11.87

Table 4. Corm yield and its parameters in second year.

Treatments	Average corm weight (g)	Average cormel weight/plant (g)	Corm diameter (cm)	Total corm yield (t/ha)	Total cormel yield (t/ha)
Chilling	78.90	5.69	6.38±0.44^{a}	10.16	0.487±0.065^{ab}
GA	72.40	4.96	5.17±0.74^{b}	8.85	0.433±0.032^{ab}
Scales removal	80.10	6.10	6.14±0.25^{a}	11.30	0.667±0.179^{a}
Smoking	62.00	4.99	6.06±0.18^{ab}	9.80	0.306±0.204^{b}
Water treatment	79.00	5.15	5.90±0.03^{ab}	9.37	0.619±0.087^{a}
Control	66.00	5.82	5.77±0.16^{ab}	9.49	0.346±0.097^{b}
F test	NS	NS	*	NS	*
CV%			10.5		39.2
Lsd			0.821		0.2464

(Tables 3 and 4). Similarly, average cormel yield and total cormel number per plot were observed maximum from chilling treatment. Chilling and scales removal were found at par while highest cormel was recorded from water treatment.

DISCUSSION

Freshly harvested corms of gladiolus do not germinate immidiately, even when planted in favorable environment (Denny, 1930). This is because of its rest period or dormancy period. Due to this rest period, gladiolus corms take certain period to germinate and flower. This dormancy period varies from variety to variety.

A high temperature treatment at 35°C for five days followed by a low temperature treatment at 0 to 5°C for 25 days for two varieties - White Gold and Cardinal Prince - was reported effective for good sprouting of corms (Tsukamoto and Yagi, 1960). They also reported that the dormancy of gladiolus corms and cormels can be broken by storing in low temperature. Our results are also in agreement with these papers because in our experiment, we found that chilling treatment with sand at 0.9°C was the most effective for sprouting the corms. According to Gerritis et al. (2003), dormancy of lilies bulb was broken by storage at low temperature (5°C) for several weeks, the longer the cold storage, the faster and more uniform leaf emergence occurred. The role of low temperature treatment on dormancy breaking was due to starch hydrolyzed during storage and sugars accumulated (Gerritis et al., 2003). Chilling treatment was the most effective dormancy breaking treatment for root/ shoot sprouting and corm yield.

Gibberellins stimulate growth, break dormancy and delay senescence. According to (Salisbury and Ross, 1978), gibeberellins overcome the seed dormancy and promote the germination. In our experiment, gibberellic acid found an effective treatment to promote the root initiation and shoot prolongation. According to Suresh et al. (2009), gladiolus corms of cultivar American Beauty dipped in the solution of GA3 at 125 ppm sprouted with less number of days (17 days) and 50% sprouting in 29 days. Tsukamoto and Yagi (1960) also reported that soaking the gladiolus corms in benzyladenine solution (20 ppm) for 24 h, followed three days later by resoaking in 100 ppm gibberellin solution resulted in good dormancy breaking and good root and shoot growth. Bhalla and Singh (2000) reported the increased plant height and spike length with 100 ppm GA3 at 5°C. Same report was reported by Negi and Raghava (1986) that application of GA3 increased plant height and number of leaves and shoots per plant as well as increased number and size of florets.

Smoking treatment was also found effective to promote sprouting and it is also easier to apply. Suh (1989) also reported that cormels treatment with smoke (1 h/ day for 3 days) before planting promoted sprouting. Our result are also in accordance with this report. Similarly, recently harvested dormant corms of Triteleia laxa Benth. Cv. Queen Fabiola to ethylene for 7 days at 20 ppm resulted in early sprouting, early flowering and more flowers and high corm and cormels weight (Han et al., 1990). Exposure of freesia corms to smokes reduce the corm dormancy as reported by Imanishi and Fprtamoer (1983). They also reported that this dormancy breaking was because these smokes contain ethylene. Exposure of freesia corms to ethylene promote sprouting (Uyemura and Imanishi, 1984); ethylene has the ability to shorten the dormant period with no adverse effects on root and shoot development (Masuda and Asahira, 1980). The ethylene releasing compound (CEPA) promotes the germination of dormant gladious cormals (Ginzburg, 1974). Bhalla and Singh (2000) also reported that ethrel (ethephon 1000 ppm) at room temperature was effective in causing early sproutings. Descaling was also reported as the dormancy breaking treatment as it stimulates the germination of dormant cormels.

Conclusion

Three treatments like chilling corms, using GA and smoking, were found effective for shortening the resting period of gladiolus. These three treatments are not only effective but easy to apply and also easily available. Among these three treatments, the chilling treatment was found superior in dormancy breaking.

So, we can conclude that corms chilled in moist sand 20 days will enhance the rooting. Thus, we can recommend the chilling treatment for early planting of gladiolus cv. Jester.

ACKNOWLEDGEMENTS

The authors are highly grateful to the Nepal Agricultural Research Council for providing the fund for this research. Persons directly or indirectly involved in this experimentation are also duly acknowledged.

REFERENCES

Bhalla R, Singh R (2000). Effect of plant bioregulators and temperature treatments on dormancy breaking of gladiolus. *National Symposium on Emerging Scenario in Ornamental Horticulture in 2000 A.D. and Beyond*, Indian Society of Ornamental Horticulture.

Denny FE (1930). Shortening the rest period of gladiolus by treatment with chemicals. Am. J. Bot., 17(6):602-613.

Gerritis MML, Miller WBM, Croes AF, Klerk GJ (2003). Effect of low temperature on dormancy breaking and growth after planting in lily bulblets regeneraged *in vitro*. Plant Growth Regulation, 40(3): 267-275.

Ginzburg C (1972). Hormonal regulation of cormel dormancy in gladiolus grandiflorus. J. Exp. Bot., 24(3): 558-566.

Ginzburg C (1974). Studies on the role of ethylene in gladiolus cormel germination. Plant Sci. Letters, 2(3):133-138.

Han SS, Halevy AH, Sachs RM, Reid M (1990). Enhancement of growth and flowering of Tritelia laxa by ethylene. J. Am. Society Hort. Sci., 115:482-486.

Imanishi H, Frotanier EJ (1983). Effects of exposing freesia corms to ethylene or to smoke on dormancy breaking and flowering. Sci. Hort., 18(4): 381-389.

Kumar PN, Raju DVS (2007). Dormancy in gladiolus: The cause and remedy- A review. Agric. Rev., 28(4).

Masuda M, Asahira T (1980). Effect of ethylene on breaking dormancy of freesia corms. Sci. Hort., 13(1): 85-92

Negi SS, Raghava SPS (1986). Gladiolus. In Ornamental Horticulture in India (eds K.L Chadha and B. Choudhary). Publications and Informations Divisiion, Indian Council of Agricultural Research, New Delhi.

Salisburry FB, Ross CW (1978). Plant Physiology, second edition. Wadsworth Publication Co. Inc. California. p. 59.

Suh JK (1989). Corm formation in gladiolus as influenced by dormancy breaking methods and cormel planting and corm harvesting dates. Korean Society for Horticultural Science, 7(1): 158-159.

Suresh KK, Chandra SR, Siva SA (2009). Effect of plant growth regulators on dormancy, corm and cormel production in gladiolus (*Gladiolus x grandiflorus* L.). J. Ornamental Hort., 12(3):182-187

Tsukamoto Y, Yagi M (1960). Dormancy of gladiolus corms VI. Effects of temperature treatment on breaking dormancy of gladiolus corms stored in a storage room and of those grown under different day length. Plant Cell Physiol., 1(3): 221-230.

Uyemura S, Imanishi H (1984). Effects of duration of exposure to ethylene on dormancy release in freesia corms. Scientia Horticulturae, 22(4): 383-390.

Yanai CN, Gautam MP, Bijl B (2007). Sector study on floriculture, June-September 2007.Nepal: Advisory Services on Export Development of Priority Sectors of Nepal. Project NEP/A1/01A, A project financed by the EU and ITC under the Asia Trust Fund. p. 45.

Physiochemical and nutritional characterization of *Vitex payos* (Lour.) Merr. (Verbenaceae): An indigenous fruit tree of Eastern Africa

James Munga Kimondo[1]*, Jacob Godfrey Agea[2], Clement Akais Okia[2], Refaat Atalla Ahmed Abohassan[3], Elizabeth T. N. Nghitoolwa Ndeunyema[4], Dino Andrew Woiso[5], Zewge Teklehaimanot[6] and Jackson Mulatya[1]

[1]Kenya Forestry Research Institute, P. O. Box 20412-00200, Nairobi, Kenya.
[2]Department of Community Forestry and Extension, Makerere University, P. O. Box 7062, Kampala Uganda.
[3]Faculty of Meteorology, Environment and Arid Land Agriculture, King Abdulaziz University, P. O. Box 80208 Jeddah, 21589 Saudi Arabia.
[4]Faculty of Agriculture and Natural Resources, The University of Namibia - Ogongo Campus, Private Bag 5520, Oshakati, Namibia.
[5]Department of Biological Science, Sokoine University of Agriculture, P. O. Box 3038, Morogoro, Tanzania.
[6]University of Bangor, Bangor Gwynedd, LL57 2UW, UK.

In the dry areas, indigenous fruits become important staples when cereals harvested are inadequate to support populations. Farmers in these areas have identified many of the handicaps in domestication but there is still need for inputs from the food industry into identification of the desirable traits and characteristics of potentially novel food. The purpose of this study was to assess the nutrient content of one edible wild fruit, *Vitex payos* that has been identified as a top priority species among the inhabitants of drylands of Kenya for domestication. The proximate, minerals and vitamin content were determined. Results showed that the fruit did contain useful quantities of potassium, manganese, phosphorus and vitamin C. Besides, sodium, magnesium and calcium were also present in minute quantities.

Key words: *Vitex payos*, indigenous fruits, dry lands, nutrients, proximate, Kenya.

INTRODUCTION

Utilization of indigenous fruits is limited in drylands of East and Central Africa (ECA) where they are abundantly available because of scarce knowledge on their nutritional values, postharvest handling (Swai and Kimata, 2005) and lack of appreciation of the quantities available. There is dearth of information on fruit processsing and value addition activities by fruit gatherers and traders; yet processing and value addition activities could serve to: (i) ensure continued availability of these fruits to the consumers even in off-season periods; (ii) improve demand and prices of these fruits as a result of good and attractive packaging; and iii) reduce the bulkiness of fruits and hence ease the transportation and storage of the products. One such resource in the drylands of Kenya is *Vitex payos* (Lour.) Merr fruit, which most subsistence farmers collect for food consumption and for subsistence sale in local markets during the fruit ripening seasons. *V. payos* is an indigenous undomesticated fruit tree that grows on farms and fallows in drylands of Eastern, Coastal and Central regions of Kenya. The fruits of *V. payos* are available between April and July during dry season. The pulp is black and mealy, though it does not stain; it is sticky and covers a fine hairy endocarp, which normally possesses four chambers containing the seed.

The endocarp is hard and requires mechanical strength to break and release seed (Mbabu and Wekesa, 2004;

*Corresponding author. E-mail: jmkimondo59@yahoo.co.uk or jmkimondo@kefri.org.

Mbora et al., 2008). The mealy pulp is usually consumed fresh, but could be used to process by-products such as jam and juice (Mbabu and Wekesa, 2004; Swai and Kimata, 2005). Pulp and nut mass are the two major constituents of the fruits of *V. payos*.

Fruits are good sources of minerals and vitamins, essential for human health (Saka and Msonthi, 1994). They have nutritional value that could provide the necessary body requirements (Beentje 1994; Mbabu and Wekesa, 2004; Ondachi, 2002; Saka, 1995; Mbora et al., 2008; Maundu et al., 1999). While these authors observed the richness of *Vitex payos* in vitamin C, Palgrave (2002) pointed out that these fruits had no vitamin C. This study was therefore carried out to evaluate the physico-chemical characteristics of *V. payos* fruits from the drylands of Kenya. The objective of this study was to determine the physical and nutritional composition of fruit pulp. The null hypothesis was that *V. payos* fruit pulp does not possess any valuable nutrients.

MATERIALS AND METHODS

Fruit characteristics

Pulp and nut mass are the two major constituents of fruits (Leakey et al., 2002, 2005). In *V. payos*, the important component is the pulp, which is the edible portion. Ten ripe fruits were randomly collected below 10 trees in Kitui and Mwingi districts, the two areas where fruits are consumed as snacks and also sold in local markets. The length and width of the fruits were measured to the nearest 0.1 mm. The fruit skin and pulp were removed at the widest diameter and length of the fruit and the nut measured for length and diameter to the nearest 0.1 mm. Assuming fruit and nut had cylindrical shape, their volumes were calculated using the formula:

$$\text{Volume} = \tfrac{1}{4}(\pi d^2 l);$$

where d is fruit/nut diameter; l is length and $\pi = 3.14$ (Foster, 2008), even though some fruits had irregular shapes. The volumes of individual fruits were computed and compared between the two districts and between individual trees.

To assess variation between fruits at different heights in the canopy, 15 ripe fruits were collected from each of the lower, middle and upper third portion of the entire canopy from 18 trees in Kitui. This was done to determine the anecdote farmers held that fruits from different parts of the canopy differed in size. Length and width of individual fruits were measured to the nearest 0.1 mm and weight to the nearest 0.1 g. The 15 fruits from every canopy layer were mixed and divided into three random subgroups of five fruits each. Each batch of five fruits had all the pulp scrapped together with the skin and weighed (fresh weight). Pulp was dried in an oven at 105°C for 12 h, weighed after cooling and expressed as percent of the fresh pulp weight. To estimate the number of fruits per kilogram, batches of 200 ripe fruits were collected from 15 randomly selected trees in Kitui. The fruits were mixed thoroughly and divided into batches of close to 1000 g and fruits in each counted.

Chemical composition of fruit pulp

Ripe fruits were collected from five random trees in Gitumbi village, Mbeere district, where fresh ripe fruits were available for immediate transport to the UK. The fruits were frozen and transported to the laboratory. The pulp was scrapped from the fruit immediately and chilled to 0°C. The frozen pulp was stored until the time of analyses. To prepare the samples for analysis, they were thawed and oven-dried at 70°C for 24 h, ground, and sieved at 200 µM. Samples were analyzed for proximate, mineral and vitamin content in the laboratory of the School of the Environment, Natural Resources and Geography, Bangor University, UK. The proximate included protein, fat, ash, total dietary fibre, carbohydrate and energy, while minerals were macronutrients: calcium (Ca), magnesium (Mg), phosphorus (P), potassium (K) and sodium (Na) and micronutrients: iron (Fe), zinc (Zn) and Manganese (Mn). Vitamins analysed were both water-soluble; ascorbic acid (vitamin C), thiamine (B^{-}), niacin (B3) and pyridoxine (B6), and fat soluble ones; α-tocopherol (vitamin E), phylloquinone (vitamin K) and β-carotene (provitamin A). The laboratory analytical methods used are described as follows.

Protein

The protein content of pulp samples was obtained by measuring nitrogen (N) content of the samples following the Kjeldahl method using a Kjeltec 2300 analyser unit (FOSS, Denmark). In brief, 200 mg of pulp sample were digested by adding 4 ml of H_2SO_4 (98%) and 2 digestive tablets, and warmed for 4 h at 30°C. The resultant nitrogen was converted into $(NH_4)_2SO_4$, which on distillation with NaOH released NH_3 in the form of ammonium ions (NH_4^+) which bonded to the SO_4^{2-} ions of the acid. After the digestion, sample solutions were placed in the Kjeltec analyser unit to determine their N content. The realised N content was multiplied by 6.25 to obtain the protein content in samples.

Fat content

The fat content was determined directly by extracting the fruit pulp with petroleum ether using Soxtec Avanti 2050 system (Foss, Denmark). Five grams of pulp sample were placed in a porous thimble, which was lodged into an extraction aluminium cup containing 80 ml of petroleum ether as a solvent and the fat was extracted by the Soxtec system. After the extraction, tubes were placed in an oven at 102°C to evaporate the remaining solvent and dry the sample. The residue in round bottom flask after solvent removal represented the fat content of the sample. Extracted fat was weighed and divided by the sample weight (5 g) to obtain the fat content (g g^{-1}).

Ash content

Two grams of pulp sample was put into crucible and weight was recorded and placed in a furnace at 600°C for 12 h. When samples were burnt, water and volatile substances were vaporized while organic substances were oxidised in the presence of oxygen. After samples were cooled, ash was weighed and the content (g g^{-1}) was calculated by dividing ash weight by the original weight of the sample.

Total dietary fibre

Dietary fibre includes polysaccharides, oligosaccharides, lignin and associated plant substances that are resistant to digestion in humans (McCleary, 2003). Therefore, dietary fibre content determination consists of the removal of digestible substances of the pulp samples and weighing the rest. To assess dietary fibre content, 1 g of pulp sample was used after fat extraction. Each sample was dissolved in 50 ml of phosphate (pH 6), after which 0.1 ml of amylase was added and the solution was incubated at 95°C for 15 min to catalyze the hydrolysis of starch into glucose. After the incubation, the solution was cooled to room temperature and its pH was adjusted to 7.5 by adding NaOH (0.275 N). Then 0.1 ml of protease was added to the solution and placed in a water bath at 60°C for 30 min to solubilise protein. At the end of this second incubation, the solution was cooled at room temperature and the pH adjusted between 4 and 4.6 by adding HCl (0.325 M). Then, 0.1 ml of Amyloglucosidase was added to the solution, which was placed again in a water bath at 60°C for 30 min to remove glycogen. By the end of this third incubation, 4 volumes of ethanol (95%) were added and the solution kept overnight at room temperature to cool. After complete precipitation, the solution was filtered and rinsed with ethanol (95%) and acetone, to extract dietary fibre. Finally, the dietary fibre was dried at 70°C in an oven overnight and then weighed to obtain the fraction of the original sample (g g^{-1}).

Carbohydrate content

Total digestible carbohydrate content was estimated by difference; its content was assessed based on the assumption that samples are constituted of ash, dietary fibre, fat, protein and digestible carbohydrate. So, when the contents of protein, fat, ash and dietary fibre are known for 1 g of sample, carbohydrate content could be calculated according to the following formula:

Carbohydrate content (g g^{-1}) = 1 - (ash content + dietary fibre content + fat content + protein content)

The calorific value calculated as kilo Joules per gram (kJ g^{-1}) of fruit pulp was determined by multiplying the carbohydrate, fibre, protein and fat contents (in grams) by 17, 8, 17 and 37 respectively (FAO, 2003).

Water-soluble vitamins

All vitamins were determined in a Varian Prostar HPLC (Varian Analytical Instruments, Walnut Creek, CA, USA) with a Phenomenex HyperClone 250 x 4.60 5-micron C18 column (Phenomenex, Torrance, CA, USA). Calibration was done by comparing with standards (Sigma-Aldrich, Saint Louis, MO, USA). All water-soluble vitamins were determined using a mobile phase of 90:10 100 mM pH 2.2 phosphate buffer (50 mM NaH_2PO_4; 50 mM H_3PO_4) with 0.8 mM sodium 1-octanesulphonate: acetonitrile. Vitamin C (ascorbic acid) was determined at a flow rate of 1.0 ml min^{-1} and absorbance measured at 240 nm. Vitamins B1 (thiamine), B3 (niacin), and B6 (pyridoxine) were determined at a flow rate of 0.8 ml min^{-1} and absorbance measured at 270 nm. Samples were extracted by maceration in 100 mM phosphate buffer at a ratio of 1 to 2 g FW to 10 ml buffer and filtered to 0.2 μM before injection.

Fat-soluble vitamins

All fat-soluble vitamins were determined using a mobile phase of 50:50 methanol:acetonitrile. The total run duration was 20 min. Flow rate was 1.5 ml min^{-1} and absorbance was measured at 210 nm for the first 7.5 min whilst vitamin E (α-tocopherol) was eluted. Flow rate was 1.5 ml min^{-1} and absorbance was measured at 254 nm for the period 7.5 to 11 min whilst vitamin K (phylloquinone) was eluted. From 11 to 20 min, flow rate was 2.0 ml min^{-1} and absorbance measured at 450 nm whilst β-carotene was eluted. Samples were extracted by maceration in cyclohexane at a ratio of *ca.* 4 g fresh weight (FW) to 50 ml cyclohexane. Volume was reduced to *ca.* 1 ml by freeze-drying before analysis.

Determination of Ca, Na and K contents

Flame photometry was used to determine Ca, Na and K contents in the pulp samples. The flame photometer (model 410) measured the light of a specific wavelength emitted when a solution of a particular element was burnt. The amount of light emitted is directly proportional to the element concentration in the solution. To prepare aqueous solutions of samples, 2 g of sample were burnt at 450°C in a furnace overnight to remove the carbon. The samples were then dissolved into 10 ml of hydrochloric acid (HCl) of 12 M concentration. The solution obtained was diluted to 10 times for K and Na and 80 times for Ca by adding distilled water. Seven standard solutions at concentrations of 0, 5, 10, 30, 50, 70 and 100 mg L^{-1} were prepared for each element (Na, Ca and K). The standards and the sample solutions were read by the flame photometer. A regression equation was derived between standard solutions and the readings of the flame photometer and the equation was used to obtain the concentration of elements in sample solutions (mg L^{-1}). Using the dilution rates, the elements content in dry samples was calculated (g g^{-1}).

Phosphorus content

Phosphorus content was determined using the colorimetric method (Ames, 1966). This method is based on the principle that phosphate ions react with ammonium molybdate to give a blue complex that has an intense absorption band at 820 nm, when reduced by ascorbic acid. The complex absorbance is proportional to phosphate concentration in the original solution and was measured using a spectrophotometer (BioTek, model PowerWave XS). Eighty times concentration solutions of fruit pulp samples were obtained as shown above. Six standard solutions (0, 10, 30, 50, 70 and 100 mg L^{-1}) of the PO_4 ions were used to determine the relation between the spectrophotometer readings with phosphate concentrations. In brief, 80 μl of sample and standard solutions were placed in a 96 wells plate, then 180 μl of Ames reagent were added at 30 second intervals and finally 30 μl of ascorbic acid (10%) were added in each well of the plate. The absorbance of the solutions was read by the spectrophotometer after 15 min at an interval of 30 s until the last well of the plate.

Regression equation between the concentration of standard solutions and the readings of the spectrophotometer was used to obtain the phosphate concentration (mg L^{-1}) in sample solutions and then the content (g g^{-1}) in dry samples was calculated using the dilution rate.

Determination of Mg, Fe, Mn and Zn content

An atomic absorption photometer (VARIAN, model SpectrAA 220FS) was used to assess Fe, Mg, Mn and Zn contents in fruit pulp samples. The principle is that each element when burnt emits a specific wavelength light whose intensity is proportional to the amount of element in the solution. Six concentrations of each element were used as standards to calibrate the photometer. Sample solutions were prepared as earlier described. Ten (10) times concentration solution was used for Fe, Mn and Zn while a ninety times (90) concentration was used for Mg. Assay tubes each containing 50 ml of sample solution were placed on a 60 wells support where the first well was a tube of water and after each five (5) sample tubes, a drift solution was intercalated to control the photometer readings accuracy. The absorbance of solutions was read by the atomic absorption photometer and expressed as elements concentration (mg L^{-1}) according to the calibration done with standard solutions. The content in dry matter (g g^{-1}) was obtained by applying the dilution rates with regard to each element solution.

Figure 1. *Vitex payos* ripe fruits showing three size width (a) and length-wise (b). S, small; M, medium; L, large fruits.

Table 1. Mean, standard error of the mean (±SEM) and range of individual fruit length, width and weight in Kitui, Kenya.

Parameter	Minimum	Maximum	Mean±SEM
Fruit wt (g)	1.3	17.7	5.1 ± 0.1
Fruit width (mm)	11.0	28.7	17.7 ± 0.1
Fruit length (mm)	13.8	30.6	21.6 ± 0.1
Fruit pulp volume (cm^3)	0.90	9.84	3.70 ± 0.13

Analysis of data

Fruit characteristics

The fruit parameters were compared between Kitui and Mwingi districts using Mann Whitney test (unequal variance), while 1-way analysis of variance (ANOVA) was used to compare different canopy levels, respectively (Dancey and Reidy, 2007).

Chemical composition

The average content of proximate, minerals and vitamin of *Vitex payos* fruit pulp were calculated from four samples with Excel for Windows 2007.

RESULTS

Fruit characteristics

There were 203 ± 5 (n = 13) fruits for every one-kilogram of fruits. Fruit length and width ranged between 13.8 to 30.6 mm (mean 21.6 ± 0.1 mm) and 11.0 to 28.7 mm (mean 17.7 ± 0.1 mm), respectively, n = 804, (Figure 1), while mean weights and pulp volume were 5.1 ± 0.1 g and 3.70 ± 1.89 cm^3, respectively (Table 1). The pulp volumes of individual fruits from different districts were not significantly different. Length, width and weight of fruits from different canopy levels were not also statistically different. These fruit parameters, including the pulp volume, however, differed significantly among individual trees, $p < 0.001$ with wide range between tree with lowest and highest values (Table 1). The average fruit length among individual trees ranged from 16.5 ± 0.2 (tree no. 14) to 27.53 ± 0.3 mm, (tree no. 6) while average fruit width ranged between 14.4 ± 0.2 to 22.6 ± 0.4 mm for the same trees. Fruit average weight among individual trees ranged from 2.4 ± 0.1 to 10.5 ± 0.4 g. The lowest and highest values of the three parameters were from two trees. However, the order of trees between the two extremes varied from one parameter to the other.

Fruit pulp was the major component with a mean weight of 2.84 ± 0.04 g which represent 57.7 ± 0.5% of total fruit weight. Pulp weights of fruits from different parts of tree canopy were not significantly different. However, like fruit weight, fresh pulp weights also differed significantly between individual trees ($F = 217.434$, $p < 0.001$) ranging from 1.23 to 5.02 g. Furthermore, the moisture content in fresh ripe fruit pulp varied from 54.9 to 91.3%, with a mean of 68.4 ± 0.4%. This was however influenced by the ripeness of individual fruits. Mean dry pulp weight was 0.88 ± 0.03 g and ranged between 0.28 and 1.76 g per fruit. Figure 2 shows six potential groups

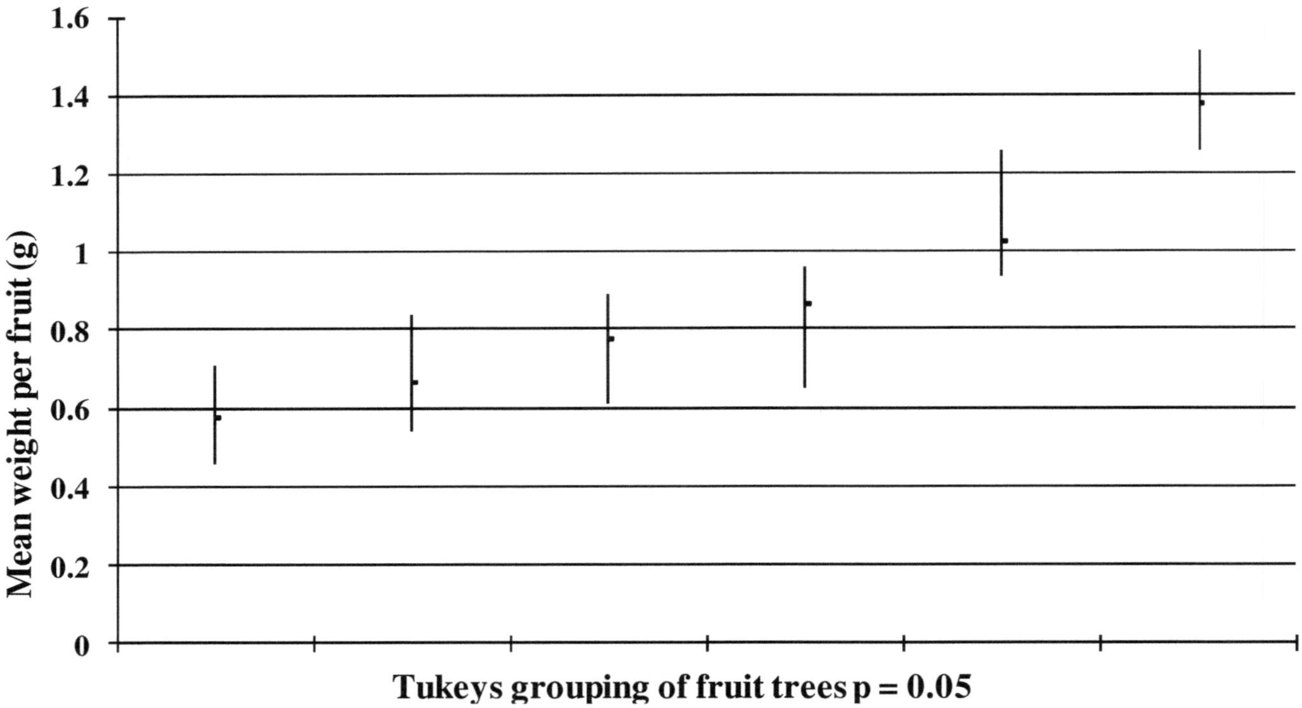

Figure 2. Tukey's test comparison for mean dry fruit pulp weight for individual trees.

Table 2. Proximate and vitamin composition of *Vitex payos* pulp.

Proximate	Mean ± sem (g 100 g^{-1} DW) (n = 4)	Vitamin	Mean ± SEM (mg 100 g^{-1} DW) (n = 3)
Water	240 ± 20	C	26.3 ± 4.9
Ash	5.3 ± 1.0	B1	n.d.
Protein	3.4 ± 0.07	B3	0.58 ± 0.2
Fat	0.44 ± 0.05	B6	0.11 ± 0.03
Fibre	60 ± 0.4	E	n.d.
Carbohydrate	30 ± 0. 3	K	n.d.
*Energy (kJ 100 g^{-1})	1064.1 ± 11.0	β-carotene	n.d.

DW, Dry weight; n.d., not detected, *assuming all fibres are metabolised (FAO, 2003).

of trees according to their average pulp dry weight. Trees with high pulp yield had over 200% more pulp than those with the least. Out of 18 trees, 8 (44.4%) produced fruits with at least twice as much pulp compared to low pulp producers (Figure 2).

Nutritional composition

Proximate compositions of the fruit pulp are shown in Table 2. The fruit pulp was rich in fibre (60 ± 0.4 g 100 g^{-1}) and moderate amount of carbohydrate (30 ± 0.3 g 100 g^{-1}). The pulp protein content was 3.4 ± 0.07 g 100g^{-1}. The fat was low in the pulp at 0.44 ± 0.04 g 100 g^{-1}, while the ash content was 5.3 ± 0.1 g 100 g^{-1}.

The fruit pulp of *V. payos* has three water-soluble vitamins, C (ascorbic acid), B3 (niacin) and B6 (pyridoxine) (Table 2). Results indicated that the fruit pulp is rich in vitamin C with 26.3± 4.9 mg 100 g^{-1} of pulp, while 0.58± 0.2 mg 100 g^{-1} of B3 and 0.11± 0.03 mg 100 g^{-1} of B6 are present (Table 2). On the other hand, vitamins B1, E, K and β-carotene (provitamin A) were not detected in the fruit pulp. Moreover, from the results presented in Table 3, the fruit pulp could provide potassium in large quantities (16 ± 0.4 mg g^{-1} DW) while it is a source of other macronutrients such as magnesium,

Table 3. Mean ± SEM (mg 100 g^{-1} DW) of mineral contents of fruit pulp.

Mineral	Mean ± SEM (n = 4)	Mineral	Mean ± SEM (n = 4)
K	1600 ± 40	Fe	11.9 ± 2.0
Mg	90 ± 3	P	11.5 ± 0.8
Na	30 ± 2	Mn	3.9 ± 0.1
Ca	20 ± 0.4	Zn	1.9 ± 0.04

DW, Dry weight.

sodium, phosphorus and calcium in smaller quantities (Table 3). The pulp contains micronutrients such as iron (11.9 ± 2.0 mg 100 g^{-1} DW), manganese (3.9 ± 0.1) and zinc (1.9 ± 0.04) (Table 3).

DISCUSSION

Black edible flesh surrounds a large nut in *Vitex payos* fruit, which contains four chambers potentially each with a seed. The pulp represents about 57% of the weight of the fresh fruit but varies greatly among individual trees. Majority of the pulp is fibre and carbohydrate (60 ± 0.44% and 30 ± 0.3% respectively). The high fibre content however causes the low energy of the pulp. Earlier studies on the species by Ondachi (2002) under the name '*Vitex doniana*', however, showed that the fruits contained more energy (1783 kJ 100 g^{-1}) than found in this study because of the high level of carbohydrate compared to the limited crude fibre. This could be due to the difference in methods of analysis used in determination of the two proximate. Fruits are generally not considered as good sources of protein; however, *V. payos* contains slightly higher protein than *Vitex doniana* with 2.8 g 100 g^{-1} (Ladeji and Okoye, 1993) and *Adansonia digitata* with 3.2 ± 0.1 g 100 g^{-1} (Osman, 2004), but lower than *Vitex mollis* which has 4.3 ± 0.11 g 100 g^{-1} (Montiel-Herrera et al., 2004). In addition, the fat content is lower than in *V. mollis*, but compares with that of *A. digitata.* The ash content compares with that of *V. mollis* at 2.9 ± 0.23 g 100 g^{-1} but was higher than for *V. doniana* and *A. digitata* at 0.3 g 100 g^{-1} (Osman, 2004).

Some of the macronutrients; (Na, K, Mg, Ca, and P) and micronutrients; (Fe, Mn, and Zn) were available in fruit pulp in quantities that could contribute significantly to the dietary requirements for humans. Their nutritional significance in the pulp is individually related to the contribution they make to the recommended dietary allowance in human beings. Potassium is the most abundant element in the pulp. The content available compares favourably with that of *V. mollis* at 1610 ± 57 mg 100 g^{-1} (Montiel-Herrera et al., 2004) and is higher than in *A. digitata,* which had 1240 ± 30 mg 100 g^{-1} (Osman, 2004). It plays an important role in the ionic balance and helps in maintaining the tissue excitability of the human body (Dhyani et al., 2007). With a daily requirement of 730 mg (Shells and Young, 1987), 50 g of fruit pulp is adequate to meet 100% of dietary requirement. K and Ca are reported as blood pressure lowering agents, thus reducing the occurrence of cardio-vascular diseases (Osborne et al., 1996).

Sodium in the pulp compares well with that of *A. digitata* but is only approximately 10% that of *V. mollis.* Sodium working together with K maintains appropriate acid-balance and is involved in enhancing nerve impulses transmission in the body (Umar et al., 2007). The Na content in the pulp is low compared to the K and this variation has significant importance in diets. According to Umar et al. (2007), during body growth, retention of protein requires a K/Na ratio within a range of 3 - 4. The ratio in *V. payos* pulp is around 53, thus indicating supplementary supply of Na could be necessary where these fruits are consumed in large quantities besides meeting the daily requirement of 1100 to 3300 mg (Shells and Young, 1987). The amount of Na available can only meet between 1 and 3% of the recommended daily allowance.

In our bodies, Mg plays an important role in circulatory diseases and Ca metabolism in bone (Ishida et al., 2000). The daily dietary requirement is 300 - 400 mg (Shells and Young, 1987) and thus 100 g of fruit pulp can therefore meet between 22 and 30%. The Ca content in *Vitex payos* is lower than in *A. digitata* with 90 ± 2 mg 100 g^{-1} (Osman, 2004) and *V. mollis* with 45 ± 3.9 mg 100 gm^{-1} (Montiel-Herrera et al., 2004) but significantly higher than in *V. doniana* which has 1 mg 100 g^{-1} (Ladeji and Okoye, 1993). Calcium is also necessary for healthy growth and maintenance of teeth and bones. With a dietary requirement of 800 -1200 mg (Shells and Young, 1987), 100 g of pulp can only supply between 1.7 and 2.5%. However, the availability of Ca and its absorption in the body depends on the Ca: P ratio (1:1) and the presence of anti-nutritional factors (Umar et al., 2007). The *V. payos* pulp however has a ratio of 2:1, thus the necessity of supplementary P in *V. payos* dependent diets to improve availability of Ca.

Phosphorus content in the fruit pulp is low with 100 g of pulp contributing between 1 and 1.4% of daily human requirement (Shells and Young, 1987). The mineral is important for bones, teeth and muscles growth and maintenance (Turan et al., 2003). Iron (Fe) is an important trace mineral and an essential nutrient that is required in the body in small quantities: 10 to 18 mg per day (Shells and Young, 1987). It is an important component of the red blood cells. It is absorbed in human bodies in large quantities in the presence of Vitamin C (ascorbic acid) (Shells and Young 1987). A 100 g of *V. payos* pulp could contribute between 66 and 100% of daily requirement and since the pulp contains Vitamin C, it is a good source especially among communities that are resource poor. Manganese, on the other hand, is required in the body in small quantities of 2.5 to 5.0 mg daily (Shells and Young, 1987). The fruit pulp with 3.9 ± 0.1 mg per 100 g of pulp, could meet 78 to 100% of the daily requirement, and therefore a good source of this nutrient. Moreover, the Zn content in *V. payos* compares well with that of *A. digitata*, (1.8 mg 100 g^{-1}) (Osman, 2004), but higher than in *V. doniana* with 0.04 mg 100 g^{-1} (Ladeji and Okoye, 1993) and lower than in *V. mollis* with 4.4 ± 0.24 mg 100 g^{-1}). The available amount of zinc could meet just above 10% of the daily requirement (Shells and Young, 1987). Zn is an essential component of many enzymes participating in metabolism of carbohydrates, lipids, proteins, and nucleic acids as well as in the metabolism of other micronutrients (USDA, 2001). It also plays a central role in supporting the immune system of body (USDA, 2001).

The energy obtained from *V. payos* pulp is lower than that from *V. mollis* (1433 kJ 100 g^{-1}) (Montiel-Herrera et al., 2004) and *A. digitata* (1340.1 ± 18.4 kJ 100 g^{-1}) (Osman, 2004). This was mainly because the latter had minimal fibre with the bulk of their pulp made up of carbohydrates with higher energy contribution than fibres. In this study, *V. payos* pulp was composed of 60% fibre, thus reducing the overall energy realized. Mbora et al. (2008) recorded an even lower carbohydrate and energy levels of 27.4 g and 63 kJ 100 g^{-1} of pulp, respectively. In their study, the fibre content was 27 g 100 g^{-1}, thus leaving a substantial amount of pulp composition unexplained.

Conclusions

Awareness of the significant contributions that wild indigenous fruits make to the diet of the inhabitants of dry areas in Kenya is increasing (Muok, 2001; Maundu et al., 1999; Mbabu and Wekesa, 2004; Ondachi, 2002; Mbora et al., 2008; Kimondo, 2010). However, the knowledge gap of the nutritional content of most of these fruits is not yet complete. The *V. payos* fruits are not endowed with all the nutrients that were analysed. The fruit is, however, a good source of some important nutrients such as potassium, phosphorus and manganese, which could be adequately supplied from consumption of these fruits as snacks. The fruits also compare favourably with others of the same genera where quantities of nutrients vary slightly from one to the other.

Meanwhile, the nutritional analysis of *V. payos* by chemical means informs us only of the potential value of these foods to those populations who may rely upon them as staples or supplements to their diet. It is important to assess the bioavailability of the essential nutrients in these plants. Such studies and the improvement of the processing of fruit pulp into various end products are contemplated.

ACKNOWLEDGEMENTS

I would like to thank Dr. Jacqualyn Eales for the technical assistance in the preparation of the samples, for analyses and for determining the various nutrients. Thanks are also due to Zewge Teklehaimanot for transporting the fresh fruits from Kenya to Wales, UK, and Peter Njeru, a field assistant in Mbeere who did the actual fruit collection on the farms. Leverhulme Trust (UK) funded the research through 'Improved Management and Utilisation of Eastern Africa Indigenous Fruit Trees' Project' with additional funds from the Government of Kenya through Kenya Forestry Research Institute.

REFERENCES

Ames BN (1966). Assay of inorganic phosphate, total phosphate and phosphatases. In: Colowick, S.P. and Kaplan, N.O., Editors, 1966. Methods Enzymol. 8:115-118.

Beentje HJ (1994). Kenya trees, shrubs and lianas. Nairobi: National Museums of Kenya 9:722.

Dancey CP, Reidy J (2007). Statistics without maths for Psychology. Prentice Hall. p. 619.

Dhyani D, Maikhuri RK, Rao KS, Kumar L, Purohit VK, Sundriya M, Saxena KG (2007). Basic nutritional attributes of *Hippophae rhamnoides* (Seabuckthorn) populations from Uttarakhand Himalaya, India. Cur. Sci. 92(8):1148-1152.

FAO (2003). Food energy – methods of analysis and conversion factors. Food and nutrition paper No. 77, Rome.

Foster MS (2008). Freeze-Frame Fruit Selection by Birds. Wil. J. Ornithol. 120(4):901-905.

Ishida H, Suzuno H, Sugiyama N, Innami S, Todokoro T, Maekawa A (2000). Nutritional evaluation of chemical component of leaves stalks and stems of sweet potatoes (*Ipomoea batatas poir*). Food Chemist. 68:359-367.

Kimondo JM (2010). The potential for optimization of Vitex payos as a dryland resource in Kenya. A PhD Thesis, University of Wales, Bangor, UK.

Ladeji O, Okoye ZSC (1993). Chemical analysis of the fruit of Vitex doniana (Verbenaceae). J. Sci. Food Agric. 63:483-484.

Leakey RRB, Shackleton SE, du Plessis P, Pate K, Lombard C (2002). Characterization of phenotypic variation in marula (*Sclerocarya birrea*) fruits, nuts and kernels in South Africa and Namibia. Winners and Losers, Final Technical Report, 2003. Project R7795, Forestry Research Programme.

Leakey RRB, Shackleton SE, du Plessis P (2005). Domestication potential of Marula (*Sclerocarya birrea* subsp. *caffra*) in South Africa and Namibia: 1. Phenotypic variation in fruit traits. Agrofor. Syst. 64:25-35.

Maundu PM, Ngugi GW, Kabuye CHS (1999). Traditional food plants of Kenya. English Press Ltd. Nairobi, Kenya. National Museums of Kenya.

Mbabu P, Wekesa L (2004). Status of indigenous fruits in Kenya. In: Chikamai B, Eyog-Matig O, Mbogga M (eds.). Review and appraisal on the status of indigenous fruits in Eastern Africa. A report prepared for IPGRI-SAFORGEN in the framework of AFREA/FORNESSA.

Mbora A, Jamnadass R, Lillesø J-PB (2008). Growing high priority fruits and nuts in Kenya: Uses and management. Nairobi. The World Agroforestry Centre. p. 61.

McCleary BV (2003). Dietary fibre analysis. Proc. Nutrition Soc. 62:3-9.

Montiel-Herrera M, Camacho-Hernandez IL, Rios-Morgan A, Delgado-Vargas F (2004). Partial physicochemical and nutritional characterization of the fruit of *Vitex mollis* (Verbenaceae). J. Food Comp. Anal. 17:205-215.

Muok B (2001). Experiences on the domestication of indigenous fruit trees in drylands. Proceedings of the regional Social Forestry Extension Seminar for Semi arid and Arid Areas. KEFRI Hqs, Muguga, Kenya. 24-27 October 2001.

Ondachi PW (2002). Nutritional studies of indigenous fruit trees in support of conservation. Capacity building in Forestry. Annual Research Report 1999-2001. KEFRI, Nairobi. p. 21.

Osman MA (2004). Chemical and nutrient analysis of baobab (*Adansonia digitata*) fruit and seed protein solubility. Plant Foods Hum. Nutr. 59:29–33.

Osborne CG, McTyre RB, Dudek J, Roche KE, Scheuplein R, Silverstein B, Weinberg MS, Salkeld AA (1996). Evidence for the relationship of calcium to blood pressure. Nutr. Rev. 54(12):365-381.

Palgrave KC (2002). Trees of Southern Africa. Struik Publishers Cape Town. p. 959.

Saka JDK (1995). The nutritional value of edible fruits: In Maghembe JA, Ntupanyama Y, Chirwa PW (eds). Improvement of indigenous fruit trees of Miombo woodlands of Southern Africa. Proceedings of a Conference held 23-27 January 1994 Mangochi, Malawi ICRAF Nairobi. pp. 50-57.

Saka JDK, Msonthi JD (1994). Nutritional value of edible fruits of indigenous wild trees in Malawi. For. Ecol. Manag. 64(2-3):245-248.

Shells ME, Young VR (1987). Modern nutrition in health and diseases. Lea & Febiger. Philadephia. PA. pp. 1487-1497.

Swai R, Kimata E (2005). Processing and value adding of indigenous fruit tree products of drylands of East and Central Africa. In: Simitu, P. (ed). Utilisation and commercialisation of drylands indigenous fruit tree species to improve livelihood in ECA. Regional workshop held in Kitui June 20-24. pp. 76-98.

Turan M, Kordali S, Zengin H, Dursun A, Sezen Y (2003). Macro and Micro Mineral Content of Some Wild Edible Leaves Consumed in Eastern Anatolia. Acta Agric. Scand. 53(3):129-137.

Umar KJ, Hassan LG, Ado Y (2007). Minerals composition of *Detarium microcarpum* grown in Kwatarkwashi, Zamfara state, Nigeria 1(2):43-48.

USDA (2001). Composition of Foods Raw, Processed, Prepared USDA Nutrient Database for Standard Reference, Release 14. Agricultural Research Service.

Physicochemical characteristics of Jamun fruit

Muhammad Shahnawaz[1]* and Saghir Ahmed Sheikh[2]

[1]Department of Food Technology, Karakoram International University, Gilgit Pakistan.
[2]Institute of Food Science and Technology, Sindh Agriculture University Tandojam, Pakistan.

Jamun fruit (*Syzygium Cuminii* L*)* processing in Pakistan is uncommon or still traditional methods are used during its processing and preservation. It becomes imperative to characterize the fruit cultivars with a view to understand the properties that may affect the design to handle its processing. This study was carried out to present basic principles of physical properties of Jamun fruit in order to facilitate its handling and processing. The morphological parameters includes weight, volume, length, diameter, shape, color, firmness/softness, edible and non-edible contents, specific gravity, juice and seed contents. Two prominent cultivars of Jamun, that is, (V1) improved and (V2) indigenous were analyzed. The improved cultivar was found superior in all parameters analyzed whereas indigenous cultivar was found substandard except seed portion which was more in it. The weight, length, width and volume of V1 was determined as 9.55 g, 3.88 cm, 2.98 cm and 7.60 ml whereas V2 was determined as 6.71 g, 2.73 cm, 2.10 cm, and 5.33 ml respectively. Likewise, edible portion was 69.10% whereas non-edible portion was 30.90% in V1. In case of V2, edible portion was determined as 39.19% whereas non-edible portion was 60.81%. These few parameters indicate that V1 is comparatively better than V2. These research findings would be beneficial references for processors, post harvest practitioners and fruit exporters.

Key words: Jamun fruit, varieties, physical and chemical properties, Sindh, Pakistan.

INTRODUCTION

Physical components of fruits refer to those fruits, whose chemical procedure determination is not almost needed. For determining most of the physical components, intact fruits are used. However, in some cases like determination of juice content, pulp color, inedible part, seed content etc, fruit samples are required to be cut into pieces for onward processing. Jamun fruit is a berry consisting of a single seed surrounded by a flashy pulp and the fruit skin. Jamun fruit are unique in that they constitute a set of properties and characteristics, which distinguish them from all major fruits. Jamun have significance as a stable food as well as an ornamental fruit plant, whilst their use in Jamun products and industrial applications could be increased. According to variety and growth conditions, Jamun fruit vary in shape, size and weight. Usually they are elliptical and ovoid though certain varieties may reach a near round shape.

Agricultural materials and food products have several unique characteristics which set them apart from engineering materials. Design of machines and process to harvest, handle and store agricultural materials and to convert these materials into food and feed requires an understanding of their physical properties (Stroshine, 1998). Physical properties are often required for the development of post harvest techniques to make them a value-added product. The physical properties of fruits are important in designing and fabricating for handling, transporting, processing and storage, and also for assessing the behavior of the product quality (Kashaninejad et al., 2006; Bart-Plange and Baryeh, 2003). Size and shape are often used when describing grains, seeds, fruits and vegetables. Shape and physical dimensions are important in screening solids to separate foreign materials and in sorting and sizing of fruits and vegetables. The product shape can be determined in terms of its sphericity and aspect ratio which affect the flow ability characteristics of the products. Size and shape determine how many fruits can be placed in shipping containers or plastic bags of a given size. The size such as (length, width, thickness and geometric mean diameter) and shape are also important in designing separating, harvesting and sizing etc. Quality differences in fruits, vegetables, grain and seeds can

*Corresponding author: E-mail. nawaz_glt@yahoo.com.

often be detected by differences in density. When fruits and vegetables are transported hydraulically, the design fluid velocities are related to both density and shape. Volumes and surface areas of solids must be known for accurate modeling of heat and mass transfer during cooling and drying. The porosity, which is the percentage of airspace in particulate solids, affects the resistance to air flow through bulk solids. Airflow resistance, in turn, affects the performance of systems designed for force convection drying of bulk solids and aeration systems used to control the temperature of stored bulk solids. Knowledge of frictional properties is needed for the designing of handling equipment. Physical properties can be used when designing and sizing machine components (Stroshine, 1998). Also the information is useful in equipment such as cleaning, sorting, grading and kernel removing. Likewise, moisture content is useful information in the drying or dehydration processes either the whole fruit or its byproducts. Fruit part fraction gives an overall idea about the composition of whole fruit and its seed which affect the yield and quality of the product. The fruit length, width and thickness etc. have also been studied for the purpose of estimating further processing (Garnayak et al., 2008; Sirisomboon et al., 2007 and Visvanathan et al., 1996).

The importance of these dimensions in determining aperture sizes and other parameters in machine design have been discussed by Mohsenin (1980) and highlighted lately by Omobuwajo et al. (1999).

Several other researchers have conducted experiments to find the physical properties of various fruits and crops. Paksoy and Aydin (2004) estimated some physical properties of squash seeds at different moisture content (6.4 to 52.9%). The role of moisture content was also studied. Owolarafe and Shotonde (2004) determined some physical properties for okra fruit at a moisture content of 11.42% (wet basis). Akar and Aydin (2005) evaluated some physical properties of gumbo fruit varieties as functions of moisture content. Karababa (2006) evaluated physical properties of popcorn kernels as a function of kernel moisture content, varying from 8.95 to 17.12% (dry basis). Kashaninejad et al. (2006) determined some physical and aerodynamic properties of pistachio nut and its kernel in order to design processing equipment and facilities. Several physical properties of pistachio nut and its kernel were evaluated as a function of moisture content in the range of 4.10 to 38.10% (wet basis). Topuz et al. (2005) determined and compared several properties of four orange varieties. Jahromi et al. (2007) determined dimensions and projected areas of date (Barhi variety) by image processing technique. Also many studies have been reported on the physical properties of agricultural crops.

There may in fact be a large number of physicochemical components of Jamun fruits that are worthy to test however, in the present study, only a few but important properties have been evaluated which are more commonly considered for their greater applicability in assessing quality, nutritive value and acceptance to consumers with the objective to determine the physical properties of jamun fruit and consider the appropriate technologies for onward processing.

METHODOLOGY

Jamun fruit was procured for the study from farmer's field near to Sindh Agriculture University, Tandojam, Pakistan. Two common cultivars typically grown in the Sindh climatic condition were selected for the studies which were chosen (V1)-Improved and (V2)-Indigenous. V1 is a newly developed cultivar in the recent years whereas V2 is a wild variety found since many decades. The sample fruits were cleaned manually to remove all foreign materials such as dust, dirt, immature fruits etc. The fruit and seed material was divided into 2 lots each and 20 samples were selected at random from each lot of fruit and seed to obtain samples for conducting the experiment. Hence, measurements of all size and shape indices as well as the fruit mass and seed mass were made thrice. As a whole, a summary of the results of determined physical and chemical parameters of fruit and seeds are collectively shown in Table 1.

Weight

The weight of a selected sample of fruits can however, be determined easily in any standard physical balance. As the Jamun fruit consists of two parts, namely, the outermost part that is edible matter and the inner portion that is the seed fraction (non-edible). The balance was calculated before measuring the fruit weight. The randomly selected representative Jamun fruit samples were placed in a pre-weighed pan then to the digital balance. The mass unit of fruit was determined by means of a digital electronic balance (Shimadzu, Japan) having an accuracy of 0.0001 to 200 g. The weigh measurements of the two fractions were made thrice to get mean values. To get correct weight it was ensured before use that the balance is properly set, placed and leveled. The weighing was done accurately and the reading was noted carefully.

Volume

The Jamun fruit volume was measured through measuring volume of displaced water directly in measuring cylinder. A measuring cylinder which was properly graduated and transparent was used keeping in mind that the representative sample of fruits could be easily and freely introduced in the cylindrical space. Before use, the measuring cylinder was thoroughly and repeatedly washed with clean running water. The clean water was then poured in it. By doing so all possible care was taken so that no air-bubble could gets trapped in the water poured. In this connection, the water was poured slowly, holding the cylinder in slanting manner. The bubbles made were expelled out by striking the outer wall of the cylinder lightly with finger. Before introducing the sample fruit in the cylinder, the poured water was left to stand for a while and the excess water was taken out by pippetting to such an extent that the water left in the cylinder comes to a known volume and this was carefully noted and this was taken as initial volume of water and denoted as (A). Then the fruit samples were slowly and carefully placed inside the cylinder containing water so that the already measured water could not splash and spill outside the cylinder. When the entire sample fruits were sunk in water, they were left to stand for a while and termed it final volume of water and denoted as (B). The total numbers of fruits were denoted as (N). Hence, the average volume of a fruit was calculated by dividing the differences in volume with the number of fruits by the following formula:

$$AV = \frac{B - A}{N}$$

Specific gravity

The relative heaviness, measured as specific gravity value is an important character of a fruit. Like other fruits, Jamun fruit's specific gravity value is also considered to be a guiding index to judge their physiological maturity. The specific gravity value is related to season of their development, maturation and ripening. The specific gravity of fruit can also be determined with the help of a hydrometer. However, in the present study, the specific gravity values of Jamun fruit samples were determined by the following formula:

$$\text{Specific gravity value} = \frac{\text{Weight of the fruit}}{\text{Volume of the fruit}}$$

Overall length

The fruit size, in terms of the two principal axial dimensions, that is length and width was measured using a Vernier-caliper with an accuracy of 0.0 to150 mm (Shanghai, China). The selected batch of Jamun fruit was measured provided that it is of such length as can be suitably held between the jaws of the calipers. By measuring the overall length, care was taken to hold the fruit in correct position, ensured that it's uppermost and the lowermost ends were in contact respectively with two jaws of the instrument.

Maximum width

To measure the width, the diameter of a fruit in its centre is emphasized. In fact, in various fruits the cross sectional diameter is the same as the maximum width; however, this is not the case for all fruits especially in Jamun fruit. Therefore, the Jamun fruit samples were considered the total width measuring the distance from the extreme points at two sides. Hence, the width of Jamun was measured in similar way as its length was measured through the same Vernier caliper.

Fruit shape

Visual observation was made in case of Jamun which is the easiest method and beyond the doubts and practical method to judge the shape of a fruit. However, a standard chart having various types of shapes commonly found in fruits was used to find the accurate shape of Jamun both cultivars, that is (V1) improved and (V2) indigenous.

Firmness/softness

The existence of firmness in Jamun fruit was measured by the following two methods:

Finger prenure/by hand-feeling: This is an easy approach where a person holds the fruit in palm and feels it by giving the desired pressure. No doubt the method is crude but practical and may be applied to have gross ideas especially when relative degree of firmness among different samples of maturation. In order to acquaint much accurate data, a panel of persons was employed instead of one person tester. The judges were M.Sc student and the faculty members of the Institute of Food Sciences and Technology, comprising both male and female. Those were divided into several groups and each group (five members) was asked to test independently by the hand feeling process. The degree of firmness or softness as assessed by each group was expressed by appropriate terms such as very hard, medium had, hard to soft, moderately soft, soft and soft to very soft and extremely soft which was already communicated to the testers before starting the test. The data obtained from each group of tester was considered together and made average to get conclusion.

By pin-drop method: This is also a crude method and may be of use to determine the relative level of softness of fruits. However, this method is applicably used for those fruits which are soft or moderately soft in texture like Jamun. To work, about 5 g sample fruit was attached at the head of a pin. The length of pin was measured 1 cm. The pin having weight at the head was left to drop vertically and freely over the surface of the Jamun fruit from 1 inch height. The length of pin penetrated into the fruit was measured. To interacting effect of air on free drop of the pin was averted by carrying out the test in a closed chamber. On the basis of penetration of pin into the fruit surface, the degree of softness was concluded (Table 1).

Color (visual estimation)

The same panel of judges who were employed to determine firmness was again asked to identify color of Jamun fruit as they were observed and narrated the color under specific terms which was already explained to them before they had started judging. Visual observations were made for determination of color and were compared with the standard color charts. The conclusion was drawn making average of the reports given by the group of judges making a comparison with a standard color dictionary.

Percentage consumable matter (pulp)

In this regard, the selected sample of Jamun fruit was first weighed. Fruits were cut to separate the non-consumable part (seed) from the consumable portion (pulp). The remaining consumable matter which adhered to the separated non-consumable part was brought back to the former part by careful scratching. Then the consumable matter was weighed. The percentage of consumable matter was calculated by the following formula:

$$\text{Percent consumable matter} = \frac{\text{Weight of consumable matter (g)}}{\text{Weight of whole fruits (g)}} \times 100$$

Percent non-edible contents (Seed)

Jamun seeds were separated from fruit pulp through a process described in pulp separation technique The pulp that adhered to the seeds was removed by washing. After washing, the water was wiped out from their surfaces with a piece of cleaned cloth and tissue paper. The seeds were then weighed and calculated asfollow to determine the percentage of seed composition:

$$\text{Percentage seed} = \frac{\text{Weight of seed matter (g)}}{\text{Weight of whole fruits (g)}} \times 100$$

Percent Juice content

After weighing the fruit, the consumable matter from them was completely taken out. A known volume of distilled water was added to that and minced thoroughly in a blender and the blender was run till blending of whole pulp was accomplished. The ground material was filtered through muslin clothe with many folds to separate the pulp pieces out completely. The volume of the filtrate was then measured in measuring cylinder. The percentage of juice in the given sample of Jamun fruit was worked out as follows:

$$\text{Percent Juice} = \frac{\text{Volume of the filtrate (ml) - Volume of distilled water used (ml)}}{\text{Weight of fruit (g)}} \times 100$$

Determination of % moisture

About 10 g of pulp sample was taken in a pre-weighed flat bottomed dish. Both empty dish and fruit sample weights were noted separately. The sample loaded dish was kept in an oven by adjusting its temperature at 100°C. Dish was removed after 4 h, cooled in the desiccators and reweighed. The dish was again placed in the oven for 3 h and was reweighed after cooling. This practice was repeated till the constant reading was obtained. The percentage moisture content was determined by the following formula:

$$\text{Moisture (\%)} = \frac{\text{Weight of (Fresh sample – Sample after drying)}}{\text{Weight of fresh sample}} \times 100$$

Determination of total solids

Total solids were estimated by deducting percentage moisture from 100 as described by James (1995).

% Total solids = 100 - % moisture

Determination of titrable acidity

Titrable acidity was determined according to the method of AOAC (2000). Each sample of the products was treated with 10 N NaOH solution using titration kit. Where 3 to 5 drops of phenolphthalein were used as an indicator. The acidity was determined in terms of lactic acid. The volume of alkali used was noted and calculated using the following formula;

$$\text{\% Titrable acidity} = \frac{\text{Normality of NaOH} \times 0.009}{\text{Weight of sample (g)}} \times 100$$

Determination of TSS

The total soluble solids (TSS) were determined as per method described by Mazumdar and Majumder (2003) using Digital-Bench-Refrectometer. Before use, the instrument was cleaned and adjusted to zero at 20°C using distilled water. An appropriate quantity of each sample was placed on the prism-plate of the refrectometer with the help of a glass rod and folding back the cover. For each sample, the instrument was calibrated by using distilled water. The reading appeared on the screen was directly recorded as total soluble solids.

Determination of pH

For determination of pH (hydrogen ion concentration) in the Jamun fruit, a method of AOAC (2000) was adopted. A digital pH meter was used. Samples solution was taken in the beaker and directly inserted the electrode into the solution. When the first reading was completed, the electrode was wiped with distilled water and dried-up with tissue paper. Hence, as a continue series, all other samples were determined accordingly.

RESULTS AND DISCUSSION

According to shape index, the Jamun fruit of V1 was concluded to be "elliptical" whereas V2 was imagined "ovoid/round shape". However, the color of V1 was recorded "reddish-purple" whereas V2 was "blackish-purple" (Table 1). The study of the shape as a physical feature of a fruit has its importance to determine whether it conforms to the standard and any deviation that may result due to interacting factors like environment, biotic agencies or application of agro-inputs to the trees bearing that fruit. Fruits belonged to a species or a variety of it has some characteristic shape of their own although variations within some limit are not considered to be an uncommon feature. The fruits commonly consumable to the human beings are found to have number of shapes (Omobuwajo et al., 1999). The selected samples of Jamun fruit were examined to determined its softness hence, V1 was ranked as "soft-to-very-soft "and V2 were "moderately-soft" in firmness (Table 1). Owolarafe and Shotonde (2004) described that firmness of a healthy fruit is linked to the degree of its physiological maturation. With progress of development, maturation and ripening either in the pre-harvest or in the post harvest condition; the fruit undergo gradual softness to a greater or a lesser extent depending on species, varieties, environment and the use of agro inputs. Chemically, the changeover from firm to soft condition of a fruit is largely due to enzymatic conversion of pectic-compounds present in them as component of cell walls. These complex polysaccharides, which are long-chain polymers of galacturonate monomers gradually transform from water-insoluble to the more and more soluble forms, that is, from protopectin to pectic acid in the fruits as they undergo development, maturation and ripening. Due to this solubilization process of pectic polymers, the fruits attain more and more softness (Stroshin, 1998).

Size of Jamun fruit cultivars found varied from 3.88 to 2.73 cm in length and 2.98 to 2.10 cm in geometric mean diameter/width in V1 and V2 respectively. The overall weight of Jamun fruit was observed 9.55 and 6.75 g for V1 and V2 respectively (Table 1). An important qualitative character of a fruit is the length that it measures. For fruits of some species including Jamun, maturity for harvest is also judged in considering the length attained by them. In the market, the consumers are even found to select out relatively long sized fruits especially for those of Jamun despite the fact it is not certain whether a

Table1. Determination of physicochemical characteristics of Jamun fruit cultivars of Sindh (mean measurements/values).

S/No.	Parameter	Improved (V1)	Indigenous (V2)
01	Shape index	Elliptical	Ovoid/round
02	Color	Reddish-purple	Blackish-purple
03	Firmness/softness	Soft to very soft	Moderately soft
04	Weight of fruit (gm)	9.55±0.685	6.71±0.520
05	Length of fruit (cm)	3.88±0.224	2.73±0.174
06	Width/diameter (cm)	2.98±0.361	2.10±0.202
07	Volume of fruit (ml)	7.60±0.361	5.33±0.176
08	Specific gravity (value)	1.26±0.120	1.25±0.048
09	Edible matter/pulp (%)	6.6 g (69.10±6.240)	2.63 g (39.19±0.416)
10	Non-edible matter/seed (%)	2.95 g (30.90±0.327)	4.08 g (60.81±0.485)
11	Length of seed (cm)	0.49±0.047	1.60±0.058
12	Width/diameter of seed	0.21±0.033	0.70±0.025
13	Juice (%)	2.11 ml (32±1.155)	0.99 ml (38±1.155)
14	Total Soluble Solids (TSS) deg Brix	13.75±0.501	15.82±0.505
15	Titrable acidity (%)	1.26±0.031	1.58±0.021
16	pH (value)	3.87±0.010	3.77±0.010
17	Moisture content in fruit (%)	81.32±0.203	80.14±0.087
18	Moisture content in seed (%)	13.31±0.262	12.34±0.021

longer fruit have any impact on its quality. Kashaninejad et al. (2006), Bart-Plange and Baryeh (2003) in their studies pointed out that width of a fruit, that is, how much it has grown across is a measure which is perhaps of greater importance than the lengthwise growth of it, in order to judge its normal and healthy growth. In some countries, the maturity for harvest of many fruits is also assessed by observing their girth-wise growth. For example, in the fruit of Jamun, width is some times considered to determine their harvest-maturity condition. The practice lies in testing whether the fruit passes or not through a ring of desired diameter. If the fruit passes out, it is considered not to have fully grown till then for which it is left to grow further instead of harvesting at that stage. Keramat et al. (2007) describe that determination of weight of fruit samples needs no emphasis in any analytical study on fruits. A part from estimating yield of a tree, weight of a fruit is considered to be an important factor in judging its compactness, maturity, juice content, levels of chemical constituents. The weight of fruit also determines its acceptance to consumers and thereby the market price of it.

The whole fruit specific gravity value was found to 1.25 in V1 and 1.26 in V2. Garnayak et al. (2008), Sirisomboon et al. (2007) and Visvanathan et al. (1996) agreed to say that volume of fruit is another important factor like that of its weight in determining quality. In the market, the consumers are also found to have a preference to select the large sized ones for many fruit including Jamun. And accordingly the price of those fruits goes higher with size. Results showed that volume varied from 5.60 to 5.33 ml in V1 and V2 respectively. To what proportion the human consumable matter is actually present in a fruit is, in fact, of practical concern to a food scientist. This varies greatly according to species, varieties under them, rootstocks used, edapho-climatic conditions of growing, agro inputs applied and so on. Thus, to a fruit scientist, the yield as human consumable matter of a fruit crop per tree or per unit area of plantation is of actual concern than the total yield of fruit. The edible matter (pulp) in V1 and V2 was 69.10% (6.6 g) and 39.19% (2.63 g) respectively. Whereas non-edible portion (seed) was 30.19% (2.95 g) in V1 and 60.81% (4.08 g) in V2 was determined. Dutta et al. (1988) stated that presence of seeds in a fruit is considered a merit or demerit. Seeds are valued at least in some fruits for certain purposes like Jamum fruit. Experimental evidence provides proof to the fact that the seedless Jamun fruits tend to be less sweet than the seeded fruits. Hence, Jamun seeds can be used for many purposes i.e. medicinal, nutritional and food byproducts etc. Also the juice percent was obtained as 32 and 38% in V1 and V2 respectively.

The chemical properties of Jamun fruit cultivars were also analyzed. Accordingly, the average dry-basis moisture content of whole Jamun fruit was found 81.32 and 80.14% in V1 and V2. Whereas, the moisture content in seeds were found 13.31 and 12.34% in V1 and V2 respectively. TSS varied as 13.75 and 15.82% in refractory index for V1 and V2 respectively (Table 1). These results are in agreement with many other researchers on Jamun aqueous extract as described in detail by Dahot et al. (1986, 1989). By describing the role of chemical constituents including moisture content, many researchers carried out experiments on various fruits and crops. Paksoy and Aydin (2004) estimated squash seeds at different moisture content (6.4 to 52.9%). Owolarafe and Shotonde (2004) determined moisture content of Okra fruit at 11.42% along with some other physical characteristics. Akar and Aydin (2005) evaluated some physical properties of gumbo fruit varieties. Karababa (2006) evaluated physical properties of popcorn kernels. Kashaninejad et al. (2006) determined some physical and aerodynamic properties of pistachio nut and its kernel.

The pH and acidity for V1 and V2 was found 3.87 and 3.77, 1.26 and 1.58% respectively. The pH values revealed that fresh extract of Jamun fruit is highly acidic and may be responsible for astringency in taste. Noomrio and Dahot. (1996) authenticated the same view point on pH of Jamun fruit. The other researchers reported that the pH value of Jamun fruit pulp is higher than other fruits like Carisa carindas 2.9 (Hasnain and Ali, 1990). Hasnain and Ali (1990) further reported that lower pH of sample is favorable for higher shelf life. Total solids are measure of the amount of material dissolved in water. This material can include carbonate, bicarbonate, chloride, sulfate, phosphate, nitrate, calcium, magnesium, sodium, organic ions, and other ions (American Public Health Association, 1998). Thus these properties may be useful in order to separate fruit and seed, processing, design equipment and facilities.

CONCLUSION

The aim of this study was to investigate the physical properties of Eugenia Jambolana fruits and seeds grown in Sindh, Pakistan. The parameters studied include size, shape, color, fruit part fraction (edible and non-edible), geometric mean diameter, specific gravity, percentage juice ratio, volume, moisture content, TSS, pH and acidity etc. The physical properties of Jamun fruit were evaluated in order to optimize better the designing of tools, equipment and machines and facilitate the overall mechanism for their processing, preservation and commercialization.

RECOMMENDATION

The present study was specific on few parameters and cultivars for physicochemical determinations of Jamun fruit grown only in Sindh region, however, the Jamun cultivars of other regions may be evaluated further with additional parameters.

REFERENCES

Akar R, Aydin C (2005). Some physical properties of gumbo fruit varieties. J. Food Eng., 66: 387-393.

American Public Health Association (1998). Standard methods for the examination of water and wastewater. 20th edition. pp. 127-131.

Bart-Plange A, Baryeh EA (2003). The physical properties of category B cocoa beans. J. Food Eng., 60: 219-227.

Dahot MU, Memon AN, Khand FD (1989). Studies on the constituents of Acacia arabia premature pods. Pak. J. Sci. Ind. Res., 32: 405-407.

Dahot MU, Memon AN, Ali SA (1986). Estimation of water soluble vitamins from fruits. J. Pharm. Uni. Kar., 5: 25-27.

Dutta SK, Nema VK, Bhardwaj RK (1988). Physical properties of grain. J. Agric. Eng. Res., 39: 259-268.

Garnayak DK, Pradhan RC, Naik SN, Bhatnagar N (2008). Moisture-dependent physical properties of jatropha seed (*Jatropha curcas L.*). Ind. Crops Prod., 27:123-129.

Hasnain A Ali R (1990). Studies on amino acids of Carissa carandas. Pak. J. Sci. Ind. Res., 33: 318-320.

Jahromi M.K, Rafiee S, Jafari A Tabatabaeefar A (2007). Determination of dimension and area properties of date (Barhi) by image analysis. International Conference on Agricultural, Food and Biological Engineering and Post Harvest Production Technology, Khon Kaen, 21-24 January, Thailand.

James CS (1995). Analytical chemistry of food. Seale-Hayne Faculty of Agriculture, Food and Land use, Department of Agriculture and Food studies, University of Polymouth, UK., 1: 96-97.

Kabas O, Ozmerzi A, Akinci I (2006). Physical properties of cactus pear (Opuntia ficus India L.) grown wild in Turkey. J. Food Eng., 73: 198-202.

Karababa E (2006). Physical properties of popcorn kernels. J. Food Eng., 72: 100-107.

Kashaninejad M Mortazavi A, Safekordi A. Tabil LG (2006). Some physical properties of Pistachio (*Pistacia vera* L.) nut and its kernel. J. Food Eng., 72 30-38.

Mohsenin NN (1980). Physical Properties of Plant and Animal Materials. Gordon and Breach Science. Publishing, New York.

Noomrio MH, Umar Dahot M (1996). Nutritive value of *Eugenia jambosa* fruit. J. Islamic Acad. Sci., 9 (1): 9-12.

Omobuwajo TO Akande EA Sanni LA (1999). Selected physical, mechanical and aerodynamic properties of African breadfruit (*Treculia africana*) seeds. J. Food Eng., 40: 241-244.

Owolarafe OK, Shotonde HO (2004). Some physical properties of fresh okro fruit. J. Food Eng., 63: 299-302.

Owolarafe OK, Olabigeand TM, Faborode MO (2007). Macro-structural characterisation of palm fruit at different processing conditions. J. Food Eng., 78 1228-1232.

Paksoy M, Aydin C (2004). Some physical properties of edible squash (Cucurbita pepo L.) seeds. J. Food Eng., 65: 225-231.

Sirisomboon P, Kitchaiya P, Pholpho T, Mahuttanyavanitch W (2007). Physical and mechanical properties of *Jatropha curcas* L. fruits, nuts and kernels. Biosys. Eng., 97(2): 201-207.

Stroshine R (1998). Physical Properties of Agricultural Materials and Food Products. Course manual. Purdue Univ. USA.

Topuz A, Topakci M, Canakci Akinci I, Ozdemir F (2005). Physical and nutritional properties of four orange varieties. J. Food Eng., 66: 519-523.

Visvanathan R, Palanisamy PT, Gothandapani L, Sreenarayanan VV (1996). Physical properties of neem nut. J. Agric. Eng. Res., 63: 19-26.

21

Anticandidal screening and antibacterial of *Citrullus colocynthis* in South East of Iran

Rasool khatibi and Jahanbakhsh Teymorri

Department of Rangeland and Watershed Management, Faculty of Natural Resources, University of Zabol, 98615, Zabol, Iran.

Resistance to current antibacterial drugs and the rise of opportunistic fungal infections are growing global concerns. Traditional medicine is a potential source of new antibacterials and antifungals. *Citrullus colocynthis* Schrad (Cucurbitaceae) endemic in Southern Iran is used in folk medicine against dermatological, gynaecological and pulmonary infections. To assess *in vitro* antibacterial and anticandidal activity of aqueous and diluted acetone extracts of *C. colocynthis* Schrad. MIC and MBC/MFC were determined for plant organs at different maturation stages. *C. colocynthis* Schrad. was harvested and its identification was verified. Aqueous and diluted acetone extracts (from the plant's roots, stems, leaves and three maturation stages of its fruit and seeds) were screened for activity against Gram-negative and Gram-positive bacteria (*Escherichia coli*, *Pseudomonas aeruginosa*, *Staphylococcus aureus* and *Enterococcus faecalis*) and various *Candida* spp. (*Candida glabrata*, *Candida albicans*, *Candida parapsilosis* and *Candida kreusei*). All extracts showed activity against all strains. The highest MICs and MBCs/MFCs were obtained from the fruit aqueous extracts (MIC 0.10 mg/ml against *C. albicans* and *C. glabrata*, 0.20 mg/ml against *E. coli* and *P. aeruginosa*), lowest activity from the root extracts. *C. colocynthis* Schrad shows antibacterial and anticandidal properties. The folk medicinal use as a broad-spectrum antimicrobial agent is validated.

Key words: *Citrullus colocynthis* Schrad, aqueous extract, acetone extract, antibacterial, anticandidal, Iran.

INTRODUCTION

The use of plants for medicinal purposes has been practiced for many centuries by a substantial proportion of Iran's population. Interest in ethnopharmacy as a source of pharmacologically active compounds has increased worldwide, particularly in the search for drugs to counter multi-resistant microorganisms. Additionally, in some developing countries, plants are the main medicinal source to treat infectious diseases due to economic conditions and availability. However, only approximately 20% of the plants found in the world have been submitted to pharmacological or biological testing, despite the substantial number of new antibiotics derived from natural or semi-synthetic resources being introduced on the market (Mothana and Lindequist, 2005). *Citrullus colocynthis* Schrad. (Cucurbitaceae), growing in arid areas, is endemic in the South of Tunisia (Pottier-Alapetite, 1981). In Iran, as in the rest of the persian gulf (Mozaffarian, 1964), the parts of plants most often used for medicinal purposes are fruits and/or seeds, though other parts of the plants can be used, for example roots to treat urinary infection (Nadkami, 1954) or leaves (Batanouny, 1999). Traditional healers seem to not pay attention to the plant's degree of maturity. The literature rarely mentions if seeds are present in preparations involving ground fruit/pulp. Modes of preparation and administration vary, even for similar indications. Common preparations use fresh, warmed or dried plant material (often ground), as well as extracts used mostly in a liquid form. Extracts are prepared either in water or in aqueous mixtures containing more lipophilic compounds (hot milk extractions, water/olive oil at various ratios) at a range from room temperature (maceration) to boiling. The use of *C. colocynthis* Shrad oil (expressed from the fruit or

*Corresponding Author E-mail: r_khatibi60@yahoo.com.

Table 1. Extraction yields (w/w%) and phytochemical screening of *Citrullus colocynthis Schrad.*

Variable	Extraction yields		Phytochemical screening of Citrullus colocynthis Schrad.																	
			Alkaloids		Coumarins		Flavonoids		Anthraq.		Cardiac glyc		Iridoids		Saponids		Steroids		Gallic tanins	
	Aq.	Ac.	Aq.	Ac.	Aq.	Ac.	Aq.	Ac.	Aq.	Ac.	Aq.	Ac.	Aq.	Ac.	Aq.	Ac.	Aq.	Ac.	Aq.	Ac.
Roots	4.558	2.630	–	–	–	–	–	–	–	–	–	–	+	–	–	–	+	+	–	–
Leaves	12.887	7.014	+	+	+	+	–	–	–	–	–	–	+	–	–	–	+	+	+	+
Stems	11.047	8.266	+	+	–	–	–	–	+	–	–	–	+	–	–	–	+	+	–	–
Immature fruit	2.758	2.611	+	+	–	–	–	–	–	–	–	–	+	–	–	–	+	+	–	–
Ripening fruit	2.907	2.980	+	+	–	–	–	–	–	–	–	–	+	–	–	–	+	+	–	–
Ripe fruit	3.041	3.075	+	+	–	–	–	–	–	–	–	–	+	–	–	–	+	+	–	–
Immature seed	2.937	3.032	+	+	–	–	+	+	–	–	–	–	–	–	–	–	+	+	–	–
Ripening seed	2.750	2.850	+	+	–	–	+	+	–	–	–	–	–	–	–	–	+	+	–	–
Ripe seed	2.212	2.275	+	+	–	–	+	+	–	–	–	–	–	–	–	–	+	+	–	–

Aq.: aqueous extract; Ac.: diluted acetone extract; Antraq.: anthraquionones; Cardiac glyc.: cardiac glycosides; +: presence; –: absence.

distilled) is not documented in Iran. This is all the more notable that family-sized oil presses and stills (Clevenger-type apparatus) for essential oils are widespread.

Ground plant material can be mixed with honey for ingestion or topical gynaecological application or with other plants for poultices (for example with *Lawsonia inermis* and *Capparis spinosa*). Methods of administration are topical, rectal or vaginal suppositories (fruit), enema, cervico-vaginal douche, and by ingestion (Boukef, 1986; Marzouk, 2008). Extreme caution should be exercised with ingestion, due to the plant's drastic laxative properties, and with contact with leaves, due to risks of syncope (for all mammals including domestic animals) (Marzouk, 2008). Use is contraindicated during pregnancy as the plant is abortifacient (Pottier-Alapetite, 1981; Delazar et al., 2006). Many of today's traditional medicinal uses of the plant are found throughout history (Eber's papyrus in ca.1550 BCE Egypt) (Riddle, 1999) over a large geographical zone from Mauritania to India, even extending outside of the plant endemic zone, to Europe (Adams et al., 2009). The current study measured in vitro antibacterial and anticandidal activities of this plant using the broth serial dilution (microdilution method). Plants were collected and identified. Then, preparation and testing were carried out on the reconstituted lyophilized extracts of aqueous and diluted acetone extracts from the roots, stems and leaves, as well as three different maturation stages of the fruit and seeds.

MATERIALS AND METHODS

Plant material

C. colocynthis Schrad plants were collected in south east of iran in zabol city. The identification was performed according to the flora of Iranica (Ghahraman, 1981) and a voucher specimen (C.C-01.01) deposited in the biological laboratory of the Faculty of Pharmacy of Monastir.

Extraction protocol

The extraction was performed on 100 g of fresh organs: Roots, stems, leaves and three different stages of maturation of fruits and seeds (immature, ripening and ripe). Yields of prepared extracts are given in Table 1.

Aqueous extract

One hundred grams of fresh organs were ground with a mixer and added to 500 ml of distilled water. The mixture was allowed to reflux for 30 min, after which the solution was allowed to cool (4 h at 3°C). The mixture was then filtered using filter paper (Whatman no.1) under the vacuum of a water pump. The filtrate obtained was lyophilized, yielding the lyophilized aqueous extract.

Diluted acetone extract

To 100 g of fresh organs were added 500 ml of an acetone-water mixture (3/4 vol. acetone completed with distilled water). The mixture was ground with a mixer, followed by reflux for 30 min, after which the solution was allowed to cool (24 h at 3°C). A precipitate formed, which was removed by filtering (Whatman paper no.1). The filtrate was concentrated by rotavapory extraction to remove the acetone. A new precipitate appeared in the concentrated liquid and was filtered out. The final filtrate was lyophilized, giving the lyophilized diluted acetone extract.

Table 2. Antibacterial MIC (mg/ml) and MBC (mg/ml) of lyophilized aqueous extracts.

Variable	*E. coli* ATCC 25922		*P. aeruginosa* ATCC 27853		*S. aureus* ATCC 25923		*E. faecalis* ATCC 29212	
	MIC	MBC	MIC	MBC	MIC	MBC	MIC	MBC
Roots	3.25	6.50	3.25	6.50	3.25	6.50	3.25	6.50
Leaves	1.63	3.25	0.81	1.63	1.63	3.25	1.63	3.25
Stem	1.63	3.25	0.81	1.63	0.81	1.63	1.63	3.25
Immature fruit	0.20	0.41	0.23	0.41	0.41	0.81	0.81	1.63
Ripening fruit	0.41	0.41	0.81	1.63	3.25	3.25	0.81	3.25
Ripe fruit	0.20	0.41	1.63	3.25	3.25	3.25	1.63	3.25
Immature seed	0.41	0.81	0.81	1.63	0.81	1.63	0.81	1.63
Ripening seed	0.81	1.63	1.63	3.25	1.63	3.25	1.63	3.25
Ripe seed	0.81	1.63	1.63	3.25	1.63	3.25	3.25	6.50

MIC positive control: Levofloxacin (*E. coli* 0.61 μg/ml, *P. aeruginosa* 0.3 μg/ml, *S. aureus* 0.3 μg/ml, *E. faecalis* 1.22 μg/ml). *E. coli: Echerichia coli; P. aeruginosa: Pseudomonas aeruginosa; S. aureus: Staphylococcus aureus; E. faecalis: Enterococcus faecalis.*

Table 3. Antibacterial MIC (mg/ml) and MBC (mg/ml) of lyophilized acetone extracts.

Variable	*E. coli* ATCC25922		*P. aeruginosa* ATCC 27853		*S. aureus* ATCC2593		*E. faecalis* ATCC29212	
	MIC	MBC	MIC	MBC	MIC	MBC	MIC	MBC
Roots	6.50	ND	6.50	ND	6.50	6.50	6.50	6.50
Leaves	3.25	6.50	0.41	0.81	0.41	0.81	0.41	0.81
Stem	3.25	6.50	0.41	0.81	1.63	3.25	0.81	1.63
Immature fruit	0.41	0.81	0.41	0.40	0.81	1.63	0.81	1.63
Ripening fruit	0.41	0.81	0.81	1.63	0.81	1.63	0.81	1.63
Ripe fruit	1.63	3.25	1.63	3.25	1.63	3.25	1.63	3.25
Immature seed	0.81	1.63	0.81	1.63	1.63	3.25	1.63	3.25
Ripening seed	1.63	3.25	0.81	1.63	1.63	3.25	1.63	3.25
Ripe seed	1.63	3.25	1.63	3.25	3.25	6.50	3.25	6.50

MIC positive control: Levofloxacin (*E. coli* 0.61 μg/ml, *P. aeruginosa* 0.3 μg/ml, *S. aureus* 0.3 μg/ml, *E. faecalis* 1.22 μg/ml). *E. coli: Echerichia coli; P. aeruginosa: Pseudomonas aeruginosa; S. aureus: Staphylococcus aureus; E. faecalis: Enterococcus faecalis.* ND: not determined.

Qualitative phytochemical screening

Each extract was screened for the presence of key families of phytochemicals (Sakar and Tanker, 1991; Trim and Hill, 1952) using the following reagents and chemicals: alkaloids with Dragendorff's reagent confirmed with Bouchardat's (I_2/MgI_2) and with Meyer's reagents ($KI/MgCl_2$), coumarins with diluted NaOH-U.V. test, flavonoids with metallic magnesium and hydrochloric acid (HCl), anthraquinones with Borntrager's reagent, cardiac glycosides with Kedde's reagent (and confirmed with Baljet's reagent), iridoids with diluted hydrochloric acid, saponids for their ability to produce suds, steroids with acetic anhydride and concentrated sulphuric acid (Liebermann reaction), tannins in general with ferric chloride (confirmed with concentrated hydrochloric acid, Bath-Smith reaction) and gallic tannins specifically with Stiasny reagent (Table 1).

Antibacterial and anticandidal activity

Microorganisms

Four reference strains were chosen for antibacterial investigation: Gram-positive cocci (*Enterococcus faecalis* ATCC 29212 and *Staphylococcus aureus* ATCC 25923) and Gram-negative bacilli (*E. coli* ATCC 25922 and *P. aeruginosa* ATCC 27853). The antifungal effect of the various *C. colocynthis* Schrad. extracts was also tested against a range of pathogenic reference yeasts (*Candida albicans* ATCC 90028, *Candida glabrata* ATCC 90030, *Candida kreusei* ATCC 6258 and *Candida parapsilosis* ATCC 22019).

MIC and MBC/MFC determinations

The minimal inhibitory concentration (MIC) preventing visible bacterial or fungal growth was measured by the broth dilution method (microdilution using 96-well microplates), following the procedure of Berche et al. (1991). All extracts stock solution were prepared by dissolution in 10% dimetyl sulfoxyde (DMSO). The plant extracts concentrations tested ranged from 0.10 to 6.50 mg/ml. The MIC of each extract was defined as the lowest concentration which inhibited either bacterial or candidal growth, after incubation at 37°C between 18 and 24 h. The minimal bactericidal concentration (MBC) and the minimal fungicidal concentration (MFC) were determined by subculture on blood agar at 37°C between 18 and 24 h. Levofloxacin was used as

Table 4. Antifngal MIC (mg/ml) and MFC (mg/ml) of lyophilized aqueous extracts.

Variable	*C. albicans* ATCC 90028		*C. glabrata* ATCC 90030		*C. kreusei* ATCC 6258		*C. parapsilosis* ATCC 22019	
	MIC	MFC	MIC	MFC	MIC	MFC	MIC	MFC
Roots	3.25	6.50	3.25	6.50	3.25	6.50	3.25	6.50
Leaves	1.63	3.25	0.81	1.63	1.63	3.25	0.81	1.63
Stem	1.63	3.25	1.63	3.25	1.63	3.25	0.81	1.63
Immature fruit	0.41	0.81	0.41	0.81	0.41	0.41	0.41	0.81
Ripening fruit	0.41	0.41	0.20	0.20	0.20	0.41	0.41	0.41
Ripe fruit	0.20	0.41	0.20	0.20	0.20	0.20	0.20	0.41
Immature seed	0.20	0.41	0.20	0.81	0.81	1.63	0.20	0.41
Ripening seed	0.81	1.63	1.63	3.25	0.81	1.63	0.81	1.63
Ripe seed	0.81	1.63	3.25	6.50	3.25	6.50	0.81	1.63

Positive control with Amphotericin B (MFC 0.5 µg/ml). *C. albicans: Candida albicans; C. glabrata: Candida glabrata; C. kreusei: Candida kreusei; C. parapsilosis: Candida parapsilosis.*

Table 5. Antifungal MIC (mg/ml) and MFC (mg/ml) of lyophilized acetone extracts.

Variable	*C. albicans* ATCC 90028		*C. glabrata* ATCC 90030		*C. kreusei* ATCC 6258		*C. parapsilosis ATCC 22019*	
	MIC	MFC	MIC	MFC	MIC	MFC	MIC	MFC
Roots	6.50	ND	3.25	6.50	6.50	ND	3.25	6.50
Leaves	0.81	1.63	0.81	1.63	0.41	0.81	0.81	1.63
Stem	1.63	3.25	0.20	0.41	3.25	6.50	0.81	1.63
Immature fruit	0.10	0.20	0.10	0.20	0.20	0.41	0.20	0.41
Ripening fruit	0.41	0.81	0.41	0.81	0.20	0.41	0.20	0.41
Ripe fruit	1.63	3.25	0.41	0.81	0.81	1.63	1.63	3.25
Immature seed	0.41	0.81	0.81	1.63	0.41	0.81	0.41	0.81
Ripening seed	0.81	1.63	0.81	1.63	0.81	1.63	0.41	0.81
Ripe seed	1.63	1.63	1.63	3.25	1.63	3.25	0.81	1.63

Positive control with Amphotericin B (MFC 0.5 µg/ml). *C. albicans: Candida albicans; C. glabrata: Candida glabrata; C. kreusei: Candida kreusei; C. parapsilosis: Candida parapsilosis.* ND: not determined.

antibacterial positive control, and Amphotericin B for the anticandidal one. Results are presented in Tables 2, 3, 4 and 5 with the highest activities for each type of extraction in various shades of grey and the two most noticeable antibacterial and anticandidal activities against different strains are shown in bold face.

RESULTS

Extraction yields

Leaves and stems have significantly much higher extraction yields than other plants parts (both for aqueous and acetone extraction). Aqueous extraction gave higher yields than the acetone extraction for roots, leaves and stems (Table 1). The extraction yields were close for both methods with immature fruits, but acetone extraction gave better yields for ripening and ripe fruits and seeds in all their stages.

Qualitative phytochemical screening

Results (reported in Table 1) were dependent upon the extraction method: aqueous extracts contained iridoids and anthraquinones not detected in diluted acetonic ones. Differences were also noted between plant organs: Alkaloids were found in all extracts except the roots, flavonoids were present only in seeds, gallic tannins and coumarins only in leaves, anthraquinones only in (aqueous) stems extracts and iridoids in all aqueous extracts but the ones from seeds. However, no extract contained saponids and all of them contained steroids.

Antibacterial activity

All the extracts tested from Tunisian *C. colocynthis* Schrad showed antibacterial activity against all tested strains. MIC and MBC were tested for concentrations ranging from 0.10 to 6.50 mg/ml (Tables 2 and 3) and all plant organs exhibited antibacterial activity against all tested strains. Gram-negative bacilli were more sensitive than the Gram-positive. The plant organs with the highest antibacterial properties were immature fruits and immature seeds (in both cases their aqueous as well as their diluted acetone extracts). The most active organs of all were immature fruits. The strongest inhibitions were obtained against *E. coli* and *P. aeruginosa* with aqueous extracts of immature fruits. For both strains, the MIC was of 0.20 mg/ml and the MBC of 0.40 mg/ml. Stem extracts were as active as those of leaves, these two organs showing generally less activity than the fruits and seeds. The lowest activity (against all strains) was observed for root extracts (MIC ranging from 3.25 to 6.50 mg/ml).

Anticandidal activity

Anticandidal activity is reported as MIC and MFC (Tables 4 and 5). With the exception of roots, all extracts showed significant antifungal activity against all tested yeasts. Overall, the best antifungal activity was against *C. albicans* and *C. glabrata* (for both aqueous and diluted acetone extracts). In terms of plant organs, the best activities were found for fruits at one or more of their ripening stages (either immature for the acetone extracts, or ripening and/or mature for the aqueous extracts).

For the aqueous extracts, mature fruits had the highest anticandidal effects in all strains (MIC 0.20 mg/ml). Immature seeds were as efficient as mature fruits for *C. albicans*, *C. glabrata* and *C. parapsilosis*. Ripening fruits (aqueous extracts) were either as strongly anticandidal as mature fruits, for *C. glabrata* and *C. krensei*, or a close second (MIC 0.40 mg/ml).

For acetone extracts, immature fruits were the most active against all strains (with the best activity against *C. albicans* and *C. glabrata*, MIC 0.10 mg/ml). Ripening fruits were either as efficient as immature ones (*C. krensei*, *C. parapsilosis*, both MIC and MFC) or had the second best efficacy against *C. albicans*. For *C. glabrata*, the stems showed the second best and very active anticandidal property. The activity of the stem diluted acetone extract against *C. glabrata* differs significantly from its activity against other strains, for which its activity ranks below that of other organs (seeds, leaves). Note also that this result is specific to the acetone extracts, as aqueous extracts from leaves showed better activity than the one from stems.

DISCUSSION

Tunisia recently increased research in Traditional Herbal Medicines following scientific confirmation of their effectiveness in treating conditions for which they were traditionally prescribed. The present investigation has explored the use of one such plant, *C. colocynthis* Schrad., endemic in the south of Tunisia, for treating infectious diseases.

The antimicrobial and anticandidal screening showed interesting activities at low concentrations: Every plant organ was active, except roots. Plant extracts are generally a crude mixture of non-active and active compounds and their MICs must be interpreted accordingly. For example, in a discussion on anticandidal properties of Canadian medicinal plant extracts (Carmeli et al., 1999), the MIC of *Epilobium augustifolium* at 0.80 mg/ml against *C. albicans* ATCC 90028 was interpreted as strong antifungal activity: The authors noted that MICs of less than 1.00 mg/ml should be interpreted as strong antifungal potential (comparable to accepted active antimicrobial single products such as cationic peptides in the 0.10 mg/ml range). MICs obtained in this study (as low as 0.10 mg/ml) are within the range of what is considered significant for plants and even purified extracts.

The study clarified a complex ethnopharmacological picture in terms of preparation: Firstly, results showed the importance of maturation: At different stages of maturation, the compositions of fruits and seeds differed, as made obvious by their extraction yields and antimicrobial and anticandidal activities. Which maturation stage was the most efficient depended on the extraction method (for example compare results of active fruits in Tables 4 and 5).For a given microorganism, the method of extraction can also change which plant organ is the most effective (for *S. aureus* and *E. faecalis*, the best antimicrobial plant organ extracts changes from immature fruits in aqueous extracts to the stronger activity in of leaves in the diluted acetone extracts). In addition, the difference in activity between the de-seeded fruits and the seeds emphasizes the need for precise recording of ethnobotanical protocols to notice if seeds are present in preparations involving "pulp" or "fruit".

For the antimicrobial activity, the plant extracts were active against both Gram-positive and Gram-negative bacteria, though more against the latest. Activity depended from the bacterial strain, the plant organ, their maturation state and the nature of the extraction.

The good MIC values (0.20 mg/ml) against *E. coli* of an extract obtained from a productive organ through an easy extraction procedure (fruit, aqueous extraction, good extraction yields) bodes well for widespread potential applications. The need to decontaminate surfaces soiled with *E. coli* is of key economic importance in the health care and industrial sectors (for example the meat packing industry). Novel anti *P. aeruginosa* activity is also of particular interest as it is the leading cause of nosocomial infections and has developed mechanisms of resistance to common antibiotic classes (Carmeli et al., 1999). In the same way, synergy between commercial drugs showed promising results for a complicated urinary infection with *P. aeruginosa* (Hayami et al., 1999) and FIBs between plants products or their various organs are getting more thoroughly investigated (Van Vuuren and Viljoen, 2008), active extracts of *C. colocynthis* Schrad. should be tested for potential synergy with other antimicrobials.

Activity cannot be imputed to one family of phytochemicals only (or its absence). Alkaloids are commonly found to have antimicrobial properties (Omulokoli et al., 1997); therefore their absence in roots could account for the lack of activity of this organ. However, alkaloids cannot be solely responsible for the activity: For the same organ, the majority of aqueous extracts are more active than the acetone ones, which could reflect preferential solubility/extraction properties of the alkaloid(s) in water versus acetone, but the diluted acetone extracts of leaves are more active than their aqueous extracts for 3 out of 4 strains. Therefore, activity is not due to the alkaloids alone. Flavonoids (detected in all the seeds extracts) are known to be synthesized by plants in response to microbial infection (Fogliani et al., 2005); therefore their potential in vitro antimicrobial effectiveness against a wide array of microorganisms should not come as a surprise, but activity of immature seeds (not containing them) and the low or near lack of activity of ripe fruit (which contain them) point to other factors at play. The same could be said about most phytochemicals detected by the qualitative chemical analysis and known from works in other plant species to have antimicrobial pharmacological activity: tannins (Ayaz et al., 2008), steroids (Khan et al., 2007), pigments (Fogliani et al., 2005; Eyong et al., 2006) and flavonoids (Fogliani et al., 2005), alkaloids (Yan et al., 2008) and iridoids (Akunvili et al., 1991). Iridoids, present in the aqueous samples only, could contribute to their better performance compared to the acetone extracts. However, iridoids are detected in the aqueous root extract but this extract showed no overall activity. Therefore, iridoids role, if any, is likely to be synergistic or potentiate other compounds.

Concerning the anticandidal activity, the present study showed MICs and MFCs significant for crude plant extracts against all strains. The potent activity against *C. albicans* (MIC 0.10 mg/ml for acetone extracts – bold face Table 5 – and 0.20 mg/ml for some aqueous extracts, Table 4) is particularly welcomed due to the fungus' frequency to cause infections, with ca. 45% of clinical fungal infections due to this strain (Gupta et al., 2004). The equally strong activity against *C. glabrata* opens the possibility to use the extracts either alone or in combination with Amphotriricin B, as has been proposed with other natural substance extracts (Mimee et al., 2005). New active drugs are particularly needed as triazoles are generally met with resistance and the few alternative treatments, such as Caspofungin, are limited by their low oral bioavailability (Boucher et al., 2004). Like for the antimicrobial activity, the anticandidal activity may be attributed, possibly in combination, to various phytochemicals detected during the extracts chemical screening and which are known to cause damage to cell membranes, causing leakage of cellular materials and ultimately the microorganisms death (Mshvildadze et al., 2000; Ghani et al., 2008; Abdel et al., 2008). The trend over all the strains to have higher activity for ripe fruits aqueous extracts versus immature fruit for the diluted acetone extract points towards various active, potentiate or antagonistic compounds present in various concentrations according to the maturation stage of the fruit and which can be preferentially extracted with various solvents. It invites and in-depth study of the most active organs contents to determine (by MICs and FICs) if further purified extracts or isolated substances, alone or in combination, could display an even higher activity and/or selectivity against specific microorganisms.

With all these wide spectrum antibacterial and anticandidal properties, *C. colocynthis* Schrad can be considered an effective antimicrobial agent to treat infectious diseases. This plant, namely its fruit extracts, demonstrated activity against some bacteria and fungi prevalent in dermatology, gynaecological and pulmonary infections. The study supported scientifically the ethnopharmacological use of the plant as an antimicrobial and anticandical agent and could account for some of the variations observed in the ethnopharmaceutical preparation methods. Therefore, the use of this plant as antimicrobial agent is validated by the results obtained in this work. Further studies are ongoing to identify the chemical compounds of these antimicrobial extracts.

REFERENCES

Abdel GSB, Weaver L, Zidan ZH, Hussein MA, Keevil CW, Brown RCD (2008). Microware-assisted synthesis and antimicrobial activities of flavonoid derivatives. Bioorg. Med. Chem. Lett., 18: 518–522.

Adams M, Berset C, Kessler M, Hamburger M (2009) . Medicinal herbs for the treatment of rheumatic disorders—a survey of European herbals from the 16th and 17th century. J. Ethnopharmacol., 121: 343–359.

Ayaz AF, Hayirlioglu-Ayaz S, Alpay-Karaoglu S, Gruz J, Valentova K, Ulrichova J , Strnad M (2008). Phenolic acid contents of Kale (*Brassica oleraceae* L. var. *acephala* DC) extracts and their antioxidant and antibacterial activities. Food Chem., 170: 19–25.

Akunvili DN, Houghton PJ, Kaman A (1991). Antimicrobial activities of the stembark of Kigelia pinnata. J. Ethnopharmacol., 12: 331–333.

Batanouny KH (1999). The Mediterranean coast dunes in Egypt: An endangered landscape. Estuar. Coasta Shelf Sci., 49: 3-9.

Berche P, Gaillard, JL, Simonet M (1991). Mesure de lactivite antibactenenne. Bactenologie, bacteries des infections humaines. Ed. Flammarion, Paris, France, pp.12-16.

Boucher WH, Groll AH, Chiou CC, Walsh TJ (2004). Newer systemic antifungal agents. Drugs, 64: 1997–2020.

Boukef MK (1986). Medecine traditionnelle in pharmacopee. Les plantes dans medecine traditionnelle tunisienne. Agence de Coopération Culturelle et technique, 165p.

Carmeli Y, Troillet N, Eliopoulos GM, Samore MH (1999). Emergence of Antibiotic-resistant *Pseudomonas aeruginosa*: comparison of risks associated with different antipseudomonal agents. Antimicrob. Agents Chemother., 43: 1379–1382.

Delazar A, Gibbons S, Kosari AR, Nazemyeh H, Modarresi M , Nahar L , Sarker SD(2006). Flavone C-glycosides and cucurbitacin glycosides from *Citrullus colocynthi.* J. F. Pharmacy. T. Univ. Med. Sci., 14: 109–114.

Eyong KO, Folefoc GN, Kuete V , Beng VP , Krohn K H , Hussain AE,

Nkengfack M, Saeftel SR (2006). Sarite and A. Hoerauf, Newbouldiaquinone A: anaphthoquinone–anthraquinone
ether coupled pigment, as AA. Agent. Phytochemistry, 67: 605–609.
Fogliani B, Raharivelomanana P, Bianchini JP, Bouraima S, Hnawia E (2005). Bioactive ellagitannins from *Cunonia macrophylla*. From New Caledonia. Phytochemistry, 66: 241–247.
Ghahraman A (1981). Flore De La Iranica, Angiospermes-Dicotyledones: Gamopetales. Ed. Imprimerie officielle de la republique Iranica, Iran, 930p.
Gupta N, Haque A, Lattif A, Narayan R, Mukhopadhyay G , Prasad R (2004). Epidemiology and molecular typing of *Candida* isolates. Mycopathologia, 158: 397–405.
Hayami H, Goto T, Kawahara M, Ohi Y (1999). Activities of β-lactams, fluoroquinolones, amikacin and fosfomycin alone and in combination against *Pseudomonas aeruginosa* isolated from complicated urinary tract infections. J. Infect. Chemother., 5: 130–138.
Khan NS, Jin Kim B, Kim HS (2007). Synthesis and antimicrobial activity of 7-fluoro-3-aminosreroids. Biol. Med. Chem. Lett., 17: 5139–5142.
Marzouk B (2008). Etude biologique et activités pharmacologiques de Citrullus colocynthis Schrad. M Thesis Biology & Health Care. Uni. Of Monastir, Monastir, Tunisia. Pp.11-16
Mimee B, Labbe C, Pelletier R, Belanger R (2005). Antifungal activity of flocculosin, a novel glycolipid isolated from pseudozyma flocculosa, Antimicrob. Agents Chemoth., 49: 1597–1599.
Mothana RA, Lindequist U (2005). Antimicrobial activity of some medicinal plants of the island Soqotra. J. Ethnopharmacol., 96: 177–181.
Mozaffarian M (1964). Medicinal Plants of Iran. Ministry Agriculture and Technology, 146p.
Mshvildadze V, Favel A , Delmas F, Elias R, Faure R, Decanosidze Q , Kemertelidze E , Balansard G (2000) . Antifungal and antiprotozoal activities of saponins from *Hedera colchica*. Pharmazie, 55: 325–326.
Nadkami AK (1954). Indian Materia Medica (3rd ed.), Popular Prakashan, Bombay, India, pp.25-32
Omulokoli E, Khan B, Chhabra S (1997). Antiplasmodial activity of four Kenyan medicinal plants. J. Ethnopharmacol., 56: 133–137.
Pottier-Alapetite G (1981). Flore De La Tunisie, Angiospermes-DicotylEdones: GamopEtales. Ed. mprimerie officielle de la republique tunisienne. Tunisia, 930p.
Riddle JM (1999). A History of Contraception and Abortion in the West, Harvard University Press, Harvard, USA, 350p.
Sakar MK, Tanker M (1991). Fitokimyla structural analysis, publications, Ankara University Faculty of Pharmacy, Ankara, Turkey, pp. 67.
Trease GE, Evans CW (1984). Trease and. Evans, Pharmacognos (12th ed.), Balliere Tindall, London, UK, London, pp.45-54
Trim AR, Hill R (1952). The preparation and properties of aucubin, asperuloside and some related glycosides, Biol. J., 310p.
Van Vuuren SF, Viljoen AM (2008). *In vitro* evidence of phyto-synergy for plant part combinations of *Croton gratissimus* (Euphorbiaceae) used in African traditional healing. J. Ethnopharmacol., 119: 700–704
Yan D, Jin C, Xiao XH, Dong XP (2008). Antimicrobial properties of berberines alkaloids in Franch *Coptis chinensis* by microcalorimetry. J. Biochem. Biophys. Method., 70: 845–849.

Spatio-tempora variation in cassava (*Manihot* sp.) in forest savannah eco-climatic zone of Nigeria

ADETAYO, A. O. and ADURAMIGBA-MODUPE, V. O.

Institute of Agricultural Research and Training, Obafemi Awolowo University, Ibadan, Oyo, Nigeria.

Multi-locational trials were conducted in four selected locations with the forest-savannah agro-ecological zone of Nigeria to investigate the spatial variability in the yield of cassava. In an attempt to compare the yield of cassava within the forest-savannah eco-climatic zone of Nigeria, four locations including Ibadan, Abeokuta, Akure and Oshogbo were selected for field trials. Four improved cassava cultivation (TMS 30572, TMS 91934, TMS 50395 and TMS 30555) were grown between 2009 and 2010 in these locations and harvested at 12 months after planting. Results obtained showed that there were significant differences in the yield of cassava in the four selected locations with Ibadan having the highest tuberous root per hectare among the four locations. TMS 30572 had the highest (52,964 t. ha^{-1}) number of tuberous root per hectare among the four varieties of cassava planted. This study revealed that cassava yield decreased with increase in rainfall distribution in the forest-savannah agro-ecological zone of Nigeria.

Key words: Cassava, tuberous root, spartial variability, rainfall distribution.

INTRODUCTION

Forest-Savannah eco-climatic zone of Nigeria covers a total land area of about 115,000 sq. km. The rainfall in the zone can be described as humid to sub-humid tropical with district dry and wet seasons. The dry season runs from early November to end of March, while the wet season is from early April to early November. There are two rainfall peaks (June and September) with a dry spell in August. The annual temperature ranged between 28 to 36°C, relative humidity is high throughout the year and ranged between 60 to 90 percent at 16.00 h. Sunshine hours are directly related to cloud cover, while average daily sunshine hours ranged from 14 h in august and 7.5 h in January.

The wellbeing of large population around the world depends on access, stability and availability of food. This is especially true in the developing world with predominant small land holders and subsistence farmers for whom both on-season and off-season agricultural labour provides the main source of food and income. It has been suggested that larger percentage of African population would be stroked with poverty as short term changes in climate will increase the stress on food production (Ahmed and Hertel, 2009).

Canbas and Olace (2009) used statistical models to highlight the importance of intra-seasonal changes in temperature and precipitation on crop production in southwestern Ontario, Canada and showed that, although precipitation and temperature variability might have impact on average yield, net crop yield will be higher in the future due to a lengthening of the growing season. Cassava (*Manihot utilisima*) is one of the most important root crops in Nigeria. It is affected by different environmental factors (Ngere, 1994). It has been discovered that, improved varieties of cassava would yield an optimum value in any environment (Dixon et al., 1994). Canter et al. (1992) reported that a genotype is adjudged to be stable if at a given location, it has little or no yield variation. This distinction is important because farmers who has to decide the cultivar to adopt is interested in the stability of the cultivars only at her location for a given yield level. The yield potential of the cultivar at any particular location is of importance to the farmer at that location.

Desert encroachment into the rainforest zones is going at a fast rate; thus, many rainforest zones are becoming dry lands, otherwise referred to as forest-savannah zone.

*Corresponding author. E-mail: wale.agromet@yahoo.com.

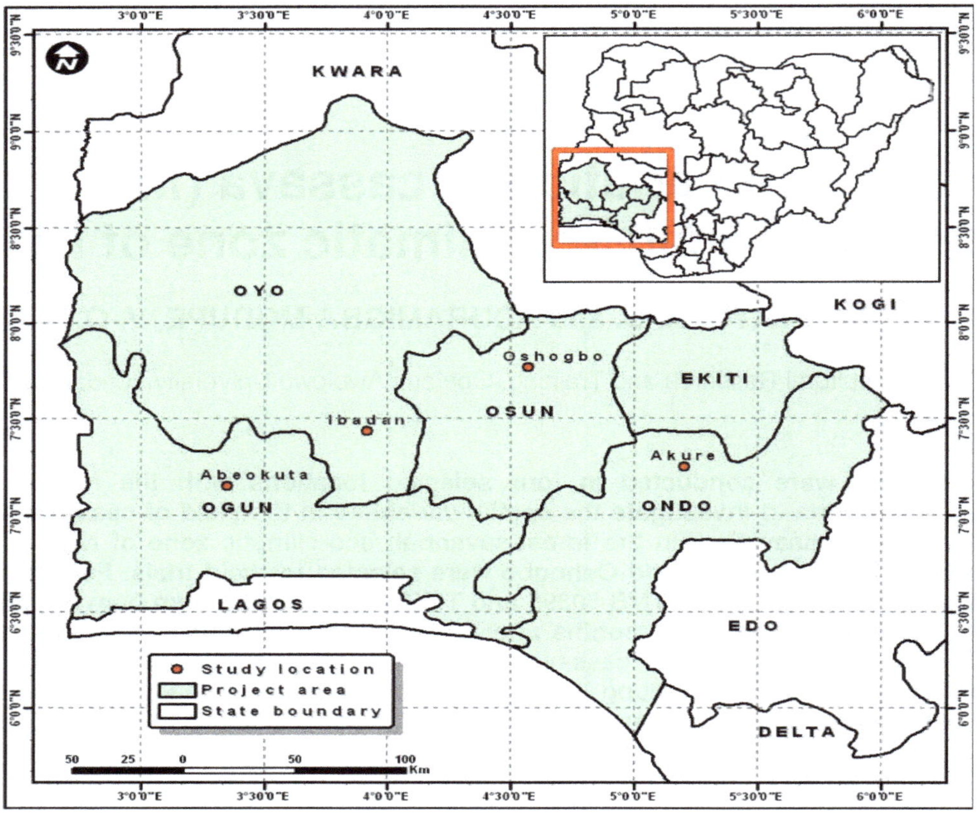

Figure 1. Map of Nigeria showing the four selected study locations.

This area has rainfall that ranges between 1500 to 2900 mm. There is a need to estimate the impact of micro-climatic variability on cassava yield and screen cassava cultivars that are high yielding and that can be recommended for farmers at particular location. The aim of this study was to compare the yield of cassava in some selected locations including Ibadan (7°24` N; 3°33`E), Abeokuta (7°9` N; 3°20` E), Akure (7°15` N; 5°11` E) and Oshogbo (7°46` N; 4°34` E) within the forest-savannah agro-ecological zones of Nigeria. These selected locations have a total annual rainfall of 1279, 1529, 1576 and 1382 mm respectively.

MATERIALS AND METHOD

The research was conducted during the 2010/2011 season at Ibadan (Institute of Agricultural Research and Training research farm), Abeokuta (Federal University of Agriculture Abeokuta research farm), Akure (Federal University of Technology agricultural Research farm) and Oshogbo (Oshun State Agricultural Development Program farm) (Figure 1). The soils of the experimental site in all the selected locations are well-drained tropical ferruginous soil classified as sandy-loam. Four improved IITA cultivars: TMS 30572, TMS 91934, TMS 50395 and TMS 30555 were planted in the four locations.

The experimental sites were cleared, ploughed, harrowed and ridged (tractor-driven). The experimental design used at each location was Randomized Complete Block Design (RCBD) with four replications. Each plot was 10 m long and 1 m apart. Stem cutting of each at 30 m long having at least 4 nodes were used as planting materials. Each plot measured 10 m and 1 m apart. Stem cuttings each of 30 cm long and having at least four nodes were used as planting material. At 12 months after planting, harvesting was done by hand. Stems were cut and tuberous roots uprooted from the soil. Data collected included number of tuberous roots per hectare and fresh tuberous root yield per hectare. These data were subjected to General Linear Model (GLM) and mean differences were determined by least significant difference (LSD).

RESULTS AND DISCUSSION

Rainfall encountered during the experiment in all the locations were higher than the amount required for the growth and development of cassava. Hence, the plant stood no risk of water stress. The results of the preliminary soil survey in all the selected locations indicated that the experimental sites were sufficient in terms of basic nutrient requirements. Mean pH, organic matter and organic carbon obtained for the sites were 6.7, 1.17 and 0.68%, respectively. Also, the mean exchangeable base Ca, Mg, K, Na and CEC were 1.36, 1.92, 0.10, 90.36 and 3.75 cmol kg^{-1}, respectively. There is no significant different in the growth indices during establishment and vegetative stages. This could be

Table 1. Average total tuberous roots number per hectare and fresh tuberous root yield (mg ha^{-1}) in 2010 and 2011.

Variety	Total tuberous roots number per hectare	Fresh tuberous root yield (mg ha^{-1})
$V_1 * P_1$	52.964	24.5
$V_1 * P_2$	48.847	19.2
$V_1 * P_3$	50.262	22.0
$V_1 * P_4$	47.562	22.4
$V_2 * P_1$	40.332	20.5
$V_2 * P_2$	39.425	18.5
$V_2 * P_3$	40.014	19.8
$V_2 * P_4$	38.813	17.8
$V_3 * P_1$	50.024	18.4
$V_3 * P_2$	40.136	18.0
$V_3 * P_3$	42.256	19.0
$V_3 * P_4$	29.468	17.5
$V_4 * P_1$	32.467	18.5
$V_4 * P_2$	29.674	17.0
$V_4 * P_3$	30.128	17.5
$V_4 * P_4$	22.245	16.0
LSD = $P \leq 0.01$	10124	5.4

V1, TMS 30572; V2, TMS 91934; V3, TMS 50395; V4, TMS 30555; P1, Ibadan; P2, Akure; P3, Abeokuta; P4, Oshogbo.

attributed to similarity in rainfall amount, temperature condition and sunshine hours in the four selected locations, which among other factors accelerated the initial sprouting ability, breaking of bud dormancy, plant vigour and plant establishment (Eke-okoro et al., 1999; Njoku and Muoneke, 2008).

Table 1 shows the results of the average total tuberous roots number per hectare and fresh tuberous root yield. Cultivar differences were observed for the parameters measured thereby reflecting the generic differences between the varieties of cassava used. Analysis of variance (ANOVA) for total number of roots per hectare and fresh tuberous root yield per hectare showed significant difference ($P \leq 0.01$) in the four locations. Cassava planted in Ibadan (P_1) had the highest for tuberous roots number and fresh cassava tuberous root yield followed by those planted in Abeokuta, whereas the lowest were recorded for those grown in Akure. This finding suggests the influence of spatial variability of climatic factors on yield of cassava. This implies that cassava yield will be enhanced by planting in certain locations since the same trend was obtained irrespective of the variety of cassava planted. This is in agreement with Eke-Okoro et al. (1999) and Simwambana et al. (1995) who reported differences in the annual productivity of cassava in Nigeria and India, respectively. The annual differences in rainfall amount in both cropping seasons predisposed the treatments to differential performances in favour of interaction of treatments and weather condition. This suggests that the productivity of cassava was driven by the differences in weather condition of rainfall and sunshine in response to climate change in our environment. The results of this study emphasized that differences in the treatments (agronomic factors) and annual weather conditions were responsible for differences in the growth and productivity of cassava. This study suggests that climatic factor in Ibadan when adequately harnessed will have the best positive influence on the yield of cassava than any of the other three locations selected.

Conclusion

Significant differences were noticed in the yield of the selected varieties of cassava in the four selected locations within the forest-savannah agro-ecological zones of Nigeria. Genotypic differences were also observed in the yield characteristics of the varieties of cassava irrespective of the planting locations. The study further revealed that yield characteristics of cassava decreased with increased rainfall distribution to a certain critical value.

REFERENCES

Ahmed NS, Hertel TW (2009). Climate volatility deepens poverty vulnerability in developing countries. Policy Research Working World Bank, Washington DC. p. 5117.

Canbas A, Olale E (2009). Crop yield response to economic site and climatic variables. Climate Change 101:599- 616.

Canter SE, Fresco LO, Jones PG, Fairbairn JN (1992). An atlas of cassava in Africa: Historical, agro-ecological and demographical aspect of distribution. CIAT, Cali, Colombia. p. 86.

Eke-Okoro ON, Okereke OU, Okeke JE (1999). Effects of weather change and planting setting on growth and productivity of cassava in South-eastern Nigeria. Afr. J. Root Tuber. Crops. 3(2):34-38.

Ngere JM (1994). Yield stability parameters for comparing cassava varieties. Proceedings of the 9th symposium of the International Society for Tropical Root Crop, Accra, Ghana. pp. 139-145.

Njoku DN, Muoneke CO (2008). Effect of cowpea planting density on growth, yield and productivity of component crops in cowpea/cassava intercropping. Syst. Agrosci. J. 7(2):106-113.

Dixon AGO, Otoo, JA, Aiedu R, Okeke JE, Hahn SK (1994). Genotype-environment interaction studies with cassava. Proceedings of the 9th symposium of the International Society for Tropical Root Crop, Accra, Ghana 20-26:130-137.

Simwambana SMC, Ekanayake IJ, Ferguson TU (1995). Effects of shading on growth and flowering behavior of four cassava genotypes. Afr. J. Root. Tuber Crops 1(1):19-25.

Effect of dressing of hot water-steeped and unsteeped seeds on longevity and seedling emergence in *Corchorus olitorious* L.

Ibrahim H., Oladiran J. A., Adeyemi R. A. and Danladi M. M.

Department of Crop Production, School of Agriculture and Agricultural Technology, Federal University of Technology, Minna, Niger State, Nigeria.

The study was conducted at the laboratory of the Department of Crop Production, Federal University of Technology Minna. Seeds of cultivars "Amugbadu" and "Oniyaya" of *Corchorus olitorius* were steeped in hot water at 97°C for five seconds (for the purpose of breaking dormancy) and subsequently air dried to about 10% moisture content (wet weight basis). They were then treated with seed-plus, Apron-star (As) or Wood ash (Wa) at the rate of 10 g per 4 kg seed, packaged in paper envelopes and stored for 20 weeks at 30°C while the unsteeped seeds served as the control. Germination ability and seedling emergence were determined before and during storage. Before storage, 90 and 95% germination levels were recorded for hot-water steeped seeds of "Amugbadu" and "Oniyaya" respectively in contrast to 25 and 33% germination for the unsteeped seeds. There was a general decline in both germinability and seedling emergence with storage period irrespective of treatments. In non-dormant "Amubgadu" seeds, viability was better maintained without seed dressing or when seeds were dressed with wood ash. In "Oniyaya", viability was poorest in As-dressed seeds. Significantly better seedling emergence was recorded in non-dormant seeds dressed with wood ash. Unsteeped (dormant) seeds germinated poorly throughout the storage period. Dormant seeds from which zero germination was recorded after 12 weeks of storage gave about 75% germinations followed by steeping in hot water. It is postulated from this observation that the poor germination of stored seeds might be due to a decline in the germination energy of the seeds which made their radicle unable to rupture the endospermic tissue and seed coat. This is an indication that the seeds were still viable at this age but required the weakening of the seed coats (which was effected by hot water-steeping) to permit germination.

Key words: Longevity, seed dressing, seedling emergence, jute mallow

INTRODUCTION

Jute mallow (*Corchorus olitorious*) leaves are consumed as pot herb in many countries of the world (Jansen et al., 2004). The leaves are rich in beta carotene, Iron, Calcium (Ca), and vitamin C. The plant has antioxidant activity with a significant vitamin E constituent and is particularly good for sick people who have difficulties in swallowing as it is slippery in texture (Moses and Ramsey, 2004). It is also said to be the second most important vegetable fibre after cotton for making sacks and cloths (Chattapadhyay et al., 2004). The role of the seed coats as a germination constraint is well known (Debeaujon and Koorneef, 2000; Koornneef et al., 2002). However, one of the positive roles played by the seed coat is to protect the embryo against detrimental agents from the environment and it also limits the detrimental activity of physical and biological agents during seed storage (Debeaujon et al., 2000). In theory, seeds with impermeable coats would present a barrier to moisture and fungi and would therefore be expected to live longer than those with permeable coats. In tandem with this view Clerkx et al. (2004) reported that mutants of Arabidopsis with seed coat alterations showed a stronger reduction in germination percentage after storage. The authors stressed that this was an indication of the importance of a functional seed coat for seed longevity. *C. olitorious* seeds are known to posses dormancy due to hard seed coats (Schippers, 2000), which can be broken by steeping seeds in boiling water at about 97°C for five seconds or by scarifying the seed coat (Oladiran, 1986). Some flavonoid pigments which are synthesized by plants and accumulated in the cells of the testa have been reported to act in defense against predators and pathogens (Dixon et al., 2002).

According to Pourcel et al. (2005), their derivatives that may be produced strengthen the testa, protect the embryo and endosperm from biotic and abiotic stresses. Observations from unpublished works have shown that when *C. olitorius* seeds are steeped in hot water, the soak water becomes brownish in colour which is an indication of the leaching of some of the coat pigments. This suggested that, seed steeping may not only soften the seed coat, it also results in a change in coat structure even when such seed may have been dried back. Damages of this sort may render the seed coat less protective resulting in faster embryo death during storage due to pathogen invasion. In situations in which damages to seed coats are not suspected, extra precautions are still being taken to protect seeds meant for planting and or storage. Proper seed dressing with appropriate chemicals that are known to be effective against wide range of pathogens (FAO, 2007), had been reported to enhance seed longevity in storage. Classen (1995) reported that, proper storage and treatment of seed with chemicals slow down deterioration in all seed species, preserving viability and vigour. Okra seeds treated with baselin (fluchoralin) at 2kg a.i/ha have been shown to exhibit higher vigour, seedling emergence and yield. Similarly, Adebisi et al. (2004) obtained significantly longer storage life by treating soyabean seeds with Apron plus, Almithio and Aldrex T. Contrary to the above however, Adebisi et al. (2003) observed that seeds treated with Apron plus, Almithio and Aldrex T did not last longer than the untreated control.

The objective of this study was to assess the effectiveness of seed treatment with wood-ash, Seed-plus (Sp) and Apron-star (As) on the germination and seedling emergence from dormant and non-dormant jute mallow (*C. olitorious*) seeds.

MATERIALS AND METHODS

The study was done between March and September 2010 using seeds of two cultivars ("Amugbadu" and "Oniyaya") of *C. olitorius* that are widely grown in Nigeria. Some seeds of the two cultivars were steeped in hot water as described by Oladiran (1986) to break dormancy; they were then spread on paper and air-dried under shade to a moisture content of about 10%. Samples of steeped (non-dormant) and unsteeped (dormant) seeds of both cultivars were dressed with As 42WS (consisting of 20% w/w thiamethoxam, 20% w/w metalaxyl-M and 20% w/w difenoconazole), Sp 30WS (consisting of Imidacloprid 10% + Metalaxyl 10% + Carbendazim 10% WS) (which are both manufactured by Shandong Xinghe Crop Technology co. Ltd Jinan) or wood-ash. The substances were applied at 10 g per 4 kg seed. The former two substances possess both fungicidal and insecticidal properties; untreated seeds served as the control. Seeds of the different treatment combinations were packaged in paper envelopes and stored at 30°C for 20 weeks. Seed samples were drawn for germination and seedling emergence tests prior to storage and at 2 weeks intervals. A sample of some dormant seeds stored for 12 weeks was steeped in hot water and then incubated to ascertain if there would be any difference in their viability level compared to the unsteeped and those stored following steeping.

For the germination test, 50 seeds of 4 replicates of each of the treatment combinations were counted and placed on layers of moist absorbent paper in plastic petri dishes and incubated at 30 °C; germination counts were taken every-other-day. The absorbent paper and seeds were kept constantly moist with distilled water. For the seedling emergence study, 4 replicates of 10 seeds from each treatment were sown into soil in plastic pots on each sampling day. The plastic pots were watered a day prior to sowing and twice weekly after sowing. All data in percentages were transformed to arcsin values before statistical analysis was done. Probit analysis (Finney, 1971) was also conducted on the germination data obtained from all treatments to determine the potential seed longevity (that is, the constant K_i), the slope, which is the rate of seed deterioration and the seed half-viability period (P_{50}), that is, the time taken for viability to drop to 50%, which according to Ellis and Roberts (1980) is a measure of absolute seed longevity.

RESULTS

Figures 1 and 2 showed the survival curves of Dormant (D) and Non-dormant (Nd) seeds of "Amugbadu" and "Oniyaya" cultivars. Before storage, 90 and 99.5% germination levels were recorded for Nd seeds of "Amugbadu" and "Oniyaya", respectively in contrast to 25 and 33% germination for D seeds. As storage progressed, there was a general decline in the germination of D and Nd seeds of both cultivars, irrespective of treatment. Both D and Nd seeds of "Amugbadu" cultivar stored better when untreated (control) or treated with wood ash (Wa) than when treated with As or Sp which values declined sharply after only 2 weeks of storage (Figure 1). The control performed better than Wa during the first four weeks of storage in Nd seeds and only at 2 weeks of storage in D seeds. Figure 2 showed the superiority of Nd, non-treated "Oniyaya" seeds over As and Sp

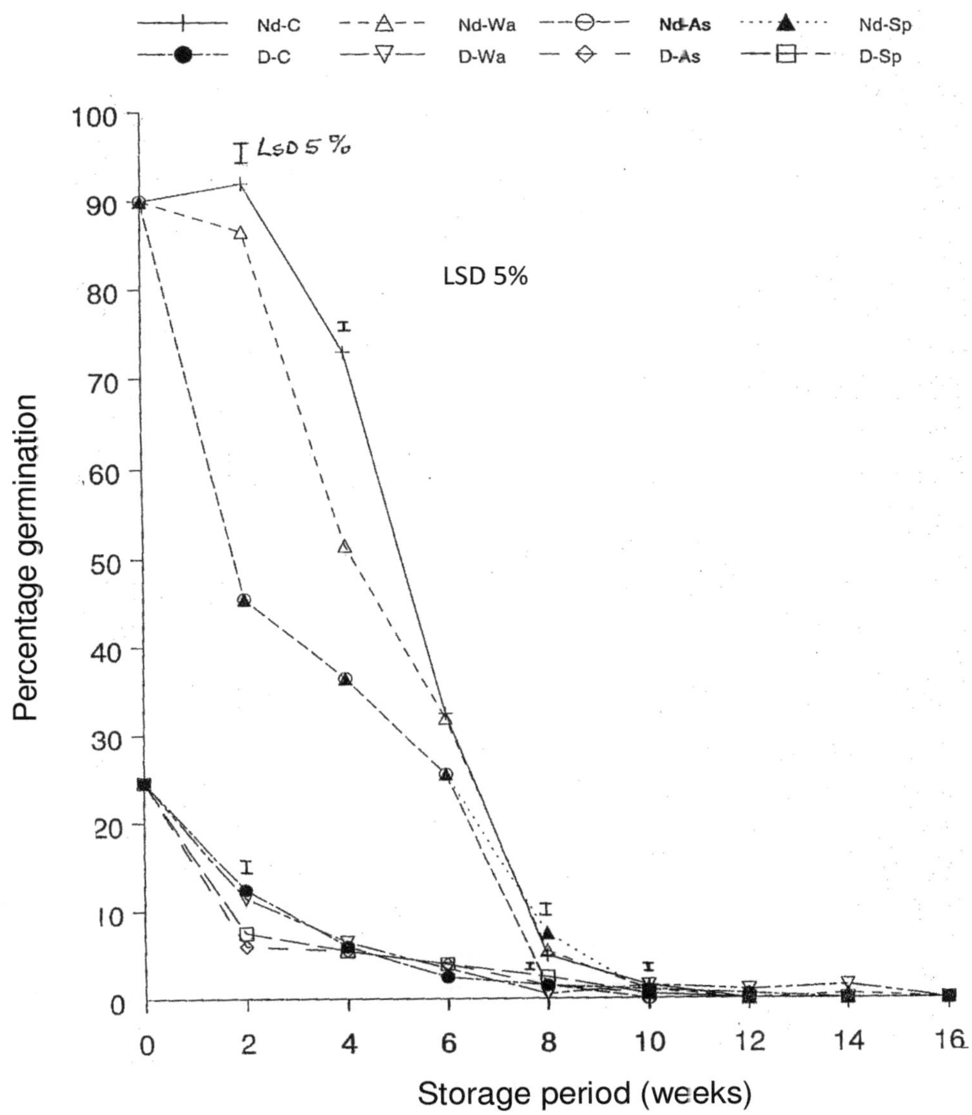

Figure 1. Germination levels of treated and untreated Dormant (D) and Non-dormant (Nd) seeds of Cv. "Amugbadu". Treatment: Control-C, Wood ash (Wa), Apron-star Ws), Seed plus (Sp).

treatments within the first 6 weeks of storage. Viability of Wa-treated seeds was also poorer than the control at 4 and 6 WAS. Performance was consistently poorest in As-treated seeds irrespective of dormancy status. The germination of As dressed seeds dropped from about 100 to about 27% within 2 weeks of storage.

Prior to storage, seedling emergence of 60 and 80% were recorded for Nd seeds of "Amugbadu" and "Oniyaya" respectively in contrast to 20 and 30% for D seeds (Figures 3 and 4). A general decline in values was recorded as from 2 weeks of storage irrespective of dormancy status and seed dressing treatment. For the first 6 weeks of storage, seedling emergence was similar in "Amugbadu" seeds irrespective of the dressing material used. Shortly thereafter and up to the end of storage however, the superiority of Wa- and Sp-dressed seeds (more especially the former), became evident. Furthermore, though no significant differences were recorded among seed treatment materials in D seeds, higher seedling emergence values were obtained in untreated seeds and those treated with Wa, compared to the As-and Sp-treated seeds.

Figure 4 shows the general superiority of Wa-treated Nd "Oniyaya" seeds over all other materials. Again, performances were poorest in seeds treated with As and Sp. Table 1 showed the estimates in respect of some storage parameters obtained from probit analysis of germination data. It reveals that treatment of seeds with Wa, Sp and As reduces the potential of seed longevity in comparison with the untreated control. Furthermore,

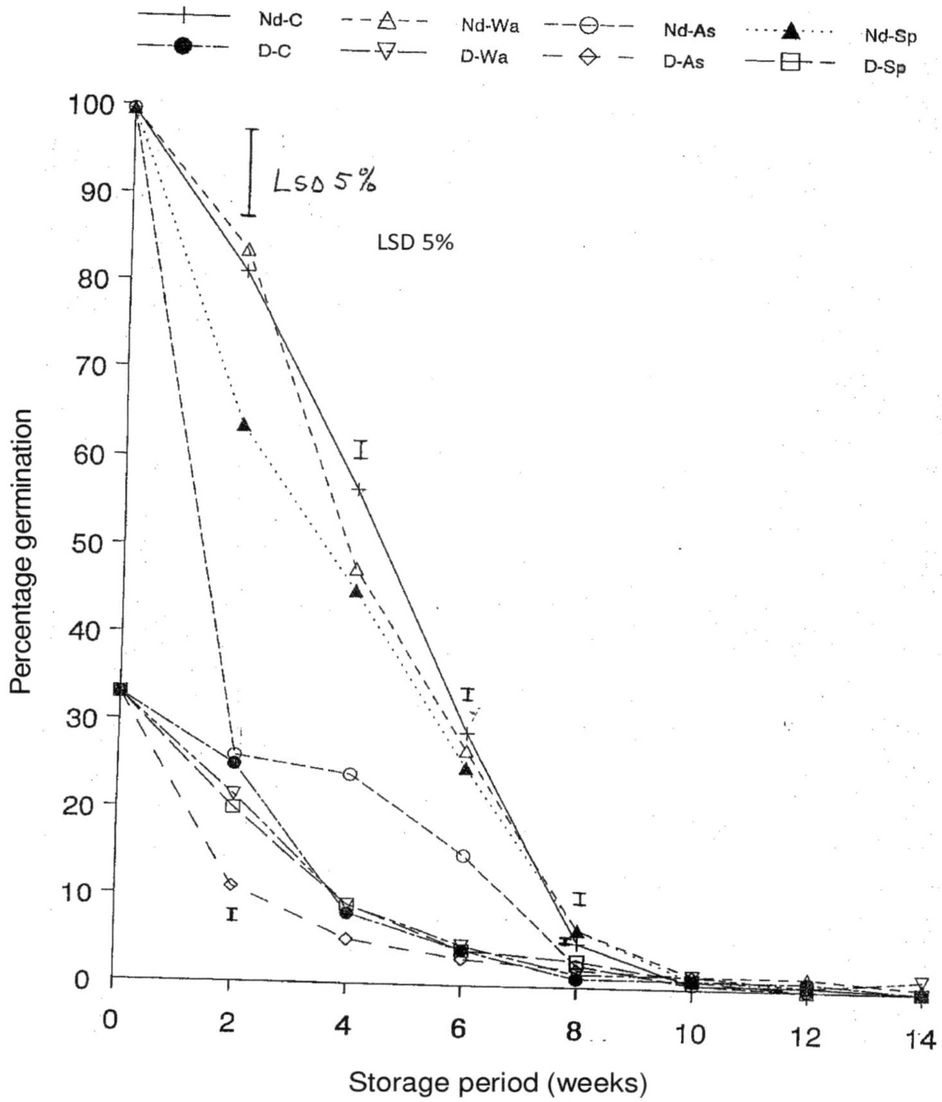

Figure 2. Germination levels of treated and untreated Dormant (D) and Non-dormant (Nd) seeds of Cv. "Oniyaya". Treatment: Control-C, Wood ash (Wa), Apron-star Ws), Seed plus(Sp).

treatment with Sp and As resulted in rapid seed deterioration which is reflected in reduced P_{50}. When a sample of the dormant seed stored for 12 weeks was tested for germination, values of 76 and 75% were recorded for "Amugbadu" and "Oniyaya", respectively compared to 0% recorded for both the unsteeped dormant and the non-dormant seeds of the two cultivars. In addition to the above, it is noteworthy that the seeds of *C. olitorius* lost vability without dormancy alleviation.

DISCUSSION

Seeds are normally dressed to protect them from pathogenic organisms and so delay seed deterioration during storage (Sekaramarthy et al., 1994). The opposite was however the case in this study. The general poor performance of seeds dressed with As and Sp might be an indication of the toxicity of these chemicals to *Corchorus*. Adebisi (1999) reported that, the dressing of soybean seed with Ap, Adrex T and Fernasan D was ineffective in the maintenance of viability beyond 2 months. Adebisi et al. (2003) also observed that, okra seeds treated with Ap and stored for 6 months had significant reduction in speed of germination. Ajala (2003) explained that, a reduction of the potency of the active ingredients of the chemical might occur during storage resulting in shorter seed life than expected. Just as Sp and As lowered both K_i and P_{50} in this study.

Adebisi et al. (2004) showed that, M-350 soybean

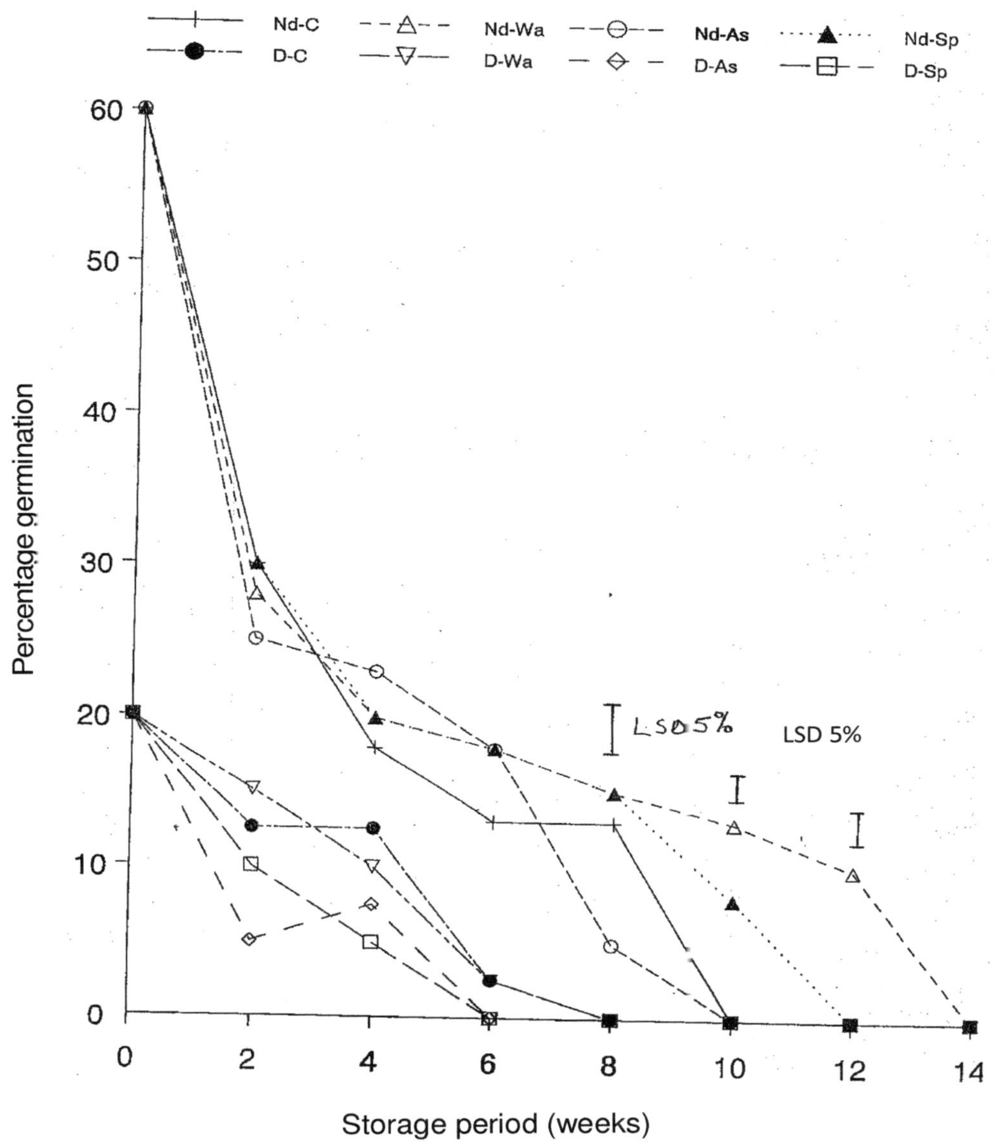

Figure 3. Seedling emergence from treated and untreated Dormant (D) and Non dormant (Nd) seed of cv "Amugbadu". Treatment: Control-C,Wood ash (Wa), Apron-star (As), Seed-plus (Sp).

variety seeds treated with Aldrex T had lower Ki and P_{50} values compared to the control. The authors also showed that different varieties of the same crop could exhibit differences in K_i and P_{50} values just as it was the case in this study. If seedling emergence is used to assess the effectiveness of the various seed dressing materials used in this study, then Wa showed some potential. Abdullahi (2006) did an analysis of the Wa obtained from a bakery in Minna, Niger State, Nigeria to determine its chemical composition. The sample contained SiO_2, Al_2O_3, Fe_2O_3, CaO, MgO, K_2O, and NaO. Wa is appreciated as a liming agent (Clapham and Zibilske, 1992) and it is known to increase soil pH in addition to the concentrations of K, Ca, B, Mg and other elements except N (Mandre, 2006).

We postulate that, the positive effect of Wa on seedlingemergence in this study might be linked probably to increase in the pH of the soil immediately around the seed. Bukvic et al. (2007) recorded the highest germination of pea seed at pH = 6 compared to higher and lower pH values. They also reported that, the pH requirement for optimum seedling growth may vary with seed age. For example, whereas the stem length of seedlings development for 33 months old seeds was greatest at pH = 6, the values obtained from 21 months old seeds was greatest at pH = 5.

In this study, the significantly high germination recorded for dormant "Amugbadu" and "Oniyaya" seeds that were steeped in hot water following storage for 12 weeks

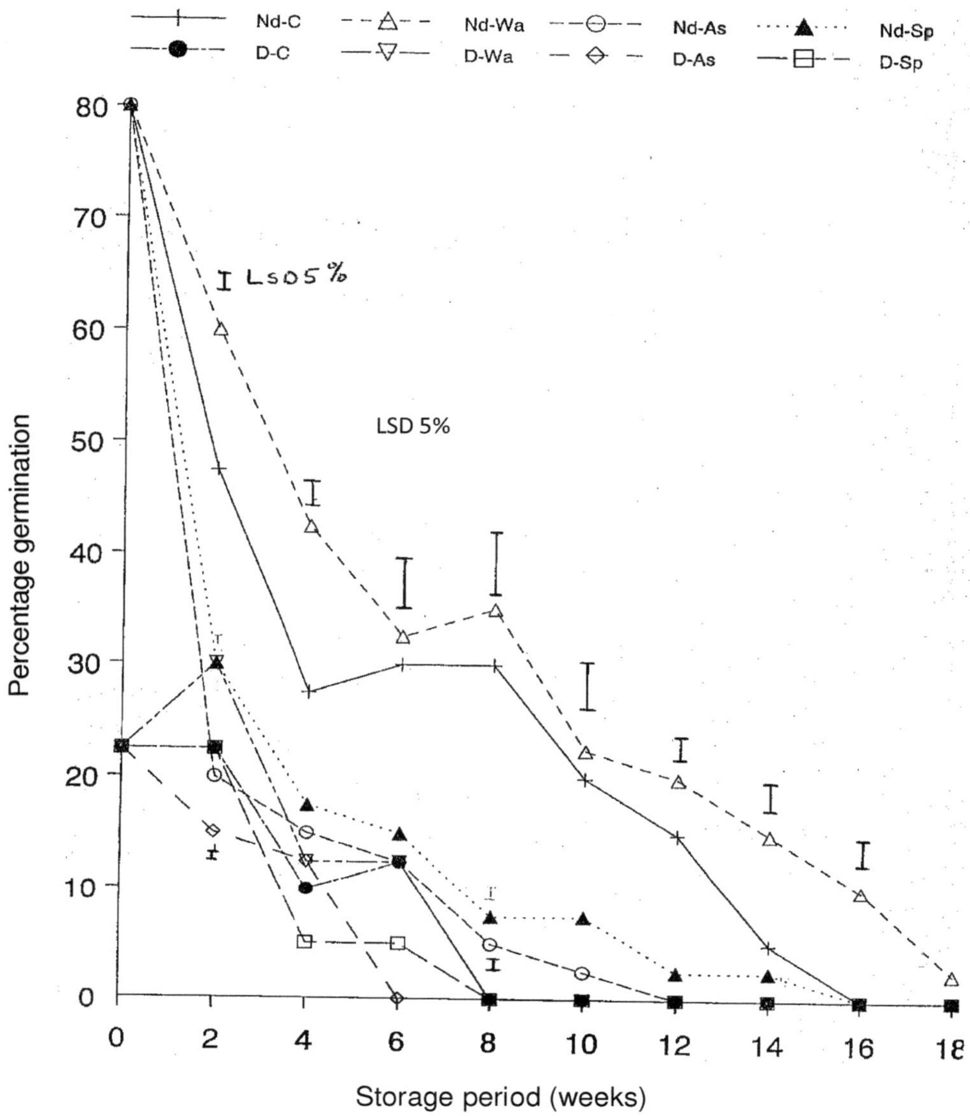

Figure 4. Seedling emergence from treated and untreated Dormant (D) and Non dormant (Nd) seed of cv "Oniyaya". Treatment: Control-C, Wood ash (Wa), Apron-star (As), Seed-plus (Sp).

Table 1. Estimates of the potential longevity (Ki) and the time taken for viability to drop to 50% (P_{50}) and R^2, the coefficient of determination of non-dormant *C. olitorius* seeds.

Crop variety	Seed dressing material	K_i	P_{50}	R_2
"Oniyaya"	Control	7.344	4.40	0.98
	Wood ash	6.850	4.80	0.97
	Apron star	6.477	3.00	0.93
	Seed plus	6.746	2.80	0.95
"Amugba"	Control	7.044	4.1	0.94
	Wood ash	6.686	2.50	0.89
	Apron star	6.163	2.50	0.89
	Seed plus	6.305	3.00	0.92

R^2 = Determination coefficient.

compared to the zero germination recorded for non-dormant seeds of the same age suggests that dormancy conferred longevity superiority on the seeds of this crop. This has also been shown to be the case in Arabidopsis by Debeaujon et al. (2000). The germination percentages of about 75% recorded for the two cultivars following the steeping of seeds at 12 WAS was however much lower than the 90 to 100% recorded for steeped seeds prior to storage. Also, the germination of 0% recorded for the unsteeped seeds of the two cultivars at 12 WAS, was much lower than the 25 to 33% obtained at the onset of storage. Two things must have been responsible for the decrease in values: One is the decline in vigour that is normally associated with seed age (Ibrahim et al., 2011) while the second could be linked to the reduction in germination energy that is known to decline as seed ages (Bukvic et al., 2007). Therefore, at 12 WAS, some seeds that though were still viable, might not have possessed enough energy to rupture the seed coats.

According to Oladiran (1986), hot water steeping might promote *Corchorus* seed germination by softening the seed coat which offers a mechanical restriction to embryo growth. An anatomical study of *Corchorus* seed revealed that the embryo is embedded in the endospermic tissue (Unpublished) which the radical would have to puncture to emerge during germination. Furthermore, Juntila (1973) and Watkins and Cantiliffe (1983) reported that, seed endosperm might constitute a mechanical restriction to radicle emergence. Aged seeds may have therefore reduced in vigour in this study and the weak embryo might therefore not be able to overcome the mechanical restriction of the seed coat and or the endospermic tissue in unsteeped seeds. Hot water steeping might have therefore, helped to alleviate the restrictions. This may need to be investigated empirically.

REFERENCES

Abdullahi M (2006). Characteristics of wood ash/OPC concrete. Leonardo Elect. J. Practices Technol. (LEJPT) 5(8):9-16.

Adebisi MA (1999). Effect of chemical seed treatments on viability and vigour in soyabean cultivars. Unpublished M. Agric. Discertation, University of Agriculture, Abeokuta, Nigeria. P. 73.

Adebisi MA, Daniel IO, Ajala MD (2004). Storage life of soybean (*Glycine max* L. Merril) seeds after seed dressing. J. Tropic. Agric. 42(1-2):3-7.

Adebisi MA, Ola JA, Akintobi DAC, Daniel IO (2003). Storage life of okra (*Abelmoschus esculentus* L. Moench.) seeds under humid tropical condition. Seed Sci. Technol. 36:379-387.

Ajala MO (2003). Influence of seed quality attributes on field emergence of pigeon pea (*Cajanus cajan* L.) and wing bean (*Psophocarpus tetragonolobus* L.). Tropic. Agric. (*Trinidad*). 80(2):118-122.

Bukvic G, Grljusic S, Rozman V, Lukic D, Lackovic D, Novoselovic D (2007). Seed age and pH of water solution effects on field pea (*Pisum sativum* L.) germination. Notulae Botanicae Horti Agrobotanici Cluj-Napoca 35(1):20-26.

Chattapadhyay SN, Pan NC, Day A (2004). A novel process of dyeing of jute fabric using reactive dye. Tech. Reports Textile Ind. India. 42:15-22.

Clapham WM, Zibilske HM (1992). Wood ash as a liming amendment. Commun. Soil Sci. Plan. 23:1209-1227.

Classen SL (1995). Seed management on agricultural research farms. IITA Research Guide 12 Training Programme. International Institute of Tropical Agriculture (IITA), Ibadan, Nigeria. P. 30.

Clerkx EJM, Vries HB, Ruys GJ, Groot SPC, Koornneef M (2004). Genetic differences in seed longevity of various Arabidopsis mutants. Physiol. Plant. 121(3):448-461.

Debeaujon I, Leon-Kloasterziel KM, Koornneef M (2000). Influence of the testa on seed dormancy, germination and longevity in Arabidopsis. Plant Physiol. 122:403-414.

Debeaujon, I, Koornneef M (2000). Gibberellin requirement for *Arabidopsis* seed germination is determined both by testa characteristics and embryonic abscicis acid. Plant Physiol. 122:425-435.

Dixon RA, Achnine L, kota P, Liu CJ, Reddy MSS, Wang LJ (2002). The phenylpropanoid pathway and plant defence-A genomic perspective. Mol. Plant Pathol. 3:371-390.

Ellis RH, Roberts EH (1980). Improved equation for the prediction of seed longevity. Annals Bot. 45:13-30.

Food and Agricultural Organization (FAO) (2007). Manual for training in seed production. In: Trivedi, R.K., Usman, I.A. and Umeh, J.C. (eds.), FAO, United Nations. P. 116.

Finney DJ (1971). Probit analysis, 3rd edition. Cambridge University Press, London. P. 33.

Ibrahim H, Oladiran JA, Habila J (2011). The effects of sarcotesta and storage temperature on the longevity of seeds of five papaya (*Carica papaya*) landraces. Int. J. Sci. Nat. 2(4):727-732.

Jansen V, Rensburg WS, Venter SL, Netshiluvhu AC, Van Dan HE, Vorster HJ, Fonde AO (2004). The role of indigenous leafy vegetables in combating hunger and malnutrition. South Afr. J. Bot. 70(1):52-59.

Juntila O (1973). The mechanism of low temperature dormancy in mature seeds of *Syringa* species. Physiol. Plant. 29:256-263.

Koornneef MM, Debeaujon I, Hilhorst H (2002). Seed dormancy and germination. Plant Physiol. 122:415-424.

Mandre M (2006). Influence of wood ash on soil chemical composition and biochemical parameters of young Scot pines. Proc. Estonian Acad. Sci. Biol. Ecol. 55:91-107.

Moses JJ, Ramsey M (2004). Nutrient components of jute and cotton material using enzyme treatment. Afr. J. Food Agric. Nutr. Dev. 22:11-21.

Oladiran JA (1986). Effect of storage, harvesting and seed treatment on germination, seedling emergence and growth in *Corchorus olitorius*. Sci. Hort. 28:227-233.

Pourcel L, Routaboul J, Kerhoas L, Caboche M, Hepiniec L (2005). Transparent testa 10 encodes a laccase-like enzyme involved in oxidative polymerization of flavinoids in Arabidopsis seed coat. Plant Cell. 17:2966-2980.

Schippers RR (2000). African indigenous vegetables. An overview of the cultivated species. Chattam, UK. Nat. Resour. Inst. P. 214.

Sekaramarthy SK, Prakash HS, Shetty SA, Prakash HS, Shetty SH (1994). Effect of thiram treatment on sorghum seed quality in relation to accelerated ageing. Seed Sci. Technol. 22:607-617.

Watkins JT, Cantliffe DJ (1983). Mechanical resistance of the seed coat and endosperm during germination of *Capsicum annuum* at low temperature. Plant Physiol. 72:146-150.

Effect of alcoholics extract on rat livers and antibacterial screening of *Citrullus colocynthis*

[1]Rasool Khatibi, [2]Mohammadreza DahmardehGhaleno and [3]Akbar Fakhireh

[1]Department of Natural Resource, University of Zabol, Iran.
[2] Department of Raneland and Watershed Managemennt, Facultyof Natural Resources Uuniversity of Zabol, 98615, Zabol,
[3] Department of Raneland and Watershed Managemennt, Facultyof Natural Resources Uuniversity of Zabol, 98615, Zabol, Iran.

Crude ethanolic extracts of fruits, leaves, stems and roots of *Citrulus colocynthis* Schrad were examined for their antibacterial potentialities against Gram positive and Gram negative bacilli. Ethanolic extracts of fruits, leaves, stems and roots were found to be active against Gram positive bacilli, viz., *Bacilus pumilus* and *Staphylococcus aureus,* while fruit and root extracts in double strength gave positive results against Gram positive bacillus *(Bacillus subtilis)*. The Gram negative bacilli viz., *Escherichia coli* and *Pseudomonas aeruginosa* showed no response, and in order for it to affect the alcoholics extract of rat liver, 50 rats were randomly divided into five groups (4 experimental and 1 controls). In the experimental groups a single daily dose of alcoholic extract of *C. colocynthis* (50, 100, 200, 400 g/kg) was administered intraperitonally. Normal saline was administered in control group. After two weeks, the rats were killed and the livers were removed and fixed with formalin (10%). Specimens were then processed and stained with H&E and Reticuline. The results indicated that there is a morphological change in liver cells including karyrrhexis, chromatolysis, and granulation of the cytoplasm. Additionally, collagen and reticular fibers were evident in liver parenchyma in high doses. Citrullus colocynthi can have toxic effects on liver cells which may induce hepatocyte necrosis and liver fibrosis. These effects were dose dependent. Further studies are necessary to clarify the issue.

Key words: *Citrullus Colocynthis*, *Bacilus pumilus*, *Staphylococcus aureus*, liver, necrosis, zabol.

INTRODUCTION

Citrullus colocynthis belongs to the family *Cucurbitaceae.* Members of this family are generally dioecious herbs which may be prostate or climbing by means of tenderils. Fruit is fleshy and many fruits are used as vegetable or as edible fruits. *C. colocynthis* is a small scarbid perennial creeping herb with prostate or climbing stem, bearing smooth spherical fruits which are mottled green when young and some what yellow when ripe (Shah and Qadry, 1985). In moderate doses a drastic hydrogogue, cathartic and diuretic; in large doses emetic and gastro-intestinal irritant; in small doses it is expectorant and alterative. Physicians use this drug extensively as a drastic purgative in ascites and jaundice and in various uterine conditions, especially in amenhcrroea. Colocynth in the form of the solid extract enters in to many of the purgative pills of modern pharmacy. It is useful in biliousness, fever, intestinal parasites, constipation, heap-tic and abdominal, visceral and cerebral congestions, dropsy, etc. Juice of the fruit mixed with sugar is a household remedy in dropsy (Anonymous, 1970).

Root is given in abdominal enlargements and in coughs and asthamatic attacks of children. For intestinal inflam-mation, tumours, etc. a powder of root is given for three days in doses of 45 grains well mixed with caster oil. A poultice of root is useful in inflammation of the breast of nursing mothers (Dastur, 1962). *C. colocynthis* (CCT) is traditionally used as an antidiabetic medication in tropical and subtropical countries (Diwan et al., 2000). This plant can induce insulinotropic (Nmila et al., 2000) and mild immunostimulating effects (Bendeddou et al., 2003). There is some evidence that it may induce side effects.

Corresponding Author E-mail: r_khatibi60@yahoo.com.

The comparative toxicity of the alcoholic extract of CCT has been studied in seven insect species in which the adult honey bee was more affected (el-Naggar et al., 1989). Sheep which were fed fresh CCT fruits and leaves (0.2 to 10 g/kg) showed signs of weeks caused death in goats (Barri et al., 1983). The other side- effects of this plant are toxic acute colitis (Golfain et al., 1989), reversible infertility (Chaturvedi et al., 2003) and hepatotoxicity in rats (Barth et al., 2002). These damages were sometimes enhanced with higher doses of CCT. As the liver is a sensitive organ and many substances including toxins accumulate in this organ and induce liver toxicity, therefore the aim of this project was to study the histopathological changes in the liver after ingestion of CCT in male rats.

MATERIALS AND METHODS

Different parts of the plant *C. colocynthis,* that is, roots stems, leaves and fruits were collected in summer 2009 and January, 2010 from the desert area near zabol, Iran. The fruits, stems, leaves and roots of *C. colocynthis* were carefully separated from weeds, soil particles, added adulterants and other extraneous matter, and were dried at room temperature for one month except fruit which took three months to dry. The fruits, stems, leave and roots were crushed to coarse power separately. 50 g of the powder of fruits, leaves, stems and roots were homogenized in 100 ml of ethyl alcohol (90%) separately in homogeniser and operated for 5 min. The process was repeated for three to four times with the same quantity of alcohol. The extract was filtered through filter paper under vacuum. The alcohol extract was evaporated through rotary evaporator under reduced pressure at 40°C.

The weights of residue obtained from 50 g powders of fruits (2.37 g), leaves (8.27 g), stems (2.24 g) and roots (5.00 g) were obtained too. The alcoholic extract of this plant in order to effect on rat livers was obtained using the percolation method.

Growth media

All the bacteria *Bacillus pumilis, Bacillus subtilis, Staphylococcus aureus, Escherichia coli* and *Pseudomonas aeruginosa* were cultivated on nutrient agar medium.

Nutrient agar medium

Beaf extract 3.0 g
Peptone 5.0 g
Sodium chloride 5.0 g
Ag 15 to 20 g
Distilled water 1000 ml
pH 7.4

All the ingredients were mixed and dissolved in distilled water then sterilized by autoclaving at 121°C (151 lb/sq. inch), for fifteen minutes (Cruickshank et al., 1975).

Preparation of Inoculum

The test organism was maintained by bi-weekly transfer on agar slants of nutrient agar medium. Growth was washed from slants with sterilized 3 ml of normal saline. This suspension was used to inoculate in wide based flask containing 200 ml of the same medium supplemented with 10 g of agar per litre (Ayoub and Yankov, 1981).

Antibacterial screening

The antimicrobial activity of extract from fruits, leaves, stems and roots against various Gram-positive and Gram-negative bacteria were observed. The organisms given below were collected from Central Drugs Testing Laboratory Zabol in south East of Iran Pharmaceuticals.

Gram-positive

1. *B. pumilis*
2. *B. subtilis*
3. *S. aureus*

Gram-negative

1. *E. coli*
2. *P. aeruginosa*

Preparation of dilution

All the glass apparatus used were sterilized in an oven at 200°C, for thirty minutes. A series of four dilutions were prepared of each extract and marked n1, n2, n3 and n4. The dilutions were prepared as follows: i) First dilution (n1) was prepared by dissolving one gram of dry extract in 5 ml of ethyl alcohol (200 mg/ml). ii) Second dilution (n2) was prepared by taking 1 ml of dilution n1 and 1 ml of ethyl alcohol (100 mg/ml). iii) Third dilution (n3) was prepared by taking 1 ml of dilution n2 and one ml of ethyl alcohol (50 mg/ml). iv) Fourth dilution (n4) was prepared by taking one ml dilution n3 and one ml of ethyl alcohol (25 mg/ml).

Preparation of assay plates

Petri dishes were sterilized in an Autoclave at 121°C (15 lb pressure/sq in. for 15 to 20 min then labeled with the name of bacterias whose inocculum were prepared. Each Petri dish was divided into four equal parts and each part was marked accordingly. One part was marked as control, while other three parts used for each sterilized in conical flask (500 ml) by autoclaving at 121°C (15 lb pressure/sq in.) for 15 to 20 min. Some quantity of media was prepared for each part of the plant and for different inoculums. 0.1 ml of bacterial inoculum was transferred in (100 ml) sterilized melted nutrient agar medium for bacterial inoculation at temperature not more than 45°C and was gently shaked to mix the inoculum. In each sterilized Petri dish 20 ml of inoculums media was poured carefully at sterilized environment to avoid contamination and then allowed to solidify at room temperature. Now small uniform and superficial holes were made by the help of sterilized borer in the center of each part of Petri dish then each hole was sealed with one drop of same melted media. One drop of extract dilution of n1, n2, n3 and n4 were dropped with the help of pipette very carefully in three holes of individual Petri dishes, respectively with the help of pipette. The fourth one hole in each Petri dish was left as control and one drop of ethyl alcohol was dropped in it.

This process was repeated for each part of the plant and also for individual microorganism. All the Petri dishes with bacterial inoculation were incubated at 37°C for 24 to 48 h. After incubation period the zones of inhibitions were measured results were noted.

Table 1. Average zone of inhibition for *Staphyloccocus aureus.*

Dilution	Zone of inhibition (cm)			
	Fruit extract	Leaves extract	Stem extract	Root extract
First	1.30	1.31	1.27	1.32
Second	1.22	1.12	1.24	1.14
Third	1.19	1.10	1.21	1.12
Fourth	1.18	1.10	1.10	1.09

Table 2. Average zone of inhibition for *Bacillus pumilus.*

Dilution	Zone of inhibition (cm)			
	Fruit extract	Leaves extract	Stem extract	Root extract
First	1.27	1.23	1.23	1.21
Second	1.26	1.16	1.20	1.17
Third	`1.203	1.163	1.19	1.11
Fourth	1.173	1.083	1.1	1.06

Table 3. Average zone of inhibition for *Bacillus subtilis.*

Dilution	Zone of inhibition (cm)			
	Fruit extract	Leaves extract	Stem extract	Root extract
First	1.02 (1.20 with	-ve	-ve	1.05 double conc.)
Second	-ve	-ve	-ve	1.02
Third	-ve	-ve	-ve	-ve
Fourth	-ve	-ve	-ve	-ve

Effect of alcoholics extract on rat livers

Fifty male rats, weighing 200 to 250 g were obtained from the Animal House of Shiraz Medical School and were maintained under standard conditions (light, temperature, humidity and free access to water and food).The rats were then selected and randomly divided into 5 groups (4 experimental and 1 control). In the experimental groups a single daily dose of the alcoholic extracts of *C. colocynthis* (50, 100, 200, 400 mg/kg) was injected intraperitoneally. Each of the experimental groups received only one particular dose for the duration of 14 days. Normal saline was administered in the control group. After 14 days, the rats were sacrificed under deep anesthesia and their livers were removed and fixed with 10% formalin. Specimens were processed routinely and sections with 5 micron thickness were prepared and stained with Hematoxiline-Eosine (H&E), Reticuline methods. The slides were studied by light microscopy by Knodell scoring system for assessing histological activity in asymptomatic chronic active hepatitis (Knodell et al., 2003) and the results were recorded. The results were analyzed using Chi-square and ANOVA tests. P-value less than 0.05 was considered statistically significant

RESULTS AND DISCUSSION

Antibacterial screening

The results obtained from the parts of *C. colocyn*this Schrad, that is, fruits, leaves, stems and roots are shown in the Tables 1, 2 and 3. A sufficient time period (24 h) was given for thorough and complete extraction but same result and quantity was obtained when extraction was made in homogenizer with ethyl alcohol (400 ml as menstrum) for 15 min and evaporated the ethyl alcohol under reduced pressure using rotary evaporator and drying the extract at 40°C in oven. Another change was made in the extraction process by using normal saline as menstrum, and soaked over night instead of ethyl alcohol. Then same series of dilutions were prepared but results were negative. Changes were also made in dilution series after first dilution of ethyl alcohol, then further dilution was prepared with sterilized water and results were negative. For experimental work the bacterial inoculum was prepared in normal saline as mentioned in method and the results were compared by changed method, that is, inoculum was prepared in sterilized water and phosphate buffer, positive results were obtained in the changed method. The same results were obtained by prolonging the incubation period from 24 to 48 to 72 h. The same zone of inhibition was obtained by changing the size of borer one number less.

During performing the experiment for checking the

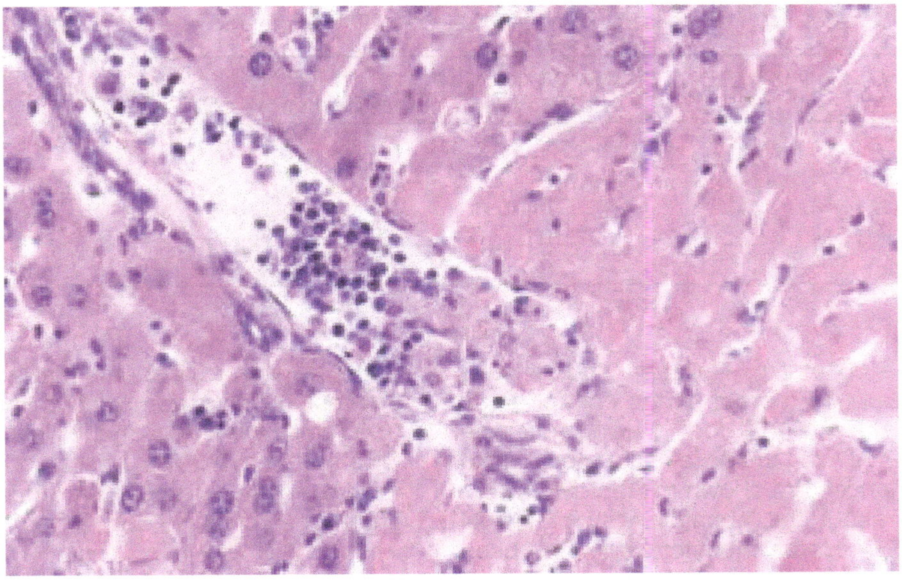

Figure 1. Liver cell necrosis in dose 400 mg/kg with H&E staining (400×).

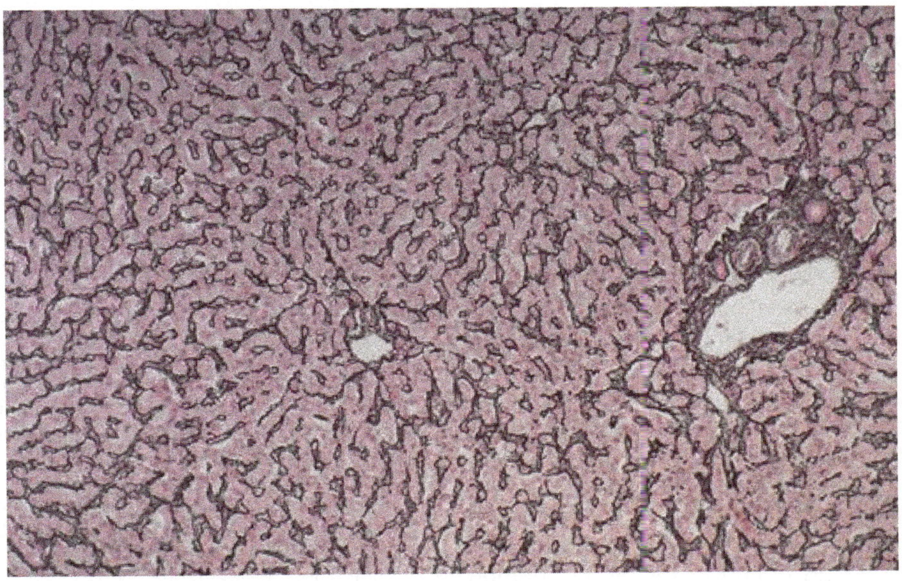

Figure 2. Reticular fiber in liver parenchyma in dose 400 mg/kg with Reticulin staining (100×).

antibacterial activity of crude extracts extreme care was taken to prevent the contamination. All the apparatus and growth media were sterilized for this purpose using oven and autoclave. A sterilized environment was made for inoculum. However, the crude extract was significantly active which shows the prominent antibacterial activity. The crude extract also shows appreciable inhibition of the growth of Staphylococcus aureus, *B. pumilus* and *B. subtilis*. The antibacterial activity against Escherichia coli and Pseudomonas aeruginosa were negative.

The conclusion which drawn from the above discussion is that crude extracts showed active response against the some strains of bacteria may be due to carbohydrates, flavonoids, glycosides and tannin which are reported in the literature.

Effect of alcoholics extract on rat livers

A morphological change in hepatocyte including karyorrehexis, chromatolysis and granulation of the cytoplasm was seen using H&E staining (Figure 1) especially

Table 4.The reaction of liver induced–CCT using H&E, PAS, staining between control and experimental groups.

Groups \ Lesion	Hepatocyt necrosis H&E staining	Fibrosis inflammatory cell	Reticuline H&E staining
Control	0	0	0
50 mg/kg	0	0	0
100 mg/kg	5	0	2
200 mg/kg	9	0	6
400 mg/kg	12	3	9

With doses of 200, 400 mg/kg. Collagen and reticular fiber were observed around more than the control group in parenchyma at a dose of 400 mg/kg using reticuline staining (Figure 2). All of these effects were dose dependent. These changes are shown in (Table 4). There is growing concern about the hepatotoxicity of herbal remedies (Larry, 1997). Herbal hepatotoxicity has been recognized for many years, but new agents are constantly being identified (Cruickshank et al., 1975). Citrullus colocynthis extract was found free of hepatotoxic effect at concentrations up to 100 μg/ml (Chaturvedi et al., 2003). But, higher concentrations seem to have some degree of hepatotoxicity. Male wistar rats that were fed diets containing 10% CCT ripe fruits showed body weight loss, inefficiency of feed utilization, diarrhea, ruffled hair and enterohepatonephrotoxicity (Adam et al., 2001). In this study, the effect of different concentrations of CCT on the liver was investigated. The results showed some histological changes in the nucleus and cytoplasm of hepatocytes. The changes observed in the nuclei included karyorrhexis and chromatolysis. The mechanism for these changes is not clear but other re-ports have showen that CCT has a damaging effect on different cells. The ethanol extract of CCT decreases the concentration of sialic acid in serum of mice. This decrease is concomitant with an increase in the unmasking of galactose residues that is recognizable by macrophages in apoptotic cell (Itzhaki et al., 2003). Therefore; it seems that CCT by decreasing sialic acid induces cell degeneration. In addition, CCT causes an increase in neutrophils (Elawad et al., 1984) which confirms the above finding. Elevation of alkaline and acid phosphates is a useful marker for diseases of the liver such as liver cirrhosis (Israeli and Bogin, 1986; Ryvniak, 1986). However, in some studies alkaline and acid phosphatase concentration was decreased by CCT (Chaturvedi et al., 2003). So, it seems that CCT has probably no effect in causing liver cirrhosis. On the other hand, it is a well-known fact that inflammation can be precursor of liver fibrosis. In our study, we observed scattered neutrophil and lymphocytes in liver parenchyma. This phenomenon could potentially lead to liver cirrhosis. CCT has been shown to increased oxidative damage (Shivakumar and Geoffrey, 2000). On the other hand CCT extract has stimulated lipid peroxidation, H_2O_2 formation and has amplified chemiluminescence in rat liver microsomes (Barth et al., 2002). Therefore, it seems that a decrease in iron and an increase in lipid peroxidation induce the generation of free radicals which damages hepatocytes. In conclusion, CCT can have toxic effects on liver cells which may induce hepatocyte necrosis and liver fibrosis. However; more research is needed to clarify the issue.

Conclusion

In conclusion, the studied plants could provide some activity against antibacterial and anti fungal; however, it is not known that which component of the extract is responsible for this effect. Further studies effect using isolated constituents instead of the whole extract should be carried out.

REFERENCES

Adam SE, Al Yahiya MA, Al Farhan AH (2001). Combined toxicity of Cassia senna and *Citrullus colocynthis* in rats. Vet. Hum. Toxicol., 43(2):70-2.

Anonymous (1970). Hamdard Pharmacopoeia of Eastern Medicine, H Nati Found.Pak. 2nd Impression, 373p.

Ayoub SMH, Yankov LK (1981). On the constituents of the peel of *Citrullus colcynthis*. P-2. Fitoterpia, 52(1): 13-20.

Barri ME, Onsa TO, Elawad AA, Elsayed NY, Wasfi IA , Abdel Bari EM, Adam SE(1983).Toxicity of five Sudanese plants to young ruminants. J. Comp. Pathol., 93(4): 559-675.

Barth A, Muller D, Durrling K (2002). *In vitro* investigation of a standardized dried extract of *Citrullus colocynthis* on liver toxicity in adult rats. Exp. Toxic. Pathol., 54(3): 223-30.

Bendeddou D, Lalaoui K, Satta D (2003). Immunostimulating activity of the hot water-soluble polysaccharide extracts of *Anacyclus pyrethrum*, *Alpinia galanga* and *Citrullus colocynthis*. J. Ethnopharma, 88(2-3): 155-60.

Chaturvedi M, Mali PC, Ansari AS (2003) . Induction of reversible antifertility with a crude ethanol extract of *Citrullus colocynthis* schrad fruit in male rats. Pharmacology, 68(1): 38-48.

Cruickshank R, Duguid JP, Marmion BP,Swain RHA (1975). Medical Microbiology. Churchill Livingstone., Edinburg. London and New York, 356p.

Dastur JF (1962). Medicinal Plants of India and Pakistan. D.B. Tarapo. Sons and Co. Pvt. Ltd. Bombay. India, 56p.

Diwan FH, Abdel-Hassan IA, Mohammed ST (2000). Effect of Saponin on mortality and histopathological changes in mice. J. East. Mediterr. Health, 6(2-3): 345-351.

Elawad AA , Abdel Bari EM, Mahmoud OM Adam SE(1984). The effect of *Citrullus colocynthis* on sheep. Vet. Hum. Toxicol., 26(6): 481-485.

El-Naggar ME, Abdel-Sattar MM, Mosallam SS (1989). Toxicity of colocynithin and hydrated colocynithin from alcoholic extract of Citrullus colocynthis pulp. J. Egypt Soc. Parasitol., 19(1):179-185.

Golfain D, Lavergne A, Galian A, Chauveinc L, Prudhomme F (1989) .

Peculiar acute colitis after ingestion of colocynth: a clinicopathological study of three cases. Gut, 30(10): 1412-1418.

Israeli BA , Bogin E (1986). Biochemical changes in liver, kidney and blood associated with common bile duct ligation. Clin. Chim. Acta. 160(2): 211-221.

Itzhaki O, Skutelsky E, Kaptzan T, Michowitz M, Huszar M, Leibovici J(2003). Ageing-apoptosis relation in murine spleen. Mech. Ageing Dev., 24(10- 12): 999-1012.

Knodell RG, Ishak KG, Black WC, Chen TS, Graig R, Kaplowitz N, Kiernan TW, Wollman J (2003). Formulation and application of a numerical scoring system for assessing histological activity in asymptomatic chronic active hepatitis. J. Hepatol., 38(4): 382-386.

Larry D (1997). Hepatotxicity of herbal remedies. Hepatology, 26(1): 47-51.

Nmila R, Gross R, Rchid H, Roye M, Manteghetti M, Petit P, Tijane Ribes G, Sauvaire Y (2000). Insulinotropic effect of *Citrullus colocynthis* fruit extracts. Planta Med., 66(5): 418-423.

Ryvniak VV (1986). Acid phosphatase distribution in the liver in cirrhosis. Bull. Eksp Biol. Med., 101(3): 363-365

Shah CS, Qadry JS (1985). A Text book of Pharmacognosy (5th Edition). BS. Shah, Prakashan, Pankore Naka, Ahmedabad, India, 284p.

Shivakumar Ch, Geoffrey C (2000). Herbal hepatotoxicity: An expanding but poorly defined problem. J. Gastroenterol. Hepatol., 15: 1093-1099.

Effect of field capacity of sacred forest soils on regeneration of *Aframomum melegueta* on Western Highlands in Cameroon

Anjah G. M.[1], Nguetsop V. F.[1], Tsombou F. M.[1] and Njoya M. T. M.[2]

[1]Department of Plant Biology, University of Dschang, P. O. Box 67, Dschang, Cameroon.
[2]Department of Biological Sciences, the University of Bamenda, P. O. Box 39- Cameroon.

This study was aimed at investigating the field capacity of three sacred forest soils of the West Region of Cameroon and their effect on the germination potential and early plant growth characteristics of fresh and dried seeds of *Aframomum melegueta.* Sample soils were collected from Bamendjo (BO), Bamendjinda (BA), and Mbing Mekoup (MB) sacred forest areas. Each area was divided into three location sites 1, 2 and 3 and each of the sites was made up of 3 collection points (P1, P2, and P3). The field capacity (FC) of each site was determined before the planting of the seeds. For each collection site, the soil of a collection point was randomly chosen to form two groups with five replicates each. In each pot, 10 fresh seeds and 10 dried seeds were planted respectively and monitored for germination. Measurements for seedling growth were spread over 14 weeks. Results showed that the field capacity (FC) of BO (71.18%) and of BA (68.64%) were significantly different from those of MB (51.66%). The germinating percentages of the seeds were very high in the soils of BO (76. 67%) and BA (61. 33%) as compared to those of MB (48%). The latent period of the seeds was longer in the soils of MB (12 days) and shorter in the soil of BO (7 days). The average height of the plants in the soils of BO (6. 75 cm) and BA (6. 05 cm) were greater than those of the plants in the soils of MB (4.61 cm). The number of leaves of the plants of BO (4), BA (4) and MB (4) were not significantly different. The leaf surface area was higher in BO (10.43 cm²) and BA (8, 30 cm²) as compared to that of MB (5. 37 cm²). This study has shown that the FC of the soils of the sacred forests of BO and BA positively affected the germination percentage and the latent period of the fresh seeds; the leaf surface area and the heights of young plants. Meanwhile, the germination rates of the fresh seeds and *A. melegueta* dried seeds were not affected by the FC of the soils.

Key words: Field capacity, Bamendjo, Bamendjinda, Mbing Mekoup, sacred forests, germination, growth, *Aframomum melegueta*.

INTRODUCTION

The forest is a living environment comprising biodiversity which allows communities to meet their socio-cultural or socio- economic needs. The forests ecosystems of Central Africa offer rural people the opportunity to always collect non-timber forest products (Honoré, 1999), among these being *Aframomum melegueta* (Jujube), a herbaceous perennial in the Zingeberaceae family. The leaves are simple and asymmetrical with a membranous ligule. The inflorescences are born at the base of the stem with reddish bracts. The fruit is a bright red berry at maturity. It is indehiscent with seeds enclosed in a white pulp.

According to Lékané (2009), *A. melegueta* is an important plant on the socio-economic and socio-cultural planning as its fruits are presented whole during various traditional rites such as in the induction of traditional leaders, successions of next-of-kin, protection, blessing rites and in traditional medicine. The fruits are also consumed as food by humans, primates and some rodents. In modern and traditional medicine, all the parts of this plant are used to cure several illnesses depending on the users and the region in which it is encountered (Rafatullah et al., 1995). According to Baumeer (1995), these fruits are sold in domestic and international markets with prizes varying between 50 and 100 FCFA per berry while a 50 kg bag costs between 14,000 and 20,000 FCFA. Honoré (1999) states that commercialization generates revenues that are intended for household economy and the sales play major economic income generation for the population at large.

A. melegueta is native to West Africa, and is found in Guinea, Nigeria, Gabon, and Cameroon (Geudje et al., 1996). In Cameroon, it is found in the Western Highlands of some sacred forests. These forests include those of Mbing Mekoup (MB), Bamendjinda (BA) and Bamendjo (BO). The populations of this plant and other *Aframomum* species are very low (Tiokeng, 2007). According to Geudje et al. (1996), the propagation of *A. melegueta* presents difficulties at the germination phase but, states that the propagation of *A. melegueta* would be more appropriate with its rhizomes. But this method proved inconclusive during our experience probably due to soil and climatic factors. According to Eyob (2008), the seeds of *A. corrorima* also present germination problems, probably due to their low permeability to water, and observed that the seeds can remain dormant for two to three years before germinating.

Niang-Diop et al. (2010), observed that the germination of seeds in natural ecosystems can be limited by a number of factors including integumentary seed dormancy. It can also be related to other factors including the soil retention capacity of soil water. Lachiheb and Neffati (2005) observed that *Aframomum* seeds could remain dormant for 2 to 3 years before germinating after seeds are planted. Indeed, a seed can germinate if the embryo has the ability to imbibe.

However, seed germination by Babou et al. (2001) is not always the only problem in the regeneration of a species, because sometimes after germination, the seedlings are subjected to several factors such as bush fires, temperature variations and predation which significantly affect the growth thereof and lead to high mortality with few, reaching maturity. This is the case with *A. melegueta* which not only presents difficulties in its germination but also shows significantly high mortality of the plantlets. The amount of water needed by a plant for its metabolic cycle may be derived from its "reservoir" which is the ground. This will depend on its ability to retain water (Segalen, 1953). Assie et al. (2010) state that, the ability of a soil to retain water depends on its texture and topography.

There is however, no existing works on the soil retentive capacity of the sacred forests of Bamendjo (BO), Bamendjinda (BA), and Mbing Mekoup (MB), and thus faced with the problem of regenerating *A. melegueta*, this study was aimed at contributing to the regeneration of this plant taking into account edaphic factors including the water retention capacity of the soil of the three sacred forests and their effects on the germination and early growth parameters.

MATERIALS AND METHODS

Study area

This study was conducted in three sacred forests located in the Western Highlands of the West Region of Cameroon. It culminates at an average altitude of 2,000 m (Kuete and Dikoume, 2000). Residual forest formations that are most often maintained because of their "sacred" nature include Bambotous Sacred forests of Mbing Mekoup located about 2,183 m above sea level at 5° 37'32 'north latitude and 10° 06'28' 'East longitude. It occupies an area of 27.7 ha in the Batcham district. Bamendjinda, another sacred forest, has small and large forest surface areas of 1.2 and 4.2 ha respectively. The study was carried out in the large. These forests are located around 1,327 m above sea level at 5° 36 '48' 'north latitude and 10° 17' 48 '' East. Bamendjo on the other hand, covers an area of 1.8 ha. It culminates to 1,376 m above sea level between 5° 33 '47' 'north latitude and 10° 17' '42' east longitude.

Climate

The Western Highlands have a temperature influenced by altitude fluctuating between 18.9 to 21.1°C (Centre Technique de la Forêt Communale, 2013). The prevailing climate is cool and moist because of their altitude, exposure and latitudinal position. According to Suchel (1972), a short dry season (4 months) from November to February and a long rainy season (8 months) from March to October. These highlands receive rainfall of about 2,000 mm/year.

Plant material: *A. melegueta* K. Shum

The plant materials for germination were the fresh and dried seeds collected from populations of the Western Highlands (Figures 1 and 2).

Methodology

The experiment had two specific objectives with the first being to investigate the field capacities of the soils in each of the sacred forest BO, BA and MB; and the second, to investigate the

Figure 1. Fresh fruits and seeds of *A. melegueta*.

performance in the germination of seeds and growth of seedlings of *A. melegueta* in these soils.

Determination of the field capacity

For each sacred forest, soil samples were collected on laid out transects of one hectare from the periphery to the interior of the forest. Each forest area (BO, BA, MB) was divided into three location sites (S1), (S2) and (S3), with each having three collection points that is, .replicates (P1, P2 and P3). Each replicate was separated from the next by a spacing distance of 50 m, and two adjacent sites were also separated by a spacing distance of 50 m (Figure 3). A machete was used to drill to a depth of 0 to 10 cm in the soil. The soil samples were stored in polyethylene bags. A GPS was used to obtain the geographical coordinates (longitude, latitude and altitude). A tape was used to measure the distances among points and between sites. The soil sample collection procedure was the same for the three forests. A total of 27 samples were collected for the three forests.

In the laboratory, soil samples of 350 g were weighed with an electronic balance of mark METTLERPM 2000, and dried. Each of the 27 samples were poured in an envelope of 23 cm x 16 cm layflats and run in an oven at a temperature of 105°C for one week, after which time a constant weight was obtained, and the dry weight (DW) of each sample measured. These were placed in cylindrical boxes of length 11 cm and diameter of 10 cm lay flats. The bottom of each box was perforated with 15 holes of 3 mm in diameter. Each box was placed on top of another of same size but not perforated. Each soil sample was washed with 350 ml of water. After 24 h soil samples were again weighed and the fresh weights (FW) noted. Both weight measurements allowed us to determine the water retentive capacity (WRC) of these soil samples and were calculated as a percentage according to the method of Braudeau and Mohtar (2005).

$$WRC = \frac{FW - DW}{DW} \times 100$$

WRC: water holding capacity, FW: fresh weight, DW: dry weight.

Seed germination

The experiment on germination was spread over eight weeks involving the collection of three soil samples from three sites within the nursery of the Forestry Department of the University of Dschang. According to Nwame (1997) Dschang is at an altitude of 1400 m, between Latitude 5°10' and 5°30' North and Longitude 9°50' and 10°20' East. There are two distinct seasons, the rainy season and the dry season whose start and end periods are not definite nowadays due to climate change. The average annual rainfall ranges from 13 mm in December to 340 mm in September. The temperatures range from between 18.9 to 21.1°C. A total of 900 seeds of *A. melegueta* divided into two groups of 450 fresh seeds and 450 dry seeds respectively were used.

Three soil samples from each site (Site 1, 2 and 3), were collected randomly. The S1 soil from P3 was used to fill 10 polyethylene bags. For S2, P6 soil was used to fill 10 bags of polyethylene and for S3, P9 was used, which gave a total of 30 pots filled per forest site and a grand total of 90 pots for the three sites from which soil samples were collected. The 90 pots filled were treated with fungicides (Furaplants, Country) to prevent fungal attack. Each batch of 10 pots filled was divided into two groups of five pots in which were sown 10 fresh seeds and 10 dry seeds of *A. melegueta* respectively.

Germination parameters measured for over 8 weeks period included: the seed germination rate expressed by the average germination time AGT (the time at which it reached 50% of sprouted seeds); latency (which is the time between planting and first germination) (LT) (Ahoton et al., 2009) and the percentage of seed germination (PG).

Measuring seedling growth

This phase of the work spread over 14 weeks. Shoot height was measured from the base to the apex of the shoot. The leaves produced were counted after every two weeks. The measurement of the length axis of the limb multiplied by the measurement of the short axis multiplied by a coefficient yielded the leaf surface area according to the formula L × W × 2/3 (Raunkiaer, 1934). Growth

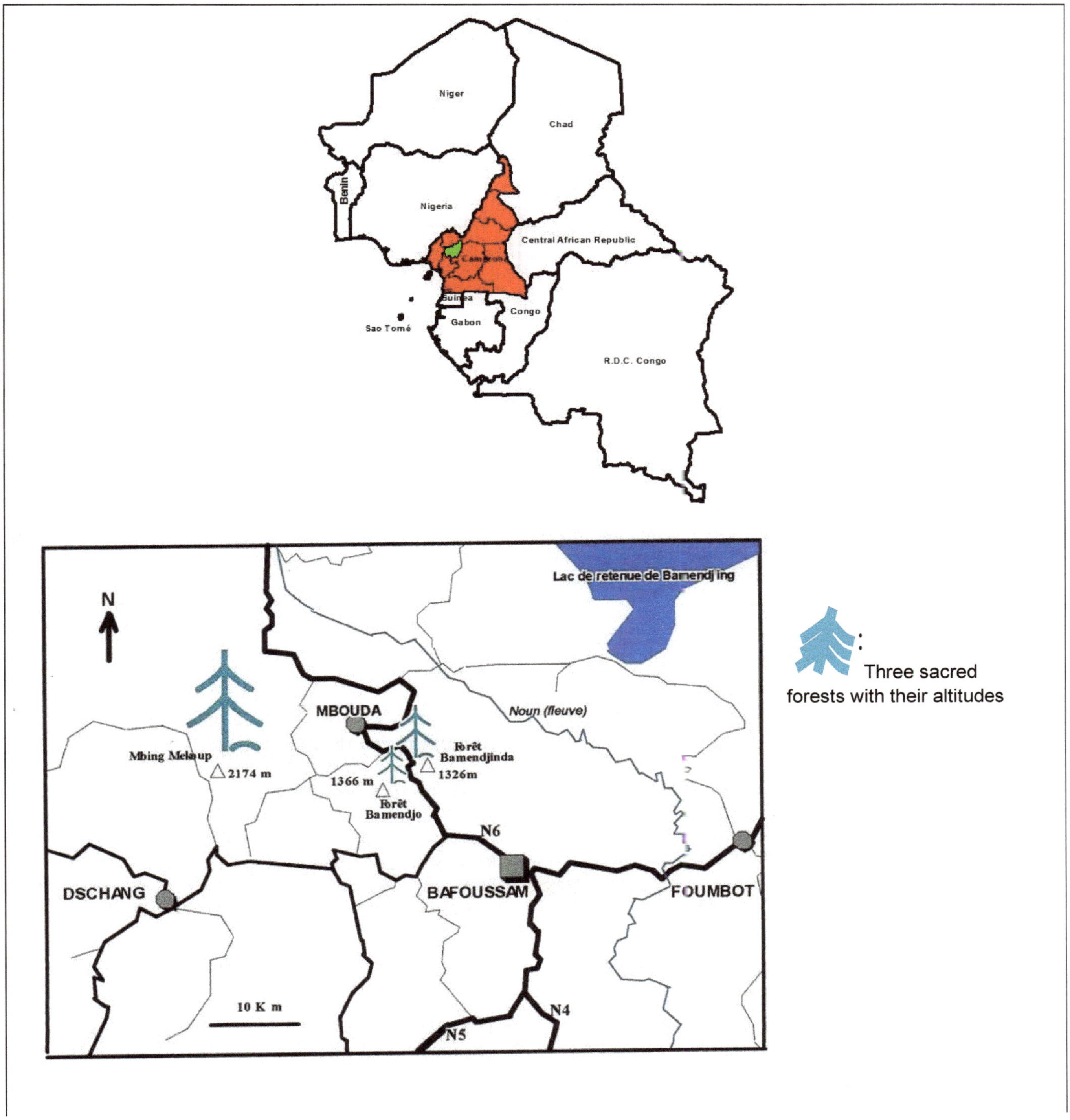

Figure 2. Location map of the study area in West and Central Africa.
Sources: Administrative maps of Cameroon, MOTHERS ISH Centre National Geographic, Yaoundé, 1984 - Lifting GPS, September 2006, April 2007 and GARMIN Database 2003.

parameters were measured as follows: the average number of leaves, the average leaf area and the average height of seedlings. These parameters were observed with four seedlings randomly selected from germination bags.

Data analysis

Data were subjected to analysis of variance (ANOVA) and means were separated using the Fischer LSD test. The Sigma Plot

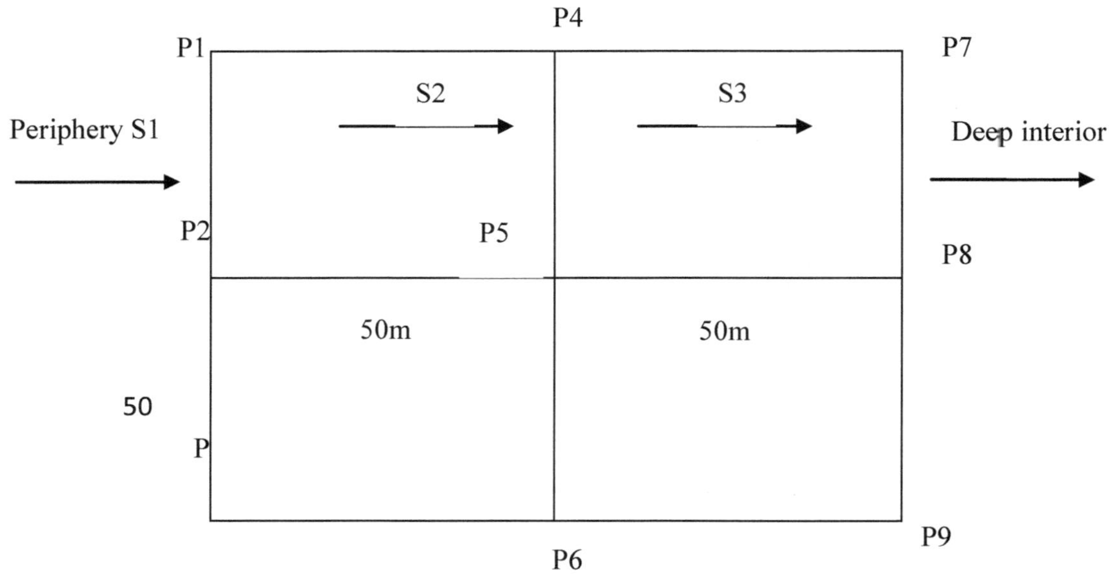

Figure 3. General scheme of the location of sites and points where soil samples were collected S1, S2, and S3 =periphery, mid-interior, and deep forest respectively; P = replicates (P1, P2, P3).

Statistical Version 11 Software was used.

RESULTS

Determination of the field or water retention capacity of the sacred forests soils

Water retention capacity was not significantly different for all sites in collection areas of BO and BA (Table 1). But all the sites of MB recorded significantly low values.

Germination rate

The seed germination rate, expressed by the average germination time (AGT) was not significantly different in the three sites of each area (Table 2).

Latency in days (TL)

The latency time was significantly ($P<0.05$) different for all collection zones, starting from MB, BA and BO in descending order. However, there was no significant ($P>0.05$) difference for the three sites of the same zone (Table 3).

Germination percent

There was no significant difference ($P>0.05$) for seed germination for Site 3 in the different locations (BO, BA, MB). All the seeds of Site 1 of BO germinated, marking a significant difference ($P<0.05$) with the other sites and locations (Table 4).

Seedling growth parameters

Average number of leaves

During the 14 weeks of growth, the mean number of leaves for each site (Table 5) varied between 3.75± 0.52 to 4.12±0.45 for all the sites in all the collection areas, but without any significant difference ($P>0.05$).

Leaf area

The mean values for leaf surface area after 14 weeks of growth varied significantly ($P<0.05$) for the different locations (Table 6) in decreasing order from BO to MB. However, there was no significant difference ($P>0.05$) for the sites within the same collection area.

Height of seedlings in different sites

The seedlings of MB were not significantly different ($P>0.05$) in height for all three sites, but were significantly ($P<0.05$) inferior in height to those of the sites in BO and BA. Very tall seedlings were found in sites 2 of BO and 3 of BA (Table 7).

Table 1. Water retention capacity in the different sites for each forest area.

Area collection	Water holding capacity (%)			
	Site 1	Site 2	Site 3	Zone
BO	72.68^{a} ±0.45	69.65^{a}±0.50	71.02^{a}± 0.42	71.18^{a}±1.50
BA	67.23^{a}±0.50	68.67^{a}±0.35	70.01^{a}±0.25	68.64^{a}±0.50
MB	50.96^{b}±0.55	54^{b}±0.52	50.03^{b}±0.53	51.66^{b}±0.54

Means ± SD values from 3 replicates; Values followed by different letter superscripts in the same column are significantly different (P <0.05).

Table 2. Germination rate of seed per area.

Area collection	AGT (weeks)			
	Site 1	Site 2	Site 3	Zone
BO	3.02^{a}±0.35	2.7^{a}±1.50	3.5^{a}±0.50	3.02^{a}±0.51
BA	2.5^{a}±0.71	3.3^{a}±0.42	2.8^{a}±0.51	2.9^{a}±0.45
MB	2.9^{a}±0.61	3.03^{a}±0.44	2.9^{a}±0.54	2.9^{a}±0.56

Means ± SD values from 5 replicates; Values followed by different letter superscripts in the same column are significantly different (P <0.05).

Table 3. Germination latency in days at various sites and collection areas.

Collecting zone	Germination latency time (days)			
	Site 1	Site 2	Site 3	Zone
BO	7.01^{c}±0.50	7.07^{c}±0.35	7.05^{c}±0.45	7.04^{c}±0.51
BA	9.03^{b}±0.65	9.06^{b}±0.56	9.05^{b}±0.46	9.08^{b}±0.36
MB	12.01^{a}±0.25	11.03^{a}±0.52	11.07^{a}±0.45	11.04^{a}±0.50

Means ± SD values from 5 replicates; Values followed by different letter superscripts in the same column are significantly different (P <0.05).

Table 4. Percentage germination (PG) in different sites and soil collection area of each zone

Area	PG (%)			
	Site 1	Site 2	Site 3	Zone
BO	100^{a}±0.00	66^{b}±1.52	64^{b}±1.55	76.67^{a}±1.53
BA	56^{b}±1.58	60^{b}±1.45	68^{b}±0.59	61.33^{b}±0.55
MB	46^{c}±0.50	46^{c}±0.51	52^{b}±0.49	48^{c}±0.54

Means ± SD values from 5 replicates; Values followed by different letter superscripts in the same column are significantly different (P <0.05).

Table 5. Average number of leaves in different sites and each area of sample collection and area.

Collection area	Number of leaves			
	Site 1	Site 2	Site 3	Zone
BO	4.10^{a}±0.35	4.12^{a}±0.45	4.09^{a}±0.51	4.11^{a}±0.53
BA	4^{a}±0.02	4^{a}±0.01	4^{a}±0.03	4^{a}±0.04
MB	3.82^{a}±0.35	3.75^{a}±0.52	3.79^{a}±0.50	3.79^{a}±0.51

Means ± SD values from 4 replicates; Values followed by different letter superscripts in the same column are significantly different (P <0.05).

Table 6. Average leaf surface area in the different sites.

Collection area	Leaf area (cm^2)			
	Site 1	Site 2	Site 3	Zone
BO	11.39^a±0.30	9.87^a±1.20	10.02^a±1.12	10.43^a±1.31
BA	8.21^b±1.51	7.88^b±1.53	8.81^b±0.51	8.3^b±1.50
MB	5.43^c±0.50	5.17^c±0.50	5.5^c±0.50	5.37^c±0.50

Means ± SD values from 4 replicates; Values followed by different letter superscripts in the same column are significantly different (P <0.05).

Table 7. Height of seedlings after 14 weeks of growth.

Collection area	Height (cm)			
	Site 1	Site 2	Site 3	Zone
BO	6.94^a±1.45	7.68^a±1.50	5.61^b±1.55	6.75^a±1.51
BA	5.41^c±1.41	5.52^c±1.40	7.22^a±1.42	6.05^a±1.43
MB	4.3^c±0.54	5.17^c±0.50	4.34^c±0.51	4.61^c±0.52

Means ± SD values from 4 replicates; Values followed by different letter superscripts in the same column are significantly different (P <0.05).

DISCUSSION

The study which investigated the field capacity of soils of three sacred forests in the Western Highlands of Cameroon and the germination and growth potential of *A. melegueta* in these soils, was worthwhile since it has gone a long way to determine at what water retentive field capacity to propagate especially, the fresh seeds of *A. melegueta.*

The soil with the optimum field or water retention capacity positively influences the germination latency and percentage of germination, as stated by Braudeau and Mohtar (2005). The field capacity is significantly important in the sacred forests of BO and BA compared to those of MB. High field capacities positively influence the germination latency of the fresh seeds of *A. melegueta*, which is significantly shorter in BO and BA compared to MB. According to Dirik (2000), these seeds have germination latency between 7 to 12 days. This is in agreement with the data obtained in this study. The seed germination rate did not appear to have been significantly affected by the soil field capacity, because it did not vary significantly among the three areas.

In the MB area, the field or soil water retention capacity did not seem to be sufficient or optimal for rapid germination. The values of the latency and germination percentages were low. These soils are not so favourable for seed germination of *A. melegueta.* The sacred forests of BO, BA and MB are located at 1.376, 1.327 and 2.183 m altitudes respectively, implying that climatic conditions are not comparable. These slower and lower germination factors at MB may be primarily a temperature effect due to the considerable higher altitude. It is also possible that weather conditions have had an influence on the physical and chemical properties of soils of the sacred forests of BO and BA which helped to get a good germination of *A. Melegueta.*

Despite the significantly large water retention capacities of the sacred forests of BO and BA, dry seeds of A. *melegueta* could not germinate after 22 weeks of observation, this, confirming the results of Lachiheb and Neffati (2005) that *Aframomum* seeds could remain dor-mant for 2 to 3 years before germinating after planting. But this does not exclude the fact that the amount of water is very important for the germination of a seed because, according to Niang-Diop et al. (2010) a seed can germinate if the embryo has the ability to imbibe.

Soil water holding capacity had a positive effect on growth of leaf surface area of seedlings. Bamendjo (BO), with higher water holding capacity significantly affected the growth in leaf area of seedlings of *A. melegueta*. The same holds for Bamendjinda (BA) which had an elevated water holding capacity and achieved growth in leaf area significantly higher compared to MB. There was however, no significant difference in number of leaves indicating that other factors such as light, temperature and composition of mineral elements could have influenced the growth of this plant (Lachiheb and Neffati, 2005). According to Theodore et al. (1971), the variations in quantity of water in the soil can affect the metabolism of a plant which may have an influence on the growth of a plant species. The water holding capacity of soils of BO and BA positively affected the growth in height of seedlings while that of MB did not. These results are similar to those of Enti (1998); who observed that high soil water holding capacity can ensure good plant growth.

Conclusion

In conclusion, soil field or water retention capacity of different sites of sacred forests of the Western Highlands affected the germination of fresh seeds of *A. melegueta*. It is also clear from this experiment that the domestication of this plant is possible with fresh seeds since the dry seeds did not germinate.

The soil water retention capacity of different sites in a sample collection area of soil and those of a collection zones also affected seedling growth of *A. melegueta*. The sacred forest of Bamendjo and Bamendjinda showed a positive effect and this affects the regeneration of *A. melegueta*. The values for the sacred forest Mbing Mekoup show that higher soil water retention capacities are needed for a better regenerate of *A. melengueta*.

Conflict of Interest

The authors declare that they have no conflict of interest.

ACKNOWLEDGEMENTS

We wish to acknowledge the moral and financial support of the Heads of the Laboratories of Applied Botany, and that of the Soils and Environmental Science, not forgetting, the authority of the Forestry Nursery Unit, all of the University of Dschang. Our appreciation is also extended to the traditional rulers of the localities of the Sacred Forests of the Western Highlands of Cameroon.

REFERENCES

Ahoton LE, Adgakpa BG, M'po Fonti M'po, Akpo LE (2009). Effets des prétraitements des semences sur la germination de *Prosopis africana* (*Caesalpianceae*). *Tropicultura*, 4:233-238.

Assie KH, Angui KTP, Danho DM, Tamia AJ, Savane I (2010). Effets des contraintes morpho- pédologique sur quelques propriétés hydrodynamiques des sols sous différents agrosysteme au (oumé) Centre –ouest de la Cote d'Ivoire. J. Appl. Bio-Sci. 34:2145-2155.

Babou AB, Sibiri GS (2001). Longevity of seeds and constraints to the survival of seedlings of *Afzelia africana* in a woodland of Burkina Faso. Ann. For. Sci. 58(69):1–11.

Baumeer M (1995). "Trees, shrubs and bushes foster in West Africa." Dakar: Enda Tiers – Monde. P 260.

Braudeau SM, Mohtar RH (2005). "Hydrostructural characteristics of two African tropical soils." Eur. J. Soil Sci. 56:375-388.

Centre Technique de la Forêt Communale (2013). Rapport d'enquête socio-economique des villages riverains a la reserve forestiere (nteingue). Centre Technique de la Forêt Communale, Yaoundé, Cameroun. Pp. 1-24. http://www.foretcommunale-cameroun.org/download/ESEForeke.pdf

Dirik H (2000). "Effect of osmotic water stress on seed germination among provenances of Cedar of Lebanon of Turkish origin." Ann. For. Sci. 57:361-367.

Enti AA (1998). "The rejuvinating Plants of Tropical Africa. Ed. FSL Scale In the North Institute, San Francisco. P 3.

Eyob SM, Appelgren, Rohloff G, Tsegaye A, Messele G (2008). "Traditional medicinal uses and essential oil composition of leaves and rhizomes of *Aframomum Corrima*". Gansen Southern Ethiopia. S. Afr. J. Bot. 74:181-185.

Geudje NM, Van DJFW, Nkongmeneck BA (1996). "Ecological Evaluation of some Non Timber Forest Products of Bipindi region. Akom II: abundance, distribution and the impact of harvesting on forests. "Tropenbos Cameroon program. (Internal document).12p

Honoré T (1999). "The Market for Non-Timber Forest Products in Central Africa in France and Belgium." Occasional Paper. 32p.

Kuete MR, Dikoume A (2000). "Space, power and conflict in the highlands of West Cameroon." Space, Society, CEREHT. University Press of Yaoundé. P 149.

Lachiheb K, Neffati ZM (2005). "Germination ability of some spontaneous halophyte grass of the Southern Tunisia." Higher Institute of Arid Regions. Medenine, Tunisia. Faculty of Sciences of Tunis, University Campus 2092 Tunisia. P 6.

Lékané TD (2009). "*Aframomum* serving the natural and cultural diversity of Grass fields of Cameroon: Valuation Issues of cultural products and sustainable management of shallows and interfluve." University of Yaoundé I Geography Department. Symposium "locate products." P 7

Niang-Diop F, Sambou LBA (2010). "Natural regeneration of *Prosopis africana* Constraints: factors affecting seed germination." Institute of Science Anta Diop. Dakar Fanna, Senegal. 4 (5):1693-1705.

Rafatullah SA, Gabal AM (1995). "Gastric and duodenal antiulcer and cytoprotective effect of *Aframomum melegueta* in rats." Int. Pharmacogn. J. 33:311-316.

Raunkiaer F (1934). The life forms of plants and statistical plant geography. Oxford and the Clarendon Press. 547p.

Segalen P (1958). Les sols de l'Ouest Cameroun. Notice sur la feuille de Mgitapon. Repport I.R.CAM. pp 94, 31.

Suchel JB (1972). "The distribution of rainfall and rainfall patterns in Cameroon. Contribution to the study of climates from Tropical Africa. "Trav. and Doc. Tropical geographies. Univ. Fed. Cam. Bordeaux, France. P 287

Theodore AE, Hsiao C, Henderson WD (1971). Immediate and Subsequent Growth Responses of Maize Leaves to Changes in Water Status. Plant Physiol. 48:631-636.

Tiokeng B (2007). "Diversity, structure, use and local conservation mode of some sacred forests in the Western Highlands of Cameroon." Master's thesis, University of Dschang. 120 P.

Growth and photosynthetic responses of *Lycoris haywardii* Traub to watering frequencies

Bao, C.S. , Zhang, P.C., Chen, C. and Zhou, H.

Hangzhou Botanical Garden, Hangzhou 310013, P. R. China.

To seek a way to solve the slow growth of bulbs of *Lycoris*, a pot experiment with three replications was conducted to evaluate the effects of watering frequency on growth, net photosynthetic rate and bulb biomass of *Lycoris haywardii* under greenhouse condition. The results revealed that leaf number, leaf length and fresh bulb weight was increased by 67, 41 and 323% respectively, under the irrigation frequency of once a month, which were significantly greater than that of other treatments. And leaf number increase rate was positively correlated (r=0.97, *P*<0.01) with the irrigation frequency, so was fresh bulb increase rate (r=0.98, *P*<0.01). However, no significant differences were observed with regard to bulb number in all the treatments. The species with a good ability of drought resistance can get by under irrigation frequency of once in every three months. The maximum net photosynthetic rate reached 18.0 μmol/m^2/s or higher, and the light saturation point was near or higher than 2000 μmol/m^2/s, therefore, a full-light management was recommended for the cultivation of *L. haywardii*. It took about a week to recover the normal photosynthetic ability after a severe drought stress. On the contrary, short-term drought stress had no more negative effect on net photosynthetic rate but remarkable light compensation effect. Higher temperature (under 22°C) was favourable for photosynthesis. Cultivation management in dormancy stage was necessary for increasing bulb yield of *Lycoris*.

Key words: Watering interval, drought resistance, photosynthesis, biomass, cultivation.

INTRODUCTION

Lycoris haywardii, a species of Amaryllidaceae, is a perennial herb with bulbs. Its leaves appear from early October to early May of the following year. Its perianth is reddish violet, a little lighter than that of *L. sprengeri*, and changes to ink-blue at the apex (Hsu et al., 1994). *Lycoris* spp. have a great ornamental value for their beautiful flowers and attractive foliage and important medicinal value in bulbs (Liu et al., 2012; Zhang and Cao, 2001). Species of *Lycoris* have a low reproduction rate, low field yield for bulb production and no technique for effective cultivation and it will take about five years to produce flowers from seedlings (Zhang and Cao, 2001). Slow growth of bulbs is a major problem in cultivation, which has not been solved effectively yet. It needs further exploratory cultural practices. We have conducted a series of studies on that problem, however, results revealed that fertilization does not increase bulb yield (Bao et al., 2012b). Species of *Lycoris* have nutrient strategies of slowly growing (Bao et al., 2012a; Chapin, 1980) and distinct eco-mechanism for retranslocation and conversion of nutrients (Bao et al., 2012a). In addition to fertilizer application, watering management is also important for cultivation. But *Lycoris* is usually regarded as a plant resistant to drought (Qin et al., 2003; Zhang

Table 1. Soil Relative Water Content (SRWC) of the sand at different stages in early 2012.

Test date	SRWC (%)				
yyyy-mm-dd	A1	A2	A3	A4	A6
2012-01-03	61.9	52.3	2.7	2.6	2.4
2012-01-31	82.1	29.4	60.8	1.2	1.1
2012-02-28	75.2	67.0	32.9	46.4	2.3
2012-04-01	33.5	14.6	6.5	8.3	1.7
2012-04-30	20.5	17.6	13.4	2.7	17.1

and Cao, 2001). Furthermore, species of *Lycoris* are in dormant stage in hot summer when plants need more water management. As a result, very little attention is paid to the water management for *Lycoris* spp. Although there are many studies on the water effect on plant growth (Flagella et al., 2002; Rawson and Turner, 1983; Shan et al., 2010; Su et al., 2007; Xia et al., 2001), very limited information is available about the relationship between water and growth of *L. haywardii*. In order to develop a reasonable watering regime and evaluate the influence of irrigation on growth of *L. haywardi*, the effect of different watering levels on growth was comparatively studied.

MATERIALS AND METHODS

This study was conducted in a greenhouse at Hangzhou Botanical Garden (120°16'E, 30°15'N), Zhejiang Province, China. Seventy-five bulbs were selected from the Garden's nursery, which were as uniform in weight (≈15.5 g each bulb in wet weight), and every five bulbs were planted in each pot (20 cm in height, 20 cm in top width and 60 cm in top length) on 16 August 2010 with river sand as a medium, which had a total nitrogen content of 0.2 g/kg, a total phosphorus content of 0.07 g/kg, a total potassium content of 4.5 g/kg, an organic matter content of 2.4 g/kg, and a pH of 5.6.

With a randomized complete block design, the experiment was performed with one single factor at five levels (or five treatments) of watering intervals, viz. A1 (once every 1 month), A2 (once every 2 months), A3 (once every 3 months), A4 (once every 4 months), and A6 (once every 6 months) and three replications. The amount of water applied each time was 3.5 L per pot to ensure sand drenched. The first watering date was on 25 September 2010. Afterwards re-watering was implemented at the end or the beginning of a month according to the design until the last watering date on 30 April 2012.

Alive leaf length and number of each clump were first recorded on 9 March 2011 when leaves almost reached the stage of most vigorous vegetative growth in the first growing season after planting, followed by the second recording on 6 May, 2011 at the late stage of leaves. The following ten recordings were conducted in the second growing season from autumn of 2011 to spring of 2012, about once every 3 w during the leaf stage of all plants. The bulbs were harvested on 4 June, 2012. Samples were dried in an oven at 85°C for 48 h after cleaning.

The net photosynthesis rate (Pn) was measured using a portable gas exchange system Li-6400XT (Li-COR, USA) at the middle-upper part of the second leaf from bottom of a randomly selected plant in each pot based on an open air path, natural atmospheric temperature and CO_2 concentration in the greenhouse. Air flowing speed and photosynthetic photon flux (*PPF*) were set at 500 μmol/s and 800 μmol/m^2/s, respectively. Totally 17 measurements were carried out from January to April 2012, just prior to re-watering and 3, 6 and 14 days after re-watering. Besides, a sand sample in 10 cm depth was taken just before re-watering to be dried in an oven at 105°C for 24 h for the measurement of soil relative water content (SRWC). Dates for soil-sampled were shown in Table 1.

Light response curves were developed using the following *PPF*: 2000, 1800, 1600, 1400, 1200, 1000, 800, 600, 400, 200, 100, 50 and 0 μmol/m^2/s at the middle-upper part of the second leaf from the bottom of a randomly selected plant for each treatment. Before recording, the system was adjusted for baseline for 30 min at 2000 μmol/m^2/s.

Data were analyzed using MS Excel 2003 and SPSS 16.0. Statistical significance was determined at $P \leq 0.05$. Differences between means were compared according to an LSD test.

RESULTS AND DISCUSSION

Watering interval and soil relative water content

SRWC is a ratio of soil water content to field saturated moisture capacity (Shan et al., 2010; Shangguan, 1997). SRWC of the medium just before re-watering at the end or the beginning of a month was shown in Table 1. The scope of SWRC of A1 was 20.5 to 82.1%. It indicated that treatment A1 could keep a higher SWRC in winter, but in early spring the interval might be too long. It was obvious that the longer a re-watering interval, the lower the SRWC. We estimated that it would be necessary to water when SRWC reached about 50%.

Impact of watering frequency on leaf number

The dynamic change in average leaf count per clump for each treatment was showed in Figure 1. It is obvious that the leaf number of treatment A1 was the highest, and A6 was the lowest. Analysis of variance (ANOVA) indicated that there was no significant difference among A1, A2 and A3, but there was significant difference between the group of A1, A2, A3 and the group of A4, A6 on 9 March 2011. Leaf number showed 3 levels on 16 November 2011 and 27 February 2012 in the second growing season, in which A1 had significantly more leaves than

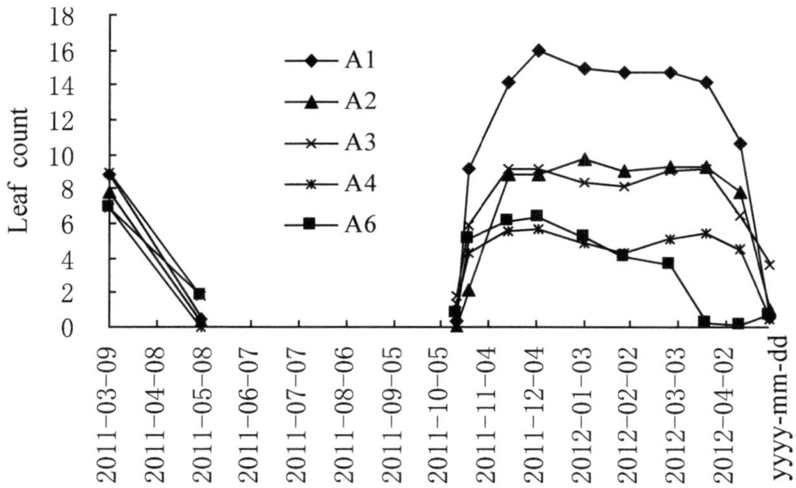

Figure 1. Dynamic changes in average leaf count per clump in different treatments.

A2 and A3, and the latter two also had significantly more leaves than A4 and A6. The ratio of the leaf number on 27 February, 2012 to that on 9 March, 2011 indicated the extent to which leaf number increased. The ratio of A1, A2, A3, A4 and A6 was 1.67, 1.19, 1.01, 0.75 and 0.52, respectively. The leaf number of treatments A1, A2 and A3 was increased by 67, 19 and 1% respectively. However, that of A4 and A6 was decreased by 25 and 48%, respectively. If watering frequencies were quantified (that is, A1, A2, A3, A4 and A6 were valued as 1, 1/2, 1/3, 1/4 and 1/6, respectively), it was known from regression analysis that the increase rate of the leaf number was significantly in positively linear correlation with watering frequency (correlation coefficient r = 0.97. *P* = 0.008), which indicated the higher the irrigation frequency, the more leaf number. The linear regression was shown as following:

Y =127.487X -54.572.

Here independent variable X was watering frequency on a monthly basis, and Y was the percentage of increased leaf number

It was illustrated that *L. haywardii* is strong in resistance to drought, because the leaf number of A3 was not decreased obviously one year after planting (Ratio = 1.01).

The difference in leaf number was not significant between A1 and A2 or A3 on 9 March, 2011. However, in the second growing season after dormancy (samples from 16 November, 2011) the difference was significant, indicating watering in the dormancy period was necessary. It may be the reason that bud differentiation requires certain moisture, for No Leaf Period (dormancy period) is the main period of leaf bud differentiation in *Lycoris* (Li and Zhou, 2005). Moist soil can ensure the bud differentiation, and generate more leaves.

Impact of watering frequency on leaf length

The dynamic change in leaf length of five treatments is shown in Figure 2. A1 had the longest average leaf length. In general, the impact on leaf length caused by water deficit was not obvious except in A6. In case of water deficit without replenishment for more than 4 months (A6), leaves started to wither from tip, but watering brought about new leaf emerging. Based on the analysis of variance of data from representative 5 December, 2011 and 27 February, 2012, leaf length in A1 was significantly greater than that in other four treatments, among which the difference were not significant with the data from 5 December 2011. Although leaf length in A1 was still significantly greater than that in other four treatments with the data from 27 February 2012, the leaf length in A2, A3 and A4 was significantly greater than that in A6. Compared with that on 9 March 2011, the leaf length of A1 on 21 March, 2012 was increased by 41%.

Dynamic change in net photosynthetic rate (Pn) and impact of watering frequency on Pn

The maximum of Pn in A1 was in January, and the minimum was in April (Figure 3). As a whole, Pn decreased over time with rising temperature and decreasing growth potential. However, Pn was positively correlated with air temperature before March when the air temperature in the leaf chamber (rTair) was below 15°C, and photosynthesis might be prone to inhibition when rTair was about higher than 22°C, for water deficit would take place if temperature is too high (Wang, 1997). It indicated that properly increasing air temperature in winter could increase photosynthesis, which was in line with that of *Lycoris aurea* (Bao et al., 2012b). We

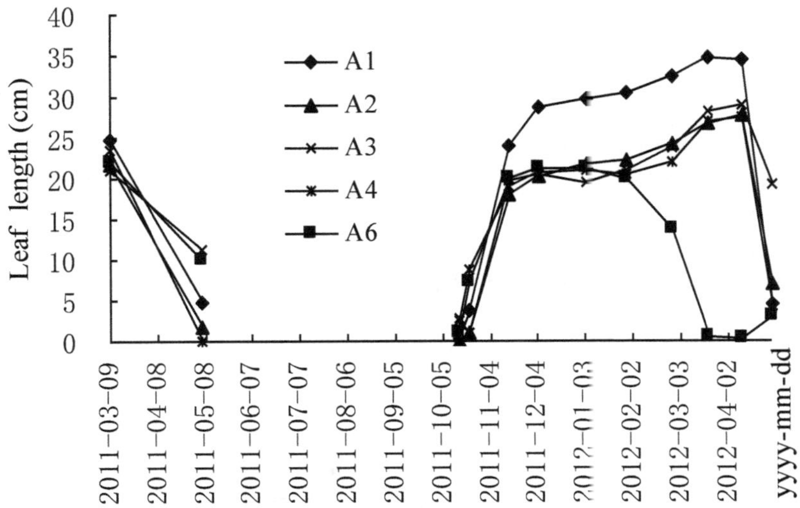

Figure 2. Dynamic changes in average leaf length per clump in different treatments.

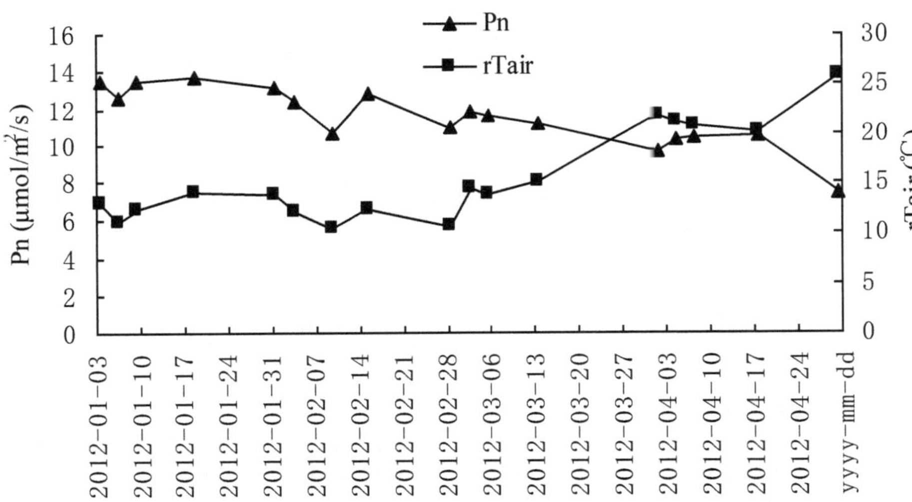

Figure 3. Dynamic changes in average Pn and rTair in Treatment A1. Pn: Net photosynthetic rate. rTair: Air temperature in the leaf chamber.

speculated that the best environmental temperature for photosynthesis should be about 22°C. And It would be better to cultivate this species in greenhouse to avoid low temperature at the leaf stage.

According to measurement, the total average Pn for A1, A2, A3, A4 and A6 was 11.55, 11.72, 11.28, 6.71 and 1.64 μmol/m^2/s, respectively. Impact of A2 on Pn was not obvious, but when watering interval was over 2 months, the effect was outstanding, such as that of A3 even in winter, and Pn appeared to be negative, but recovered to normal within six days after watering (Table 2). Pn could recover to normal photosynthesis even in treatment A4. It demonstrated that *L. haywardii* was in strong tolerance of drought and could maintain alive when relative lack of water and recovered photosynthesis after watering. As for A6, because leaves died, few new leaves appeared a week after watering, and the leaf was so long that it could be used for measurement of Pn half a month later when Pn reached a normal level again. On 30 April, 2012, RSWC of A1, A2, A3, A4 and A6 was 20.5, 17.6, 13.4, 2.7 and 17.1%, respectively, and the corresponding Pn was 7.48, 9.41, 10.34, 0.04 and 10.11 μmol/m^2/s. It indicated that *L. haywardii* had a good physiological capability for photosynthetic recovery, with which resistance to drought such as intensity of drought tolerance, recovery rate and recovery extent after watering was stronger than that of other plants, such as wheat, maize, *Leymus chinensis,* and sunflower (Liang et al., 2009; Lin et al., 2008;

Table 2. Re-watering effect on Pn recovery in five treatments from January to April 2012[z].

Days before or after re-watering	Pn (μmol/m²/s)				
	A1	**A2**	**A3**	**A4**	**A6**
The day just before re-watering	10.95	8.57	-0.09	0.33	(NL)
3 days after re-watering	11.81	11.49	10.57	7.86	(NL)
6 days after re-watering	11.52	11.77	13.61	13.04	(LTS)
14 days after re-watering	12.08	10.71	13.55	12.23	11.86

[z] Pn: Net photosynthetic rate; NL: No leaf; LTS: The leaf was too short to measure Pn

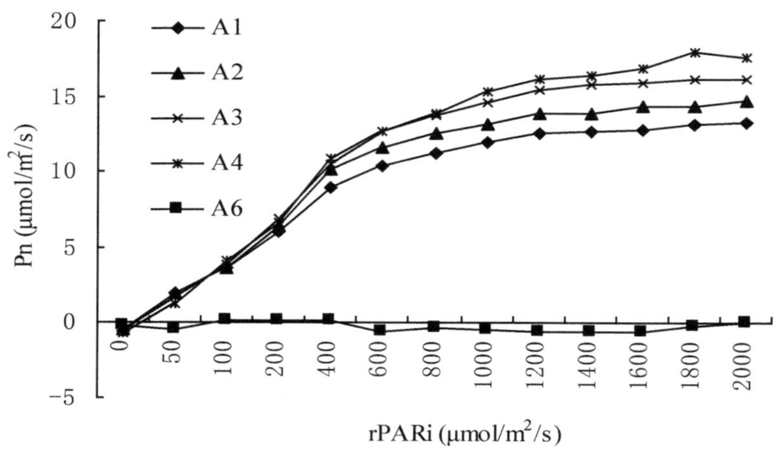

Figure 4. Light response curves of treatments on 14 February 2012. Pn: Net photosynthetic rate. rPARi: active photosynthetic radiation.

Shangguan, 1997; Yagoub et al., 2010).

Watering interval and light response

Treatment A6 had had no watering for 4 months on 14 February 2012, and the leaf was in wilting status, and total Pn was negative (Figure 4.). Photosynthesis of other treatments was normal (watered after 6 weeks for A3 and watered after 2 weeks for the others). The normal curve shape was like a parabola in line with that of *L. aurea*, *L. chinensis*, *L. longituba* and *L. sprengeri* (Bao et al., 2012b; Liu et al., 2012; Meng et al., 2008). The photosynthetic ability had no positive correlation with watering frequency. According to the average, photosynthetic capacity was in the following order: A4>A3>A2>A1. Lower watering frequency had a higher utilization of irradiance. That meant photosynthetic capacity was increased by watering after drought stress because of the light compensatory effect (Liu et al., 2004) We could infer that *L. haywardii* was heliophilous because of its light resistance. Light-inhibition phenomenon had not appeared at the active photosynthetic radiation (rPARi) level of 2000 μmol/m²/s, so full-light management was suggested in cultivation in certain conditions if temperature was not too high. The maximum of Pn in this test was 18.0 μmol/m²/s (from treatment A4).

Impact of watering frequency on bulb count and biomass

Results from ANOVA revealed that there was no significant difference in bulb number, but significant difference for dry bulb weight (Table 3). The ratio of the present fresh weight (PFW) to the initial fresh weight (IFW) represents an increase in biomass. Treatments A1, A2 and A3 were increased by 323, 164 and 47% in fresh bulb weight, respectively. However, that of A4 and A6 was decreased by 21 and 52% in fresh bulb weight, respectively. This indicated that fresh biomass was also positively correlated with watering frequency if quantifiedly regressed with watering frequency as leaf number (r=0.98, P=0.003). Watering once every 3 months could be regarded as a critical interval value. It could be seen that the wet bulb weight still increased at this watering frequency, and stopped increasing when the interval was longer than 3 months for bulbs became lighter and lighter. Additionally, the single average bulb biomass had a significant difference among treatments. The weight of a single bulb was negatively correlated

Table 3. Average fresh bulb weight ratio, bulb number and dry bulb weight and multiple comparison[y].

Treatment	PFW/IFW[x]	Bulb number	Dry bulb weight (g)
A1	4.23	8.7[a]	109.87[a]
A2	2.64	7.0[a]	70.67[b]
A3	1.47	8.7[a]	38.03[c]
A4	0.79	5.7[a]	23.82[d]
A6	0.48	7.3[a]	8.44[e]

[y]Different letters behind values in the same column indicate significantly different ($P \leq 0.05$). [x]PFW: Present fresh weight. IFW: Initial fresh weight.

with the watering interval. It is partly because the bulb gradually sacrificed the outside epimatium to maintain the survival of the inner short shoot, partly because the bulb tended to split into smaller new bulbs to survive under dry stress.

Our study revealed that within a certain range the more watering frequency, the more leaf number, leaf length and bulb biomass, which demonstrated the importance of water to plant (Yagoub et al., 2010). It was necessary to strengthen watering management even in no leaf period for *Lycoris* although *L. haywardii* is a drought-resisting species because of high water content in bulb. Since temperature was low in winter and evaporation was weak, watering once a month could ensure plant growth. But in early autumn or early spring, when temperature occurs to be a little higher, watering frequency should be higher. Roughly speaking, timely watering should be carried out when SRWC reached about 50%. While in a high temperature situation dry stress should be avoided to ensure leaf bud differentiation and leaf sprout. Since the minimum watering interval was once a month in this study, further study was needed to determine if there exists a shorter and better watering pattern.

ACKNOWLEDGEMENT

This study was supported by West Lake Scenic Area Management Committee of Hangzhou City, Zhejiang Province, P.R. China.

REFERENCES

Bao CS, Zhang PC, Zhang HZ, Jiang Y, Li NL (2012a). Seasonal changes of biomass allocation and micronutrient contents in *Lycoris longituba*. J. Northeast For. Univ. 40(9):34–38.

Bao CS, Zhang PC, Zhang HZ, Xu YQ, Jiang Y (2012b). Effect of fertilization in sand bed on growth and photosynthesis of *Lycoris aurea*. J. Nanjing For. Univ. (Natural Science Edition) 36(5):61–65.

Chapin FS (1980). The mineral nutrition of wild plants. Ann. Rev. Ecol. Sys. 11:233-260.

Flagella Z, Rotundo T, Tarantino E, Di Caterina R, De Caro A (2002). Changes in seed yield and oil fatly acid compotation of high oleic sunflower (*Helianghus annuus* L) hybrids in relation to the sowing date and water regime. Eur. J. Agron. 17:221–230.

Hsu PS, Kurita S, Yu ZZ, Lin JZ (1994). Synopsis of the genus *Lycoris* (Amaryllidaceae). SIDA 16(2):301–331.

Li AR, Zhou J (2005). Study of the growth cycle and development of *Lycoris chinensis* leaves. Chin. Bull. Bot. 22(6):680–686.

Liang ZJ, Tao HB, Zhao HZ, Liu HM, Wang P, Wei X (2009). Photosynthetic physiological recovery of maize after water deficit at seedling stage. Acta Agriculturae Boreali-Sinica 24(2):117–121.

Lin XL, Xu ZZ, Wang YH, Zhou GS (2008). Modeling the responses of leaf photosynthetic parameters of *Leymus chinensis* to drought and re-watering. Acta Ecologica Sinica 28:4718–4724.

Liu GS, Guo AH, Ren SX, An SQ (2004). Compensatory effects of re-watering on summer maize threatened by water stress at seedling period. Chin. J. Ecol. 23(3):24–29.

Liu K, Tang CF, Zhou SB, Wang YP, Zhang D, Wu GW, Chang LL (2012). Comparison of the photosynthetic characteristics of four *Lycoris* species with leaf appearing in autumn under field condition. Photosynthetica 50(4):570–576.

Meng PP, Ge Y, Cao QJ, Chang J, Pan P, Liu C, Lu YJ, Chang SX (2008). Growth and photosynthetic responses of three *Lycoris* species to levels of irradiance. HortScience 43(1):134–137.

Qin WH, Zhou SE, Wang HY, Wang H (2003). Advance in *Lycoris* Herb. J. Anhui Normal Univ. (Nat. Sci.) 26(4):385–390.

Rawson HW, Turner NC (1983). Irrigation timing and relationship between leaf area and yield in sunflower. Irrig. Sci. 4(3):167–175.

Shan YF, Wang LK, Ma YS, Song CH (2010). Effect of different water deficit on yield, water use efficiency and economic benefit of sunflower. J. Northeast Agric. Univ. 41(7):70–74.

Shangguan ZP (1997). The response of photosynthesis to different kind of water deficits in wheat leaves. Acta Agriculturae Boreali-occidentalis Sinica 6(4):38–41.

Su D, Sun GF, Zhang JZ, Jiang CD, Yu QB, Gu DF, Dong R (2007). Effects of soil water stress on growth and biomass distribution of *Sedum aizoon* and *Sedum spectabilis*. Acta Horticulturae Sinica 34:1317–1320.

Wang H (1997). Effect of water deficit on net photosynthesis rate of winter Wheat. Chin. J. Ecol. 16(1):1–6.

Xia GJ, Yan YL, Cheng SM, Gao SJ, Luo Y (2001). Research on compensatory effects to water deficits on dryland winter wheat. Agric. Res. Arid Areas 19(1):79–82.

Yagoub SO, Osman AAM, Abdesalam AK (2010). Effect of watering intervals and weeding on growth and yield of Sunflower (*Helianthus annuus* L). J. Sci. Tech. 11(2):52–56.

Zhang L, Cao FL (2001). Advance on the cultural technology in genus *Lycoris*. Acta Agriculturae Universitatis Jiangxiensis 23(3):375–378.

Site factor on nutritional content of *Arundinaria alpina* and *Oxytenanthera abyssinica* bamboo shoots in Ethiopia

Sisay Feleke

Ethiopian Institute of Agricultural Research, Forest Products Utilization Research Coordination, P. O. Box 2322, Addis Ababa, Ethiopia.

The objective of this study was to characterise some of the nutritional values of the two bamboo species grown and consumed in western parts of Ethiopia. The bamboo shoots of each the two species were collected from each of the three sites and analyzed for their nutritional and mineral composition. The average moisture content of the bamboo shoots was more than 90% for both species but the fat content as determined on dry weight basis indicated a significant variation ranging from 0.6 to 2.2% for *Oxytenanthera abyssinica* and from 0.6 to 1.5% for *Arundinaria alpina.* Highland bamboo *(A. alpina)* shoots characterized with high nitrogen content (5.40%) implying that they may have high content of protein, but the lowland bamboo (*O. abyssinica)* was low in nitrogen content (3.10%). The average ash contents of the highland bamboo shoots collected from Tekur Incheny, Enjibara and Masha site were 14.20, 15.50 and 17.20%, respectively. The average nutrient elements in the shoots of *O. abyssinica* were in the order of K>N>P>Ca>Mg, whereas in *A. alpina* bamboo shoots observed to have slight variation in the arrangement of these elements with the order of K>N>P> Mg > Ca.

Key words: *Arundinaria alpina, Oxytenanthera abyssinica,* bamboo shoots, mineral elements, nutrients.

INTRODUCTION

Bamboo is a perennial grass with over 1200 species in 50 genera of bambuosidaea and grows mainly in the tropics and subtropics (Zhang et al., 2002). In different parts of the world, it is used as a source of raw material for fodder, construction materials, paper production, laminated boards, energy, food, beverage and medicine (Choudhury et al., 2010). The bamboo shoots have been used as food for centuries in Asia. Their protein content is as high as 1.49 to 4.04% with 17 kinds of essential amino acids accounting for 35% of the total protein. They are also rich in other mineral nutrients which are highly valuable for human being (Hunter and Feng'e, 2000; Choudhury et al., 2010).

Ethiopia has two bamboo species: The African alpine Bamboo *Arundinaria alpina* (K. Schumann Lin) which grows between 2200 and 4000 m above sea level (asl) and the lowland bamboo *Oxytenanthera abyssinica* (A. Richard Munro) which grows between 1200 and 1800 m asl (Sylvia, 1995; Embaye et al., 2005). In Ethiopia the common use of bamboo resource is limited to fencing, house construction (tukul), waving, household furniture and utensils but there are the experiences of bamboo shoots consumption in the western and south western parts of Ethiopia. In general, the most common use of bamboo is for the construction purposes. The area coverage of bamboo in the country is reported to vary from eight hundred thousand to one million ha (FRA, 1985; Embaye et al., 2003) but currently this figure may be reduced due to the expansion of mechanised agriculture in bamboo growing parts of the country. Recently, in response to the growing interest of utilizing the existing bamboo resource, some research activities have been going on in areas such as product characterization and different management options (Embaye et al., 2003, 2005; Seyoum et al., 2006, 2007).

In Ethiopia Bamboo shoots (BS) usually emerge after the spring rain, namely between April and June without disregarding its dependence on location and species. The quality of bamboo shoots for food depends on the biological characteristics of the bamboo such as species, shooting time, duration of shooting, the rate of shoot growth and the size of shoots.

Bamboo shoots are known for their low fat and cholesterol content and high content of protein, mineral and dietary fibre. Reports show that some species of BSs like other root crops such as cassava, have some cyanogenic glycoside composition which decrease following harvesting of the material and during the time of food preparation (Anonymous, 2004; Agatemor, 2009; Choudhury et al., 2010, 2012).

Bamboo has an immense potential in realizing the food security mechanism of the country. To achieve such objective, the indigenous knowledge of bamboo shoots consumption which accumulated over centuries can be easily introduced to other bamboo growing and potential areas. However, dissemination of such indigenous knowledge can be effective only when preceded by some improvement works which include taste and preparation ease that lead to getting fast adoption rate in the community. The traditional methods of preparing food from bamboo shoots by the indigenous people (ethnic community) of south western and western parts of Ethiopia demands some improvement work, especially with regard to nutritional analysis, the removal of the natural toxicants or bitterness prior to consumption and storability.

The objective of this paper was to identify and raise awareness on some of the parameters of the nutritional values of bamboo shoots of *O. abyssinica* and *A. alpina* species that used for food in south western and western parts of Ethiopia.

MATERIALS AND METHODS

Site description

The samples of bamboo shoots are collected from the following six sites. Masha forest which located in southwest Ethiopia (7° 30'N; 35° 30'E) has an altitude that ranges from 2400 to 3000 m asl, mean annual rainfall of over 2300 mm and annual temperature range 16 to 20°C (Embaye et al., 2005). Enjibara is located in northern parts of the country (10°56' N and 36°56'E) at elevation of 2560 m asl. Tikur Inchini is situated in western parts of the country (9°19' N and 38°24' E) and has an altitude of 2700 m asl. Pawe found in northwest Ethiopia (11° 09' N and 36° 03' E) and has an altitude of 1120 m asl and mean annual rainfall of 1587 mm. Didhessa is located in the south-western of the country (8°41' N and 36°24' E) at the altitude of ~ 1300 m asl. Assosa is located in the western region of the country (10°7' N and 34°37' E) with an altitude of 1510 m asl.

Sample preparation

Lowland bamboo shoots *(O. abyssinica)* were collected from Asossa, Pawe and Dhidhessa while the highland bamboo shoots (*A. alpina*) collected from Enjibara, Masha and Tikur Enchiny. Shoot samples of both bamboo species were transported to the laboratory under liquid nitrogen to avoid spoilage of the shoots.

The outer cover of the hard sheath, the bottom and top portions of the shoots were removed by hand and the remaining middle portions (the edible part of the shoots) which account for 45% of the emerging ones were washed three times in cold water, sliced at the thickness of 5 mm and then dried at a temperature of 45°C using oven drier for 40 h.

Moisture content

The moisture content of the bamboo shoots were determined by drying of the samples in an oven at 105 ± 2°C.

Fat extraction

The bamboo shoots were extracted with hexane in a Soxtec apparatus (a product of Germany) and the average reading of six replicate for each location were taken (Greenfield and Southgate, 2003).

Ash determination

A sample of the bamboo shoots was first heated on a burner in air to remove its smoke, and then burned in a furnace at 550°C. The ash content was expressed as a percentage ratio of the weight of the ash to the oven dry weight of bamboo shoots.

Nitrogen content

Nitrogen analysis was undertaken using Kehdjal Distillation Apparatus. The amount of nitrogen was calculated by the amount of acid added that change the colour of the distilled solution. The protein content was calculated using the nitrogen conversion factor (NCF) of 6.25 (Anonymous, 1998).

Mineral content

Ash obtained from samples of bamboo shoots were dissolved in concentrated sulfuric acid. Then the solution was used for the determination of the studied minerals by Atomic Absorption Spectrometer except for Phosphorous. Wet ash method was used for the determination of Phosphorous. Finaly the amount of Phosphorous was determined using UV-VIS spectrometer at 400 nm wave length (AOAC, 2000).

Statistical analysis

Results were expressed as the mean of three replicates. A one-way

Table 1. Mean nutritional values of *Arundinaria alpina*.

Components	Location		
	Injibara	Masha	Tikur Incheny
Nitrogen (%)	5.83[a]	6.22[a]	4.15[b]
Fat (%)	1.1[a]	0.64[b]	1.5[a]
Moisture (%)	93.3[a]	93.0[a]	91.9[a]
Ash (%)	15.6[b]	17.1[a]	14.2[c]
P (ppm)	9660[a] (0.97%)	9903.33[a] (0.99%)	7113.33[b] (0.71%)
K (ppm)	90833.33[b] (9.08%)	97716.67[a] (9.77 %)	85350.00[c] (8.54%)
Ca (ppm)	2440.00[a] (0.24%)	2853.30[a] (0.29%)	1633.33[b] (0.16%)
Mg (ppm)	1890.00[a] (0.19%)	1850.00[a] (0.19%)	1386.67[b] (0.14%)
Fe (ppm)	72.67[b]	147.00[a]	41.67[c]
Mn (ppm)	25.33[a]	28.33[a]	11.00[b]
Zn (ppm)	107.00[b]	210.67[a]	97.67[b]

Means with similar letters for comparison between different sites are not significantly different at $p = 0.05$.

analysis of variance (ANOVA) was used to compare the mean using the least significant difference at 5% level.

RESULTS AND DISCUSSION

In Masha and Assosa areas where the bamboo shoots emerge between April and June the indigenous communities use it as a food during spring season of the year. The shoots are immediately picked within the first week of their development as they have a growth rate of up to 1 m per day for some species (Anonymous, 2004). Before serving, the shoots are boiled and drained the first water to reduce the bitterness or the toxic chemicals of the shoots which could be the cynogenic glycosides (Hunt and Feng'e, 2000; Agatemore, 2009; Choudhury et al., 2010, 2012).

In both shoot samples Ethiopian bamboo species the fat extract has a light yellow colour. The bamboo shoots of *A. alpina* were low in fat content with an average of 1.09% on dry weight basis (dwb). The statistical analysis of the fat content showed that the samples of Masha site was lower in fat content which differ significantly from the contents of the samples of the other two locations (Table 1). The average fat content of the lowland bamboo shoots collected from Assosa, Dhedhessa and Pawe sites was 1.44% on dwb. The average fat content of shoots collected from Pawe area was significantly different from those of the other two locations. The low fat content in bamboo shoots is also reported by others and this is one of the reasons that make the bamboo species as a preferred food items (Choudhury et al., 2010, 2011).

The mean ash content of highland bamboo shoots collected from the three sites range from 14.20 to 17.10% dwb. The ash content of the highland bamboo shoots was significantly varied with locations. The lowest ash amount of the shoot was 14% dwb and the highest was 18% dwb for samples of Tekur Incheny and Masha, respectively. The high amount of ash reveals the mineral richness of the biological material which serves as a source of mineral elements required in the nutrition. Whereas the average ash content for the lowland bamboo shoots collected from Assosa, Dhidhessa and Pawe were 11.17, 12.03 and 11.63% dwb, respectively. The lowest ash content in the lowland bamboo was obtained from Assosa (10.87% dwb) and the highest was obtained from Dhidhessa (12.38% dwb) areas. The ash amount of the lowland bamboo from Dhidhessa site showed a significant difference from that of Assosa sample (Table 2). High amount of ash also reported on bamboo shoots of *Fargesia yunnanensis* (Wang et al., 2009). High ash content of biological material is also observed in barks and annual crops, indicating that it is a phenomenon for fast growing plants, which can accumulate high amount of mineral materials in the growing season (Raveendran et al., 1995; Azarov et al., 1999). Similar with the BSs of *Bambusa balcoa, Bambus polymorpha, Dendrocalmus hamiltonii, D. strictus, D. gigantus, Phyllostachys praecox* the ash content of the studied species are high but their mineral contents depend on the species, location and season of harvesting (Zhang et al., 2011; Chouduhry et al., 2011). The high amounts of ash content in the bamboo shoots enable them to hold the macro- and micro-elements making them preferable.

The average nitrogen content of the bamboo shoots of *A. alpina* from Tekur Incheny, Injebara and Masha areas were in the range of 4.15 to 6.22% dwb. The lowest amount was 4.15% dwb and the highest was 6.22% dwb from Tekur Incheny and Masha areas, respectively. The nitrogen content of the bamboo shoots from Tekur Incheny site was low and significantly different from those of the other two locations (Figure 1). The average nitrogen content in *A. alpina* bamboo shoots are 8 to 10 times richer than the 1st to 3rd year old culms of Masha area (Embaye et al., 2005). Such high variation in nitrogen

Table 2. Mean nutritional values of *Oxytenanthera abyssinica*.

Components	Location		
	Assoa	**Dehdhesa**	**Pawe**
Nitrogen (%)	3.30^{a}	3.99^{a}	2.96^{a}
Fat (%)	1.5^{a}	2.2^{a}	0.6^{b}
Moisture (%)	91.65^{a}	92.57^{a}	92.54^{a}
Ash (%)	11.17^{b}	12.03^{a}	11.63^{ab}
P (ppm)	5686.67^{c} (0.57%)	6166.67^{b} (0.62%)	7453.33^{a} (0.75%)
K (ppm)	66285^{b} (6.63%)	70283.33^{a} (7.03%)	71533.33^{a} (7.15%)
Ca (ppm)	933.33^{c} (0.09%)	1200.00^{b} (0.12%)	1460.00^{a} (0.15%)
Mg (ppm)	1903.33^{a} (0.19%)	2020.00^{a} (0.20%)	1986.67^{a} (0.20%)
Fe (ppm)	33.67^{c}	45.00^{b}	88.00^{a}
Mn (ppm)	19.67^{a}	19.33^{a}	18.67^{a}
Zn (ppm)	71.67^{c}	88.33^{b}	106.67^{a}

Means with similar letters for comparison between different sites are not significantly different at p = 0.05.

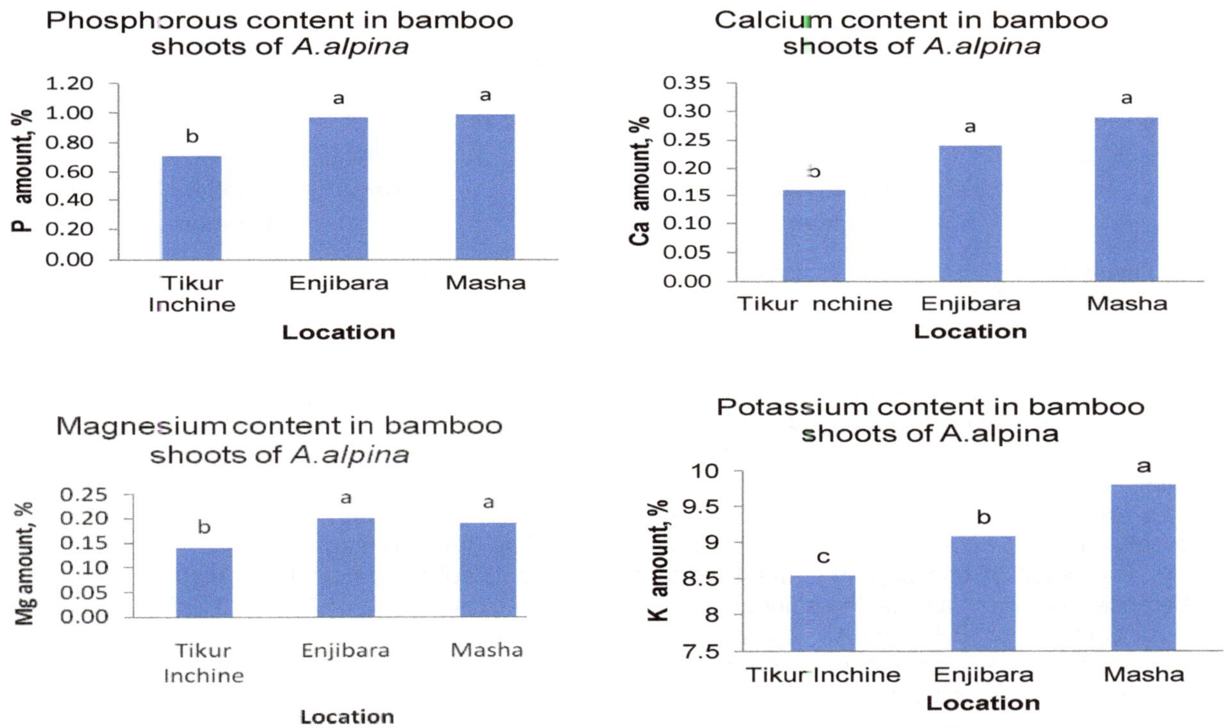

Figure 1. Mineral contents in bamboo shoots of *A. alpina* from different locations. Bars with the same letter are not significantly different at p = 0.05 level.

content is attributed to the active and fast growing stage of the plant which requires more nitrogen in the chlorophyll make up, cell differentiation of the growing bamboo shoots and biosynthesis of nitrogen containing compounds. The average nitrogen content for lowland bamboo shoots was 3.29, 3.99 and 2.95% dwb for samples from Assosa, Dhidhessa and Pawe areas, respectively. There was no significant difference in nitrogen content between the lowland bamboo shoots collected from the three locations. When the nitrogen content in the shoots multiplied by the factor 6.25 the total protein amount ranges between 2.88 and 4.13% of the fresh BS which is in the range of other BS of Asian species (Wang et al., 2009; Chouduhury et al., 2010, 2011).

The average phosphorous content of the *A. alpina* bamboo was 0.71, 0.97 and 0.99% for samples collected from Tekur Incheny, Injebara and Masha sites,

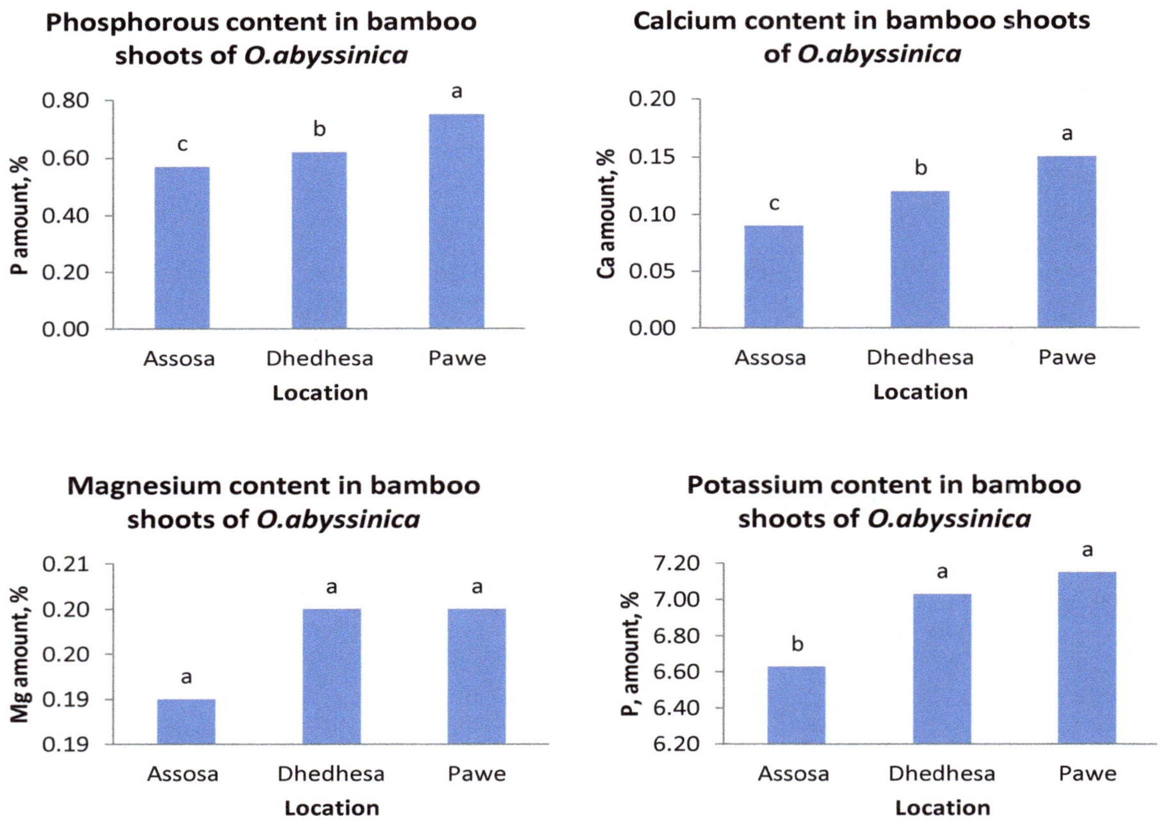

Figure 2. Mineral contents in bamboo shoots of *O. abyssinica*. Bars with the same letter are not significantly different at p = 0.05 level.

respectively. The phosphorous amount in the highland bamboo shoots from Tekur Incheny was significantly different from Injebara and Masha locations with its low content. Similarly, the average phosphorous content of the samples of lowland bamboo shoots of Assosa, Dhidhessa and Pawe were 0.57, 0.62 and 0.75%, respectively. The result of phosphorus content indicates strong differences between all the three locations. The phosphorous content of BSs of *A. alpina* and *O. abyssinica* is much higher than those of Bambusa and Dendrocalamus species (Choudhury et al., 2010, 2011).

The average calcium content of the bamboo shoots of *A. alpina* was analyzed and found the lowest content in the samples of Tekur Incheny site which was statistically different from those of Injebara and Masha areas. The average calcium contents were 0.16, 0.24 and 0.29% for bamboo shoot samples of Tekur Incheny, Injebara and Masha areas, respectively (Table 1). The low calcium content of the samples of Tekur Incheny is related to the low ash content of the sampled material. The average calcium content of the lowland bamboo shoot samples of *O. abyssinica* collected from Assosa, Dhidhessa and Pawe areas were 0.09, 0.12 and 0.15%, respectively. There was significant difference between the sites in average calcium content the lowest calcium content in the samples of Assosa area (Table 2). In both studied species, the calcium content accounts for less than 1% of the mineral content unlike the perennial plant wood which consists more than 50% (Fengel and Wegner, 1984; Azarov et al., 1999).

The average magnesium contents of *A. alpina* bamboo shoots collected from Tekur Incheny, Injebara and Masha was in the range of 0.14 to 0.19%. The highland bamboo shoots of Tekur Incheny site exhibited lowest amount which was significantly different from those of Injebara and Masha locations as indicated in Figure 1. On the other hand, the average magnesium content in the shoots of *O. abyssinica* collected from Assosa, Dhidhessa and Pawe areas was in the range of 0.19 to 0.20% with no significant difference between locations (Figure 2). The amount of magnesium in these bamboo shoots were 10 fold higher than the leafy vegetables of *Amaranthus spinosus, Hibiscus species, Solanum macrocarpon* consumed in Northern Ghana (Amagloh and Nyarko, 2012) and equal to *Eragrostis tef* an Ethiopian staple food called "Injera" made of it (Seyfu, 1997; Parish, 2006).

The average potassium contents in the bamboo shoots of *A. alpina* were 8.53, 9.08 and 9.77% for the samples of Tekur Incheny, Injebara and Masha areas respectively.

The statistical analysis for the average potassium content showed significant differences among all three sites. Whereas the average potassium content in *O. abyssinica* bamboo shoots collected from Assosa, Dhidhessa and Pawe areas were 6.63, 7.02 and 7.15%, respectively. The average potassium content of the samples of Assosa site was found to be lower and significantly different from those of the other locations. There are also reports which reveal similar high amounts of potassium in bamboo shoots of other species (Shanmughavel and Francis, 2003; Choudhury et al., 2011). In general the bamboo shoots exhibit higher amount of potassium than *A. spinosus, Hibiscus species, Solanum macrocarpon* leafs consumed in Ghana (Amagloh and Nyarko, 2012) and higher amount than teff (*Eragrostis tef*) grain (Seyfu, 1997; Parish, 2006).

The mean iron content in the *A. alpina* bamboo shoots of all the three sites was significantly different from each other with 41, 75 and 145 ppm for samples collected from Tekur Incheny, Injebara and Masha sites respectively. The iron content in the bamboo shots of *A. alpina* in Masha area is 2 to 5 folds higher than those of the other locations. The average iron content in *O. abyssinica* bamboo shoots collected from Assosa, Dhidhessa and Pawe were 33.67, 45 and 88 ppm respectively. The analysis for iron content for lowland bamboo shoots was significantly different from each other. The iron content of the bamboo shoots of *O. abyssinica* from Pawe area is 2 to 3 folds higher than the other locations. The iron content in the two bamboo shoots is higher than the leafy vegetables used in Northern Ghana and almost equal to the amount in teff (*Eragrostis tef*) grain (Seyfu, 1997; Parish, 2006; Amagloh and Nyarko, 2012).

Conclusion

The analysis of bamboo shoots of *A. alpina* collected from the three locations showed that the bamboo shoots collected from Tekur Incheny area was the lowest in all analyzed mineral content. The lowland bamboo shoots of *O. abyssinica* except for the magnesium, all samples differ in their content of mineral elements and the bamboo shoots from Assosa area showed the lowest values. Analysis of both shoot samples of bamboo species nutrient elements were in the order of K>N>P>Ca>Mg for *O. abyssinica* bamboo shoots and K>N>P> Mg > Ca for the *A. alpina.* In general the result of this study and the experiences of southwest Ethiopia strongly indicate the great potential of bamboo shoot for food security mechanism of the country.

ACKNOWLEDGMENT

The author thank the Ethiopian Agricultural Research Institute for its Financial support, and the offices of Rural and Agricultural Development of the districts targeted for materials, to his assistants Ato Kinfe Tesfaye and Ato Ayeshem Tebeje and also extend his thanks to the anonymous reviewers for their valuable and constructive comments on previous version of this paper.

REFERENCES

Agatemor C (2009). Cooking time and steeping time effects on the cyanide content and sensory quality of cassava slices. Electronic J. Environ. Agric. Food Chem. 8:189-194.

Anonymous (1998). Food composition table for use in Ethiopia. Part IV. P. 32.

Anonymous (2004). Cyanogenic Glycosides in Cassava and Bamboo Shoots: A Human Health Risk Assessment. Technical Report Series P. 28.foodstandards.gov.au/_srcfiles/28_Cyanogenic_glycosides.pdf.

Association of Officials American Chemists (AOAC) (2000). Official analytical methods. 17th edition, AOAC International, Gaithersburg.

Azarov VI, Burov AV, Obolenskaya AV (1999). Wood chemistry and synthetic polymers (RUSSIAN). LTA, St. Petersburg, Russia, P. 627.

Choudhury D, Sahu JK, Sharma GD (2011). Bamboo shoot based fermented food products: a review. J. Scientific Industrial Res. 70:199-203.

Choudhury D, Sahu JK, Sharma GD (2012). Bamboo shoot: Microbiology, biochemistry and technology of fermentation: a review. Indian J. Trad Knowledge 11:242 – 249.

Choudhury D, Sahu JK, Sharma GD (2010). Biochemistry of Bitterness in Bamboo Shoots. Sci. Technol. 6:105-111.

Embaye K, Christersson L, Ledinc S, Weiha M (2003). Bamboo as bioresource in Ethiopia: management strategy to improve seedling performance (*Oxytenanthera abyssinica*). Bioresour. Technol. 88:33–39.

Embaye K, Weiha M, Ledinc S, Christersson L (2005). Biomass and nutrient distribution in a highland bamboo forest in southwest Ethiopia: implications for management. Forest Ecol. Management. 204:159-169.

Fengel D, Wegener G (1984). Wood: Chemistry, Ultrastructure, Reactions. Walter de Gruyter, Berlin, P. 613.

Forest Resource Assessment (FRA) (1985). Global Forestry Resource Assessment. ftp://ftp.fao.org/docrep/fao/010/a1243e/a1243e02.pdf.

Greenfield H, Southgate DAT (2003). Food composition data: Production, management and use. 2nd edition, FAO, Rome: P. 288.

Hunter I, Feng'e Y (2000). Cyanide in Bamboo Shoots. http://www.inbar.int/publication/txt/INBAR_Working_Paper_No39.htm

Raveendran K, Anuradda Ganesh, Kartic CK (1995). Influence of mineral matter on biomass pyrolysis characteristics. Fuel 74:1812-1821.

Seyoum K, Paridah MT, Wong En Ding, Rahim Sudin (2006). The effect of site on the variability on the suitability of Ethiopian highland bamboo (*Yushinia alpina*) for particleboard production. Ethiopian J. Nat. Resour. 8:251-269.

Seyoum K, Paridah MT, Wong EN Ding, Rahim S (2007). Influence of Culm age particle size and board density on the performance of particleboard made from the Ethiopian highland bamboos (*Yushina alpina*). J. Bamboo. Rattan. 6:239-250.

Shanmughavel P, Francis K (2003). Biomass accumulation and nutrient distribution in Dendrocalamus hamiltonii http://www.fao.org/docrep/article/wfc/xii/0269-b3-shanmughavel.htm.

Sylvia P (1995) Poaceae (Gramineae). In: Inga Hedberg and Sue Edwards (Eds). Flora of Ethiopia and Eritrea, Addis Ababa 7:420.

Wang S, Pu Xiaolan, Ding Y, Wan X (2009). Aboveground biomass and bamboo shoot nutrients of high altitude bamboos (*Fargesia yunnanensis* Hsueh et Yi) from different sites in Yunnan province, China. J. Trop. Agric. 47:48-53.

Zhang Jin-jie, Rong Ji, Hu Ya-qin, Chen Jian-chu, Ye Xing-qian (2011). Effect of three cooking methods on nutrient components and antioxidant capacities of bamboo shoot (*Phyllostachys praecox* C.D. Chu et C.S. Chao). Biomed. Biotechnol.12:752-759.

Amagloh FK, ES Nyarko (2012). Mineral nutrient content of commonly consumed leafy vegetables in northern Ghana. Afr. J. Food Agric. Nutr. Develop. (AJFAND) 12(5):6397– 9408 (online).

Seyfu K (1997). Tef. Eragrostis tef (Zucc.) Trotter. Promoting the conservation and use of underutilized and neglected crops. 12. Institute of Plant Genetics and Crop Plant Research, Gatersleben/International Plant Genetic Resources Institute, Rome, Italy. P. 49.

Parish C (2006). Whole Grains and the Gluten-Free Diet. Practical Gastroenterology. Series 2:66-78.

Performance of some promising genotypes of Glory lily (*Gloriosa superba* L.) against leaf blight disease

R. Chitra*, P. Muthulakshmi and K. Rajamani

Horticultural College and Research Institute, Tamil Nadu Agricultural University, Coimbatore-641 003, Coimbatore, India.

The study was conducted on performance of eighteen genotypes of glory lily (*Gloriosa superba* L.) against leaf blight disease under Tamil Nadu conditions. Experiment was laid out in a randomized block design with three replications at the Medicinal Plants Unit, Botanical Garden, Tamil Nadu Agricultural University, Coimbatore. All genotypes expressed significant variations for percent disease index and varied from 21.73 to 64.48%. The genotype GS 07 and genotype GS 05 exhibited the lowest index (21.73%) and highest index (64.48%), respectively. The percent disease index exhibited highly positive significant correlation both at phenotypic and genotypic levels for polyphenol oxidase. But it had negative significant association with peroxidase, total phenol content and catalase activity. The genotype GS 07 exhibited the lowest PDI value and this genotype considered as tolerant to leaf blight.

Key words: Genotypes, glory lily, leaf blight, peroxidase, polyphenol oxidase, catalase.

INTRODUCTION

Gloriosa superba (Liliaceae) is a native of tropical Africa and is found growing naturally in many countries of tropical Asia including Bangladesh, India, Sri Lanka, Malaysia and Myanmar. It is one of the major medicinal plants in India cultivated for its seeds which are exported to developed countries for pharmaceutical use. In India, it is usually found in Himalayan foot-hills, Central India, Tamil Nadu, Andhra Pradesh and Bengal. Seeds and tubers contain valuable alkaloids *viz*., colchicine and colchicoside as the major constituents, which are used to treat gout and rheumatism. Due to the action of colchicoside on spindle fibre formation during cell division, the plant has been identified as a potential anti-cancerous drug. In the Indian Systems of Medicine, the tubers are used as tonic, antiperiodic, antihelmenthic and also against snake bite (Gupta et al., 2005).

Leaf blight disease in Glory lily as observed during 2004 to 2006 in different areas of India and losses in *Gloriosa* were estimated to be 65 to 80% per annum (Maiti et al., 2007). The disease caused by the fungus, *Curvularia lunata* and being an air borne fungus, infects the leaves initially turning the leaf colour into light yellow. Later small black spots are seen which spread to the entire leaf surface. The incidence was reported more during cloudy days coupled with high humidity (Maiti et al., 2007). Plants produce different antimicrobial compounds and peroxidases when pathogen invades them (Sharma et al., 1990). Paramasivam and Arumugam (1991) identified following problems with regard to the cultivation of this crop such as i) scarcity and high cost of planting material, ii) seed tuber has a prolonged period of dormancy, iii) problems in cross pollination and poor seed set and iv) disease incidence particularly leaf blight is high. In view of all these observations, an attempt has been made in the present investigation to compare the performance of leaf blight disease and peroxidase and polyphenol oxidase activities in the elite genotypes of glory lily grown in Tamil Nadu, India with a view to identify the superior genotypes.

*Corresponding author. E-mail: chitra.varadharaj@gmail.com.

MATERIALS AND METHODS

Eighteen genotypes of *Gloriosa* were grown in the Medicinal Plants Unit, Botanical Garden, Tamil Nadu Agricultural University, Coimbatore during 2007 to 2008 in Randomized Block Design with three replications. Each plot consist of three rows of 5 m length with inter and intra row spacing of 150 and 30 cm, respectively and also having 17 plants / plot. The recommended agronomical practices

Table 1. List of genotypes and percent disease index of *Gloriosa superba.*

S/N	Name of the germplasm	Genotypes / treatment	Percent disease index
1	Nallampalayam cultivated	GS 01	44.34
2	Kallimanthayam cultivated	GS 02	60.22
3	Sathyamangalam wild	GS 03	42.87
4	Aruppukotai wild	GS 04	48.92
5	Aruppukotai cultivated	GS 05	64.48
6	Kankayam cultivated	GS 06	61.54
7	Kallimanthayam wild	GS 07	21.73
8	Ottanchadram cultivated	GS 08	45.50
9	Moolanur cultivated	GS 09	57.70
10	Jeyankondam cultivated	GS 10	42.18
11	Udangudi cultivated	GS 11	40.38
12	Viralimalai cultivated	GS 12	39.64
13	Pudukottai cultivated	GS 13	54.68
14	Andhra cultivated – I	GS 14	30.24
15	Andhra wild	GS 15	33.25
16	Z-Melur cultivated	GS 16	35.83
17	Poondurai wild	GS 17	38.25
18	Andhra cultivated -II	GS 18	52.77
Mean			45.25
SE(d)			0.11
CD (0.05%)			0.22

and plant protection measures were followed to ensure a normal healthy crop. Observations were taken on five randomly selected plants from each genotype for PDI and biochemical traits.

The percentage of disease incidence was calculated along with severity grade by 0-9 scale, where 0, no disease incidence; 1, <1% leaf area affected; 3, 1-10% leaf area affected; 5, 11-25% leaf area affected; 7, 26-50% leaf area affected and 9, >50% leaf area affected (Prasad, 1982). The enzyme activities of peroxidase, catalase, polyphenol oxidase and total phenol were also estimated from all the genotypes.

Polyphenol oxidase (PPO) activity (abs $min^{-1}g^{-1}$)

PPO in tubers was assayed using the modified method of Mayer et al. (1965). Standard reaction mixture contained 1.5 ml of 0.1 M phosphate buffer (pH 6.5), 0.5 ml of the enzyme extract and 0.5 ml of 0.01 N catechol. The changes in the absorbance were recorded at 495 nm at 30 s interval for 3 minutes and expressed as abs $min^{-1}g^{-1}$.

Catalase activity (unit min^{-1} g^{-1} fresh wt)

Catalase activity of tubers was determined by adopting the method of Diby and Sarma (2005). 50 ml of the enzyme extract was taken in sodium phosphate buffer (pH 7.0) (2 ml) and 500 ml of H_2O_2 was added. To this, was added 1 ml of distilled water and the absorbance was read at 230 nm for 3 min with a time interval of 15 s and the enzyme activity expressed as unit min^{-1} g^{-1} fresh wt.

Peroxidase activity (Δ OD 430 nm $min^{-1}g^{-1}$)

The peroxidase activity of tubers was determined by the method Suggested by Perur (1962). Reaction mixture consists of 1.0 ml of enzyme extract, 3 ml of 0.05 M pyrogallol and 0.5 ml of 1% H_2O_2 was added. The enzyme activity expressed as change in OD at 430 nm min^{-1} g^{-1}.

Total phenol content (mg g^{-1})

The total phenol content of the tuber sample was estimated by the method employing folin ciocalteu reagent (Malick and Singh, 1980) and expressed as mg g^{-1} of fresh weight.

All the data regarding the disease incidence and enzyme activities were statistically analyzed by Panse and Sukhatme (1978) and correlation coefficients at phenotypic and genotypic levels were worked out by adopting the standard methods of the analysis as suggested by Johnson et al. (1955).

RESULTS AND DISCUSSION

All genotypes expressed significant variations for percent disease index which varied from 21.73 to 64.48% (Table 1). The genotype GS 07 and genotype GS 05 exhibited the lowest index (21.73%) and highest index (64.48%), respectively. Seven genotypes (GS 02, GS 04, GS 05, GS 06, GS 09, GS 13 and GS 18) exhibited significantly higher index than the general mean value of 45.25%, while ten genotypes (GS 01, GS 03, GS 07, GS 10, GS 11, GS 12, GS 14, GS 15, GS 16 and GS 17) exhibited significantly lower index than the general mean. The genotype GS 07 exhibited the lowest percent disease incidence (PDI) value and this genotype considered as tolerant to leaf blight. Wider variations in *Phytophthora*

Table 2. Mean performance of *G. superba* genotypes for peroxidase, catalase, polyphenol oxidase and total phenol activities.

GP	Peroxidase activity	Catalase activity	Polyphenol oxidase activity	Total phenol
GS 01	0.68	2.20	0.036	0.62
GS 02	0.64	1.70	0.036	0.62
GS 03	0.68	1.70	0.036	0.60
GS 04	0.96	2.60	0.024	0.63
GS 05	0.20	1.40	0.032	0.88
GS 06	0.88	2.20	0.080	0.33
GS 07	2.24	8.20	0.012	1.34
GS 08	0.44	1.67	0.048	0.50
GS 09	0.92	2.40	0.032	0.72
GS 10	1.04	2.60	0.020	0.78
GS 11	1.48	5.60	0.012	1.11
GS 12	1.04	2.60	0.020	0.87
GS 13	1.02	2.60	0.024	0.79
GS 14	1.12	3.10	0.020	1.00
GS 15	1.24	4.10	0.020	1.06
GS 16	1.12	3.10	0.020	1.06
GS 17	1.24	3.10	0.020	1.07
GS 18	1.40	4.77	0.016	1.11
Mean	1.02	3.09	0.028	0.84
SE(d)	0.007	0.05	0.001	0.03
CD (0.05%)	0.014	0.11	0.003	0.07

blight incidence (14.3 to 78.5%) was reported in the several genotypes of *Colocasia esculenta* (Mitra et al., 2007).

Vidyasekaran (1997) reviewed that peroxidase and poly phenol oxidase activity would be more in the plants infected by pathogen and has great role in inhibition on the pathogenesis. Vidyasekaran (2000) reported that phenolics may not accumulate to fungi toxic levels but pathogens may produce toxins that suppress the host polyphenol oxidases. The highest peroxidase and catalase activity of 2.24 Δ OD 430 nm min^{-1} g^{-1} and 8.20 ΔA at 240 nm g^{-1} min^{-1} respectively was observed in the genotype GS 07 and lowest value of 0.20 Δ OD 430 nm min^{-1} g^{-1} and 1.40 ΔA at 240 nm g^{-1} min^{-1}, respectively was recorded in the genotype GS 05 (Table 2).

Accumulation of phenolic compounds in host parasite reaction is the general phenomenon of resistance and breakdown of these compounds determined the degree of resistance (Farkas and Kirlay, 1962; Sindhan and Parashar, 1984). Post infectional increase of phenols might be due to the tendency of phenols to accumulate at the site of infection which is involved in the defense mechanisms of plants through the interference in the metabolic activities of pathogens (Farkas and Kirlay, 1962). Phenolic substances and their oxidation products, quinones have been reported to play an important role in imparting resistance against pathogens (Kosuge, 1966; Goodman et al., 1967).

Host enzyme like polyphenol oxidase, catalase and peroxidase play an important role in disease resistance. These enzymes are responsible for degradation and synthesis of phenolics and quinones respectively. Peroxidases have been implicated in the regulation of cell wall elongation, phenol oxidation, polysaccharide cross-linking, indole-3-acetic acid (IAA) oxidation, cross linking of extension monomers, oxidation of hydroxyl-cinnamyl alcohols onto free radical intermediates and wound healing (Vidhyasekharan, 1997).

The phenotypic and genotypic correlation co-efficient were analyzed to understand the direction and magnitude of relationship between biochemical and percent disease index among the genotypes. In general, the genotypic correlation co-efficient was slightly higher than the phenotypic correlation co-efficient (Table 3). The percent disease index exhibited highly positive significant correlation both at phenotypic and genotypic levels for polyphenol oxidase. But it had negative significant association with peroxidase, total phenol content and catalase. Regarding inter correlation of the biochemical characters, the peroxidase activity had negative highly significant association with polyphenol oxidase and also had highly significant positive association with total phenol content and catalase. Polyphenol oxidase showed negative correlation with total phenol content and catalase. But total phenol content had positive highly significant association with catalase.

Knowledge about association of various characters help in selection of genotypes and also suggests the

Table 3. Correlation among host enzymes activity and percent disease index of 18 genotypes of *G. superba*.

Character		Peroxidase	Poly phenol oxidase	Total phenol	Catalase	Percent disease index
Peroxidase	P	1.000	-0.822**	0.754**	0.941**	-0.541**
	G	1.000	-0.822**	0.758**	0.941**	-0.541**
Poly phenol oxidase	P		1.000	-0.798**	-0.632**	0.544**
	G		1.000	-0.802**	-0.633**	0.544**
Total phenol	P			1.000	0.643**	-0.566**
	G			1.000	0.646**	-0.569**
Catalase	P				1.000	-0.485*
	G				1.000	-0.485*
Percent disease index	P					1.000
	G					1.000

*, Significant at 5% level; **, significant at 1% level.

advantage of a selection method for more than one character at a time, which could be explained that improvement of one character results in the simultaneous improvement of all other positively related characters. In disease resistance breeding programmes, knowledge of association between the incidence of these diseases and other characters is a must particularly when the breeder wishes to keep a balance between disease resistance and agronomic traits.

In this investigation, percent disease incidence (PDI) registered highly significant positive correlation with poly phenol oxidase and exhibited significant negative correlation with peroxidase, total phenol and catalase. These results were similar to earlier findings in chillies by Markose (1996), tomato by Jamou (2002), pea by Luhova et al. (2006) and taro by Mitra et al. (2007). Considering all the aspects it can be concluded that the genotypes GS 07, GS 14, GS 15, GS 16, GS 17, GS 12, GS 11 and GS 10 considered as tolerant to leaf blight.

REFERENCES

Diby P, Sarma YR (2005). *Pseudomonas fluorescens* mediated systemic resistance in black pepper (*Piper nigrum* L.) is driven through an elevated synthesis of defence enzymes. Archives Phytopathol. Plant Protect. 38:139-149.

Farkas GL, Kirlay Z (1962). Role of phenolic compound in physiology of plant disease and disease resistance. J. Phytopathol. 44:104-105.

Goodman RN, Kiraly Z, Zitlin M (1967). The biochemistry and physiology of infectious plant diseases. D. Van Nostrand Company. Inc. Princeton, New Jersey. p. 354

Gupta LM, Rana RC, Raina R, Meenakshi G (2005). Colchicine content in *Gloriosa superba* L. J. Res, 4:238-241.

Jamou AD (2002). Studies on tospovirus of soybean and tomato and their detection by serodiagnostic techniques. M.Sc. Thesis, Tamil Nadu Agricultural University, Coimbatore.

Johnson HW, Robinson WE, Comstock RF (1955). Genotypic and phenotypic correlations in soyabeans and their implication in selection. Agron. J. 47:447-483.

Kosuge T (1966). Role of pheolics substances in resistance of plants pathogens. Ann. Rev. Phytopathol. 7:195-222.

Luhova L, Lebeda A, Kutrova E, Hederervova D, Pec P (2006). Peroxidase, catalase, amine oxidase and acid phosphatase activities in *Pisum sativum* during infection with *Fusarium oxysporum* and *F. solani*. Biol. Plantarum. 50:675-682.

Maiti CK, Surjit S, Amal KP, Krishnendu A (2007). First report of leaf blight disease of *Gloriosa superba* L.caused by *Alternaria alternata* (Fr.) Keissler in India. J. Gen. Plant Pathol., 73:377-378.

Malick CP, Singh MB (1980). *In*: Plant Enzymology and Histo Enzymology. Kalyani Publishers, New Delhi. p. 286.

Markose BL (1996). Genetic and biochemical basis of resistance of bacterial wilt in chilli Ph.D. Thesis, Kerala Agricultural University, Vellanikkara, Thrissur. p. 140.

Mayer AM, Haul E, Shaul RB (1965). Assay of catechol oxidase, a critical comparison of method. Phytochemistry 5:783-789.

Mitra S, Sinha B, Pal H, Tarafdar J (2007). Comparative Studies on Morphological Characters, Yield, Nutritional Status and Isozymes Activity of Some Elite Genotypes of Taro (*Colocasia esculenta* (L.) Schott var. *antiquorum*) Grown in West Bengal. Acta Hort. 752:219-224.

Panse VG, Sukhatme PV (1978). Statistical Methods for Agricultural Workers. ICAR, New Delhi. pp. 134-192.

Paramasivam P, Arumugam R (1991). Survey of Gloriosa superba cultivation – a preliminary report. Scientific worker's conference, College of Horticulture, Periyakulam. July, 1991. pp. 24-25

Perur MG (1962). Measurement of peroxidase activity in plant tissues. Curr. Sci. 3:17-18.

Prasad SM (1982). National survey for diseases of tropical tuber crops. Regional centre of CTCRI, Bhubaneswar, (India). p. 49

Sharma ON, Mehrotra RS, Mahajan R, Bajaj KL (1990). Activities of some enzymes in tomato roots susceptible and resistant to root knot nematode, *Meloidigyne incognita* (Kofcid and White) Chitwood. J. Plant Sci. Res. 6:80-81.

Sindhan GS, Parashar RD (1984). A comparative study of pea varieties resistant and susceptible to powdery mildew by leaf curl and mosaic disease. Prog. Hort. 20:185-189.

Vidyasekaran P (1997). Fungal pathogenesis in plants and crops, molecular biology and host defence mechanisms. *In*: Marcel Dekker (ed.), Incorporation, Madison Avenue, New York 10016 270:465-493.

Vidyasekaran P (2000). Physiology of disease resistance in plants. Vol. II. Boca Raton: CRC. p. 128

Effect of different levels of N and P on ratoon banana (*Musa* spp. AAA)

K. S. Navaneethakrishnan[1], M. I. S. Gill[1] and S. Ramesh Kumar[2]

[1]Department of Fruit Science, Punjab Agricultural University, Ludhiana-141004, Punjab, India.
[2]Department of Horticulture, Vanavarayar Institute of Agriculture, Manakkadavu, Pollachi-642 103, Tamil Nadu Agricultural University, Tamil Nadu, India.

In India, banana is the important and highly remunerative fruit crop grown in tropical and subtropical region. It represents the second most important fruit crop after mango in India. Banana is grown throughout Asian countries but it is commercially cultivated on a large scale in the Indo-china region. A field experiment was carried out in the College Orchard, College of Agriculture, Punjab Agricultural University, during 2009. The objective of the study was to determine the proper rate of N and P fertilizers that would maximize the potential yield of commercially-grown banana Grand naine. Different levels of chemical fertilizers were applied, *viz.,* N (150, 200, 250 and 300 g/tree), P (60 and 90 g/tree) in addition to non-fertilized control (no N and P applied) treatment. The results indicated that vegetative growth parameters (that is, pseudostem height, pseudostem girth, number of leaves, leaf area index) of fertilized plants neither were significantly different nor produced better vegetative growth when compared to non-fertilized control plants. The fertilizer treatment T5 consisting of N (200 g N in 5 split doses and 60 g P_2O_5) numerically influenced the yield and fruit characteristics like bunch weight, number of hands per bunch, hand weight and number of fingers per hand of 'Grand naine' banana as compared to other treatments. Hence, this fertilizer dose of N and P was recommended for cultivation of 'Grand naine' banana in northern region (Punjab) of India.

Key words: Fruit yield, bunch weight, banana, nitrogen and phosphorus.

INTRODUCTION

Banana (*Musa* spp.) belongs to the family Musaceae and originated from South East Asia. It is the fourth important global food commodity and is grown in 4.5 million ha area with an annual production of about 73 million tons in the world (Anonymous, 2005). Banana is the leading fruit crop of India, ranks third in area (0.647 million ha) and first in production (23.20 million tons) among fruit crops (Anonymous, 2008) and grown in variety of soils provided with soil moisture, deep, well drained, friable loamy soil with adequate organic matter is ideal for its cultivation. In plains it is grown in well drained black cotton soil. The medium black and lighter soils of East Khandesh are also suitable. Banana is one of the fruit which has a restricted root zone. Therefore, two soil factors are important namely depth and drainage. In banana it is essential to induce quick growth and produce more leaves with larger area. So it is a gross feeder of nutrients and responds well to nitrogen and phosphorus.

As Indian soils are deficient in nitrogen and phosphorus - these two plant nutrients together with organic manure play an important role to get good crop returns. The choice and dosage of nutrients to be applied depends on the cultivar, inherent soil fertility, stage of plant growth, climate etc.

The better vegetative growth ensures better bunch development. High fertilization requirement of banana is mainly due to its rapid and vigorous growth and high fruit yield. Grand Naine is a popular variety of banana grown mostly in all export oriented countries of Asia, South America and Africa. This is a superior selection of Giant Cavendish which was introduced to India in 1990's. Due to many desirable traits like excellent fruit quality, immunity to fusarium wilt etc, it has proved better variety (Singh and Chundawat, 2002). However, its requirement of nutrients through organic sources along with inorganic fertilizers is not well documented. Keeping this in mind, the present investigation was undertaken to find out the suitable combination of nitrogen (N) and phosphorus (P) fertilizers for Grand Naine banana. Banana being a new crop for Punjab, requisite information on nutrient management for optimum growth and fruiting is lacking. So there is an urgent need was felt to generate this valuable information for optimum production of banana under sub-tropical conditions of Punjab. Thus, the present studies were planned to gather the much needed information on nutrient management in banana.

MATERIALS AND METHODS

The present investigation was carried out at New Orchard of the Department of Horticulture, Punjab Agricultural University, Ludhiana during 2009. Nutrients (Nitrogen and Phosphorus) were applied to the ratoon crop. All plants were given uniform cultural practices, except the fertilizers treatments.

Layout of the experiment

Treatments

Main treatments - 2 (Phosphorus at 60 g and 90 g/plant)
Sub treatments - 6 (Nitrogen detailed below)
Number of replications - 3, Treatment unit - 8
Total number of experimental plants - 6 × 2 × 3 × 8 = 288
Design: Split plot design

Treatment details

T1 - 150 g N (5 split doses) + 60 g P_2O_5
T2 - 150 g N (5 split doses) + 90 g P_2O_5
T3 - 200 g N (4 split doses) + 60 g P_2O_5
T4 - 200 g N (4 split doses) + 90 g P_2O_5
T5 - 200 g N (5 split doses) + 60 g P_2O_5
T6 - 200 g N (5 split doses) + 90 g P_2O_5
T7 - 250 g N (4 split doses) + 60 g P_2O_5
T8 - 250 g N (4 split doses) + 90 g P_2O_5
T9 - 250 g N (5 split doses) + 60 g P_2O_5
T10 - 250 g N (5 split doses) + 90 g P_2O_5
T11 - 300 g N (5 split doses) + 60 g P_2O_5
T12 - 300 g N (5 split doses) + 90 g P_2C_5

Nitrogen application was in the form of urea
Phosphorus application was in the form of single superphosphate
Potassium application was in the form of muriate of potash

Time of application

Nitrogen - Urea in 4 split doses - April, May, June, and July; Urea in 5 split doses - May, June, July and August.
Phosphorus - Single superphosphate was applied at the time of planting along with farmyard manure. Potassium - Muriate of potash 200 g/plant was applied uniformly in 5 split doses at monthly intervals starting from June onwards (till September).Different NP combinations comprised of six levels of Nitrogen and two levels of Phosphorus.

Observations

Pseudostem height

The pseudostem height was measured from June to September at monthly intervals by using a steel measuring tape from the base of pseudostem to the emergence point of the peduncle and was recorded in centimeters.

Pseudostem girth

The pseudostem girth was measured from June to September with a steel measuring tape from 90 cm above the ground and was recorded in centimeters.

Leaf area index (Length × Breadth)

The length and breadth of leaves was recorded by using a steel measuring tape and expressed in centimeters. Leaf area was calculated with formula L × B × Constant factor. Constant factor 0.8 was developed by using calculated area with the help of graph paper.

Span of harvesting

The date of harvesting of each plant was judged by the criteria given by Dhillon et al. (2002), that is, the fruits were harvested when fruit colour changed from deep green to green and ridges on the skin change from angular to round.

Bunch weight

The bunch of each plant was weighed and recorded in kilograms per plant.

Number of hands per bunch

The number of hands per bunch was recorded and an average of eight plants was pooled.

Hand weight

The 2nd and 3rd hand of the bunch was weighed and recorded in

Table 1. Effect of various combinations of N and P fertilizers on pseudostem height (cm) of banana ratoon crop *cv* Grand Naine.

Treatment	June	July	August	September
T1 - 150 g N (5 split doses) + 60 g P_2O_5	41.00	49.93	111.95	200.75
T2 - 150 g N (5 split doses) + 90 g P_2O_5	40.00	48.40	111.35	199.37
T3 - 200 g N (4 split doses) + 60 g P_2O_5	43.24	52.56	118.50	209.48
T4 - 200 g N (4 split doses) + 90 g P_2O_5	42.50	51.08	116.86	207.00
T5 - 200 g N (5 split doses) + 60 g P_2O_5	43.80	53.16	119.58	210.64
T6 - 200 g N (5 split doses) + 90 g P_2O_5	42.44	51.91	118.52	208.45
T7 - 250 g N (4 split doses) + 60 g P_2O_5	42.09	51.38	117.16	207.02
T8 - 250 g N (4 split doses) + 90 g P_2O_5	41.30	50.91	115.49	205.57
T9 - 250 g N (5 split doses) + 60 g P_2O_5	41.93	50.58	115.71	204.19
T10 - 250 g N (5 split doses) + 90 g P_2O_5	40.27	49.42	113.30	200.46
T11 - 300 g N (5 split doses) + 60 g P_2O_5	41.85	51.36	117.00	206.28
T12 - 300 g N (5 split doses) + 90 g P_2O_5	41.18	49.34	115.16	202.57
CD at 5% (N × P)	NS	NS	NS	NS

kilograms.

Number of fingers per hand

The number of fingers per hand was recorded as an average of 3 randomly selected hands that is, upper, lower and middle.

Finger length

Finger length was recorded by measuring the length of mature fruits with scale. Fingers were taken from randomly selected hands.

Statistical analysis

The data recorded during the course of studies were statistically analyzed as per split plot design method (Chao and Lincoln, 1969).

RESULTS AND DISCUSSION

Effect of nitrogen and phosphorus fertilizers on growth parameters

The effect of various combinations of N and P treatments on vegetative growth was observed in terms of pseudostem height, girth, number of leaves and leaf area index. The uniform sized plant selection was done in the month of May and the data were recorded at monthly intervals starting from June till September, one month after the application of fertilizers. The results obtained are presented and discussed below:

Pseudostem height

The data pertaining to the effect of N and P fertilizers on pseudostem height is presented in Table 1, the various combinations of N and P treatment means was found to be non significant in all the months from June to September. However it is clear from the data that the Treatment T5 (200 g N in 5 split doses and 60 g P_2O_5) recorded maximum pseudostem height, but it was at par with all other treatment combination means.

It is also evident from the Table 2 that the pseudostem height is significantly influenced by various N Treatments. In the month of June, the mean maximum pseudostem height (43.12 cm) was observed with the application of 200 g N in 5 split doses (N3), followed by 200 g N in 4 split doses (N2) with the mean height of (42.87 cm). During the month of July, application of 200 g N in 5 split doses (N3) had significantly higher mean pseudostem height (52.53 cm) than all the other N Treatments. Likewise, in the months of August and September maximum mean pseudostem height (119.05 cm) and (209.54 cm) respectively, was recorded by application of 200g N 5 split doses (N3). The applications of different P Treatments (Table 3) also indicate significant effect on pseudostem height. It is clear from the data that the Treatment P1 (60 g P_2O_5) per plant had resulted in significantly higher mean pseudostem height from the month of June to September (42.32, 51.66, 116.65 and 206.39 cm), respectively, as compared to the Treatment P2 (90 g P_2O_5).The result of the experiment also did not record much difference in height of banana plants due to application of P at the rate of 60g and 90g P_2O_5 per plant, but lower dose of 60 g P_2O_5 was significantly better than 90 g P_2O_5.

Pseudostem girth

The data on the effect of different doses of N and P Treatments and their combinations on pseudostem girth are presented in Tables 4, 5 and 6. The application of

Table 2. Effect of various doses of applied N fertilizers on pseudostem height (cm) of banana ratoon crop *cv* Grand Naine.

N (g/tree)	June	July	August	September
N1 - 150 g N (5 split doses)	40.50	49.16	111.65	200.06
N2 - 150 g N (4 split doses)	42.87	51.87	117.68	208.24
N3 - 200 g N (5 split doses)	43.12	52.53	119.05	209.54
N4 - 250 g N (4 split doses)	41.70	51.29	116.32	206.29
N5 - 250 g N (5 split doses)	41.10	50.00	114.50	202.33
N6 - 300 g N (5 split doses)	41.52	50.80	116.08	204.43
CD 5%	0.45	0.60	0.78	1.49

Table 3. Effect of two different doses of applied P fertilizers on pseudostem height (cm) of banana ratoon crop *cv* Grand Naine.

P (g/tree)	P1 (60 g P_2O_5)	P2 (90 g P_2O_5)	CD 5%
June	42.32	41.28	0.87
July	51.66	50.22	1.32
August	116.65	115.11	1.53
September	206.39	203.90	1.66

Table 4. Effect of various combinations of N and P fertilizers on pseudostem girth (cm) of banana ratoon crop *cv*Grand Naine.

Treatment	June	July	August	September
T1 - 150 g N (5 split doses) + 60 g P_2O_5	20.54	25.05	39.91	53.19
T2 - 150 g N (5 split doses) + 90 g P_2O_5	19.89	24.90	39.55	52.56
T3 - 200 g N (4 split doses) + 60 g P_2O_5	21.42	26.86	43.13	58.35
T4 - 200 g N (4 split doses) + 90 g P_2O_5	20.87	26.11	42.38	57.20
T5 - 200 g N (5 split doses) + 60 g P_2O_5	21.78	27.82	44.25	58.54
T6 - 200 g N (5 split doses) + 90 g P_2O_5	20.96	26.32	43.04	57.40
T7 - 250 g N (4 split doses) + 60 g P_2O_5	21.11	26.58	42.41	58.09
T8 - 250 g N (4 split doses) + 90 g P_2O_5	20.52	25.36	41.10	56.50
T9 - 250 g N (5 split doses) + 60 g P_2O_5	20.86	26.31	42.43	56.76
T10 - 250 g N (5 split doses) + 90 g P_2O5	20.63	25.01	40.01	54.44
T11 - 300 g N (5 split doses) + 60 g P_2O_5	21.31	26.57	41.92	57.41
T12 - 300 g N (5 split doses) + 90 g P_2O_5	20.10	26.03	40.26	55.73
CD at 5% (N × P)	NS	0.38	0.88	NS

Table 5. Effect of various doses of applied N fertilizers on pseudostem girth (cm) of banana ratoon crop *cv* Grand Naine.

N (g/tree)	June	July	August	September
N1 - 150 g N (5 split doses)	20.22	24.97	39.73	52.87
N2 - 150 g N (4 split doses)	21.14	26.48	42.75	57.77
N3 - 200 g N (5 split doses)	21.37	27.07	43.64	57.97
N4 - 250 g N (4 split doses)	20.81	25.97	41.75	57.29
N5 - 250 g N (5 split doses)	20.74	25.66	41.22	55.60
N6 - 300 g N (5 split doses)	20.70	26.30	41.09	56.57
CD 5%	0.45	0.27	0.62	0.57

Table 6. Effect of two different doses of applied P fertilizers on pseudostem girth (cm) of banana ratoon crop *cv* Grand Naine.

P (g/tree)	P1 (60 g P_2O_5)	P2 (90 g P_2O_5)	CD 5%
June	21.17	20.49	0.19
July	26.53	25.62	0.19
August	42.34	41.06	0.58
September	57.05	55.63	0.77

various combinations of N and P Treatments had shown a significant effect on the pseudostem girth in the months of July and August, but in the month of June and September it was found to be non-significant.

However, in the month of July, different N and P Treatments and their combinations significantly influenced the pseudostem girth. The maximum pseudostem girth (27.82 cm) was recorded in treatment combination 200 g N 5 split doses and 60g P_2O_5 (T5) which was at par with combination 200 g N 4 split doses and 60g P_2O_5 (T3) with the height (26.86 cm) significantly higher than all the other combinations. The minimum pseudostem girth (24.90 cm) was recorded in NP Treatment T2 (150 g N 5 split doses and 90 g P_2O_5). Similarly in the month of August, the maximum pseudostem girth (44.25 cm) was recorded with the application of 200 g N 5 split doses and 60g P_2O_5 (T5) followed by Treatment 200g N 4 split doses and 60g P_2O_5 (T3) having pseudostem girth of 43.13 cm. The pseudostem girth of these treatments was significantly higher than all the other treatment combinations. The minimum pseudostem girth (39.55 cm) was recorded in T2 (150 g N 5 split doses and 90 g P_2O_5).

The data presented in Table 5 reveal that the application of different doses of N significantly influenced the pseudostem girth during June. The application of 200g N in 5 split doses (N3) per plant had maximum mean pseudostem girth (21.37 cm) however, it was at par with the treatment were 200 g N in 4 split doses (N2) was applied. The lowest mean pseudostem girth was recorded in Treatment N1 (150 g N in 5 split doses). It is quite evident from the data that in the month of July, the Treatment N3 (200 g N in 5 split doses) had the highest mean pseudostem girth (27.07 cm) which was at par with Treatment N2 (200 g N in 4 split doses) which had mean pseudostem girth of 26.48 cm. Similarly in the months of August and September the Treatment N3 (200 g N in 5 split doses) had the maximum mean pseudostem height that is, 43.64 cm, and 57.97 cm respectively. The lowest mean pseudostem girth was recorded in Treatment N1. The data presented in the Table 6 reveal that the applied P Treatments also has a significant effect on pseudostem girth, but among the phosphorus treatments, the Treatment P1 (60 g P_2O_5) had significantly higher mean pseudostem girth (21.17, 26.53, 42.34 and 57.05 cm) respectively, in all the months than the Treatment P2 (90 g P_2O_5). Tirkey et al. (2003) also reported similar results, as the growth rate of banana during May and June (initial stages) were not affected by the different nitrogen levels, split applications of NPK and their interactions. They further explained that non-significant difference in growth parameters during early months may be due to slow nutrient uptake at initial growth phase due to the reason that tissue cultured plants take time to get acclimatized under field conditions. The significant results at later stages may be due to better utilization of nutrients.

Number of leaves per plant

Data on number of leaves per plant is presented in Tables 7, 8 and 9. Like other growth parameters (height and girth) the effect of applied N and P Treatment combinations significantly influenced the number of leaves per plant during the months of June, August and September. However, in the month of July, it was found to be non-significant. In the month of June the maximum number of leaves per plant (9.10) were recorded (Table 7) in treatment where 200 g N 5 split doses and 60 g P_2O_5 (T5) was applied, however the number of leaves obtained where at par with the treatment where 200 g N 4 split doses and 60 g P_2O_5 (T3) were applied, while the minimum number of leaves per plant (7.30) were recorded in T2 (150 g N 5 split doses and 90g P_2O_5) combination. In the month of August, the maximum number of leaves per plant (13.05) were recorded in Treatment T5 (200 g N 5 split doses and 60g P_2O_5) followed by Treatment T6 (200 g N 5 split doses and 90 g P_2O_5) and the minimum number of leaves per plant (11.70) were recorded in T2 (150 g N 5 split doses and 90 g P_2O_5).

Similarly in the month of September the maximum number of leaves per plant (15.55) were recorded in Treatment T5 (200 g N 5 split doses and 60 g P_2O_5) which was however at par with T6 (200 g N 5 split doses and 90 g P_2O_5) and the minimum number of leaves per plant (13.68) were recorded in Treatment T1 (150 g N 5 split doses and 60 g P_2O_5). The perusal of data presented in Table 8 revealed that the various doses of N have significant influence on number of leaves per plant in all the months. During the month of June, the

Table 7. Effect of various combinations of N and P fertilizers on number of leaves per plant of banana ratoon crop *cv*Grand Naine.

Treatment	June	July	August	September
T1 - 150 g N (5 split doses) + 60 g P_2O_5	7.38	9.17	12.16	13.68
T2 - 150 g N (5 split doses) + 90 g P_2O_5	7.30	9.02	11.70	14.08
T3 - 200 g N (4 split doses) + 60 g P_2O_5	9.06	9.97	12.68	15.26
T4 - 200 g N (4 split doses) + 90 g P_2O_5	8.33	9.56	12.76	15.26
T5 - 200 g N (5 split doses) + 60 g P_2O_5	9.10	10.38	13.05	15.55
T6 - 200 g N (5 split doses) + 90 g P_2O_5	8.31	9.94	12.96	15.50
T7 - 250 g N (4 split doses) + 60 g P_2O_5	8.12	9.85	12.40	15.32
T8 - 250 g N (4 split doses) + 90 g P_2O_5	7.89	9.71	12.53	14.82
T9 - 250 g N (5 split doses) + 60 g P_2O_5	7.91	9.71	12.30	14.81
T10 - 250 g N (5 split doses) + 90 g P_2O_5	7.75	9.52	12.35	13.98
T11 - 300 g N (5 split doses) + 60 g P_2O_5	7.82	9.78	12.45	14.86
T12 - 300 g N (5 split doses) + 90 g P_2O_5	7.88	9.66	12.50	14.91
CD at 5% (N × P)	0.27	NS	0.24	0.30

Table 8. Effect of various doses of applied N fertilizers on number of leaves per plant of banana ratoon crop *cv* Grand Naine.

N (g/tree)	June	July	August	September
N1 - 150 g N (5 split doses)	7.34	9.09	11.93	13.88
N2 - 150 g N (4 split doses)	8.69	9.77	12.72	15.26
N3 - 200 g N (5 split doses)	8.70	10.16	13.00	15.52
N4 - 250 g N (4 split doses)	8.00	9.78	12.46	15.07
N5 - 250 g N (5 split doses)	7.83	9.61	12.32	14.39
N6 - 300 g N (5 split doses)	7.85	9.72	12.47	14.88
CD 5%	0.19	0.15	0.17	0.21

Table 9. Effect of two different doses of applied P fertilizers on number of leaves per plant of banana ratoon crop *cv* Grand Naine.

P (g/tree)	P1 (60 g P_2O_5)	P2 (90 g P_2O_5)	CD 5%
June	8.23	7.91	0.18
July	9.81	9.57	0.73
August	12.50	12.46	NS
September	14.91	14.76	NS

Treatment N3 (200 g N in 5 split doses) had maximum mean number of leaves per plant (8.70) which however was at par with Treatment N2 (200 g N in 4 split doses) with (8.69) mean number of leaves per plant. Whereas the lowest mean number of leaves per plant (7.34) were recorded in Treatment N1 (150 g N in 5 split doses). Similarly in the month of July, Treatment N3 (200 g N in 5 split doses) had significantly higher mean number of leaves per plant (10.16). The lowest mean number of leaves (9.09) were recorded in Treatment N1 (150 g N in 5 split doses). Table 8 further revealed that in the months of August and September the Treatment N3 (200 g N in 5 split doses) had the higher mean number of leaves per plant (13.00 and 15.52), respectively. Whereas, the lowest mean number of leaves per plant (11.93 and 13.88) were recorded in the Treatments N1 (150 g N in 5 split doses), in the months of August and September, respectively.

The data pertaining to the effect of P fertilizers on number of leaves per plant is presented in the Table 9. It is clear from the data that in the early months viz., June and July the P treatments was significant, but in the later months it was found to be non-significant. The data further revealed that in both these months, Treatment P1 (60 g P_2O_5) had significantly higher mean number of leaves (8.23 and 9.81) per plant.

Table 10. Effect of various combinations of N and P fertilizers on leaf area index (cm^2) of banana ratoon crop *cv* Grand Naine.

Treatment	June	July	August	September
T1 - 150 g N (5 split doses) + 60 g P_2O_5	2018	3660	5605	6873
T2 - 150 g N (5 split doses) + 90 g P_2O_5	1961	3586	5491	6801
T3 - 200 g N (4 split doses) + 60 g P_2O_5	2619	4476	6283	7324
T4 - 200 g N (4 split doses) + 90 g P_2O_5	2562	4312	6218	7214
T5 - 200 g N (5 split doses) + 60 g P_2O_5	2718	4623	6393	7423
T6 - 200 g N (5 split doses) + 90 g P_2O_5	2702	4518	6389	7414
T7 - 250 g N (4 split doses) + 60 g P_2O_5	2365	4303	6104	7161
T8 - 250 g N (4 split doses) + 90 g P_2O_5	2268	4243	5983	7044
T9 - 250 g N (5 split doses) + 60 g P_2O_5	2047	3768	5796	6936
T10 - 250 g N (5 split doses) + 90 g P_2O_5	1965	3755	5825	6943
T11 - 300 g N (5 split doses) + 60 g P_2O_5	2176	3960	5934	7112
T12 - 300 g N (5 split doses) + 90 g P_2O_5	2114	3919	5891	6998
CD at 5% (N × P)	NS	NS	NS	NS

Table 11. Effect of various doses of applied N fertilizers on leaf area index (cm^2) of banana ratoon crop *cv* Grand Naine.

N (g/tree)	June	July	August	September
N1 - 150 g N (5 split doses)	1989	3623	5548	6837
N2 - 150 g N (4 split doses)	2590	4394	6250	7269
N3 - 200 g N (5 split doses)	2710	4570	6391	7418
N4 - 250 g N (4 split doses)	2316	4273	6044	7103
N5 - 250 g N (5 split doses)	2006	3761	5810	6939
N6 - 300 g N (5 split doses)	2145	3939	5912	7055
CD 5%	95.6	97.4	102.1	85.9

Table 12. Effect of two different doses of applied P fertilizers on leaf area index (cm^2) of banana ratoon crop *cv* Grand Naine.

P (g/tree)	P1 (60 g P_2O_5)	P2 (90 g P_2O_5)	CD 5%
June	2324	2262	59.1
July	4131	4055	NS
August	6019	5966	NS
September	7138	7069	59.3

Leaf area index

The data pertaining to leaf area index are presented in Tables 10, 11 and 12. The different N and P combinations did not significantly influence this trait. However, the maximum leaf area index (2718, 4623, 6393 and 7423 cm^2) was recorded by application of Treatment T5 (200 g N 5 split doses and 60 g P_2O_5) however, it was at par with all other treatment combinations.

The data presented in Table 11, has revealed that the application of different doses of N have a significant effect on leaf area index in all the months. In the month of June, the Treatment N3 (200 g N in 5 split doses) resulted in significantly higher mean leaf area index (2710 cm^2) followed by Treatment N2 (200 g N in 4 split doses) with mean leaf area index of 2590 cm^2. Whereas the lowest leaf area index (1989 cm^2) was recorded in Treatment N1 (200 g N in 5 split doses). In the month of July, Treatment N3 (200 g N in 5 split doses) had the highest mean leaf area index (4570 cm^2) followed by the Treatment N3 (200 g N in 5 split doses) with mean leaf area index of 4394 cm^2. Whereas the lowest leaf area index of (3623 cm^2) was recorded in Treatment N1 (150 g

Table 13. Effect of various combinations of N and P fertilizers on Bunch weight (kg) and Mean no of hands per bunch of banana ratoon crop *cv* Grand Naine.

Treatment	Bunch weight (kg)	Mean no. of hands per bunch
T1 - 150 g N (5 split doses) + 60 g P_2O_5	15.88	7.19
T2 - 150 g N (5 split doses) + 90 g P_2O_5	15.37	7.08
T3 - 200 g N (4 split doses) + 60 g P_2O_5	17.57	10.08
T4 - 200 g N (4 split doses) + 90 g P_2O_5	17.17	9.41
T5 - 200 g N (5 split doses) + 60 g P_2O_5	18.11	10.61
T6 - 200 g N (5 split doses) + 90 g P_2O_5	17.65	9.98
T7 - 250 g N (4 split doses) + 60 g P_2O_5	16.83	9.09
T8 - 250 g N (4 split doses) + 90 g P_2O_5	16.51	8.50
T9 - 250 g N (5 split doses) + 60 g P_2O_5	16.08	7.51
T10 - 250 g N (5 split doses) + 90 g P_2O_5	15.79	7.30
T11 - 300 g N (5 split doses) + 60 g P_2O_5	16.42	8.11
T12 - 300 g N (5 split doses) + 90 g P_2O_5	16.12	7.62
CD at 5% (N × P)	NS	NS

Table 14. Effect of various doses of applied N fertilizers on Bunch weight (kg) and Hands per bunch in banana ratoon crop *cv* Grand Naine.

N (g/tree)	Bunch weight (kg)	Hands per bunch
N1 - 150 g N (5 split doses)	15.63	7.13
N2 - 150 g N (4 split doses)	17.37	9.75
N3 - 200 g N (5 split doses)	17.88	10.29
N4 - 250 g N (4 split doses)	16.67	8.80
N5 - 250 g N (5 split doses)	15.93	7.40
N6 - 300 g N (5 split doses)	16.27	7.86
CD 5%	0.22	0.35

N in 5 split doses). Similarly in the months of August and September, Treatment N3 (200 g N in 5 split doses) resulted in higher mean leaf area index of (6391 and 7418 cm^2, respectively) was followed by N2 (200 g N in 4 split doses) with mean leaf area index of (6250 and 7269 cm^2). Whereas the lowest leaf area index (5548 and 6837 cm^2) was recorded in Treatment N1 (200 g N in 5 split doses). The perusal of data presented in Table 12 revealed that the P treatments were non-significant in the month of July and August, but it had shown a significant effect in June and September. In the months of June and September, among phosphorus treatments the maximum leaf area index (2324 and 7138 cm^2) has been recorded in the Treatment P1 (60 g P_2O_5) as compared to the Treatment P2 (90 g P_2O_5).

These results showed a close relationship between temperature and number of leaves and leaf area index. During the months of July and August the range of temperature remained closer to the optimum growth temperature that is, 27°C which along with humidity were most favorable for the better growth of banana leaves. Arumugam and Manivannan (2001) also reported similar results with (200 g) nitrogen per plant.

Effect of nitrogen and phosphorus on fruit yield parameters

Bunch weight (kg/plant)

The data pertaining to the effect of different N and P Treatment and their combinations on bunch weight is presented in the Tables 13, 14 and 15. This revealed that N and P treatments significantly affected the yield but the effects of NP combinations were not-significant. The perusal of data presented in Table 5a, indicated that the plants with Treatment N3 (200 g N in 5 split doses) had significantly higher mean bunch weight (17.88 kg/plant) followed by the Treatment N2 (200 g N in 4 split doses) with bunch weight of 17.37 kg/plant and was significantly higher than all other treatments. The minimum mean bunch weight (15.63 kg/plant) was recorded in Treatment N1 (150 g N in 5 split doses). In case of phosphorus Treatments P1 (60 g P_2O_5) had significantly higher mean bunch weight (16.81 kg/plant) than P2 (90 g P_2O_5).

The increase in bunch weight by increasing N dose from (150 to 200 g/plant) could be attributed largely to increased growth, higher bunch weight, more number of

Table 15. Effect of two different doses of applied P fertilizers on Bunch weight (kg) and Hands per bunch in banana ratoon crop *cv* Grand Naine.

P (g/tree)	P1 (60 g P_2O_5)	P2 (90g P_2O_5)	CD 5%
Bunch weight (kg)	16.81	16.43	0.32
Hands per bunch	8.76	8.31	0.25

fingers per hand and more number of hands per bunch, which might have been increased by timely availability of the nutrients at critical growth stages, which enhanced photosynthesis that lead to accumulation of more carbohydrates and other metabolites ultimately translocations towards the fruit tissue. Nitrogen is a constituent of nucleoproteins, amino acids, amines, amino sugars, polypeptides and many other organic compounds which are associated with all vital plant processes like photosynthesis, respiration, building plant structure and fatty acid synthesis etc. Where a yield decrease had occurred, it was due to an adverse effect of higher N dose on the availability of other nutrients, notably P or due to the toxicity caused by the release of excessive amounts of some other elements such as copper etc.

The results of the present study are in accordance with the findings of Naresh and Sharma (2004) reported that N application up to 240 g N/plant increased yield and above this dose it reduced the fruit yield significantly, due to possible toxic effects of N at higher levels in Jahajee Banana. P increased the growth and foliar status of tree which improved ATP formation thus providing more physiological efficiency to the tree which indirectly helped in better yield.

Number of hands per bunch

The data pertaining to number of hands per bunch is presented in Tables 13, 14 and 15. The data indicates that interaction between N and P Treatments is non-significant. The highest number of hands per bunch (10.61) were recorded by the application of Treatment T5 (200 g N 5 split doses and 60 g P_2O_5), while the lowest number of hands per bunch (7.08) were recorded in T2 (150 g N 5 split doses and 60 g P_2O_5). The data presented in Table 14. Further revealed that with the Treatment N3 (200 g N in 5 split doses) the significantly highest mean number of hands per bunch (10.29) were obtained followed by the Treatment N2 (200 g N in 4 split doses) with (9.75) mean number of hands per bunch. While the lowest mean number of hands per bunch (7.13) were recorded in Treatment N1 (150 g N in 5 split doses). In phosphorus treatments, P1 (60 g P_2O_5) resulted in significantly higher mean number of hands per bunch (8.76) and fingers per hand than Treatment P2 (90 g P_2O_5). The results on above said yield attributing characters like number of fingers per hands and number of hands per bunch in this experiment are in agreement with findings of Naresh and Sharma (2004) reported that N dose from 60 to 240 g/plant was effective in increasing the number of fingers per hand and number of hands per bunch and reduction in these characters were noticed beyond this nitrogen dose in Jahajee banana. This increase in yield attributing traits may be due to the availability of N during critical growth phase and increase in N supply resulted in translocation of more N to the top and this resulted in vigorous protein synthesis, utilizing the photosynthates produced by leaves for production of fingers and hands.

Hand weight (kg/bunch)

The effects of applied N and P on hand weight (kg/bunch) are presented in the Table 16. The data has revealed that the applied N and P Treatment combinations had not significantly influenced the hand weight (kg/bunch). It is evident from the data presented in Table 17. that with Treatment N3 (200 g N in 5 split doses) maximum mean hand weight of (2.14 kg/bunch) was obtained which however was at par with Treatment N2 (200 g N in 4 split doses) and N4 (250 g N in 5 split doses) with hand weight of (2.02 and 1.90 kg/bunch, respectively). The minimum mean hand weight of (1.65 kg/bunch) were recorded in Treatment N1 (150 g N in 5 split doses). Among phosphorus treatments as presented in Table 18 the Treatment P1 (60 g P_2O_5) had significantly higher mean hand weight of (1.92 kg/bunch).

Number of fingers per hand

The effect of applied N and P on number of fingers per hand is presented in the Table 16. The data revealed that the applied N and P treatment combinations did not significantly affect number of fingers per hand. However, it is evident from the data presented in Table 17, that various N doses have shown a significant effect on number of fingers per hand. The significantly higher mean number of fingers per hand (19.58) were recorded in Treatment N3 (200 g N in 5 split doses) followed by the Treatment N2 (200 g N in 4 split doses) with 18.99 fingers per hand. The minimum mean numbers of fingers per hand (15.14) were recorded in Treatment N1 were150 g N in 5 split doses was applied. The data presented in Table 16 reveal that P has a significant

Table 16. Effect of various combinations of N and P fertilizers on Hand weight (kg), Mean no fingers per hand and finger length (cm) of banana ratoon crop *cv* Grand Naine.

Treatment	Hand weight (kg)	Mean no of fingers per hand	Finger length (cm)
T1 - 150 g N (5 split doses) + 60 g P_2O_5	1.66	16.51	18.25
T2 - 150 g N (5 split doses) + 90 g P_2O_5	1.64	15.77	17.92
T3 - 200 g N (4 split doses) + 60 g P_2O_5	2.04	19.19	19.24
T4 - 200 g N (4 split doses) + 90 g P_2O_5	2.01	18.80	18.74
T5 - 200 g N (5 split doses) + 60 g P_2O_5	2.20	19.75	20.30
T6 - 200 g N (5 split doses) + 90 g P_2O_5	2.08	19.40	19.40
T7 - 250 g N (4 split doses) + 60 g P_2O_5	1.96	18.14	19.00
T8 - 250 g N (4 split doses) + 90 g P_2O_5	1.84	17.92	18.36
T9 - 250 g N (5 split doses) + 60 g P_2O_5	1.76	16.98	17.85
T10 - 250 g N (5 split doses) + 90 g P_2O_5	1.64	16.07	17.59
T11 - 300 g N (5 split doses) + 60 g P_2O_5	1.88	17.74	18.55
T12 - 300 g N (5 split doses) + 90 g P_2O_5	1.78	17.21	17.98
CD at 5% (N × P)	NS	NS	NS

Table 17. Effect of various doses of applied N fertilizers on Hand Weight (kg), Fingers per hand and Finger length

N (g/tree)	Hand weight (kg)	Fingers per hand	Finger Length (cm)
N1 - 150 g N (5 split doses)	1.65	16.14	18.08
N2 - 150 g N (4 split doses)	2.02	18.99	18.99
N3 - 200 g N (5 split doses)	2.14	19.58	19.85
N4 - 250 g N (4 split doses)	1.90	18.03	19.68
N5 - 250 g N (5 split doses)	1.70	16.52	17.72
N6 - 300 g N (5 split doses)	1.83	17.47	18.26
CD 5%	0.57	0.37	0.37

Table 18. Effect of two different doses of applied P fertilizers on hand weight (kg), Fingers per hand and Finger Length (cm) in banana ratoon crop *cv* Grand Naine.

P (g/tree)	P1 (60 g P_2O_5)	P2 (90 g P_2O_5)	CD 5%
Hand weight (kg)	1.92	1.83	0.16
Fingers per hand	18.05	17.53	0.82
Finger length (cm)	18.86	18.33	NS

effect on mean number of fingers per hand. Among phosphorus treatments, P1 (60 g P_2O_5) recorded significantly higher mean number of fingers per hand (18.05).

Finger length

The data on finger length as affected by different N and P treatments and their combinations is presented in Table 16. From perusal of data, it is clear that the finger length was significantly affected by different N treatments but P however the interaction between N and P combinations failed to show any significant effects. It is evident from the data presented in Table 17, that the different levels of applied N significantly influencec the finger length. The highest mean finger length (19.85 cm) was recorded by the application N3 (200 g N in 5 split doses) which was at par with N4 (250 g N in 4 split doses) having mean finger length of (19.68 cm) and significantly higher than the all other N treatments. The minimum mean finger length (17.75 cm) was recorded in Treatment N5 (250 g N in 5 split doses). The highest finger length with Treatment N3

(200 g N in 5 split doses) than all other N and P combinations may be due to significant effect of this treatment on vegetative characters like pseudostem height, girth, number of leaves, leaf area index etc which might have augmented photosynthesis and formation of more carbohydrates, thereby resulted in better finger size.

Conclusion

Application of four levels of N and two levels of P fertilizer greatly influence vegetative growth studied, which is pseudostem height, pseudostem girth, number of leaves and leaf area index. Application of fertilizers influenced the reproductive characteristics of 'Grand Naine' banana including total bunch weight, number of hands per bunch, hand weight and number of fingers per hand. However, long-term studies are needed to determine the residual effects of fertilizer application on soil and plants as well as its interaction with other factors such as irrigation, desuckering and management practices.

REFERENCES

Anonymous (2005). Food and Agriculture Organization (www.fao.org.com).

Anonymous (2008). National Horticultural Board Database (www.nhb.gov.in).

Arumugam S, Manivannan K (2001). Response of *in vitro* raised banana cv. Robusta to different levels of N and K application. South Indian Hort. 49:362-363.

Chao R, Lincoln L (1969). Statistics methods and analysis 2nd ed. McGraw Hill, New Delhi, pp. 116-17.

Dhillon WS, Kamboj JS, Bal JS (2002). Maturity indices and harvesting of fruit crops. Post-harvest handling of fruits and vegetables PAU, Ludhiana P. 26.

Naresh B, Sharma A (2004). Effect of different nitrogen doses and their split applications on growth, yield and quality of Jahajee banana South Indian Hort. 52:35-40.

Singh HP, Chundawat BS (2002). Improved technology of banana. Ministry of Agriculture, Government of India. pp. 1-46.

Tirkey T, Pandey SD, Agrawal S (2003). Studies on the response of Nitrogen levels and split application of N, P, K on growth, yield and quality of tissue culture raised banana cv. Dwarf Cavendish. Orrisa. J. Hort. 31:45-50.

Impact of bio-inoculants on seed germination and plant growth of guava (*Psidium guajava*)

D. V. Pathak, Surender Singh and R. S. Saini

Chaudhary Charan Singh Harayana Agricultura University (CCSHAU) Regional Research Station, Bawal (Haryana), India.

An experiment was conducted to study the impact of bio-inoculants on seed germination and plant growth of guava at CCSHAU Regional Research Station, Bawal, during the period 2007 to 2008. The bio-inoculants tested were *Azotobacter chroococcum*, phosphate solubulising bacteria (PSB), plant growth promoting bacteria (PGPR) and mycorrhiza. Their impact on seed germination, plant height and other plant growth parameters was studied in presence of farmyard manure (FYM) as well as vermicompost. During the 2007 period, maximum percent seed germination (34.2) was observed in the treatment having FYM + PGPR or FYM + *A. chroococcum* at 40 days (DAS); followed by PGPR (29.2) and vesicular arbuscular mycorrhizae (VAM) (25.8) treatments. While during the 2008 period, highest seed germination (51.1%) was observed in the treatment having FYM + PGPR or FYM + *A. chroococcum* at 40 DAS; followed by the treatments having FYM + PGPR + PSB + *A. chroococcum* or vermicompost + PSB + *A. chroococcum* (48.9%). Number of leaves per plant observed at 150 DAS were maximum in the treatment having FYM + VAM (18.8). Plant height at 150 DAS was maximum in FYM treatment having all the three bio-inoculants (31.5 cm). However, these values were quite comparable to each other in FYM as well as in vermicompost filled plastic bags.

Key words: Bio-inoculants, seed germination, guava (*Psidium guajava*), phosphate solubulising bacteria (PSB), plant growth promoting bacteria (PGPR), vesicular arbuscular mycorrhizae (VAM), *Azotobacter chroococcum*.

INTRODUCTION

Due to rising cost of chemical fertilizers and their adverse effects on soil health, an economically attractive and alternate potential source of plant nutrients should be exploited. The excessive use of these chemical fertilizers adversely affects human health resulting in dreadful diseases like cancer, hypertension and other abnormalities. Further, to sustain the productivity, the bio-inoculants can supplement them to certain extent in various food crops. But it is not the common practice in various horticultural crops. The seed coat of most of the fruit crops is very hard. To break the seed dormancy, either some chemical treatment or long incubation period is required. These bio-inoculants can be helpful in breaking the seed dormancy by producing various plant growth substances. Hence, the present investigations were undertaken to study the response of different bio-inoculants in combination with either farmyard manure (FYM) or vermicompost on seed germination and plant growth in guava (*Psidium guajava*).

MATERIALS AND METHODS

The experiment was conducted at the Experimental Farm, CCSHAU Regional Research Station, Bawal, during the years 2007/2008 and 2008/2009. The experiment was laid out in randomized block design with six replications of each treatment

Table 1. Effect of bio-inoculants on seed germination and plant height in guava.

Treatment	Seed germination (%)		Plant height (cm)	
	2007	2008	2007	2008
FYM alone	20.8	37.7	18.8	27.5
FYM + PGPR	25.8	51.1	29.7	28.7
FYM + VAM	29.2	45.5	28.0	29.2
FYM + PSB	20.0	44.4	18.3	27.9
FYM + *Azoto.*	24.0	51.1	22.3	28.2
FYM + PSB+ *Azoto.* + PGPR	34.2	48.9	19.2	31.5
VC alone	21.0	23.3	15.5	20.3
VC + PGPR	22.5	46.6	27.7	25.0
VC + VAM	21.7	37.5	26.3	26.7
VC + PSB	26.3	37.7	16.3	23.3
VC + *Azoto.*	25.2	48.9	18.0	27.4
VC + PSB+ *Azoto.* + PGPR	26.7	28.9	25.2	28.4
Mean	24.8	41.8	22.1	27.0
CD at 5%	2.04	3.41	1.87	2.17

both with vermicompost as well as farm yard manure. For filling up of polythene bags, the loamy sand soil and vermicompost/ FYM were mixed together in 1:1 ratio. The different treatment combinations were T_0: uninoculated treatment; T_1: *Azotobacter chroococcum*; T_2: phosphate solubilizing bacteria (PSB); T_3: vesicular arbuscular mycorrhizae (VAM); T_4: plant growth promoting bacteria (PGPR) (*Pseudomonas maltophilia* PM4); T_5: PGPR+PSB+ *A. chroococcum*. In total, there were twelve treatments; six each of vermicompost + soil and FYM + soil. The seeds of guava were surface sterilized with 0.2% mercuric chloride solution for five minutes and then washed 2 to 3 times with sterilized distilled water, coated with charcoal based inoculants and then, twenty seeds per replication for each treatment were imcregnated into the polythene bags. The bio-inoculants used in the study were collected from Department of Microbiology, CCSHAU, Hisar, and grown in their respective media under aseptic conditions.

Glomus fasciculatum, species of VAM fungi was used as VAM inoculant. The inoculum consisted of soil, spores, hyphae from chopped root fragments of pearl millet. Ten gram of the inoculum was mixed with the top soil in each polythene bag before sowing of guava seeds. The observations were recorded in terms of percent seed germination at different time intervals up to 40 days of sowing. Then, plant saplings were thinned down to one healthy sapling per bag and observed for plant height, number of leaves per plant and dry shoot weight at 150 days of sowing.

RESULTS AND DISCUSSION

During the period 2007/2008, maximum percent germination (34.2) was observed in the treatment having FYM + PSB + *Azotobacter* + PGPR; followed by FYM + VAM (29.2). Percent seed germination was slightly better in the treatments having FYM over their respective vermicompost treatments; however, the difference between them was non-significant. During 2008/2009, the maximum germination (51.1) was recorded in FYM + PGPR and FYM + *Azotobacter*, closely followed by FYM + PSB + *Azotobacter* + PGPR and vermicompost+ *Azotobacter* (48.9%) (Table 1). Plant height was also stimulated by different bio-inoculants in combination with farm yard manure as well as vermicompost during the period of investigation. However, PSB alone did not contribute much on different plant growth parameters.

VAM inoculation with FYM as well as with vermicompost positively affected number of leaves per plant during both years (Table 2). The similar trends were followed when dry weight of shoot was recorded after 150 days (DAS). In general, VAM culture and coinoculation of PSB, PGPR and *Azotobacter* stimulated plant growth parameters more positively as compared to single inoculation or untreated control. The response with FYM was slightly better over their respective vermicompost treatments.

Various reports in horticultural crops indicated that bio-inoculants either individually or in combination had synergistic effect on plant growth. The dual inoculation of *Azotobacter* and *G. fasciculatum* had more positive response in peach seedlings as compared to single inoculation or control (Godara et al., 1998). Sharma et al. (2002) reported that VAM fungi enhanced nutrient uptake and level of plant growth substances in apple seedlings. Subbiah (1990) also reported that when adequate amount of farmyard manure added to the soil with biofertilizers, it improved biofertilizer efficiency and ultimately nutrient status of the soil. Similar increase in growth of fruit plants with biofertilizers has also been reported by Sharma and Bhutani (1998). Increase in the growth of pecan seedlings could be attributed to the combined effect of biofertilizers on nutrient uptake and plant growth (Joolka et al., 2004). The possible reason for better plant growth and germination can be attributed to maximum and early bacterization near root zone which induce germination by inducing root inducing substances (Wani et al., 1988). Similar reports have been made by

Table 2. Effect of bio-inoculants on other plant growth parameters in guava.

Treatment	Number of leaves/plant		Dry shoot weight (g)	
	2007	2008	2007	2008
FYM alone	11.5	15.42	0.38	0.56
FYM + PGPR	21.8	17.17	1.17	0.77
FYM + VAM	22.5	18.80	0.95	0.83
FYM + PSB	15.5	15.75	0.53	0.79
FYM + *Azoto.*	17.2	18.20	0.48	0.67
FYM + PSB + *Azoto.* + PGPR	14.8	17.75	0.95	0.92
VC alone	13.3	14.24	0.42	0.48
VC + PGPR	19.0	17.33	1.25	0.62
VC + VAM	20.5	15.67	1.08	0.68
VC + PSB	15.0	18.20	0.59	0.72
VC + *Azoto.*	14.0	16.50	0.57	0.84
VC + PSB + *Azoto.* + PGPR	14.5	17.83	0.82	0.82
Mean	16.6	16.9	0.76	0.72
CD at 5%	1.49	1.35	0.12	0.17

Nath and Korla (2000) in ginger.

REFERENCES

Godara RK, Awasthi RP, Kaith NS (1998). Interaction effect of VA-mycorrhizae and Azotobacter inoculation on growth and macronutrients of peach seedlings. Hary. J. Hort. Sci. 27:235-240.

Joolka NK, Singh RR, Sharma MM (2004). Influence of biofertilizers, GA3 and their combinations on groth of pecan seedlings. Ind. J. Hort. 61:226-228.

Nath B, Korla BN (2000). Studies on effect of biofertilizers in ginger.Ind. J. Hort. 57:168-171.

Sharma SD, Bhutani VP (1998). Response of apple seedlings to VAM, Azotobacter and inorganic fertilizers. Hort. J. 11:1-8.

Sharma SD, Bhutani VP, Awasthi RP (2002). Effect of vesicular-arbiscular mycorrhizae and phosphorus on leaf and soil mineral nutrient status of apple seedlings. Ind. J. Hort. 59:140-144.

Subbiah K (1990). Nitrogen and Azospirillum interaction on fruit yield and nitrogen use efficiency in tomato. South Ind. Hort. 38:342-344.

Wani SP, Chandrapalaiah S, Zambre MA, Lee KK (1988). Association between N2 fixing bacteria and pearl millet plants-Response, mechanisms and resistance. Plant Soil 110:284-302.

31

Stability analysis of components characters in cowpea (*Vigna unguiculata* (L.) Walp)

El-Shaieny A. A. H.[1] , Y. Y. Abdel-Ati[2], A. M. El-Damarany[3] and A. M. Rashwan[1]

[1]Faculty of Agriculture, South Valley University, Qena, 83523 Egypt.
[2]Faculty of Agriculture, Minia University, Egypt.
[3]Faculty of Agriculture, Sohag University, Egypt.

Stability of yield and its attributes were assessed for nineteen genotypes over twelve environments (two seasons 2009 and 2010 × six planting dates), to determine the quantitative responses of cowpea genotypes. The interaction between genotypes and environments (G×E) were significant for all the characters studied characters except pod length, hundred seed weight and weight of pods per plant. The longest pods and heaviest hundred seeds weight were produced by genotype TVU 21, IT82C-116, providing the highest number of seeds per plant. Whereas, Sudany genotypes gave the highest number of pods per plant and heaviest seeds per plant, Blackeye Crowder genotypes had the heaviest pods per plant and total dry seed yield. The best season and planting date are fall season, third planting date (August, 15th) for most studied traits. The stable genotypes were Chinese Red, IT81D1064, IT85F2205 and Sudany for total dry seed yield.

Key words: Sowing dates, stability parameters, genotype × environment, selection, grain yield.

INTRODUCTION

Cowpea (*Vigna unguiculata* (L.) Walp) is one of the most ancient crops known to man. In Egypt, cowpea is a popular vegetable crop. The total area under cultivation of this crop was estimated at 9155 feddans (feddan= 4200 m^2) for dry seed in 2008 with a mean production of 980 kg/fed. Also, the area that produced green pods was 10064 feddans with a mean of 5.19 ton/fed (Department, Agriculture, Statistics, Ministry of Agriculture, Giza, Egypt). Stable performance of cowpea genotypes across contrasting environments is essential for the successful selection of stable and high yielding varieties (Dashiell et al., 1994; Ariyo, 2000; Ahmed et al., 2005; Yousaf and Sarwar, 2008). Combination of genotypes stability with high yield is an important criteria for selecting high yielding and stable genotypes. Therefore, a number of techniques that simultaneously coupled with high yield and stability of performance have been proposed. The regression technique (Eberhart and Russell, 1966) has been used. In this technique, the response of genotypes to a given environment is considered. G x E cannot be avoided, in fact, it is an important limiting factor for testing the efficiency of any breeding programme. The occurrence of large genotype × environment (G × E) interaction affects the recommendations of the breeders in selecting genotypes for specific environment. Genotype × environment analysis is used to provide unbiased estimates of yield and other agronomic characteristics and to determine yield stability or the ability to withstand both predictable and unpredictable environmental variation (Kamdi, 2001).

Table 1. Source, seed color and growth habit of the tested cowpea genotypes.

Genotype	Seed color	Growth habit
1. Dokii 331	White with black eye	Determinate
2. Kaha 1	Yellowish-white	Determinate
3. Cream 7	Yellowish-white	Determinate
4. IT91K-118-20	Light Brown	Determinate
5. IT93K2045-20	Light Brown	Determinate
6. TVU-21	White with red eye	Indeterminate
7. IT82D-889	Light Brown	Determinate
8. Chinese Reds	Dark Brown	Indeterminate
9. IT81D1064	Dark Brown	Determinate
10. IT85F-2205	Light Brown	Determinate
11. IT90K-1020-6	Light brown	Determinate
12. Blackeye Crowder	White with black eye	Determinate
13. IT82C-16	Dark Brown	Determinate
14. IT82-812	Light Brown	Indeterminate
15. Sudany	Black	Indeterminate
16. Cream 12	Yellowish-white	Determinate
17. Monarch Blackeye	White with black eye	Determinate
18. Azmerly	White with black eye	Determinate
19. Black Crowder	Black	Indeterminate

The regression coefficient (b_i) and genotype mean yield were used together as measure of adaptation (Bilbro and Ray, 2000). Genotype with b = 1.0 was considered as adapted to al environments, genotype with b < 1.0 was considered adapted for low yielding environments and genotype with b > 1.0 was considered as better adapted for high yielding environments, depending upon the genotype mean yield. De Rocha et al. (2007a) found that TE97-321G-4, EVX-92-49E and EVX-63-10E cowpea lines were highly adaptable, but only the last one was highly predictable. The BRS Guariba cultivar as well as EVX-92-49E and TE97-321G-4 lines best expressed their genetic potential in environments of high yield. Taiwo (2007) reported that IT 98K-1111-1, IT 86D-1010, IT 86D-719, IT 93K-452 and IT 97K-503-1 were identified to be of a high fodder yield and stable genotypes performance across performance environment. Ajeigbe et al. (2008) found that IT98K-506-1, IT97K-1113-7, IT97K-1069-6, IT97K-1092-2, IT97K-1069-5, IT98K-131-2 and IT97K-568-18 produced higher grain and fodder yielders than the other varieties. The objective of this investigation were to assess the magnitude of G×E interaction as well as the relative performance and stability of 19 cowpea genotypes under abiotic (heat) stress of Upper Egypt environmental conditions, to identifying the most stable genotypes for this stress.

MATERIALS AND METHODS

Study sites and experimental design

The field experiments were conducted at Faculty of Agriculture Farm, South Valley University, Qena Governorate, Egypt, during the growing seasons of 2009 and 2010. The material used in this study and sources of the investigated genotypes are shown in Table 1. These nineteen genotypes were evaluated in summer and fall seasons of 2009 and 2010. In each season, the genotypes were arranged in a Randomized Complete Block Design (RCBD) with three sowing dates viz, March, 15th, 30th and April, 15th in the summer seasons of 2009 and 2010, and July, 15th, 30th and August, 15th in the fall seasons of 2009 and 2010. Each genotype was represented by single row and was repeated three times, the length of the row was 3 m, 60 cm apart and plants spaced 20 cm from each other. Then, different agricultural production practices that is, fertilization and pest management were applied as per the commercial cowpea production in Egypt.

Data collection

The measured traits included

(1) Pod length (cm): Ten normal and fully dry pods for each genotype from each plot were taken to determine dry pod length and the average were recorded.
(2) Number of pods/plant: Average pod number of ten plants for each genotype from each plot was estimated.
(3) Number of seed per pod: Recorded from 10 pods per plant at harvesting time and the average was estimated.
(4) Hundred seed weight (gram): Average weight of the ten samples for each genotype in each plot was determined.
(5) Average seed weight/plant (gram): Ten plants from each genotype were taken from each plot to determine the weight of seeds/plant (gram) and the average was recorded.
(6) Average pod weight (gram): Ten normal and fully dry pods for each genotype from each plot were taken to determine dry pod weight and the average were recorded.
(7) Total dry seed yield (ton/fed.): Estimated as the weight of the dry seed per plot.

Data from all plots were subjected to analysis of variance (Steel and Torrie, 1980). Stability parameters were worked out according to (Eberhart and Russell, 1966).

Table 2. Means of pod length of the nineteen genotypes under twelve environments.

Genotype	Pod length (cm)												
	E1	E2	E3	E4	E5	E6	E7	E8	E9	E10	E11	E12	Mean
Dokii 331	13.73	12.50	14.8	13.33	12.43	14.40	14.10	14.70	14.80	13.43	14.97	15.47	14.06
Kaha 1	12.73	12.63	14.00	12.33	11.77	12.40	13.10	13.70	13.47	13.94	14.05	14.05	13.18
Cream 7	15.4	15.17	16.47	15.00	15.10	14.73	15.77	16.37	16.13	16.00	17.00	17.00	15.85
IT91K118-20	10.97	11.73	12.57	12.17	12.33	11.57	12.70	12.20	13.97	11.90	12.00	12.23	12.20
IT93K2045-20	14.00	13.86	15.00	14.14	14.07	14.80	14.80	15.00	15.77	15.20	15.90	15.57	14.84
TVU 21	18.40	17.90	19.40	18.00	17.97	18.20	17.10	17.50	18.47	18.83	19.00	19.80	18.38
IT81D-889	15.20	14.63	16.27	14.80	14.57	15.87	15.70	15.50	15.60	16.23	17.00	16.63	15.67
Chinese Red	12.13	11.60	13.20	11.73	11.53	12.80	13.47	13.10	13.87	13.13	13.33	13.87	12.81
IT81D1064	12.80	12.67	13.87	12.40	12.60	13.47	14.27	13.77	15.87	14.37	15.50	14.50	13.84
IT85F2205	13.63	13.13	14.70	13.23	13.07	14.30	14.85	14.98	15.30	14.50	15.80	15.37	14.41
IT90K1020-6	11.07	11.50	12.13	10.67	11.10	11.73	11.93	12.03	13.80	11.93	12.67	12.80	11.95
Blackeye Crowder	12.73	11.50	13.80	12.33	12.43	13.40	13.77	13.70	14.80	13.10	14.40	14.47	13.37
IT82C-16	17.73	16.50	18.80	17.33	16.77	18.40	18.00	17.10	18.17	18.10	19.60	19.47	18.00
IT82- 812	12.20	11.60	13.27	11.80	12.20	12.87	14.53	13.37	13.93	13.73	13.37	13.93	13.07
Sudany	11.00	9.53	10.00	9.67	9.93	9.17	10.43	10.73	10.80	11.43	12.23	12.00	10.58
Cream 12	15.73	14.50	16.8	15.33	14.43	16.40	15.63	16.00	16.30	16.03	16.63	17.17	15.91
Monarch Blackeye	10.90	10.17	11.97	10.50	10.10	11.57	11.93	11.87	12.63	12.27	12.60	12.63	11.60
Azmerly	15.73	15.50	16.8	15.33	15.43	16.40	14.90	15.70	16.07	16.13	16.57	17.17	15.98
Black Crowder	15.07	14.17	16.13	14.67	14.10	15.73	15.00	15.37	16.00	15.40	16.00	16.40	15.34
Environmental mean	13.74	13.20	14.74	13.41	13.26	14.12	14.31	14.35	15.04	14.51	15.19	15.29	14.26

E1 = Summer season 2009, First date E5 = Fall season 2009, Second date E9 = Summer season 2010, Third date, E2 = Summer season 2009, Second date, E6 = Fall season 2009, Third date, E10 = Fall season 2010, First date, E3 = Summer season 2009, Third date E7 = Summer season 2010, First date E11 = Fall Third 2010, Second date E4 = Fall season 2009, First date E8 = Summer season 2010, Second date E12 = Fall Third 2010, Third date.

RESULTS AND DISCUSSION

The combined analysis variance (Table 9) revealed that highly significant differences among genotypes (G), environments (E) as well as interaction between genotypes and environments (G×E) for most of the studied traits. These results indicated that cowpea genotypes responded differently to the diverse environmental conditions, differences were due to the genetic variations among genotypes and environmental factors and climatic conditions, among others. Similar results were obtained by Teixeira et al. (2007) and Akande and Balogun (2009). The mean response of each trait is outlined below.

Pod length (cm)

The mean performance of genotypes is presented in Table 2. The average pod length of the 19 genotypes over all environments ranged from 18.38 cm (TVU 21) to 10.58 cm (Sudany). The delay in planting date increased the fresh pod length in all seasons. These results agree with those reported by Ali et al. (2004) and Rashwan (2010) who found that the delay in plating date until Dec. 15 increased the fresh pod length. Partitioning the genotype × environment interaction mean square (Table 10) that (G×E) mean squares were estimated with insignificant value. The stability parameters ($\bar{x}$, bi and s^2d) of the individual genotypes are illustrated in Table 11.

Table 3. Means of number of seeds per pods of the nineteen genotypes under twelve environment.

Genotype	Number of seeds per pods												
	E1	E2	E3	E4	E5	E6	E7	E8	E9	E10	E11	E12	Mean
Dokii 331	9.33	10.00	10.00	9.33	10.33	10.00	10.67	11.00	10.33	10.33	11.00	10.67	10.25
Kaha 1	7.67	8.67	8.33	7.67	8.67	8.33	9.67	9.33	8.33	8.33	10.33	8.33	8.64
Cream 7	11.00	12.00	12.33	11.00	12.00	12.33	12.67	10.00	11.00	11.33	10.67	11.00	11.44
IT91K118-20	8.00	10.33	11.00	10.33	11.00	11.33	12.00	11.33	11.33	11.33	10.33	11.33	10.81
IT93K2045-20	10.33	11.33	12.00	11.67	11.33	12.00	10.67	9.00	10.00	10.67	10.67	10.67	10.86
TVU 21	10.67	11.67	10.67	10.67	11.33	10.67	12.67	11.33	12.00	12.00	10.67	12.00	11.36
IT81D-889	9.33	10.33	9.00	9.33	10.67	9.00	11.00	11.00	11.00	11.00	10.00	11.00	10.22
Chinese Red	8.33	9.33	8.00	8.33	10.00	8.00	11.00	10.00	10.00	10.33	10.00	10.33	9.47
IT81D1064	8.33	9.33	8.67	8.33	9.33	8.67	9.33	8.33	9.00	9.67	9.00	9.33	8.94
IT85F2205	8.33	9.33	8.67	8.33	9.33	8.67	10.33	8.00	9.33	9.67	10.00	10.33	9.19
IT90K1020-6	7.67	8.33	9.00	7.67	9.33	9.00	11.00	10.33	11.00	9.33	9.33	8.67	9.22
Blackeye Crowder	10.33	9.33	10.67	10.33	10.67	10.67	12.67	12.67	13.33	11.00	9.67	11.33	11.06
IT82C-16	12.67	13.67	13.67	12.67	13.67	13.67	9.33	9.00	9.33	13.00	12.00	13.00	12.14
IT82- 812	8.33	9.33	9.00	8.33	9.33	9.00	10.33	9.33	10.00	10.33	9.00	9.67	9.33
Sudany	9.33	7.67	7.67	9.00	8.00	8.33	9.33	10.00	10.67	9.67	8.67	9.67	9.00
Cream 12	11.00	11.33	12.00	7.33	12.00	12.00	11.67	12.33	8.00	11.33	11.33	11.33	10.97
Monarch Blackeye	7.67	8.67	7.67	7.67	8.67	7.67	11.33	10.00	10.67	9.67	11.33	9.67	9.22
Azmerly	9.33	10.33	10.33	9.33	10.33	10.33	11.33	11.00	11.33	10.33	11.00	10.67	10.47
Black Crowder	10.33	11.33	11.00	10.33	11.33	11.33	11.33	11.00	11.33	11.33	12.00	11.67	11.19
Environmental mean	9.37	10.12	9.98	9.35	10.39	10.05	10.96	10.26	10.42	10.56	10.37	10.56	10.20

11. All genotypes except IT85F2205 and Monarch Blackeye exhibited non significant stability parameters from unity and zero for regression coefficient (bi) and deviation from regression (s^2d), respectively. The genotypes IT91K118-20, Sudany and Kaha 1 appeared to be stable and exhibited below average response to different environments, ($b_i < 1$) their genotypes were considered to perform relatively better in less favorable environments. The genotypes Cream 7, TVU 21, IT81D-889 and Azmerly could be considered good inserted gave a pod length more than the overall of average genotypes besides their stability. The genotypes IT93K2045-20, Black Crowder, Azmerly and IT81D-889 might be considered superior, because they should be the tallest pod length when compared with the average overall genotypes besides their stability. Similar results were reported by Akande and Balogun (2009), Teixeira et al. (2007) and Sarutayophat et al. (2007).

Number of seeds per pod

Results in Table 3 showed that average number of seeds per plant of genotypes overall environments ranged from 12.14 (IT82D-16) to 8.64 seeds per pods for genotype (Kaha 1), with an average of (10.20) seeds per pods for all genotypes. These are in accordance with the finding of Rajput (1994) who observed that sowing on 10th March recorded significantly more number of pods per plant (16.0), seeds per pod (13.2), seed yield (12.1 q ha-1), stover yield (24.7 q ha-1) and harvest index (34.5%) compared to sowing in 18th February and 30th March. The highest number of seeds per pod was that of genotype IT82D-16, in the third planting date (August 15th and April 15th) at two seasons (summer and fall), while, the lowest was for genotypes Kaha 1, in the first planting date in fall season (July, 15th).

Table 4. Means of number of pods per plant traits of the nineteen genotypes under twelve environments.

Genotype	Number of pods per plant												
	E1	E2	E3	E4	E5	E6	E7	E8	E9	E10	E11	E12	Mean
Dokii 331	54.33	54.67	56.33	55.67	57.33	56.00	53.33	53.67	54.67	57.00	58.67	57.67	55.78
Kaha 1	34.00	34.67	36.33	34.67	37.00	35.67	31.67	32.00	33.00	35.33	38.67	38.67	35.14
Cream 7	53.67	55.00	55.33	54.67	56.67	55.00	52.00	52.33	53.33	55.67	58.00	57.67	54.94
IT91K118-20	48.33	49.33	50.33	49.33	51.33	50.00	46.33	46.67	47.67	50.00	53.00	52.67	49.58
IT93K2045-20	41.33	42.33	43.33	42.67	44.33	43.00	40.33	40.67	41.67	44.00	45.67	45.67	42.92
TVU 21	43.00	44.33	44.67	44.00	46.00	44.00	41.67	42.00	43.00	45.33	47.00	46.67	44.31
IT81D-889	37.00	37.33	39.33	38.00	40.33	38.33	36.33	36.67	37.67	40.00	41.33	41.00	38.61
Chinese Red	49.00	50.33	50.67	51.00	52.00	51.00	47.67	48.00	49.00	51.33	54.00	53.33	50.61
IT81D1064	35.67	35.67	38.00	37.67	38.67	38.33	35.00	35.33	36.33	38.67	40.67	39.33	37.44
IT85F2205	44.67	45.33	47.00	45.67	47.67	46.33	44.00	44.33	45.33	47.67	49.33	48.67	46.33
IT90K1020-6	34.67	35.00	38.00	37.00	38.67	38.00	36.33	37.33	37.00	35.33	40.00	39.00	37.19
Blackeye Crowder	62.67	63.33	64.67	64.00	65.33	65.00	61.00	61.67	62.00	63.67	68.00	65.00	63.86
IT82C-16	34.33	35.00	37.67	35.67	38.00	36.67	36.00	36.67	36.67	34.00	38.67	37.67	36.42
IT82- 812	38.33	38.67	41.00	40.33	41.00	41.00	39.00	39.33	40.00	37.67	43.33	41.00	40.06
Sudany	66.33	67.33	67.67	66.67	68.67	67.33	65.67	66.33	66.67	66.67	70.33	67.33	67.25
Cream 12	51.00	51.67	52.67	51.67	53.33	51.67	51.00	51.67	51.67	51.00	54.67	52.67	52.06
Monarch Blackeye	33.67	34.67	35.00	34.67	36.00	35.00	33.33	34.00	34.00	34.00	38.00	35.00	34.78
Azmerly	57.00	57.67	58.67	58.33	59.33	58.33	57.00	57.67	57.67	57.33	61.33	58.67	58.25
Black Crowder	55.67	56.00	57.33	57.00	57.67	57.33	55.67	56.33	56.33	55.33	60.00	57.33	56.83
Environmental mean	46.04	46.75	48.11	47.30	48.91	47.79	45.44	45.93	46.51	47.37	50.56	49.21	47.49

The differences among the tested genotypes (G) were highly significant; also, environmental (E) effect and the interactions between genotypes and environments (G×E) were highly significant as shown in Table 9. Most of this interaction was in a linear function with the environmental values as indicated by greater magnitude of the G×E (linear) mean squares (5.51) in comparison with the estimated value for E+ (G×E) mean squares (3.04), which appeared also highly significant. These results were presented in Table 10. These results appeared to be in harmony with those obtained by Torres et al. (2008) and Akande and Balogun (2009).

The stability parameters ($\bar{x}$, bi and s^2d) of the individual genotypes are illustrated in Table 11. The results indicated that all genotypes values were non-significant except IT93K2045-20, Chinese Red and Monarch Blackeye were significant, genotypes Dokii 331, IT93K2045-20 and Black Crowder were considered specially adapted to unfavorable environments because the regression coefficient of theses genotypes less than one ($b_i < 1$) while, genotypes IT91K118-20, IT82C-16 and Blackeye Crowder could consistently performed better under favorable environments because their regression coefficient (bi) were more than one. The genotypes IT93K2045-20, Dokii 331, Cram 7, and Black Crowder might be consider superior because they gave high mean values for number of seeds per pods above the grand mean, besides their stability. These results is in agreement with those obtained by Damarany (1994b), Ushakumari et al. (2002), Dahiya et al. (2007a, b, c) and Singh et al. (2007).

Number of pods/plant

Average number of pods per plant of genotypes overall environments ranged from (67.25) for genotype Sudany to (34.78) pods per plant for genotype Monarch Blackeye, with an average of (47.49) pods per plant for all genotypes, data are presented in Table 4. The highest number of pods per plant was for genotype Sudany at fall season at second planting date (July, 30th), in both seasons, while, the lowest was for genotype Monarch Blackeye at summer season at first planting date (March, 15th), in both seasons. The significance of genotype by environment interaction in regional variety trials or in selection for wide adaptation has been reviewed by other workers (Becker and Leon, 1988; Crossa et al., 1990; Cooper and DeLacy, 1994). Other studies (Allen and Allen, 1981; Singh and Rachie, 1985; Damarany, 1994a; Ishiyaku et al., 2005) pointed out the existence of significant genotypic differences in cowpea for yield and agronomic traits. However, most of the studies were conducted under single location or controlled environments that might underestimate the environmental as well as genotype by environment interaction.

Results illustrated in Tables 9 and 10 showed that the differences among all genotypes (G) and environments (E) were highly significant. Also, the interactions between genotypes and environments (G×E) were highly significant. Also, highly significant effect of E (linear) was reported, indicating that the studied trait was highly influenced by the combination of environment. G×E (linear) item was highly significant, suggesting that cowpea genotypes were different in their response to environments. Similar results were reported by Teixeira et al. (2007) and Torres et al. (2008).

The estimated stability parameters ($\bar{x}$, bi and s^2d) of the studied genotypes for number of pods per plant indicated that Sudany, Cream 12, Azmerly and Monarch Blackeye genotypes were stable ($b_i < 1$) with high mean values, while, IT90K1020-6, IT82C-16 and IT82-812 genotypes were stable with the mean values lower than the grand mean. On the other hand, Dokii 331, Cream 7, IT91K118-20 and Chinese Red were unstable ($b_i > 1$) and could consistently do better in favorable environments. These results are presented in Table 11. Similar results were obtained by Ushakumari et al. (2002) and Dahiya et al. (2007b).

Hundred seed weight (gram)

Average hundred seed weight (gram) of genotypes overall environments ranged from 22.16 (gram) for genotype TVU 21 to 11.63 (gram) hundred seed weight (gram) for genotype Chinese Red, with an average of 14.89 (g) hundred seed weight (gram) for all genotypes. The data were presented in Table 5. These results are in agreement with that obtained by Damarany (1994b), Dahiya et al. (2007b, c), Peksen (2007) and De Rocha et al. (2007b). The highest hundred seed weight was that of genotype TVU 21, in the third planting date at fall season, while, the lowest was for genotypes Chinese Red, in the third planting date (April, 115th), in summer season. The stability parameters ($\bar{x}$, bi and s^2d) of the individual genotypes are illustrated in Table 11. The results indicated that all genotypes values were non-significant except IT81D-889 and IT82C-16 were highly significant, genotypes Azmerly, IT81D-889, Blackeye Crowder and Black Crowder were considered specially adapted to unfavorable environments because the regression coefficient of theses genotypes less than one ($b_i < 1$) while, genotypes Dokii 331, IT91K118-20, IT82C-16 and IT85F2205, Blackeye Crowder could consistently performed better under favorable environments because their regression coefficient (bi) were more than one. The genotypes IT82D-889 and Azmerly might be consider superior because they gave high mean values for hundred seeds weight above the grand mean, besides their stability. These results are in agreement with those obtained by De Rocha et al. (2007b and Akande and Balogun (2009).

Average seed weight/plant (gram)

The performance of tested genotypes is presented in Table 6. The results indicated that average weight of seeds per plant of the various genotypes ranged from 67.81 g for (Sudany) to 37.03 g for (Kaha 1), with an average of 48.05 g for all genotypes. The heaviest weight of seeds per plant 69.23 and 68.80 g was found for (Sudany) in summer season, at third plating date, in both seasons, respectively. While, the lightest of 35.80 g was found for (Kaha 1) genotype in fall season at first planting date. These results are in agreement with that obtained by Ushakumari et al. (2002), and Dahiya et al. (2007b, c).

The joint regression analysis of variance is presented in Table 9. The differences among the tested genotypes (G) were highly significant; also, environmental (E) effect and the interactions between genotypes and environments (G×E) were highly significant as shown in Table 10. Most of this interaction was in a linear function with the environmental values as indicated by greater magnitude of the G×E (linear) mean squares in comparison with the estimated value for E+ (G×E) mean squares, which appeared also highly significant. These results appeared to be in harmony with those obtained by Dahiya et al. (2007a, b).

The estimated stability parameters ($\bar{x}$, bi and s^2d) of the studied genotypes for average seed weight indicated that Cream 7, Azmerly, Blackeye Crowder, Dokii 331 and Black Crowder genotypes were stable ($b_i < 1$) with high mean values, while, Kaha 1, and IT85F2205 genotypes were stable with the mean values lower than the grand mean. On the other hand, Sudany, Monarch Blackeye, IT82-812 and IT82C-16 genotypes were unstable ($b_i > 1$) and could consistently do better in favorable environments (Table 11). Similar results were obtained by

Table 5. Means of hundred seeds weight per plant traits of the nineteen genotypes under twelve environments.

Genotype	Hundred seed weight(g)												
	E1	E2	E3	E4	E5	E6	E7	E8	E9	E10	E11	E12	Mean
Dokii 331	18.40	18.40	18.40	20.60	19.82	20.09	19.10	18.73	18.47	19.70	20.37	20.00	19.34
Kaha 1	11.93	11.70	11.93	13.90	13.12	13.63	12.13	12.87	12.00	12.90	13.47	13.30	12.74
Cream 7	12.87	12.80	12.87	15.00	14.22	14.56	13.47	13.93	12.93	14.00	14.60	14.40	13.80
IT91K118-20	12.43	12.53	12.43	14.73	13.96	14.13	12.90	13.33	12.50	13.73	14.03	14.13	13.40
IT93K2045-20	16.03	15.70	16.03	17.90	17.12	17.73	15.73	15.53	16.10	17.07	16.87	17.30	16.59
TVU 21	21.70	21.33	21.70	23.53	22.76	23.39	21.27	20.67	21.77	22.53	22.30	22.93	22.16
IT81D-889	13.50	13.10	13.50	15.30	14.52	15.19	13.67	14.23	13.57	14.30	14.63	14.70	14.18
Chinese Red	10.80	10.60	10.80	12.80	12.02	12.49	11.00	11.77	10.87	12.10	12.13	12.20	11.63
IT81D1064	10.93	10.93	10.93	13.13	12.36	12.63	10.70	10.67	11.00	12.13	11.83	12.53	11.65
IT85F2205	15.77	15.43	15.77	17.63	16.86	17.46	15.13	14.47	15.83	16.63	16.27	17.03	16.19
IT90K1020-6	12.60	12.50	12.60	14.70	13.92	14.29	12.90	13.97	12.67	13.70	14.03	14.10	13.50
Blackeye Crowder	12.80	12.90	12.80	15.10	14.32	14.49	13.20	13.73	12.87	14.10	14.33	14.50	13.76
IT82C-16	15.47	15.67	15.47	17.87	17.09	17.16	15.90	15.60	15.53	16.87	17.03	17.27	16.41
IT82- 812	12.70	12.37	12.70	14.57	13.79	14.39	12.73	14.27	12.77	13.57	13.87	13.97	13.47
Sudany	11.73	11.40	11.73	13.60	12.80	13.43	11.23	11.10	11.80	12.60	12.37	13.00	12.23
Cream 12	11.37	11.30	11.37	13.50	12.70	13.06	11.47	11.67	11.43	12.50	12.60	12.90	12.16
Monarch Blackeye	16.43	16.37	16.43	18.57	17.77	18.13	17.17	16.97	16.50	17.57	18.30	17.97	17.35
Azmerly	17.90	17.43	17.90	19.63	18.83	19.59	17.27	17.50	17.97	18.63	18.40	19.03	18.34
Black Crowder	13.10	13.00	13.10	15.20	14.40	14.79	13.80	13.63	13.17	14.20	14.93	14.60	13.99
Environmental mean	14.13	13.97	14.13	16.17	15.39	15.82	14.25	14.45	14.20	15.20	15.39	15.57	14.89

Ushakumari et al. (2002) and Dahiya et al. (2007b).

Average pods weight/plant (gram)

The performance of tested genotypes is presented in Table 7. The results indicated that average weight of seeds per plant of the various genotypes ranged from 82.00 (g) for Blackeye Crowder to 50.36 (g) for Chinese Red, with an average of 64.45 (g) for all genotypes. The heaviest weight of pods per plant 85.67 (g) was found for Blackeye Crowder in fall season, at second plating date, while, the lightest of 45.33 (g) was found for Chinese Red in summer season at third planting date. These results are in agreement with that obtained by Hazra et al. (1999), and De Rocha et al. (2007b).

The joint regression analysis of variance is presented in Table 10. The differences among the tested genotypes (G) were highly significant, also, environmental (E) effect, while, the interactions between genotypes and environments (G×E) were insignificant, as shown in Table 9. Indicating the presence of genetic variability among these genotypes and the suitability of stability analysis. These results appeared to be in harmony with those obtained by Chattopadhyay et al. (2001) evaluated twenty cowpea genotypes for stability in yield and its components such as number of pods per plant, pod length, and pod weight and revealed that the significant genotype and environment interaction was observed for all characters except pod length.

The estimated stability parameters ($\bar{x}$, bi and s^2d) of the studied genotypes for average pods weight are presented in Table 11. All genotypes exhibited insignificant stability parameters from unity and zero for regression coefficient (bi) and deviation from regression (s^2d), the results indicated that Dokii 331, TVU 21, Blackeye Crowder, Black

Table 6. Means of weight of seeds per plant traits of the nineteen genotypes under twelve environments.

Genotype	Average weight of seeds per plant (cm)												
	E1	E2	E3	E4	E5	E6	E7	E8	E9	E10	E11	E12	Mean
Dokii 331	52.53	56.57	56.17	54.73	56.23	56.20	54.73	56.23	55.87	55.27	56.77	56.03	55.61
Kaha 1	38.03	37.73	36.90	35.80	37.47	37.10	35.82	37.47	36.77	36.33	38.00	36.93	37.03
Cream 7	51.70	56.23	56.10	54.77	55.97	56.27	54.77	55.97	55.93	55.30	56.50	56.10	55.47
IT91K118-20	44.70	50.37	51.10	49.37	50.13	50.87	49.37	50.13	50.53	49.90	50.67	50.70	49.82
IT93K2045-20	46.40	44.43	43.13	41.47	44.00	43.13	41.47	44.00	42.80	42.00	44.53	42.97	43.36
TVU 21	52.67	47.10	44.80	43.53	46.77	45.17	43.53	46.77	44.83	44.07	47.30	45.00	45.96
IT81D-889	44.77	40.60	39.17	37.83	40.27	39.03	37.83	40.27	38.70	38.37	40.80	38.87	39.71
Chinese Red	46.03	51.13	50.97	50.07	50.63	51.20	50.07	50.63	50.87	50.60	51.17	51.03	50.37
IT81D1064	43.43	38.80	36.93	35.77	38.17	36.60	36.00	38.40	36.50	36.30	38.70	36.43	37.67
IT85F2205	47.50	47.17	45.77	44.57	46.30	45.70	44.90	46.63	45.70	45.17	46.90	45.57	45.99
IT90K1020-6	48.10	38.30	36.17	34.57	37.50	35.87	34.90	37.83	35.87	34.97	37.90	35.73	37.31
Blackeye Crowder	62.67	66.17	66.20	64.63	65.47	65.83	64.97	65.80	65.83	65.03	65.87	65.70	65.35
IT82C-16	47.73	38.03	36.00	34.27	37.37	35.73	34.60	37.70	35.73	34.67	37.77	35.60	37.10
IT82- 812	51.40	42.00	40.10	38.10	41.37	39.87	38.43	41.70	39.87	38.50	41.77	39.73	41.07
Sudany	68.30	67.20	69.23	67.20	66.53	68.80	67.53	66.87	68.80	67.60	66.93	68.67	67.81
Cream 12	50.27	53.20	53.20	51.37	52.60	52.97	51.70	52.93	52.97	51.77	53.00	52.83	52.40
Monarch Blackeye	54.10	38.53	35.93	34.47	37.97	35.73	34.80	38.30	35.73	34.87	38.37	35.60	37.87
Azmerly	56.93	59.07	59.00	57.40	58.50	58.63	57.73	58.83	58.63	57.80	58.90	58.50	58.33
Black Crowder	55.77	57.60	57.30	55.13	56.60	56.87	55.47	56.93	56.87	55.53	57.00	56.73	56.48
Environmental mean	49.63	48.96	48.11	46.58	48.41	47.98	46.77	48.60	47.83	47.05	48.89	47.83	48.05

Crowder and Azmerly genotypes were stable ($b_i < 1$) with high mean values, while, Kaha 1, Cream 7 and IT81D1064 genotypes were stable with the mean values lower than the grand mean. On the other hand, IT82C-16 and IT82-812 genotypes were unstable ($b_i > 1$) and could consistently do better in favorable environments. Similar results were obtained by Hazra et al. (1999) and De Rocha et al. (2007b).

Total dry seed yield (ton/fed.)

Results in Table 8 showed that total dry seed yield of genotypes overall environments ranged from 0.989 (ton/fed.) for (Blackeye Crowder) to 0.328 (ton/fed.) dry seed yield for (IT81D1064), with an average of 0.706 (ton/fed.) dry seed yield for all genotypes. These are in accordance with the finding of Kurubetta (2006), he found that time of sowing influenced significantly the seed yield per plant. June second fortnight sowing recorded significantly higher seed weight (14.40 g plant^{-1}) compared to July first fortnight (8.51 g plant^{-1}) and July second fortnight (6.12 g plant^{-1}) sowing. However, July first fortnight was significantly superior to July second fortnight sowing, Damarany (1994c), Torres et al. (2008), Yousaf and Sarwar (2008) and Akande and Balogun (2009). The highest dry seed yield 1.200 (ton/fed) was that for (Azmerly), in fall season at third planting date (August, 15th), while, the lowest 0.270 (ton/fed.) was for IT81D1064 in the summer season at third planting date (April, 15th).

The joint regression analysis of variance is presented in Table 10. The differences among the tested genotypes (G) were highly significant, also, environmental (E) effect, while, the interactions between genotypes and environments (G×E) were non-significant, as shown in Table 9. These results indicated that cowpea genotypes responded differently to different environmental conditions, suggestion the importance of assessment of

Table 7. Means of weight of pods per plant traits of the nineteen genotypes under twelve environments.

Genotype	Average weight of pods per plant (gram)												
	E1	E2	E3	E4	E5	E6	E7	E8	E9	E10	E11	E12	Mean
Dokii 331	70.33	70.00	69.33	75.33	75.67	76.00	70.00	68.67	69.67	72.33	76.33	76.00	72.47
Kaha 1	59.00	59.00	58.00	64.00	64.33	64.67	59.00	57.67	58.00	61.33	65.33	64.67	61.25
Cream 7	59.33	59.00	58.33	64.33	64.67	65.00	59.00	57.67	58.67	61.33	65.33	65.33	61.50
IT91K118-20	57.67	57.67	56.67	62.67	63.00	63.33	57.67	56.33	56.67	60.00	64.00	65.33	60.08
IT93K2045-20	63.67	62.00	62.67	68.67	69.00	69.33	62.00	60.67	64.33	64.33	68.33	69.00	65.33
TVU 21	77.33	78.00	76.33	82.33	82.67	83.00	78.00	76.67	75.67	80.33	84.33	84.67	79.94
IT81D-889	61.33	61.00	60.33	66.33	66.67	67.00	61.00	59.67	60.67	63.33	67.33	69.33	63.67
Chinese Red	47.33	48.33	46.33	52.33	52.67	53.00	48.33	47.00	45.33	50.67	54.67	58.33	50.36
IT81D1064	56.00	54.00	55.00	61.00	61.33	61.67	54.00	52.67	57.00	56.33	60.33	60.67	57.50
IT85F2205	56.67	56.33	55.67	62.00	62.00	62.00	56.33	54.33	56.00	58.67	63.00	61.00	58.67
IT90K1020-6	69.00	68.33	68.00	74.33	74.33	74.33	68.33	66.33	68.67	70.67	75.00	74.33	70.97
Blackeye Crowder	80.33	78.67	79.33	85.67	85.67	85.67	78.67	77.67	81.00	81.00	85.33	85.00	82.00
IT82C-16	69.00	68.33	68.00	73.67	74.33	75.00	68.33	67.33	68.67	70.67	74.67	77.67	71.31
IT82- 812	68.33	66.00	67.33	73.00	73.67	74.33	66.00	65.00	69.67	68.33	72.33	74.00	69.83
Sudany	52.00	51.67	51.00	56.67	57.33	58.00	51.67	50.67	51.33	54.00	58.00	59.33	54.31
Cream 12	56.00	54.67	55.00	60.67	61.33	62.00	54.67	53.67	56.33	57.00	61.00	61.67	57.83
Monarch Blackeye	61.67	60.67	60.67	66.33	67.00	67.67	60.67	59.67	61.67	63.00	67.00	69.67	63.81
Azmerly	67.67	66.33	66.67	72.33	73.00	73.67	66.33	65.33	68.00	68.67	72.67	73.33	69.50
Black Crowder	71.00	69.67	70.00	75.67	76.33	77.00	69.67	68.67	71.33	72.00	76.00	75.67	72.75
Environmental mean	63.35	62.61	62.35	68.28	68.68	69.09	62.61	61.35	63.09	64.95	69.00	69.74	65.43

genotypes under different environments in order to identify the best genetic make up for particular environment. These results appeared to be in harmony with those obtained by Nwofia et al. (2007), Padi (2007), Peksen (2007), De Rocha et al. (2007 a,b), Sarutayophat et al. (2007) and Taiwo (2007).

The stability parameters ($\bar{x}$, bi and s^2d) of the individual genotypes are illustrated in Table 11. All genotypes except IT93K2045-20, Dokii 331, Chinese Red and Cream 12 exhibited highly significant stability parameters from unity and zero for regression coefficient (bi) and deviation from regression (s^2d), respectively. The genotypes Dokii 331, Cream 12, IT81D-889, Chinese Red, IT81D1064, IT85F2205 and Sudany appeared to be stable and exhibited below average response to different environments, ($b_i < 1$), their genotypes were considered to perform relatively better in less favorable environments. Genotypes Cream 7, IT91k118-20, TVU 21 and Blackeye Crowder were unstable because regression coefficient (bi) more than one. The genotypes Dokii 331 and Cream 12, could be considered good inserted gave dry seed yield more than the overall of average genotypes besides their stability. Similar results were reported by Patel et al. (2005), conducted the experiment in loamy sandy soil with cowpea which revealed that sowing in 2nd March recorded significantly higher seed and haulm yield compared to sowing in 15th February, 17th March and 2nd April, (Gurushara and Sharma, 2004; Jena, 2003; Singh and Singh, 2000; Obiadalla-Ali et al., 2007; Rashwan, 2010).

Conclusion

The study identified considerable degree of genotypic differences and average stability for yield in cowpea when tested under various environments. The best genotypes were Dokii 331 and Cream 12. These genotypes were most stable that would be suitable for growth parameters under the test

Table 8. Means of total dry seed yield traits of the nineteen genotypes under twelve environments.

Genotype	Total dry seed yield (ton/fed.)												
	E1	E2	E3	E4	E5	E6	E7	E8	E9	E10	E11	E12	Mean
Dokii 331	0.950	0.920	0.920	0.950	1.000	1.100	0.968	0.929	0.890	0.960	0.988	1.101	0.973
Kaha 1	0.610	0.598	0.592	0.750	0.800	0.852	0.628	0.599	0.569	0.735	0.787	0.858	0.698
Cream 7	0.730	0.660	0.610	0.800	0.850	0.900	0.720	0.680	0.594	0.796	0.836	0.886	0.755
IT91K118-20	0.600	0.600	0.550	0.800	0.820	0.902	0.680	0.550	0.520	0.744	0.793	0.950	0.709
IT93K2045-20	0.610	0.570	0.549	0.680	0.700	0.740	0.601	0.575	0.550	0.657	0.708	0.757	0.642
TVU 21	0.700	0.670	0.660	0.850	0.900	0.950	0.711	0.680	0.640	0.836	0.895	0.955	0.787
IT81D-889	0.340	0.330	0.300	0.322	0.380	0.400	0.358	0.310	0.300	0.328	0.380	0.400	0.346
Chinese Red	0.400	0.380	0.360	0.480	0.500	0.552	0.400	0.380	0.360	0.477	0.490	0.549	0.444
IT81D1064	0.320	0.300	0.280	0.330	0.350	0.390	0.328	0.299	0.270	0.326	0.343	0.401	0.328
IT85F2205	0.650	0.600	0.550	0.730	0.750	0.792	0.640	0.589	0.569	0.698	0.760	0.799	0.677
IT90K1020-6	0.360	0.300	0.300	0.345	0.375	0.402	0.342	0.300	0.331	0.328	0.385	0.400	0.347
Blackeye Crowder	0.910	0.880	0.860	1.050	1.100	1.150	0.930	0.887	0.830	1.040	1.080	1.161	0.990
IT82C-16	0.850	0.800	0.750	0.950	0.980	1.003	0.853	0.790	0.755	0.946	0.985	1.000	0.889
IT82- 812	0.810	0.780	0.760	0.260	0.270	0.290	0.850	0.810	0.689	0.245	0.277	0.298	0.528
Sudany	0.590	0.550	0.510	0.623	0.650	0.680	0.583	0.539	0.519	0.597	0.655	0.697	0.599
Cream 12	0.850	0.810	0.790	0.890	0.900	0.943	0.901	0.800	0.750	0.847	0.890	0.978	0.862
Monarch Blackeye	0.870	0.830	0.800	1.000	1.040	1.100	0.881	0.820	0.799	0.987	1.073	1.097	0.941
Azmerly	0.800	0.700	0.700	1.000	1.100	1.200	0.773	0.732	0.699	1.000	1.072	1.190	0.914
Black Crowder	0.880	0.860	0.810	1.070	1.100	1.152	0.910	0.851	0.810	1.064	1.112	1.147	0.980
Environmental mean	0.675	0.639	0.613	0.731	0.767	0.816	0.687	0.638	0.602	0.716	0.764	0.822	0.706

Table 9. Combined analysis of variance for studied traits of 19 genotypes under various environments.

SOV	d.f	pod length (cm)	No. of seeds per pods	No. of pods per plant	hundred seeds weight (g)	weight of seeds per plant (g)	weight of pods per plant (g)	total dry seed yield Kg/fed.
Environments (E)	11	36.41**	12.69**	130.58**	34.46**	48.53**	598.04**	738283.36 **
Replication/ E	24	24.65	2.13	10.56	20.82	34.68	47.56	1416.64
Genotypes (G)	18	157.15**	39.02**	3661.90**	295.29**	3319.79**	2548.05**	1820938.13 **
G × E	198	0.79NS	2.50**	1.74**	0.23NS	18.63**	1.57NS	30879.28 **
Error	432	1.14	0.78	1.48	0.45	2.25	2.87	1249.18

* Significant at $P < 0.05$; ** highly significant at $P < 0.01$.

Table 10. The joint regression analysis of variance for the studied traits.

SOV	d.f	pod length (cm)	No. of seeds per pods	No. of pods per plant	Hundred seeds weight (g)	weight of seeds per plant (g)	weight of pods per plant (g)	total dry seed yield Ton/fed.
Genotypes (G)	18	157.24**	39.02**	3661.90**	295.29**	3319.78**	2548.05**	1820898.68**
E + (G×E)	209	2.28**	3.04**	8.52**	2.03**	20.20**	32.96**	41725.26**
E (linear)	1	340.45**	139.64**	1436.38**	379.09**	533.83**	6578.29**	3591930.96**
G×E (linear)	18	1.03^{NS}	5.51**	4.56**	0.08^{NS}	90.06**	0.73^{NS}	222547.50**
Pooled deviation	190	0.62^{NS}	2.09**	1.38^{NS}	0.23^{NS}	10.88**	1.57^{NS}	5909.44**
Pooled error	432	1.14	0.78	1.48	0.45	2.25	2.87	1249.18

*Significant at P < 0.05; ** highly significant at P < 0.01.

Table 11. Genotype average performance over 12 environments, and stability parameters of 19 cowpea genotypes.

Genotype	Pod length (cm)			No. of seeds per pods			No. of pods per plant			hundred seeds weight (g)			weight of seeds per plant (g)			weight of pods per plant (g)			total dry seed yield Kg/fed.		
	$\bar{x}$	Bi	s^2d	$\bar{x}$	Bi	s^2d	$\bar{x}$	Bi	s^2d	$\bar{x}$	Bi	s^2d	$\bar{x}$	Bi	s^2d	$\bar{x}$	Bi	s^2d	$\bar{x}$	Bi	s^2d
Dokii 331	14.06	1.20^{NS}	0.59	10.25	0.96^{NS}	0.33	55.78	1.07^{NS}	0.72	19.34	1.05^{NS}	0.18	55.61	0.01*	4.52	72.47	0.96^{NS}	0.35	0.973	0.82^{NS}	2127.16
Kaha 1	13.18	0.94^{NS}	0.50	8.64	1.05^{NS}	1.22	35.14	1.45**	1.52	12.74	0.95^{NS}	0.13	37.03	0.81**	0.09	61.25	0.95^{NS}	0.87	0.698	1.41**	1786.88
Cream 7	15.85	0.91^{NS}	0.50	11.44	0.37^{NS}	2.06	54.94	1.20^{NS}	1.16	13.80	0.97^{NS}	0.20	55.47	-0.23*	5.56	61.50	0.97^{NS}	0.25	0.755	1.37**	975.67
IT91K118-20	12.20	0.41^{NS}	1.43	10.81	1.65^{NS}	1.42	49.58	1.35*	1.13	13.40	1.03^{NS}	0.11	49.82	-0.62*	8.48	60.08	1.02^{NS}	0.98	0.709	1.90**	2313.73
IT93K2045-20	14.84	0.92^{NS}	0.13	10.86	-0.51*	2.31	42.92	1.12^{NS}	0.97	16.59	1.02^{NS}	0.20	43.36	1.51**	0.26	65.33	1.00^{NS}	1.36	0.642	0.98^{NS}	322.70
TVU 21	18.38	0.64^{NS}	1.26	11.36	1.17^{NS}	0.62	44.31	1.07^{NS}	1.16	22.16	1.03^{NS}	0.51	45.96	2.40**	4.43	79.94	0.98^{NS}	2.67	0.787	1.57**	2099.50
IT81D 880	16.67	0.06^{NS}	0.36	10.22	1.37^{NS}	0.08	30.61	1.00^{NS}	1.15	14.10	0.00**	0.10	39.71	1.03*	2.42	83.07	1.05^{NS}	0.74	0.348	0.44**	791.74
Chinese red	12.81	1.10^{NS}	0.24	9.47	1.87*	0.98	50.61	1.24^{NS}	1.06	11.63	0.96^{NS}	0.12	50.37	-0.53**	5.86	50.36	1.11^{NS}	8.01	0.444	0.93^{NS}	573.68
IT81D1064	13.84	1.38^{NS}	0.76	8.94	0.76^{NS}	0.30	37.44	1.14^{NS}	1.09	11.65	1.09^{NS}	0.19	37.67	1.97*	3.85	57.50	0.98^{NS}	2.66	0.328	0.51**	272.77
IT85F2205	14.41	1.23*	0.21	9.19	1.31^{NS}	0.84	46.33	1.09^{NS}	0.90	16.19	1.07^{NS}	0.77	45.99	0.98**	0.10	58.67	0.94^{NS}	1.59	0.677	1.16*	657.26
IT90K1020-6	11.95	1.00^{NS}	0.63	9.22	1.82^{NS}	1.65	37.19	0.87^{NS}	3.35	13.50	0.94^{NS}	0.27	37.31	3.10*	16.55	70.97	0.99^{NS}	0.57	0.347	0.44**	940.41
Black eye Crowder	13.37	1.26^{NS}	0.37	11.06	1.27^{NS}	3.88	63.86	1.22^{NS}	0.90	13.76	1.03^{NS}	0.10	65.35	-0.12**	3.09	82.00	0.97^{NS}	1.64	0.990	1.55**	1296.05
IT82D-16	18.00	1.16^{NS}	0.78	12.14	-1.49^{NS}	9.47	36.42	0.67^{NS}	3.75	16.41	1.11**	0.11	37.10	3.11*	15.50	71.31	1.07^{NS}	1.26	0.889	1.27**	1890.71
IT82- 812	13.07	1.04^{NS}	0.95	9.33	1.28^{NS}	0.26	40.06	0.83^{NS}	2.73	13.47	0.87^{NS}	0.48	41.07	3.11*	13.79	69.83	1.02^{NS}	3.40	0.528	-2.88**	82418.73
Sudany	10.58	0.94^{NS}	1.53	9.00	0.47^{NS}	2.84	67.25	0.74*	0.95	12.23	1.02^{NS}	0.31	67.81	-3.42*	73.98	54.31	1.01^{NS}	0.48	0.599	0.82**	238.80
Cream 12	15.91	1.04^{NS}	0.44	10.97	1.28^{NS}	7.23	52.06	0.65**	0.82	12.16	1.03^{NS}	0.01	52.40	0.11*	2.76	57.83	0.99^{NS}	0.57	0.862	0.83^{NS}	1649.81
Monarch black eye	11.60	1.26**	0.14	9.22	2.36*	2.59	34.78	0.75*	0.87	17.35	1.00^{NS}	0.16	37.87	4.26*	42.68	63.81	1.06^{NS}	0.79	0.941	1.55**	1822.09
Azmerly	15.98	0.65^{NS}	0.72	10.47	1.20^{NS}	0.37	58.25	0.73*	0.78	18.34	0.97^{NS}	0.28	58.33	0.20**	1.47	69.50	0.99^{NS}	0.57	0.914	2.57**	6115.00
Black Crowder	15.34	0.96^{NS}	0.25	11.19	0.80^{NS}	0.29	56.83	0.73^{NS}	1.29	13.99	0.99^{NS}	0.15	56.48	0.50^{NS}	1.39	72.75	0.95^{NS}	1.04	0.980	1.76**	3986.68
Mean		14.26			10.20			47.49			14.89			48.05			65.43			0.706	
L.S.D of G. M		0.528			0.434			0.602			0.331			0.743			0.839			17.503	
S. E (bi)		0.240			0.866			0.246			0.062			1.790			0.046			1.089	

localities or other similar environments.

Conflict of Interest

The authors have not declared any conflict of interest.

REFERENCES

Ajeigbe HA, Singh BB, Emechebe AM (2008).Field evaluation of improved cowpea lines for resistance to bacterial blight, virus and striga under natural infestation in the West African Savannas. Afr. J. Biotechnol. 7(20):3563-3568.

Akande SR, Balogun MO (2009). Multi-locational evaluation of cowpea grain yield and other reproductive haracters in the forest and southern guinea Savanna agroecologies of Nigeria. Electron. J. Environ. Agric. Food Chem. 8:526-533.

Ali Y, Aslam Z, Hussain F, Shakur A (2004). Genotype and environmental interaction in cowpea (*Vigna Unguiculata*-L) for yield and disease resistance. Int. J. Environ. Sci. Technol. 1(2):119-123.

Allen ON, Allen EK (1981). The Leguminosae: a source book of characteristics, uses and nodulation. The University of Wisconsin Press, Madison.

Ahmed S, Zargar MA, Ali T (2005). Genetic variability, heritability, genetic advance for seed yield and component traits in cowpea. Natl. J. Plant Improvement 7:85-87.

Ariyo OJ (2000). Effectiveness and relative discriminatory abilities of techniques measuring Genotype × environment interaction and stability in Okra. Euphytica 7:99-105.

Becker HC, Leon J (1988). Stability analysis in plant breeding. Plant Breed. 101:1-23.

Bilbro JD, Ray LL (2000). Environmental stability and adaptation of several cotton cultivars. Crop Sci. 16:821-840.

Chattopadhyay A, Dasgupta T, Hazra P, Som MG (2001). Phenotypic stability for green pod yield and its components in cowpea (*Vigna unguiculata* L. Walp). Indian Agric. 45:141-146.

Cooper M, DeLacy IH (1994). Relationships among analytical methods used to study genotypic variation and genotype by interaction interactions in plant breeding multi-environment experiments. Theor. Appl. Genet. 88:561-572.

Crossa J, Gauch HGJ, Zobel RW (1990). Additive main effects and multiplicative interaction of two international maize cultivar trials. Crop Sci. 30:493-500.

Dahiya OP, Singh D, Mishra SK, (2007a). Genetic divergence in cowpea (*Vigna unguiculata* L. Walp). J. Arid Legumes 4:62-65.

Dahiya OP, Singh D,Mishra SK (2007b). Correlation and path co-efficient analysis in cowpea (*Vigna unguiculata* (L.) Walp.). J. Arid Legumes 4:127-129.

Dahiya OP, Singh D,Mishra SK (2007c). Genotype × environment interaction and stability studies in cowpea (*Vigna unguiculata* (L.) Walp.). J. Arid Legumes 4:69-71.

Damarany AM (1994a). Estimates of genotypic and phenotypic correlation, heritability and potence of gene set in cowpea (*Vigna unguiculata* (L.). Walp). Assiut J. Agric. Sci. 25:1-8

Damarany AM (1994b). Testing and screening of some cowpea (*Vigna unguiculata* (L) Walp). Genotypes under Assiut conditions. Assiut J. Agric. Sci. 25(4):9-19

Damarany AM (1994c). Estimates of genotypic and phenotypic correlation, heritability and potency of gene set in cowpea (*Vigna unguiculata* [L.] Walp). Assiut J. Agric. Sci. 25:1-8.

Dashiell KE, Ariyo OJ, Ojo DK (1994). Genotype x Environment interaction and simultaneous selection for high yield and stability in soybeans. Ann. Appl. Biol. 124:133-139.

De Rocha MM, Filho FRF, Ribeiro VQ, de Carvalho HWL, Filho JB (2007a). Yield adaptability and stability of semi-erect cowpea genotypes in the Brazil Northeast Region. Pesq. Agropec. Bras. 42:1283-1289.

De Rocha MM, Filho FRF, Ribeiro VQ, de Carvalho HWL, Filho JB (2007b). Preliminar evaluation of cowpea genotypes for green-bean. Rev. Cien. Rural 12:153-156.

Eberhart SA, Russell WA (1966). Stability parameters for comparing varieties. Crop Sci. 6:36-40.

Gurushara NP, Sharma BB (2004). Effect of planting date, seed rate and row spacing on yield and yield attributes of bold seeded mungbean during spring/summer season. Indian J. Pulses Res. 17:45-46.

Hazra P, Chattopadhyay A, Pandit MK (1999). Genetic variability in three cultigroups of cowpea. J. Interacademicia 3:263-268.

Ishiyaku MF, Singh BB, Craufurd PQ (2005). Inheritance of time to flowering in cowpea (*Vigna unguiculata* (L.) Walp.). Euphytica 142:291-300.

Jena JC (2003). A short note on influence of date of sowing in vegetable pod yield of French Bean under Tarai Zone of West Bengal and Orissa. J. Horticult. 31:112-113.

Kamdi RE (2001). Relative stability, performance and superiority of crop genotypes across environments. J. Agric. Biol. Environ. Stat. 6:449-460.

Kurubetta KD (2006). Effect of time of sowing, spacing and seed rate on seed production potentiality and quality of fodder cowpea [*Vigna unguiculata* (l.) Walp]. M.Sc. Thesis. Department of Agronomy College of Agriculture, Dharwad University of Agricultural Sciences.

Nwofia GE, Okocha PI, Ene-Obong EE (2007). Evaluation of cowpea genotypes for yield and yield components in humid condition. J. Sustainab. Agric. Environ. 9:32-41.

Obiadalla-Ali, HA, Salman AMA, El-Hady MAHA (2007). Screening some local and introduced cowpea cultivars for dry-seed yield and resistance to *Callosobruchus maculatus* (F.). Ann. Agric. Sci. 52:197-212.

Padi FK (2007). Genotype environment interaction and yield stability in a cowpea-based cropping system. Euphytica 158:11-25.

Patel IC, Patel BS, Patel MM, Tikka SBS (2005). Effect of varieties, levels of irrigation and dates of sowing on yield and monetary returns of summer cowpea under North Gujarat agro-climatic condition. Indian J. Pulses Res. 18:217-218.

Peksen E (2007). Yield performance of cowpea (*Vigna unguiculata* L. Walp.) cultivars under rainfed and irrigated conditions. Int. J. Agric. Res. 2(4):391-396.

Rajput AL (1994). Response of cowpea (*Vigna unguiculata*) to rhizobium inoculation, date of sowing and phosphorus. Indian J. Agron. 39(4):584-587.

Rashwan AMA (2010). Estimation of some genetics parameters using six populations of two cowpea hybrids. Asian J. Crop Sci. 2:261-267.

Sarutayophat T, Nualsri C, Santipracha Q, Saereeprasert V (2007). Characterization and genetic relatedness among 37 yardlong bean and cowpea accessions based on morphological characters and RAPD analysis. Songklanakarin J. Sci. Technol. 29:591-600.

Singh J, Bhardwaj BL, Singh P (2007). Genetic diversity analysis in grain cowpea over environments. Crop Improv. 34:175-178

Singh AK, Singh SS (2000). Effect of planting dates, nitrogen and phosphorus levels on yield contributing factors in French bean. Legume Res. 23:33-36.

Singh SR, Rachie KO (1985). Cowpea Research, Production and Utilization. Books on Demand, Chichester, West Sussex, UK. 488p.

Steel RCD, Torrie JH (1980). Principles and procedures of statistics. McGraw-Hill Book Co., Inc., New York, U.S.A. P 633.

Taiwo AO (2007). Studies on stability and interrelationships among stability parameters for fodder yield in cowpea (*Vigna unguiculata* [L.] Walp). Agric. J. 2(1):77-81.

Teixeira NJP, de Machado CF, Filho FRF, de Rocha MM, Gomes RLF (2007). Grain yield, yield components and their interrelationship in genotypes of cowpea [*Vigna unguiculata* (L.) Walp.] Of determinate growth habit Editora UFV. Univ. Fed. Vicosa Rev. Ceres 54:375-383.

Torres SB, de Oliveira FN, de Oliveira RC, Fernandes JB (2008). Yield and morphology of cowpea accessions in Mossoro, Rio Grande do Norte State, Brazil. Hortic. Bras. 26:537-539.

Ushakumari R, Backiyarani S, Dhanakoodi CV (2002). Character contribution to diversity in cowpea. Legume Res. 23:122-125.

Yousaf A, Sarwar G (2008). Genotypic × environmental interaction of cowpea genotypes. Int. J. Environ. Res. 2:125-132.

32

Morphological variability of the fruiting branches in *Argania spinosa*: Effects of seasonal variations, locality and genotype

ZAHIDI A. , BANI-AAMEUR F. and EL MOUSADIK A.

Laboratory of Biotechnologies and Valorization of Naturals Ressources, Faculty of Sciences, Ibn Zohr University, BP 8106 Agadir 80000 Morocco.

The argan tree, is a member of the tropical family Sapotaceae, is an endemic of south western Morocco appreciated for its edible, high nutritional oil, extracted from the kernels of the fruit. The total number of fruiting branches (F), with one (F1), two (F2), three (F3) and four or more fruits (F4) in ten units of four different types of branches were observed for three consecutive seasons in three localities in south west Morocco. The twigs of the season and those less than two seasons have not fruited even if they have flourished. In contrast, the twigs more than two seasons and the main branches bear fruits. Fruit production in argan tree is largely dependent on temperatures and rainfalls during the cycle of flowering and fruiting which covers about 16 months. Prolonged drought during the flowering season is manifested by a significant reduction of the fruiting branches and number of fruits on twigs during the fruit ripening season. Contribution in the phenotypic variance of the climatic season and tree x environment interaction were very significant (18.5 and 52.9%). Broad sense heritabilities were low and ranged between 0 and 14.4%. Differentiation between the three populations for the fruiting branches is not established. However, most trees from Argana and Ait Melloul were most fruit bearing. Argan is especially valued by its fruit and oil, this work shows the existence of significant potential to improve fruiting in this species, which is in the wild state, by the choice of plus genotypes and the optimization of fruit production techniques for the argan domestication as a fruit tree for oil production.

Key words: *Argania spinosa*, diversity, fruit, fruiting branches, repeatability, multivariate analysis.

INTRODUCTION

Fruit trees have the potential to contribute towards food security, nutritional health and income generation and mitigate environmental degradation in developing countries (Jamnadass et al., 2009; Cuni-Shanchez et al., 2011; Simbo et al., 2012). Plant growth and productivity is hampered by environmental conditions, such as water scarcity, recurrent aridity and others. Under these conditions, few species were capable to stand to adverse situation maintaining some productivity. Such is the case of *Argania spinosa* in arid and semi-arid areas of North Africa, able to provide a diversity of resources that are the basis of economy for the local population (Zunzunegui et al., 2010). This multi-purpose tree is often described as an endangered species since several physical and anthropogenic factors reduce the density and surface of the arganeraie ecosystem (Msanda et al.,

2005). The argan tree is best known for its adaptation to drought and oil extracted from kernels of harvested fruit from trees in the wild state. However, the dried pulp, meal and leaves are sources of food for livestock (Sandret, 1957; Ehrig, 1974; M'Hirit, 1989, Prendergast and Walker, 1992; Maurin, 1992). Its exploitation is always in the economy picking mode. However, argan oil constitute up to 25% of fat consumed in the region. It is the subject of a commercial flow through Morocco and starts even if required at the international level for uses in dietetics and cosmetics. The multiple uses of the argan tree, especially the interest of oil combined with resistance to drought, make it a good candidate for domestication as a fruit tree for oil production and genetic improvement for arid areas (Bani-Aameur and Benlahbil, 2004; Ait Aabd et al., 2011).

Most of the reports available on the fruit were devoted to the chemical composition of argan oil, but very little research has focused on the fruit productivity in particular. The yield of fresh fruit varies according to tree, environment and climate season. It is 500 kg / ha / year per hectare on average (M'Hirit, 1989) and about 15 kg / tree (Rahali, 1989). The total production of ripe fruit in hot and dry season varies within wide limits according to trees at the Ait Melloul (Bani-Aameur, 2002a). Dried fruit yields are between 1.52 and 22.4 kg / tree / year. In addition, the frequency of fruit-bearing trees, fruit, pulp and kernel weights was highly variab Bani-Aameur le depending on season, trees and trees x environment interaction (Ferradous et al., 1996). These authors also reported that a minimum of 100 mm of rainfall recorded in autumn of fruit ripening promotes good fruiting. However large variability of flowering intensity was observed among climatic years, sites, tree genotypes and types of twigs. In any case, the peak of flowering occurs in spring (Bani-Aameur, 2002a). Small fruits on tree start to grow from October (Metro, 1952). But in February, fruits grow very quickly. In July, the fruit maturation was almost complete. The young fruits from flowering this season remain incompletely developed until the first rains next autumn. Thus, the flowering-fruiting cycle cover a period of nine to 16 months depending on trees (Bani Aameur et al., 1998; Benlahbil and Bani-aameur, 1999).

Some trees are able to have fruit once per season in March (early tree) or June (late tree), while other trees were able to flourish twice and then produce early and later fruits on the same individual (Ferradous et al., 1996; Bani Aameur et al., 1998). In early trees, the ripening of fruit from flowers fertilized in autumn of the last season occurs in May (Ferradous et al., 1996). In late trees, fruit maturation from flowers fertilized in spring of the last campaign occurs in August. While in intermediate trees, fruits are highly variable in size; their maturation is spread between spring and summer. All fruits from fertilized flowers do not persist until maturity, but a drop more or less important interested young fruit, ripe fruit and fruit whose maturation process is interrupted. The percentages of losses expressed in number of fruit varied from 3 to 39% depending on the trees (Bani Aameur et al., 1998). On the same tree, there are different branches and twigs with variable age and size. Twig of the season, twig less than two seasons, twig more than two seasons and the main branches growing on the carpenter branches (Zahidi et al., 1995). All these twigs and main branches bear flowers in very variable proportions (Ferradous et al., 1996; Bani-Aameur, 2000); we aimed to know what types can bear the fruits at maturity, and to establish the relationship between seasonal variations in temperatures and rainfall, the locality and the tree genotype and fruiting in three populations of argan in southwestern Morocco.

MATERIALS AND METHODS

Plant material and measurements

The experiment concerned trees was located at Ait Melloul at 35 m altitude in the Souss plain, Argana at 620 m altitude on southern slopes of High Atlas Mountains and Ait Baha (AB) at 50 km from the Atlantic ocean at 550 m altitude on the northern slopes of Anti Atlas mountains south west of Morocco. Thirty trees randomly selected and characterized for several morphological characters of fruit, kernel, flower, pollen, branching and foliation were observed in each site (Ferradous et al., 1996; Zahidi and Bani-Aameur, 1999a, b; Bani-Aameur and Benlahbil, 2004). Observations occur during three consecutive seasons, the first season was dry and warm; the second season was very wet with a relatively warm autumn, but winter and spring were cold. The third season is characterized as wet and hot, with gaps relatively high between the minimum and maximum temperatures (Figure 1).

Among the twigs and main branches facing South because of its large flowering (Bani-Aameur, 2002a), we observed at the end of April for three consecutive seasons the following characters (Figure 2): Among the 10 twigs of the season labeled (green twigs) we counted: Total number of the fruiting twigs (F); number of twigs with one fruit (F1); number of twigs with two fruits (F2); number of twigs with three fruits (F3); number of twigs with four or more fruits (F4). The same operation is performed for the ten twigs less than two seasons labeled (red color), ten twigs more than two seasons (lignified) and 10 principal branches (lignified with different ages and dimensions).

Variability characterization

The variance components and the relative percentage of the variance related to different factors in the total variance were estimated using the model in Table 1:

$$\sigma^2 T = \sigma^2 A + \sigma^2 l + \sigma^2 A \times l + \sigma^2 a/l + \sigma^2 A \times a/l + \sigma^2 e$$

Where, σ^2 T, Total variance (phenotypic variance); σ^2 A, variance related to seasonal variations (season factor); σ^2 l, variance due to locality; σ^2 A x l, variance due to season x locality interaction; σ^2 a/l, variance due to tree / locality (genotype); σ^2 A x a/l, variance related to genotype x environment interaction (season x tree / locality); and σ^2e: variance due to error. The percentage of the variance of each factor in phenotypic variance per each site was calculated using the model in Table 1.

$$\sigma^2 Ts = \sigma^2 A + \sigma^2 a + \sigma^2 A \times a + \sigma^2 e$$

Where, σ^2Ts, Total variance by site; σ^2A, variance related to season; σ^2a, variance due to tree; σ^2A x a, variance due to season x tree interaction; σ^2e: variance by site due to error.

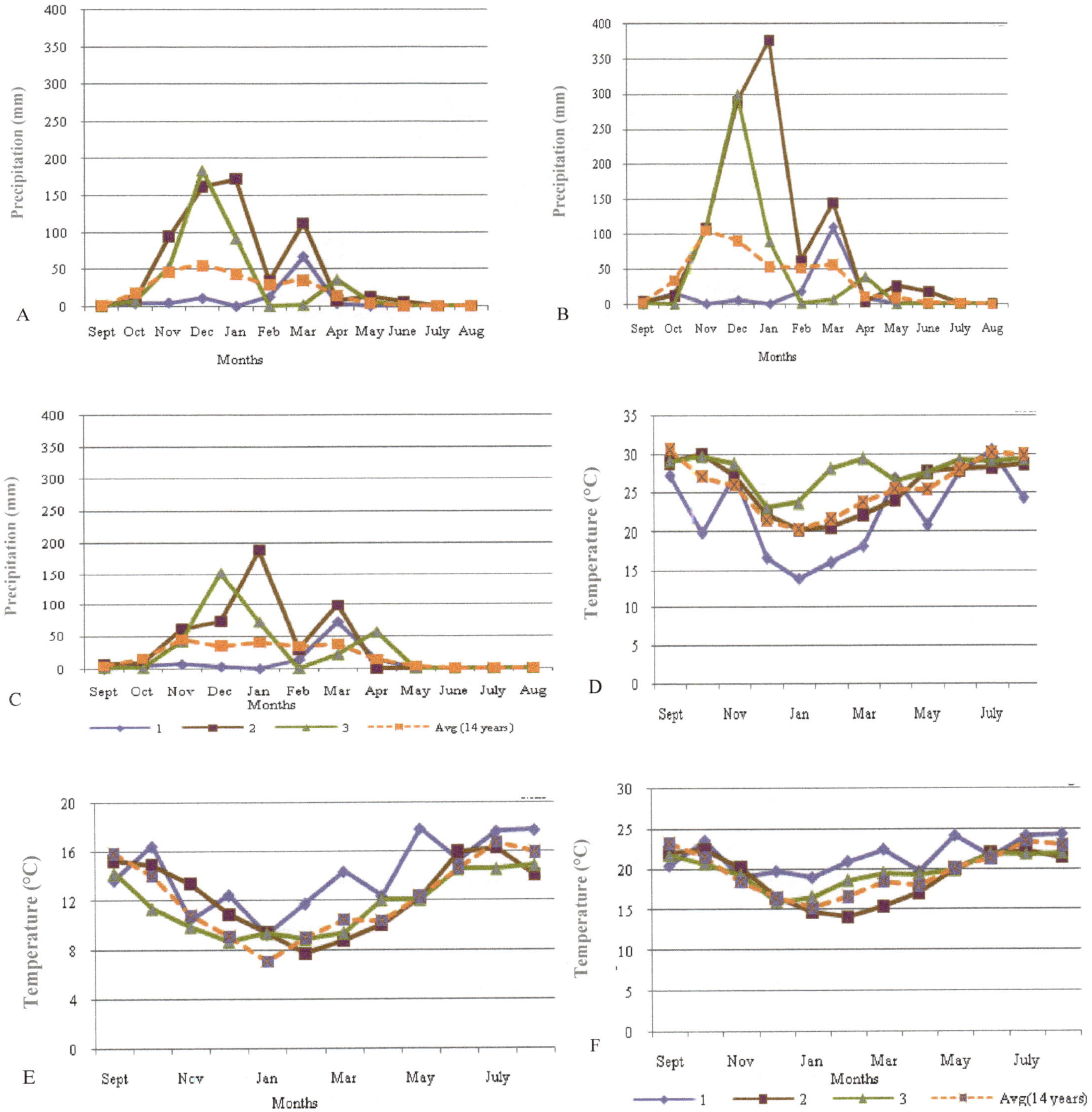

Figure 1. Climatic data from meteorological stations: mean monthly precipitation (mm), maximum, minimum and mean temperatures recorded at Ait Melloul (AM), Argana (AR) and Ait Baha. A, AM; B, AR; C, AB; D, max; E, min; F, avg.

Repeatability (broad sense heritability estimated by the ratio of variance tree / locality to the total phenotypic variance) was calculated according to the formula below given that trees are not repeated between sites and in each site (Pfahler et al., 1996; Bani-Aameur et al., 2001):

$$r^2 = 100 \times (\sigma^2 a/l \; / \; \sigma^2 a/l + \sigma^2 A \times a/l + \sigma^2 e)$$

Where the sum $\sigma^2 a/l + \sigma^2 A \times a/l + \sigma^2 e$) represents the total phenotypic variance in the three site and ($\sigma^2 a/l$) constitute the

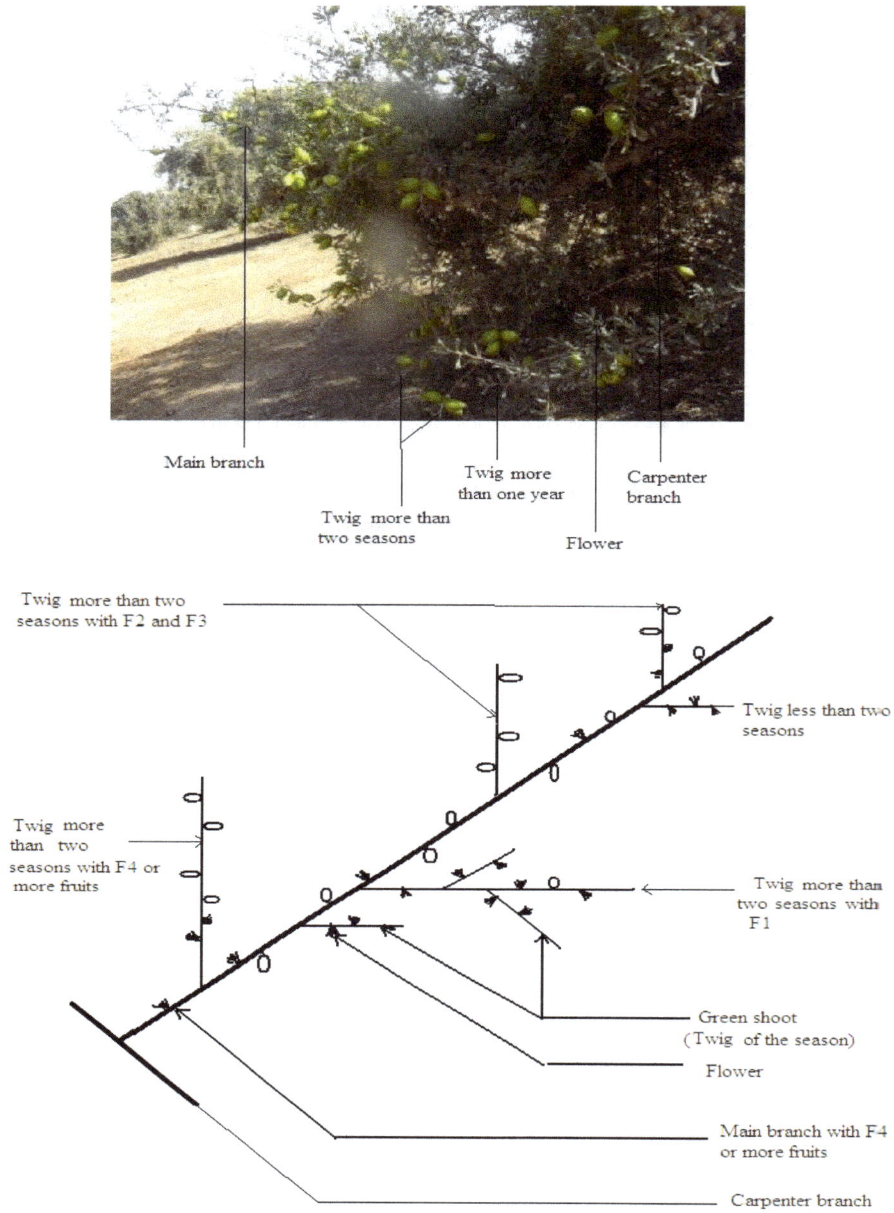

Figure 2. Morphological characters of the fruiting branches observed in Ait Melloul, Ait Baha and Argana during three consecutive seasons.

genetic variance. The repeatability per site was estimated using the following model:

$r^2 = 100 \times (\sigma^2 a \ / \ \sigma^2 a + \sigma^2 A \times a + \sigma^2 e)$

Where the sum ($\sigma^2 a + \sigma^2 A \times a + \sigma^2 e$) represents the total phenotypic variance per locality and ($\sigma^2 a$) variance related to tree.

Data analysis

An analysis of variance with four factors in hierarchical model was adopted (Table 1). Genotype (tree / locality) is hierarchical to locality factor because trees are not repeated between sites. Climatic season, locality and type of branch were crossed. The least significant difference test (LSD α = 5 %) of equality of means was used to compare differences between means (Steel and Torrie, 1960; Dagneli, 1984; Sokal and Rohlf, 1995). Factorial discriminate analysis (AFD) was performed on annual averages of each tree in order to examine the simultaneous contribution of all parameters studied in discriminating trees and sites (Frontier, 1981; Bernstein et al., 1988).

Dendogram was built using clustering method UPGMA "pair-group method unweighetd arithmetic average". Statistical treatments were performed using Statitcf, Statistix software and Ntsys version 1.40 (Rohlf, 1988).

Table 1. Expectations of mean squares and estimated variance components for morphological characters in the three localities.

Source of variation	DF	Mean square	Expectations of mean squares
Global			
Season	2	CM A	$\sigma^2e + 2\sigma^2Aal + 60\sigma^2Al + 180\sigma^2A$
Locality	2	CM I	$\sigma^2e + 2\sigma^2Aal + 60\sigma^2Al + 6\sigma^2al + 180\sigma^2l$
Tree / locality	87	CM al	$\sigma^2e + 2\sigma^2Aal + 6\sigma^2al$
Season x locality	4	CM Al	$\sigma^2e + 2\sigma^2Aal + 60\sigma^2Al$
Season x tree / locality	174	CM Aal	$\sigma^2e + 2\sigma^2Aal$
Error	270	CM e	σ^2e
By locality			
Season	2	CM A	$\sigma^2e + 2\sigma^2Aa + 60\sigma^2A$
Tree	29	CM a	$\sigma^2e + 2\sigma^2Aa + 6\sigma^2a$
Season x tree	58	CM Aa	$\sigma^2e + 2\sigma^2Aa$
Error	90	CM e	σ^2e

Table 2. Analysis of variance of number of main branches and twigs more than two seasons fruit bearing (F), with one (F1), two (F2), three (F3) and four or more fruits (F4) observed in the three localities.

Source of variation	DF	Mean square				
		F	F1	F2	F3	F4
Season	2	2806.8*	579.3^{ns}	282.9**	81.46*	26.2^{ns}
Locality	2	749.01^{ns}	248^{ns}	27.8^{ns}	32.5**	28.6**
Type of branch	1	90.7^{ns}	1.13^{ns}	12.03^{ns}	12.3^{ns}	12.9^{ns}
Tree / locality	87	14.25^{ns}	5.5^{ns}	2.03^{ns}	1.81**	1.5**
Season x locality	4	234.6**	125.03**	8.88**	8.5**	8.45**
Season x type of branch	2	52.64*	6.6^{ns}	2.88**	6.89**	7.85**
Locality x type of branch	2	11.27^{ns}	2.07^{ns}	0.76^{ns}	1.6*	2.26*
Season x locality x type of branch	4	6.34**	1.82^{ns}	0.86^{ns}	0.9^{ns}	0.82^{ns}
Season x tree / locality	174	12.14**	5.96**	1.76**	1.31**	1.12**
Season / locality x type of branch	87	1.83^{ns}	1.05^{ns}	0.61^{ns}	0.36^{ns}	0.5*
Season x tree / locality x type of branch	174	1.34^{ns}	1.13^{ns}	0.57**	0.38**	0.39**
Error	540	1.45	1.26	0.4	0.25	0.21

^{ns}Not significant; *: significant at 5%; **, significant at 1%.

RESULTS

Variability characterization

Type of branch

The principal branches and twigs more than two seasons have borne fruits for three seasons, but twigs of the season and those less than two seasons have not borne fruits at maturity even if they have flowered. The type of branch was not significant for all traits (F, F1, F2, F3 and F4) (Table 2). Type of branch x climatic season interaction is significant for total number of fruiting the twigs (F), number of twigs to two (F2), three (F3) and four or more fruits (F4), but not significant for number of twigs with one fruit (F1). Locality x type of branch interaction was significant for F3 and F4. Locality x climatic season x type of branch interaction was significant only for F. Type of branch x tree / locality interaction (genotype x type of branch) was significant for F4, while climatic season x type of branch x tree / locality interaction was highly significant for F2, F3 and F4.

During the second season (2nd) characterized as low fruiting and even during the season at intermediate fruiting (1st), no difference was found between the main branch and twig over than two seasons for F1, F2, F3 or F4. But, in season of high fruiting, twigs more than two seasons have borne more fruits (72%) than the main branch (57.3%). They bear more than one fruit in 58.9% of cases against 48.2% for main branch (Table 3).

In all three localities, no difference was observed for F, F1 and F2 between the main branches and twigs more

Table 3. Average number of branches (main branch and twigs more than two seasons) with fruits (F), one (F1), two (F2), three (F3) and four or more fruits (F4) per season observed in three localities.

Type of branch	Season	Main branch	Twig more than two seasons
Twig or main branch with fruits (F)	1	4.13	4.17
	2	0.78	1.02
	3	5.73 b	7.19 a
	Average	3.55	4.13
Twig or main branch with one fruit (F 1)	1	2.82	2.47
	2	0.54	0.73
	3	3.01	2.97
	Average	2.12	2.06
Twig or main branch with two fruits (F 2)	1	0.66	0.8
	2	0.16	0.24
	3	1.72 b	2.14 a
	Average	0.85	1.06
Twig or main branch with three fruits (F 3)	1	0.42	0.55
	2	0.06	0.05
	3	0.74 b	1.28 a
	Average	0.41	0.63
Twig or main branch with four or more fruits (F 4)	1	0.2	0.32
	2	0.01	0.0
	3	0.3 b	0.82 a
	Average	0.17	0.38

Means followed by letters are significantly different at 5%.

Table 4. Average number of branches (main branch and twigs more than two seasons) with fruits (F), one (F1), two (F2), three (F3) and four or more fruits (F4) observed in Ait Melloul (AM), Argana (AR) and Ait Baha (AB).

Type of branch	Main branch				Twig more than two seasons			
	AM	AR	AB	Average	AM	AR	AB	Average
With fruit (F)	4.18	4.37	2.09	3.55	4.99	5.13	2.26	4.13
With one fruit (F 1)	2.87	2.25	1.25	2.12	2.93	2.02	1.23	2.06
With two fruits (F 2)	0.92	1.04	0.59	0.85	1.17	1.31	0.7	1.06
With three fruits (F 3)	0.33^b	0.7^b	0.22^b	0.41	0.58^a	1.01^a	0.28^a	0.62
With four or more fruits (F 4)	0.67^a	0.4^b	0.02^b	0.16	0.3^b	0.77^a	0.07^a	0.38

Means followed by different letters are significantly at 5%.

than two seasons (Table 4). In Argana and Ait Baha, the twigs more than two seasons have more fruits, four or more fruits in greater proportions than the main branch. In Ait Melloul, twigs more than two seasons had formed three fruits in greater proportions, but the main branches had formed more than four fruits. In the three sites, main branches and twigs more than two seasons are capable of producing at least one or two fruits, but the young twigs (twig more than two seasons) have a higher production potential than older branches (main branches). All trees have not borne the same number of twigs with three, four or more fruits. In trees (1, 4, 5, 14, 23 and 29) from Ait Melloul, (2, 5, 6, 13, 17, 20, 19 and 27) from Argana and (2, 11 and 20) from Ait Baha, the twigs more than two seasons have more fruits than the main branches. While among the trees (20, 21 and 25) of Ait Melloul (11, 18 and 28) of Argana, and (21, 22, 30) of Ait Baha, the main branches were more fruiting than the

Table 5. Average number of branches (main branches and twigs more than two seasons) with fruit (F), one (F1), two (F2), three (F3) and four or more fruits (F4) per season and locality.

Type of branches	Season \ locality	Ait Melloul	Argana	Ait Baha	Average
Number of branches with fruits (F)	1	5.94^{b}	5.81^{b}	0.7^{b}	4.15^{b}
	2	1.38c	1.01c	0.32c	0.9c
	3	6.44^{a}	7.43^{a}	5.51^{a}	6.46^{a}
	Average	4.59	4.75	2.18	3.84
Number of branches with one fruit (F1)	1	4.5^{a}	2.87^{a}	0.56^{b}	2.64
	2	1.07c	0.7^{b}	0.16c	0.64
	3	3.13^{b}	2.84^{a}	3.0^{a}	2.99
	Average	2.89	2.13	1.24	1.99
Number of branches with two fruits (F2)	1	0.93^{b}	1.18^{b}	0.08^{b}	0.73^{b}
	2	0.27c	0.23c	0.1^{b}	0.2c
	3	1.93^{a}	2.12^{a}	1.74^{a}	1.93^{a}
	Average	1.04	1.18	0.64	0.95
Number of branches with three fruits (F3)	1	0.37^{b}	1.04^{b}	0.05^{b}	0.5^{b}
	2	0.05c	0.06c	0.06^{b}	0.06c
	3	0.95^{a}	1.43^{a}	0.64^{a}	1.01^{a}
	Average	0.46^{b}	0.84^{a}	0.25c	0.52
Number of branches with four or more fruits (F4)	1	0.07^{b}	0.7^{b}	0^{b}	0.26
	2	0^{b}	0c	0.03^{b}	0.01
	3	0.48^{a}	1.06^{a}	0.11^{a}	0.55
	Average	0.18^{b}	0.59^{a}	0.04c	0.27

Means followed by different letters are significantly at 5%.

twigs more than two seasons.

Climatic season

The climatic season is significant for F, F2 or F3 (Table 2). It is not significant for F1 and F4. The fructification is higher during the third season compared to the first and second seasons (Table 5). Indeed, during the humid season (3rd) following a very humid campaign, 65% of main branches and the twigs more than two seasons have fructified. Whereas during very dry season (1st) following a dry campaign, 41.5% of total of the twigs or main branches observed have borne fruits, while during the campaign very humid (2nd) following a very dry season, about 9% of main branches and the twigs more than two seasons have fructified. Among these fruiting the twigs, 45.5% in the 3rd season, 29.6% in the 1st and 28.8% in 2nd season had borne two or three fruits.

Locality

Locality is highly significant for F3 and F4, but not significant for F, F1 and F2 (Table 2). Locality x climatic season interaction is highly significant for all traits. In Argana site, number of twigs or main branches to three and four or more fruits is higher than in Ait Melloul, and Ait Baha (Table 5). Reducing the number of fruits on the branches is probably a reaction to variations of temperatures and rainfalls. This reduction was more pronounced in Ait Baha, more arid site than in Ait Melloul to mild temperatures and Argana the most humid site especially during the 1st and 2nd season (Table 5). These effects are manifested by a remarkable reduction in the number of fruiting branches, since 7% of the total branches during the dry season and only 3.2% in very humid season following a dry campaign have fruited in Ait Baha. In the three stations, in season at low and intermediate fruiting, most (over 50%) of branches bear fruits. In contrast, in season at high fruiting, 62.1% in Argana, 52.2% at Ait Melloul and 45.1% Ait Baha having more than one fruit.

Genotype

Tree / locality (genotype) is highly significant for F3 and F4 (Table 2). Indeed some trees as (6 and 10) of Ait Melloul, (6, 7, 11, 17, 26, 27 and 28) of Argana (4 and 21) from Ait Baha were able to produce more fruiting branches (main branches or twigs more than two seasons) with three and with four or more fruits and therefore more fruits. So, these trees are of high potential

Table 6. Frequencies of trees that produced fruits at Ait Melloul, Argana and Ait Baha.

Type of branch	Season \ locality	Ait Melloul		Argana		Ait Baha		Average	
		Number of trees	Frequency (%)	Number of trees	Frequency	Number of trees	Frequency (%)	Number of trees	Frequency (%)
Branches with one fruit (F1)	1	28	93.3	27	90	9	30	21.3	71
	2	13	43.3	7	23.3	2	6.7	7.3	24.4
	3	30	100	30	100	30	100	30	100
Branches with two fruits (F2)	1	15	50	25	83.3	4	13.3	14.7	48.9
	2	13	43.3	6	20	2	6.7	7	23.3
	3	30	100	30	100	30	100	30	100
Branches with three fruits (F3)	1	6	20	21	70	2	6.7	9.7	32.2
	2	3	10	5	16.7	2	6.7	3.3	11.1
	3	28	93.3	30	100	27	90	28.3	94
Branches with four or more fruits (F4)	1	1	3.3	13	43.3	0	0	4.7	16
	2	0	0	0	0	2	6.7	0.7	2.2
	3	21	70	25	83.3	8	26.7	18	60

for fruit production and then can serve as germplasm in a breeding program and for domestication as a fruit tree for the production of argan oil. Season x tree / locality interaction is highly significant for all traits (Table 2). Trees in the three sites have reacted differently with respect to seasonal variations of temperatures and rainfalls (Table 6). This differential response was reflected by the frequency of trees whose fruiting branches have presented one fruit, two, three and four or more fruits. Thus, during the very humid season (2), following a very dry season, 6.7% of total trees at Ait Baha, 23.3% in Argana and 43.3% in Ait Melloul were fruitful. While during the humid season (3), which followed a very humid campaign all trees have borne fruits. During the first season, which followed a dry campaign, about 93.3% in Ait Melloul and Argana and 30% in Ait Baha have fruited.

Variance components

The relative percentage of variance due to climatic season in the total variance is high more than 50% for F and F2, but relatively low to moderate (8% to 28%) for the others characters (Figure 3). Climatic season effect is more pronounced at Ait Baha than in Argana and Ait Melloul for all characters except number of branches (twigs more than two seasons and main branches) to four or more fruits. The percentages of total variance per site varied between 44.1 and 83.4% in Ait Baha, 26.1 and 70.9% in Argana, and between 35.4 and 61.5% in Ait Melloul. The relative contribution of variance related to locality and locality x season interaction in the phenotypic variance is relatively low and ranged from 3.6 and 13.2% except the number of branches with one fruit (22.7%) The contribution of variance due to genotype x environment interaction (season x tree / locality) in phenotypic variance is greater for all characters. It varied from 18.5% for F and 42.8% for F4 (Figure 3). Percentages remarkable of genotype x environment interaction are mainly related to the importance of season x tree interaction in Ait Melloul and Argana compared to Ait Baha. Thus, at Ait Melloul and Argana where seasonal variations are less important, season x tree interaction explains 30.1 to 59.9% in Ait Melloul and about 0 to 44.4% in Argana. By cons, at Ait Baha, the driest site, this contribution varied from 8.6 to 57.1%. The relative contribution of variance associated to genotype (tree/locality) in the phenotypic variance is low and ranged from 0% for F1 and 6.2% for F4 (Figure 3). The same observation is noted in each locality, the percentage of total variance attributed to tree factor is also low (0% for F, F2, F3 and 7.7% for F4). Highest repeatabilities (8.9 and 9.8%) were observed for F3 and F4, while for the other characters, the repeatabilities were low (0 and 4.8%) (Table 7). Low repeatability (overall and by station) recorded for the fruiting branches reflect the crucial role of seasonal variation in the fruits productivity in argan tree.

Variability distribution

The total number of the fruiting branhes (F) is correlated in different degrees with F1, F2 and F3 during the dry season (1st), and during the very humid season (2nd) (Table 8). While during the humid season (3rd), F is highly correlated with F2, F3 and F4. The correlation coefficients of F1 were low in very dry season but higher with F2, F3 in very humid season and F3, F4 in humid

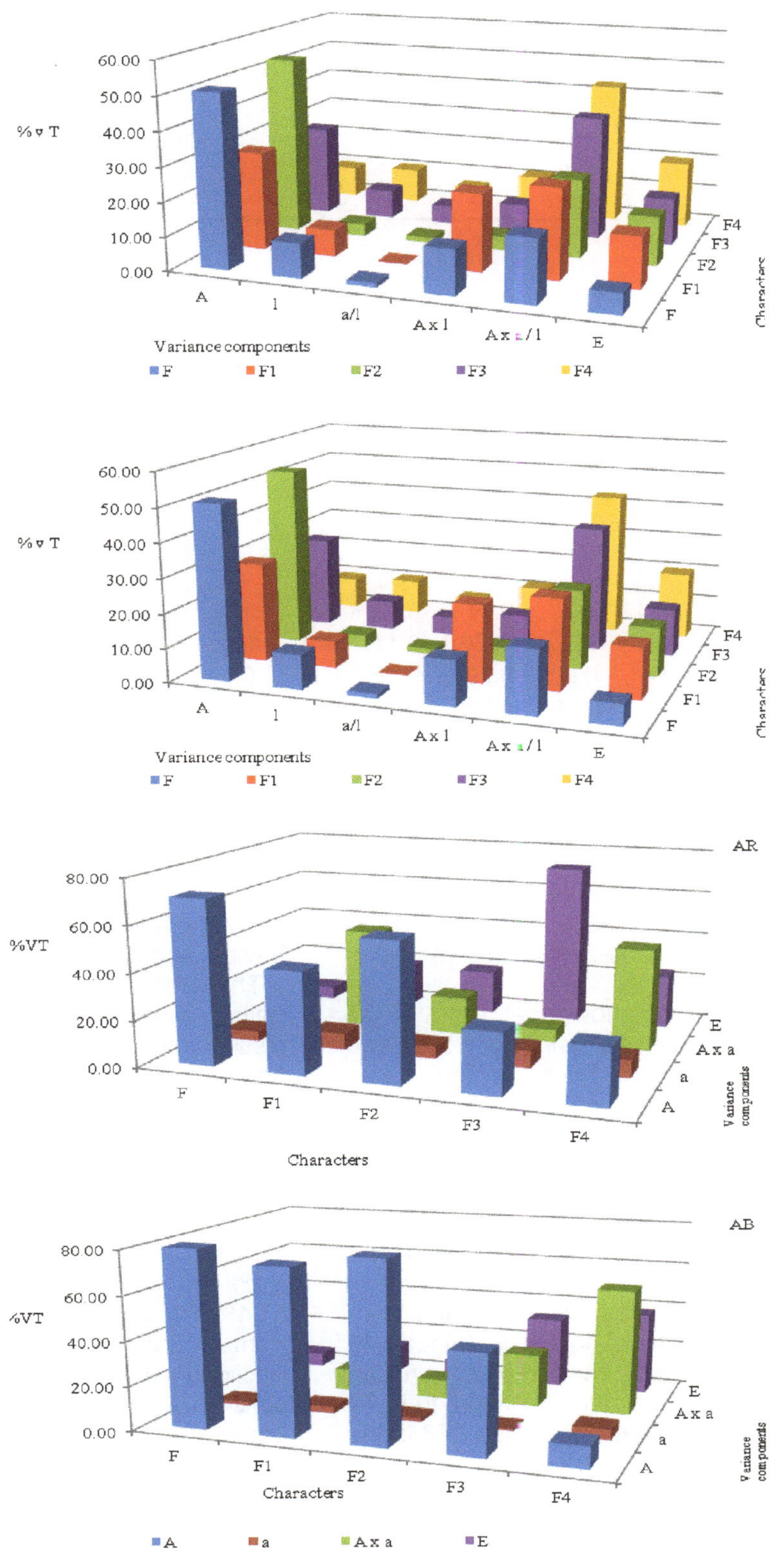

Figure 3. Percentage in phenotypic variance of variance components for main branches and twigs more than two seasons with fruit (F), with one (F1), two (F2), three (F3) and four or more fruits (F 4) observed in Ait Melloul (AM), Argana (AR) and Ait Baha (AB). (A, Season, l, locality, a / l, tree / locality, a, tree, A x l, season x locality; A x a / l, season x tree / locality; e, error).

Table 7. Repeatabilities in percentage for the fruiting branches observed in the three localities.

Character	F	F1	F2	F3	F4
Global	4.8	0.0	4.04	9.8	8.9
Ait Melloul	0.0	1.03	0.0	0.00	5.4
Argana	14.4	0.0	12.7	10.1	10.2
Ait Baha	10.2	11.8	8.9	0.0	0.0

Table 8. Matrix of correlations for the fruiting branches observed in the three localities during the three consecutive seasons.

Characters	Fs1	Fs2	Fs3	F1s1	F1s2	F1s3	F2s1	F2s2	F2s3	F3s1	F3s2	F3s3	F4s1	F4s2	F4s3
Fs1	1.00														
Fs2	0.15	1.00													
Fs3	0.43	0.21	1.00												
F1s1	0.75	0.12	0.23	1.00											
F1s2	0.16	0.97	0.18	0.14	1.00										
F1s3	-0.01	-0.02	-0.1	0.14	0.01	1.00									
F2s1	0.67	0.13	0.25	0.23	0.16	0.02	1.00								
F2s2	0.11	0.93	0.21	0.1	0.84	-0.03	0.08	1.00							
F2s3	0.32	0.17	0.73	0.22	0.166	-0.11	0.19	0.18	1.00						
F3s1	0.58	0.1	0.4	-0.03	0.09	-0.19	0.56	0.09	0.2	1.00					
F3s2	-0.03	0.69	0.14	0.01	0.53	-0.06	-0.06	0.77	0.07	0.01	1.00				
F3s3	0.32	0.15	0.82	0.15	0.11	-0.49	0.13	0.15	0.49	0.37	0.13	1.00			
F4s1	0.43	-0.01	0.35	-0.06	-0.03	-0.23	0.18	0.01	0.25	0.67	-0.03	0.24	1.00		
F4s2	-0.13	0.34	-0.1	-0.1	0.21	0.06	-0.07	0.45	-0.03	-0.08	0.7	-0.11	-0.05	1.00	
F4s3	0.33	0.21	0.78	0.08	0.19	-0.53	0.2	0.2	0.4	0.44	0.15	0.78	0.37	-0.1	1.00

season. The values of F, F1, F2, F3 and F4 obtained in the dry season were not correlated with the values of humid seasons.

Discriminate factorial analysis shows that 100% of the total variance could be explained using only the two canonical components. First CP1, explaining about 69.5% of variation, was linked to F, F2 and F3 in the three seasons, to F1 in the 1st and 2nd season, and F4 in the 2nd and 3rd season. Thus individuals to high fruiting such as (1, 2 and 3) of Ait Melloul, (1, 3 and 23) of Argana and (27) of Ait Baha were projected on the negative side of the first axis. But, individuals with low fruiting such as (8 and 15) of Ait Melloul (4 and 24) of Argana and great number of trees from Ait Baha were projected on the negative side of this axis (Figure 4). Second CP2 that was responsible for 30.5% of variation is linked to F1 in 3rd season and F4 in 1st season, to F, F4 and F2 in 3rd season and in varying degrees to F3 in the three seasons (Table 9).

The ordering of trees revealed that genotypes are not grouped according to their origins since respectively 58.9% (53/90 trees) from Ait Melloul, 48.8% (44/90 trees) from Argana are among trees most fruiting and which produce three and four fruits. But more than 60% of trees from Ait Baha are among genotypes low fruiting. We can therefore conclude that there is no differentiation of the three populations for the fruiting branches. Ait Melloul and Argana sites are relatively far from Ait Baha, while Ait Melloul and Argana are nearer (Table 10).

The dendrogram generated based on all morphological traits, showed a similar pattern. Two groups are distinguished in a Euclidean distance 3.2 (Figure 5). The first group is divided in a Euclidean distance of 2.74 in a first class containing M15, R24 and B22 characterized by low fruiting, and a second class containing 13.9% from Ait Melloul, 16.7% from Argana and 69.4% from Ait Baha. The second group is divided in a Euclidean distance of about 2.63 into two subgroups. The first subgroup includes 39.4% from Ait Melloul, 51.5% from Argana, and 9.1% from Ait Baha. The second subgroup contains 61.1% from Ait Melloul, 33.3% from Argana and 9.1% from Ait Baha. Sites classification shows two groups at a Euclidean distance about 2.56 (Figure 6). A first group consists Ait Baha and a second group containing Ait Meloul and Argana. This classification is not the result of geographical isolation. Argana characterized by cold winter, and Ait Melloul with mild temperatures are generally not differentiated from Ait Baha known for its drought summer.

DISCUSSION

The main branch and the twigs more than two seasons

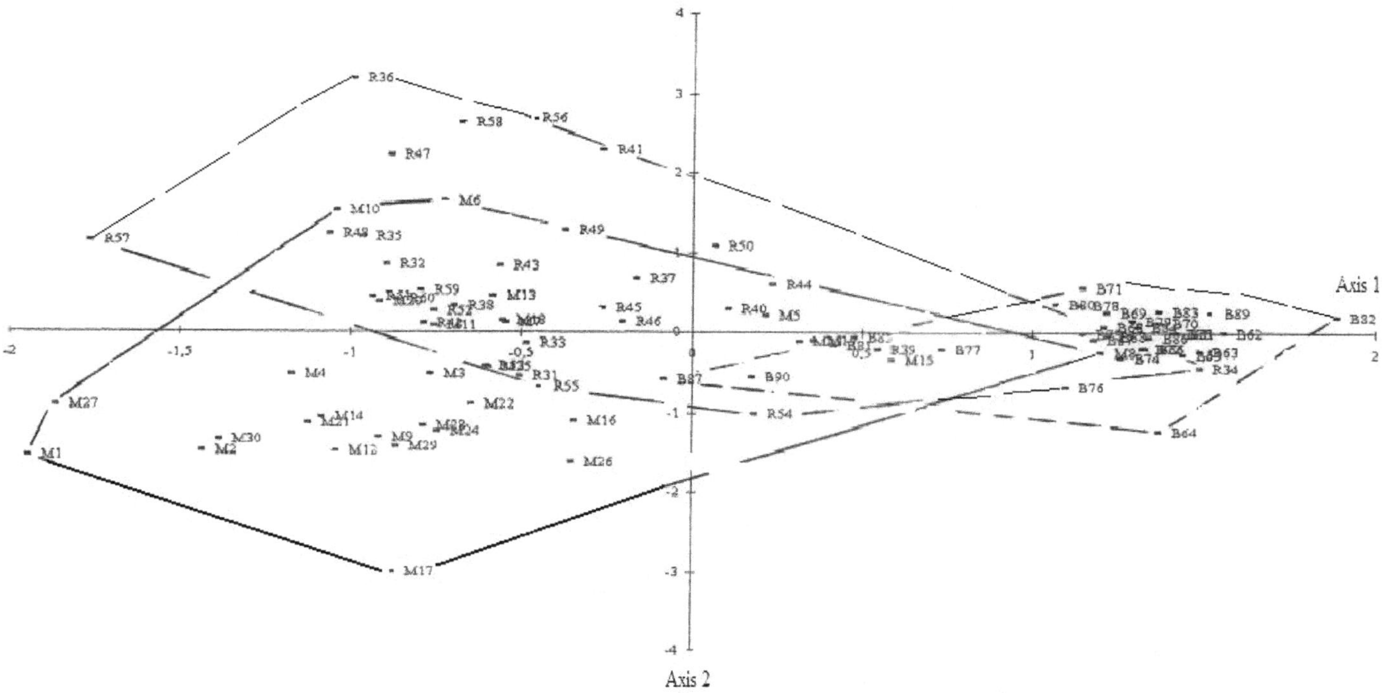

Figure 4. Projection of individuals from the three populations on the plane defined by the first two canonical components.

Table 9. Correlations between canonical components and characters of the fruiting branches observed in the three localities.

Variable	CP1	CP2
Fs1	-0.99	0.1
Fs2	-0.97	-0.24
Fs3	-0.8	0.61
F1s1	-0.95	-0.31
F1s2	-0.95	-0.31
F1s3	-0.04	-0.99
F2s1	-0.95	0.32
F2s2	-0.99	-0.15
F2s3	-0.81	0.58
F3s1	-0.67	0.74
F3s2	0.6	0.81
F3s3	-0.73	0.69
F4s1	-0.48	0.88
F4s2	0.99	-0.11
F4s3	-0.72	0.69
Eigenvalues	0.68	0.3
Explained Percentages (%)	69.5	30.5
Cumulative percentages (%)	69.5	100

CP1, First canonical component; CP2, second canonical component.

have borne the fruits during the three campaigns, but the twigs of the season and those less two seasons have not borne the fruits at maturity even if they have flowered. The type of branch (main branches or twigs more than two seasons) does not affect the fruiting of argan tree. However, the fruiting is strongly influenced by the season, locality and tree. The twigs more than two seasons are more fruiting than the main branches in

Table 10. Mahalanobis distance between Ait Melloul, Argana and Ait Baha for characters of the fruiting branches.

Locality	Ait Melloul	Argana	Ait Baha
Ait Melloul	0		
Argana	1.35	0	
Ait Baha	1.92	1.81	0

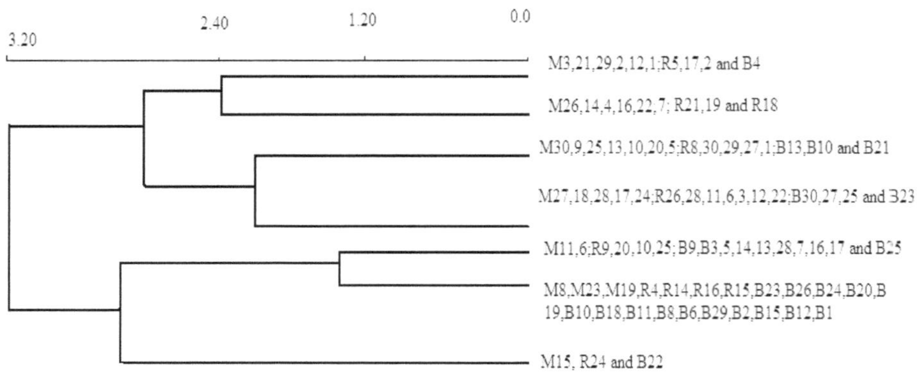

Figure 5. Dendrogram obtained when clustering individuals of Ait Melloul (M), Argana (R) and Ait Baha (B) on the basis of characters of the fruiting branches characters.

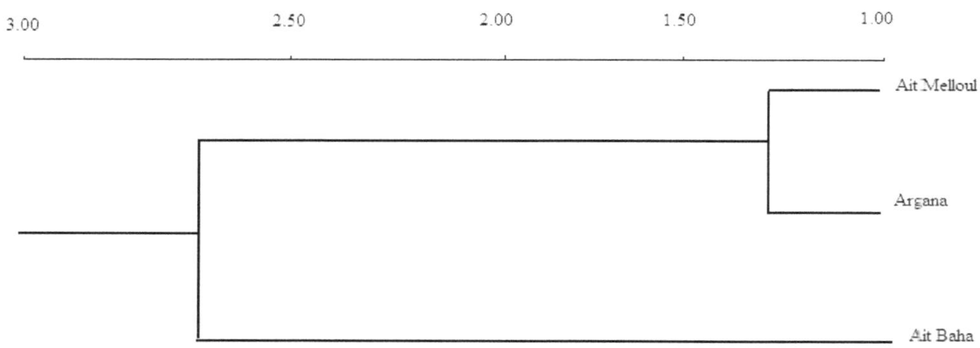

Figure 6. Classification of localities Ait Melloul (AM), Argana (AR), and Ait Baha (AB) for characters of the fruiting branches in argan tree.

favorable seasons. It seems that this difference has a trophic origin in relation to the age of the branches. Thus, for the regulation of the fruiting in argan tree, pruning-lightening operations by removing some old branches that grow on the carpenter branches reduces nutrient competition and ensure regularity of fruit production in argan as is practiced in other fruit species (Andales et al., 2006; Tworkoski and Glenn, 2010). For the establishment of orchards, this operation must be coupled with an appropriate irrigation in case of drought, at least during the flowering period since 100 mm of rainfall recorded in autumn of fruit ripening promotes good fruiting as reported by Bani-Aameur (2002a) and Bani-Aameur (2002a). Those operations must take into account yield components (number of fruiting branches, number of flowers, number of fruits per flower).

In a given season, in the three localities, all trees do not bear fruits. The effect of drought occurs partly by a reduction in number of branches with fruits and secondly by reducing the number of fruit on the twigs. The flowering-fruiting cycle cover a period of 9 to 16 months depending on trees (Bani Aameur et al., 1998; Benlahbil and Bani-aameur, 1999). Thus, in case of drought during the flowering period, most fruiting branches in the next season produce mainly one or two fruits and secondarly three fruits. But, if the flowering season is humid, fruiting in the following season will be more important and number of branches bearing fruits will be also higher. Significant decrease in the number of the fruiting branches and the number of fruits on twigs during the very humid season (2nd) is related to minimum flowering observed during the previous season characterized as dry and warm. Indeed, during the dry seasons, the flowering was late (March), it concerned only 50 to 70% of trees in the three sites and the number of glomerules ranged from 0.18 to 24 units. By cons, during the humid seasons, all trees have flowered, and the number of glomerules ranged from between 10 and 74 units (Benlahbil and Bani-aameur, 1999).

In addition, the percentage of losses expressed in number of fruits due to physiological drop (young fruits), which interrupted the process of maturation and ripe fruits ranged from 3 to 39% depending on the tree in hot and dry season (Bani Aameur et al., 1998). It appears that the adjustment of fruiting was a response to unfavorable conditions by a reduction in number of fruits on the branches. If the effects of the climatic season were manifested in the three localities by a reduction of fruiting during the second campaign, it appears that Ait Baha site was the most affected than Ait Melloul and Argana during the dry season (1st) and during the very humid season (2nd). Thus, at Ait Baha, this reaction is manifested by a very limited number of trees producing fruits (9 trees in first season, and 2 trees in second season) and a reduction in number of fruit-bearing branches. These observations confirm the findings reported by Ferradous et al. (1996) for the frequency of trees bearing fruits, weight of fruit, kernel and pulp where the effects of climatic year was perceived at Ait Baha. This station will be considered as a medium for selection of resistant genotypes to drought.

Trees in the three populations have reacted differently to seasonal variations of temperatures and rainfalls. Some individuals from Ait Baha (4 and 21), (6 and 10) of Ait Melloul and (6, 7, 11, 17) of Argana have borne the fruits in dry seasons or in humid seasons. But other trees have not borne fruits in the same conditions. Those behaviors have been observed previously since frequencies of trees that borne fruits differ mainly at Ait Melloul and Argana (Ferradous et al., 1996) confirm the impotance of genotype, in addition to seasonal variation in determining the fruiting in Argan tree. In argan, there are two categories of genotypes. Some genotypes are able to produce fruits even under unfavorable conditions. Other genotypes may only bear the fruits if temperatures and rainfalls are in favor of the flowering and ripening fruits. Trees from Ait Baha are the most affected by these changes of environmental conditions.

The relative percentage of variance related to seasonal variations in the phenotypic variance is higher than that observed for fruits characters (0.71% and 11.2%). But, this contribution related to locality and season x locality interaction was relatively low as reported by Bani-Aameur et al. (2001) (0.7 to 4.2%) except the fruit color (64.8%) and for characters of simple leaves (0.5 and 17.9%) (Zahidi, 2004). These results confirm the idea that argan tree shows a high adaptive plasticity with respect to his living environment, as has been noticed in other plant species (Sultan, 2000; Mückschel and Otte, 2003; Ait Aabd et al., 2011).

Genotype x environment interaction (season x tree / locality) contribution in the phenotypic variance is remarkable for the studied characters of the fruiting branches. This result is also observed for characters of fruit (10.4 and 14.7%) (Bani-Aameur et al., 2001). But, genotype (tree / site) contribution in the total variability is very low. These values are low compared to those obtained in sour cherry germplasm collected from the most important growing regions in Serbia. The highest degree of variability was observed number and composition of the fruiting branches, fruit set and yield (Rakonjac et al., 2010); and for characters of fruit (7.5 and 43.9%) (Bani-Aameur et al., 2001). Repeatabilities observed for the fruiting branches are much lower than those recorded for simple leaves characters except leaf dry weight (21.4 and 56.9%) (Zahidi, 2004), for fruit characters (8.02% for number of almonds and 93.28% for oil content (Bani-Aameur et al., 2001; Ait Aabd et al., 2011).

In the fruiting branches, intra-population variability (difference between trees in the same locality) is more important than inter-population variability (difference between localities). In addition, Euclidienne distance calculated based on characters of the fruiting branches is similar to that obtained for fruit and kernel (3.2) (Ferradous, 1995), but low in walnut (*Julans regia* L.) on the basis of leaf (4.6) and fruit (6.4) characters (Malvolti et al., 1994). So, differentiation of the three populations is not established since classification of individuals does not coincide with the groups that belong to the sites. This classification is not the result of geographic isolation; Ait Melloul and Argana with different climatic characteristics are not differentiated from Ait Baha. This result is confirmed by low contribution of variance related to locality (σ^2 inter-populations + σ^2geographical) in the phenotypic variance for the fruiting branches. But a large heterogeneity between trees is observed because approximately 93.3% of trees from Ait Baha, 13.3% of Ait Melloul and 6.7% of Argana are ranged in group of small producers. While about 93.3% of trees from Argana, 86.7% of Ait Melloul and 6.7% of Ait Baha (4 and 21) were among the fruit producers.

Dinis et al. (2011) suggested that annual climate conditions influence significantly the fruits and leaf characters. In addition, the morphological and phonological differences among ecotypes were not related to the small genetic differences, but were simply phenotypic adaptations to different climatic conditions. Both trees from Ait Baha, and some genotypes from Ait Melloul and Argana can produce fruits even in an arid environment will be used as germoplasm for domestication of argan as a fruit tree for oil production.

Conclusion

The main branch and the twigs more than two seasons have borne the fruits during the three campaigns, but the twigs of the season and those less two seasons have not borne the fruits at maturity even if they have flowered. The fruiting in argan tree is dependent on temperatures and rainfalls especially during the flowering season. For the establishment of orchards, the choice of efficient genotypes, pruning-lightening operations coupled with an appropriate irrigation in case of drought, at least during the flowering period should be taken into account. Differences observed for characters of the fruiting branches between trees and between localities indicate that an important genetic variation exists between individuals within each site. Ait Baha site is less far to Ait Melloul and Argana, but having some good genotypes with a high production potential even in unfavorable conditions. This variability can be exploited for the selection of desirable genotypes for breeding programme. Moreover, this result has practical implications for genetic management of resource for futur domestication programs of argan as oil-producing tree which is still in the wild state.

ACKNOWLEDGEMENTS

We gratefully acknowledge anonymous reviewers and office journal which provided helpful comments that greatly improved the manuscript. We thank the Morocco-Germany Co-operative Project 'Conservation Project and Development the argan forest' (PCDA-GTZ) and the project Pars-Agro 128 of the Morocan Ministery of Scientific Research for financial support.

REFERENCES

Ait Aabd N, El Ayadi F, Msanda F, El Mousadik A (2011). Evaluation of agromorphological variability of argan tree under different environmental conditions in Morocco: implication for selection. Int. J. Biodiver. Conserv. 3(3):73-82.

Andales A, Wang J, Sammis TW, Mexal JG, Simmons LJ, Miller DR, Gutschick VP (2006). A model of pecan tree growth for the management of pruning and irrigation. Agric. Water Manage. 84(1-2):77-88.

Bani-Aameur F (2000). Phenological phases of *Argania spinosa* (L. Skeels) flower. Forest Genetics 7:333-338.

Bani-Aameur F, Ferradous A, Dupuis P (2001). Fruit and stone variability in three argan (*Argania spinosa* (L.) Skeels) sites. Forest Genetics 8:39-45.

Bani-Aameur F (2002a). *Argania spinosa* (L.) Skeels flowering phenology. Genetic Resources Crop Evolution 49:11-19.

Bani-Aameur F, Benlahbil S (2004). Variation in RAPD markers of *Argania spinosa* trees and their progenies. Forest Genetics 11(3-4):337-342.

Benlahbil S, Bani-Aameur F (1999). Pollination of the argan tree is mostly entomophilous. In: Symposium of international plant resources: Argan and plants in arid and semi-arid areas. Faculty of Sciences, Agadir 23-25 April: pp. 119-120.

Bernstein IH, Teng GK, Garbin CP (1988). Applied Multivariate Analysis. Springer-Verlag New York Berlin Heidelberg London Paris Tokyo, P. 508.

Cuni-Shanchez A, De Smedt S, Haq N, Samson R (2011). Variation in baobab seedling morphology and its implications for selecting superior planting material. Scientia Horticulturae 130:109-117.

Dagneli P (1984). Théorie et Méthodes Statistiques. Applications Agronomiques. Tome II, 2Ed. Les Presses Agronomiques A.S.B.L. de Gembloux (Belgique), 464 pages.

Dinis LT, Peixoto F, Pinto T, Costa R, Bennett RN, Gomes-Laranjo J (2011). Study of morphological and phenological diversity in chestnut trees ('Judia' variety) as a function of temperature sum. Environ. Exper. Bot. 70(2-3):110-120.

Ehrig FR (1974). The argan character, ecology and economic maintenance of a significant tertiary in Morocco. Ptermanns geographical releases, 118(2):117-125.

Ferradous A (1995). Genetic diversity of some morphological characters of fruit and kernel of the argan tree (*Argania spinosa* (L.) Skeels). Third cycle thesis of higher education, Faculty of Sciences, University Ibn Zohr. Agadir. 180p + appendices.

Ferradous A, Bani-Aameur F, Dupuis P (1996). Stationnel climate, phenology and fruiting of the argan tree. Act. Inst. Agron. Vet. Hassan II. 17(1):51-60.

Frontier S (1981). Statistical Methods. Applied to Biology, Medicine and Ecology. Eds Masson, Paris, New York, Barcelona, Milan. P. 246.

Jamnadass R, Lowe, AJ, Dawson I (2009). Molecular markers and the management of tropical trees: the case of indigenous fruits. Trop. Plant Biol. 2:1-12.

Maurin R (1992). Argan oil, *Argania spinosa* (L.) Skeels Sapotaceae. French Rev. Fatty Substances 56:139-146.

Malvolti ME, Fineschi S, Pigliucci M (1994). Morphological integration and genetic variability in *Juglans regia* L. J. Heredity 85:389-394.

Metro A (1952). Preliminary observations on argan tree in Oued Cherrate and Dar Askraoui for genealogical selections. Ann. Forest. Rech. (Rabat), Annual Report: pp. 201-215.

M'hirit O (1989). Argan: a forest and fruit tree to multiple uses. Continuous Forest Formation, theme "argan". Forest Research Station. Rabat. 13-17 March. pp. 31-58.

Msanda F, El Aboudi A, Peltier JP (2005). Biodiversity and biogeography of the Moroccan argan tree. Cahiers Agricultures. 14(4):357-364.

Mückschel C, Otte A (2003). Morphometric parameters: an approach for the indication of environmental conditions on calcareous grassland. Agriculture, Ecosyst. Environ. 98(1-3):213-225.

Pfahler PL, Pereira MJ, Barnett RD (1996). Genetic and environmental variation in anther, pollen and pistil dimensions in sesame. Sex Plant Reprod. 9:228-232.

Prendergast HDV, Walker CC (1992). The argan: multipurpose tree of Morocco. The Kew Magazine. pp. 9:76-85.

Rahali M (1989). The production of argan forest. Continuous Forest Formation, theme "argan". Forest Research Station, Rabat, 13-17 March. pp. 31-58.

Rakonjac V, Akšić MF, Nikolić D, Milatović D, Čolić S (2010). Morphological characterization of 'Oblačinska' sour cherry by multivariate analysis. Scientia Horticulturae 125(4):679-684.

Rohlf FJ (1988). NTSYS-pc Numerical Taxonomy and Multivariate Analysis System. Version 1.80 Exeter Publ., Ltd, Setauket, New York.

Sandret F (1957). The argan pulp: Chemical composition and fodder value, change during maturation. Ann. For. Rech. (Rabat) 4:153-177.

Simbo DJ, De Smedt S, Den Bilcke NV, De Meullenaer B, Camp JV, Uytterhoeven V, Tack F, Samson R (2012). Opportinities for domesticating the African baobab (*Adonsonia digitata* L.): multi-trait fruit selection. Agroforest Syst, DOI: 10.1007/s10457-012-9568-7.

Sokal RR, Rohlf FJ (1995). Biometry, the Principales and Practice of Statistics in Biological Research. 3Ed. W. H. Freeman and Company New York. P. 887.

Steel RGD, Torrie JH (1960). Principles and Procedures of Statistics. 1Ed. McGraw-Hill Book Company, Inc. New York Toronto London. P. 481.

Sultan SE (2000). Phenotypic plasticity for plant development, function and life history. Plant Sci. 5(12):537-542.

Tworkoski TJ, Glenn DM (2010). Long-term effects of managed grass competition and two pruning methods on growth and yield of peach trees. Scientia Horticulturae 126(2):130-137.

Zahidi A, Bani Aameur F, Dupuis P (1995). Characterization of branching of the argan tree. Acts of International Symposium: Forest Facing Desertification: the case of argan. Faculty of Sciences Agadir:pp. pp. 36-52.

Zahidi A (1997). Phenology, typology and genetic variability of branching and foliation traits *Argania spinosa* (L.) Skeels. Third cycle thesis, Faculty of Sciences, University Ibn Zohr. Agadir. 177p + appendices.

Zahidi A, Bani-Aameur F (1999a). Leaf phenology of *Argania spinosa*. In: Symposium of international plant resources: Argan and plants in arid and semi-arid areas. Faculty of Sciences, Agadir 23-25 April: pp. 173-176.

Zahidi A, Bani-Aameur F (1999b). Phenology of branching in argan tree. In: Symposium of international plant resources: Argan and plants in arid and semi-arid areas. Faculty of Sciences, Agadir 23-25 April: pp. 177-179.

Zahidi A (2004). Genetic diversity of leaf, growth and branching of stem and root in *Argana spinosa* (L.) Skeels seedlings. Doctorate Thesis. Faculty of Sciences, University Ibn Zohr. 180p + appendices.

Zunzunegui M, Ain-Lhout F, Jáuregui J, Díaz Barradas MC, Boutaleb S, Álvarez-Cansino L, Esquivias MP (2010). Fruit production under different environmental and management conditions of argan, *Argania spinosa* (L.) Skeels. J. Arid Environ. 74:1138-1145.

Growth variability in *Argania spinosa* seedlings subjected to different levels of drought stress

Zahidi A. , Bani-Aameur F. and El Mousadik A.

Laboratory of Biotechnologies and Valorization of Naturals Resources, Faculty of Sciences, Ibn Zohr University, BP 8106 Agadir 80000, Morocco.

Argan tree, the only representative species of the tropical family Sapotaceae in Morocco is distributed in arid and semi-arid areas. Tolerance to drought remains poorly described for this species; we applied five levels of drought stress and monitored growth variables and biomass production of seedlings in pot and in field propagated from seeds of eight genotypes from three geographical origins. Drought stress decreased seedlings height, basal diameter, leaf number, leaf area, and biomass production and water content but increased root / shoot length. Even under moderate or severe drought stress (1/4 and 0 FC), some genotypes (mother-trees) sustained shoot growth, they have taproots exceeding 50 cm, which are accompanied by a large number of lateral roots, more leaves and great leaf areas. But others showed an increase below-ground length and biomass indicating a higher root / shoot ratio under drought stress conditions. So, in order to enhance the survival rate and conserving rate of seedlings planting, appropriate human intervention is required to reduce the damage to Argan seedlings resulting from drought.

Key words: *Argania spinosa*, water stress, aridity, shoot, root, seedling biomass production.

INTRODUCTION

Argan tree, the only representative species of the tropical family Sapotaceae in Morocco is distributed in arid and semi-arid areas. It plays major roles against soil erosion, desertification and preservation of biodiversity and in the daily life of local populations (Emberger, 1924; M'Hirit et al., 1998; Msanda et al., 2005). Currently, it is proven that the adult tree is drought resistant, survives in its area for many years (Emberger, 1925, 1955; Boudy, 1952) and produces branches, leaves, flowers and fruit with a rainfall exceeding 100 mm (Ferradous et al., 1996; Zahidi, 1997a, b; Bani-Aameur, 2001). In this ecological environment quite fragile, degradation of Argan is a dynamic process progressive owing to increasing population, high demand for agricultural and overgrazing. Therefore, conservation and regeneration of this tree are a regional priority for protection against the increased desertification (Bani-Aameur, 2007). Argan tree is regenerated by seed and release. Regeneration by seedlings allows preservation of genetic diversity of this species. However, fruit harvesting, livestock grazing of few seedlings issued from germination of some remaining kernels (Boudy, 1950, 1952), action of aridity on seed germination and difficult conditions of seedlings survival after germination are factors limiting natural regeneration from seed (Zahidi and Bani-Aameur, 1998; Bani Aameur and Alouani, 1998).

Artificial regeneration is the only possibility that can ensure the survival and maintain an appropriate level of genetic diversity of the species. If germination has been very successful, since percentage of germinated seeds was more than 80% (Bani-Aameur and Alouani, 1999), testing in field of transplanted seedlings grown in nursery was always doomed to failure. Traditionally, observed reforestation failures can be explained by the difficulty of seedling establishment after transplantation. Some techniques to improve transplantation rate have been tested on a small scale (Harrouni et al., 1995). They sometimes lead to improved success rates; however, they will remain below an acceptable level. In arid or semi-arid areas, during periods of drought, most plants are likely to be under stress, but differences in the degree of stress can have important implications for plant survival. One of the most general types of stress experienced by plants is water limitation. As a result of water limitation, plants express various responses and have developed a wide diversity of drought tolerance mechanisms from both morphological and physiological aspects (Blum, 1983; Loss and Siddique, 1994; Li et al., 2007). In field, the very definition of drought accepted by many authors includes a reduction in plant growth due to decrease in average precipitation amounts (Tucker and Goward, 1987). Water shortage can cause plants to reduce their metabolic activity, causing a decrease in photosynthesis, carbon fixation and ultimately growth (Younis et al., 1993; Li et al., 2007). Some studies have shown that drought stress can affect the growth of plant organs differently, which may result in the alteration of the morphological features of the plants (French and Turner, 1991).

The change in root to shoot dry mass ratio has been considered as one of the mechanisms involved in the adaptation of plants to drought stress. In order to diminish consumption and increase absorption of water, plants in dry conditions often decrease their growth rate and biomass production, and contribute more biomass to roots, so that they could maintain a higher root to shoot ratio (Yin et al., 2005; Villagra and Cavagnaro, 2006). Drought stress reduces both root and shoots growth. However, root growth seems to be less affected. Drought stress often leads to a decrease in leaf dry mass ratio (leaf dry mass/plant total dry mass) in many species (Turner, 1997). Several studies have concerned the adult tree and reported that Argan is xerophytic and can support low and irregular rainfall (Boudy, 1952; Ferradous et al., 1996). After a prolonged drought biomass production stops and the tree loses completely leaves, flowers and fruits (Bani-Aameur, 1997; Zahidi, 1997). Stomatal regulation by closure of stomata contributes very little to avoid drought (El Aboudi et al., 1991). Very few studies have focused on seedlings, according to Zahidi and Bani-Aameur (1997a, b) root length can reach 27 cm before the cotyledons opening. In seedlings grown in nursery conditions for one year, root length varied between 138.4 and 10.7 cm, stem height between 69.3 and 6.5 cm. Seedlings grown for 16.5 months were longer four times than those grown for 4.5 months, but nine times greater than those grown only for 3 months. After six months, seedlings irrigated daily at field capacity showed a gain of 9.7 cm in stem height (Harrouni et al., 1995). The decrease in water regime leads to a reduction in seedling leaf number.

In addition, in non-irrigated seedlings, root was twice longer than those of irrigated seedlings after nine months (Kaabouss, 1992; Harrouni et al., 1995). However, to ensure extension and conserving genetic diversity, especially in areas where drought can be a very important environmental factor which limit the increase of survival rate of transplanted seedlings; evaluating seedlings responses to various degrees of water stress is the key to speed up forest restoration of Argan. No specific information is available on the tolerance of seedlings to drought or on variability of growth in the difficult conditions of their natural environment.

Our objective is to test morphological responses of Argan seedlings issued from eight selected genotypes grown for 14 months in nursery and in field subjected to five water regimes.

MATERIALS AND METHODS

Plant material and growth conditions

Seeds of eight selected trees of *Argania spinosa* were collected from three geographical origins; that is, Ait Melloul Argana and Ait Baha with ecological characterization (Bani-Aameur and Zahidi, 2005; Zahidi et al., 2013). Before germination, fruit were kept in cold for one month, and then scarified as described by Bani-Aameur and Alouani (1999) before sowing (March 20 first season) in vats containing sand in a mini-greenhouse. After emergence, seedlings were transplanted (May 20 first season) in pots containing 1/3 of peat, 1/3 of sand and 1/3 of Argan loam and placed in open air at the Faculty of Sciences, Agadir. Transplantation was done with five seedlings per pot reduced to two seedlings per pot after two months of seedling growth.

In early summer (June 6th first season), seedlings were irrigated daily by five water regimes [1 field capacity (1 FC = 200 ml), 3/4 FC (150 ml), 1/2 FC (100 ml), 1/4 FC (50 ml) and 0 FC (0 ml)].

Measurements and calculations

At the end of each month, seedling height or stem length (LT cm) and basal diameter (DT cm) were recorded. The seedlings were harvested at the end of the experiment (June 6th second season), the following characters were measured: LT: Seedling height (cm), NF: leaf number on the main stem, NE: spines number on the main stem, RS: number of secondary on the main stem, RT: tertiary shoots on the main stem, LR: root length (cm), RC: number of lateral roots. Each seedling was then divided into roots, stems and leaves and they were dried in an oven for at least 48 h at 70°C to constant weight for biomass determination (g). PFT: Fresh stem mass, PST: dry stem mass, PFR: fresh root mass, PSR: dry root mass, RPF: root / shoot fresh mass ratio, RPS: root / shoot dry mass ratio, PFF: leaf fresh mass, PSF: leaf dry mass, FWC: leaf water content, TWC: stem water content and RWC: root water content, were determined by difference between fresh and dry mass as follows

$$FWC = PFF - PSF; \quad TWC = PFT - PST; \quad RWC = PFR - PSR$$

In order to determine leaf area (SF in cm^2), the first five units were observed in April 9th (second season). Leaves were photocopied on paper, surface of 1 cm^2 of paper (P) and the image of leaf (Fi) was weighed in grams. Leaf area was calculated using the specific

Table 1. Variance analysis for stem diameter and stem length in Argan seedling grown for 14 months under five water regimes.

Source of variation	DF	Mean square	
		Stem diameter (cm)	Stem lenght (cm)
Block	3	0.017ns	205.74**
Water regime	3	5.72**	27571**
Duration of regime	10	2.732**	6783.3**
Mother-tree	7	0.457**	4027.2**
Water regime × duration	30	0.05**	172.74**
Water regime × mother-tree	21	0.095**	536.58**
Duration of regime × mother-tree	70	0.001ns	7.63ns
Water regime × duration of regime × mother-tree	210	0.002ns	10.62ns
Water regime × duration of regime × mother-tree × seedling	352	0.98ns	42.18ns
Error	2096	0.008	53.31

DF: Degree of freedom; ns: not significant; *: significant at 5%; **: significant at 1%.

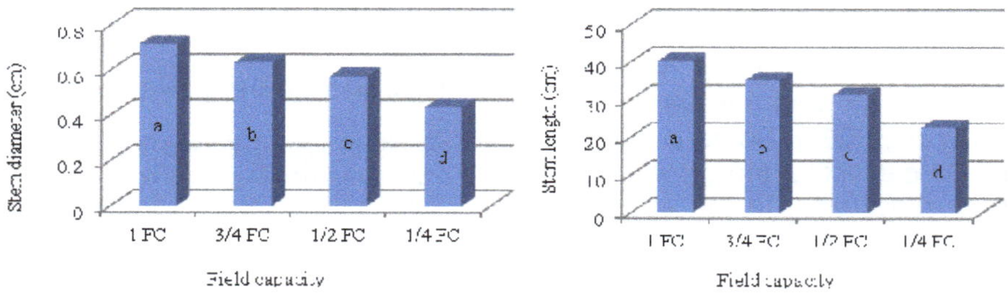

Figure 1. Stem diameter and stem length in Argan seedling grown for 14 months under five water regimes (letters indicate significant differences using LSD comparison).

surface area as proposed by Mosseddaq (1988) (S = Fi / P).

Field test

Preparation of seedlings is the same as the previous test. After germination, seedlings were grown in three types of containers measuring 17 cm in height: plastic sachets, jute sachets and seedling trays as is the case of forestry nurseries. Seedlings issued from eight genotypes grown in nursery for seven months at the Faculty of Sciences, Agadir were subjected to five water regimes [1 field capacity (1 FC = 200 ml), 3/4 FC (150 ml), 1/2 FC (100 ml) and 1/4 FC (50 ml) and 0 FC (0 ml)]. Those seedlings were then transplanted in two study plots located in the communal forests belonging to two adjacent cities: Ait Hamadi (Ouled Teima in February 2 second season) near Agadir and in Onagha near Essaouira (February 9 second season) (Water and Forests Services). Seedling height (LT) and basal diameter (DT) were recorded five times: on July 19, August 19, September 19, October 19, November 19 the second season, respectively.

Statistical analysis

In the first test, we adopt randomized complete block as experimental design. In each of the four blocks, 40 combinations (five water regime and eight varieties) are randomly distributed. Each pot is a variety and a water regime with two replications. In the second test, the experimental design is a Split-plot (Vessereau, 1988). Main water regime consists of eight mother-tree seed sources distributed in 36 plots and randomly allocated to each of the two blocks. The type of container in which Argan seedlings were grown is subsidiary water regime. Analysis of variance with four factors is adopted. All factors (block, mother-tree, water regime and observation date) were crossed. When significant differences were noted, LSD (least significant difference method) test was used to determine differences (Dagnelie, 1984). Calculations were performed using Statistix software.

RESULTS

During the pot test

Seedlings under 0 FC (field capacity) (0 ml) have dried completely after one month of stopping irrigation.

Stem characters

Water regime: Water regime is highly significant for the stem diameter and length during the test duration and at the end of the experiment (Table 1). The decrease in water amount leads to a reduction of stem diameter and stem length. Values varied respectively between 0.72 and 40.2 cm at field capacity (1 FC) to 0.43 and 22.5 cm at 1/4 FC (Figure 1).

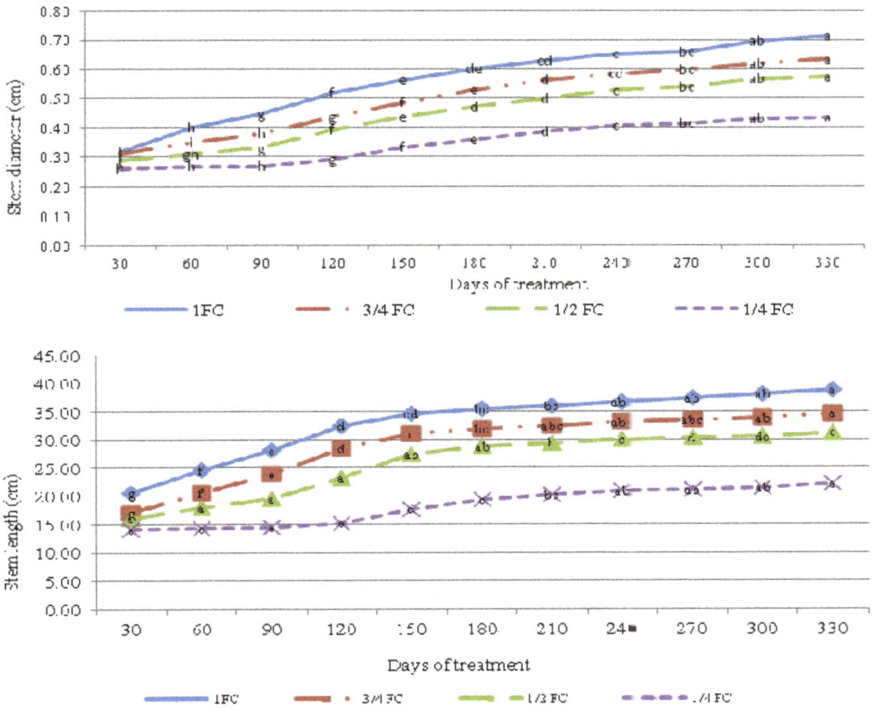

Figure 2. Stem diameter and length in Argan seedling grown for 14 months under five water regimes.

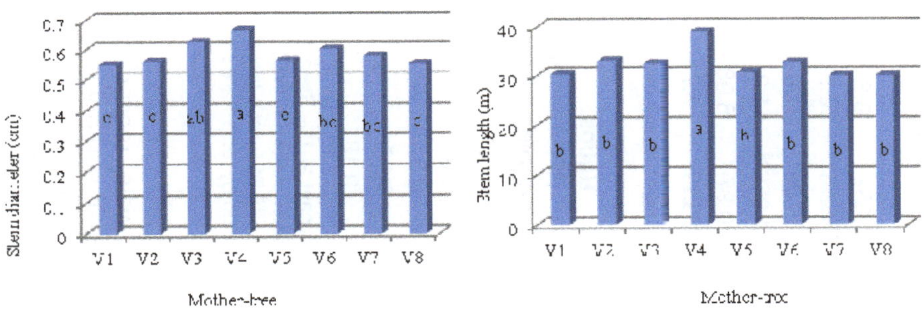

Figure 3. Stem diameter and length (cm) in Argan seedling grown for 14 months under five water regimes by mother-tree.

Duration of water regime: Duration of water regime and water regime × duration interaction influenced significantly stem diameter and length in Argan seedlings (Table 1). In seedling grown for 14 months, stem diameter remains very low at ¼ FC and ranged from 0.26 to 0.44 cm, whereas it increases in the other three water regimes and reaches 0.7 cm. Stem length increases rapidly during the first five months for water regimes 1, 3/4 and 1/2 FC and becomes slower after the fifth month (Figure 2). For 1/4 FC, stem length remains stable during the first four months, seedlings remain dwarf since length reaches only 22.5 cm at the end of the experiment.

Mother-tree: Mother-tree × water regime interaction and mother-tree were highly significant for the stem diameter and length in Argan, showing genetic variability for seedling reaction to water stress (Table 1). Diameter varied from 0.55 cm in seedlings of family V1 and V2 to 0.67 cm in family V4 (Figure 3). Stem length varied from 39.1 cm in seedlings of mother-tree V4 and 30.03 cm for V7 and V8.

At the end of experiment

Water regime

Water regime was highly significant for stem diameter

Table 2. Variance analysis for the stem, root and leaf characters in Argan seedlings grown for one year subjected to four water regime.

Source of variation	DF	Mean square												
		Stem characters							Root characters					
		DT	LT	RS	RT	PFT	PST	TWC	LR	RC	PFR	PSR	RWC	
Block	3	0.015^{ns}	63.32^{ns}	8.86^{ns}	3.46^{ns}	7.73^{ns}	1.74^{ns}	2.14^{ns}	1119.4^{ns}	4.80*	2.17^{ns}	0.23^{ns}	1.13^{ns}	
Water regime	3	0.886**	3582.1**	12.50^{ns}	3.22^{ns}	343.1**	112.17**	63.35**	7925.6**	11.28*	966.16**	266.10**	219.17**	
Mother-tree	7	0.054**	289.95**	24.02**	0.80^{ns}	44.07**	19.4**	5.14**	3824.1*	0.75*	6.56*	2.76*	1.14*	
Water regime × mother-tree	21	0.016^{ns}	74.056^{ns}	11.31^{ns}	1.49^{ns}	11.07*	4.49*	1.56^{ns}	1648.1^{ns}	3.26**	10.63^{ns}	3.09^{ns}	2.55^{ns}	
Water regime × mother-tree × seedling	32	0.019^{ns}	50.97^{ns}	6.98^{ns}	1.43^{ns}	5.56^{ns}	2.17^{ns}	0.82^{ns}	1312.8^{ns}	1.11^{ns}	9.76^{ns}	2.65^{ns}	2.38^{ns}	
Error	173	0.014	73.27	7.71	1.67	6.32	2.41	0.97	1476.6	1.63	10.128	3.14	2.11	

Source of variation	DF	Mean square							
		Ratios			Leaf characters				
		RPF	RPS	RL (LR/LT)	NF	PFF	PSF	FWC	NE
Block	3	0.02^{ns}	0.04^{ns}	0.82^{ns}	10640^{ns}	1.99^{ns}	0.42^{ns}	0.78^{ns}	2247.2**
Water regime	3	2.93**	1.64**	4.64**	1654400**	146.69**	20.20**	58.09**	10514**
Mother-tree	7	2.66**	2.33*	7.61**	37321*	1.74^{ns}	0.26^{ns}	0.74^{ns}	1371.8**
Water regime × mother-tree	21	0.39*	0.31*	1.32^{ns}	64615*	1.68^{ns}	0.24^{ns}	0.77^{ns}	300.88^{ns}
Water regime × mother-tree × seedling	32	0.20^{ns}	0.17^{ns}	1.22^{ns}	34657^{ns}	1.93^{ns}	0.32^{ns}	0.77^{ns}	166.38^{ns}
Error	173	0.21	0.18	1.52	36343	1.41	0.26	0.52	274.23

DF: Degree of freedom, ns: not significant, *: significant at 5%; **: significant at 1%. Stem diameter (DT), stem length (LT), number of secondary on the main stem (RS), number of tertiary shoots on the main stem (RT), fresh stem mass (PFT), dry stem mass (PST), stem water content (TWC); root length (LR), number of lateral roots (RC), fresh root mass (PFR), dry root mass (PSR), root water content (RWC); root / shoot fresh mass ratio (RPF), root / shoot dry mass ratio (RPS), ratio length (RL = LR/LT); leaf number on the main stem (NF), leaf fresh mass (PFF), leaf dry mass (PSF), leaf water content (FWC) and spines number on the main stem (NE); Means followed by letters are significant..

and length, for spines number (EP), fresh (PFT), dry biomass weight (PST) and stem water content (TWC) at the end of experiment (Table 2). It was not significant for number of secondary, tertiary shoots on the main stem. Drought stress diminished biomass production, stems water content and reduced spines number (Figure 4).

Mother-tree effect

Water regime × mother-tree × seedling interaction was not significant for all stem traits. Mother-tree × water regime interaction was significant only for fresh and dry mass of the stem, which shows the individual response of each seedling to each water regime. Mother-tree factor was significant for all characters, except number of tertiary shoots (Table 2). Great variability was observed between all genotypes (mother-trees) for these characters. Dry mass varied from 4.5 g for mother-tree V4 to 2 g in seedlings of V1. The fresh mass in genotype V4 (7 g) was twice larger than that in seedlings from V1 (3.2 g). Stem water content varied from 2.5 g for mother-tree V4 to 1.2 g in seedlings of V1. Number of secondary shoots varied between 4.3 for mother-tree V3 to 1.5 in seedlings of V1 (Figure 5). Spines number varied from 62.6 in mother-tree V4 and 42.9 in seedling from mother-tree V1.

Root characters

Water regime

Water regime is highly significant for root length (LR), number of lateral roots (NRL), fresh (PFR), dry mass (PSR), root water content (RWC), fresh

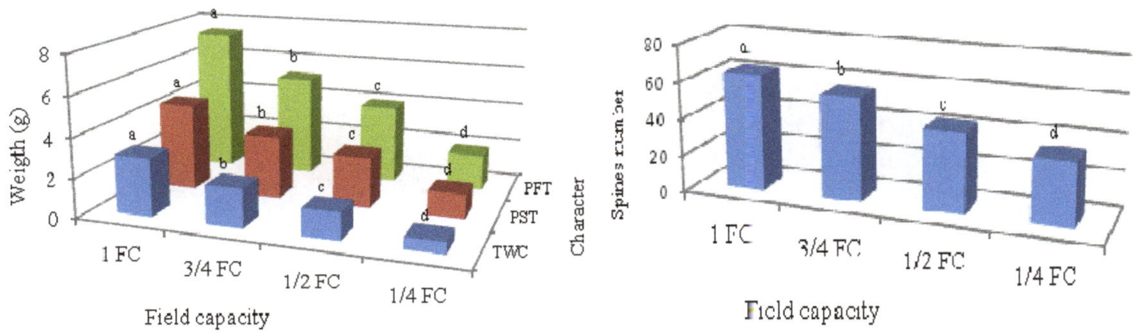

Figure 4. Fresh, dry weight and stem water content and spines number in Argan seedlings grown for 14 months by water regime.

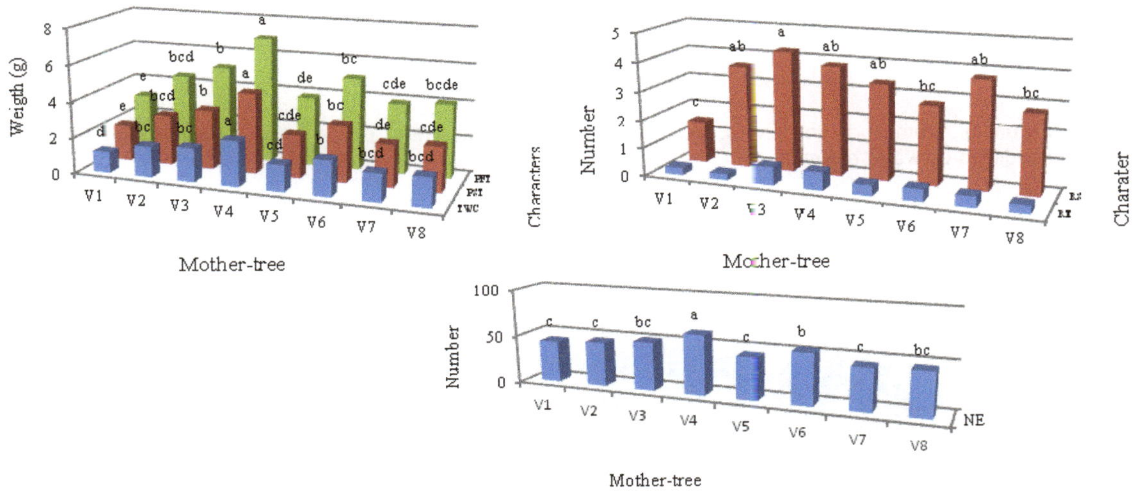

Figure 5. Fresh, dry mass, water content and number of secondary, tertiary shoots and spines on the main stem by mother-tree.

(RPF) and dry (RPS) mass ratio and length ratio (RL) (Table 2). Root length varied from 49.9 cm at ¼ FC to 73.4 cm at 1 FC, it is about twice the stem length (Figure 6). The decrease in the amount of irrigation water leads to a reduction in lateral roots number, root water content and root fresh mass. The fresh mass is at field capacity six times greater decreases also than that obtained in ¼ FC. Root dry biomass at field capacity is five times larger than that recorded at ¼ FC. The fresh and dry weight ratios varied from 1.7 and 1.6 at field capacity to 1.1 and 1.2 at 1/4 FC. Root / stem ratio of length (RL) varied from 2.4 at ¼ FC to 1.9 at 1 FC.

Mother-tree

Mother-tree × water regime was not significant for all root characters except number of lateral roots. This shows genetic variability in Argan seedling for water stress. Mother-tree × water regime × seedling interaction is not significant for all root traits. The production of root fresh, dry biomass and water content depends only on water regime. Mother-tree factor was significant for root length (LR), root / stem ratio of length (RL), fresh (RPF) and dry (RPS) mass ratio (Table 2). Thus, seedlings from genotype V1 have longer roots (82.3 cm) than seedlings from V4 which shows the shortest roots (53.13 cm) (Figure 7). Water stress induces the development of root to allow the seedling in search of water in depth. Thus, root / stem ratio varied from 2.9 in seedlings from mother-tree V1 to 1.4 in descendants of mother V4. Fresh and dry mass ratio varied from 2 and 1.9 in seedlings from V1 to 0.92 and 0.89 for descendants of V4.

Leaf characters

Water regime

Water regime is highly significant for leaf number on the

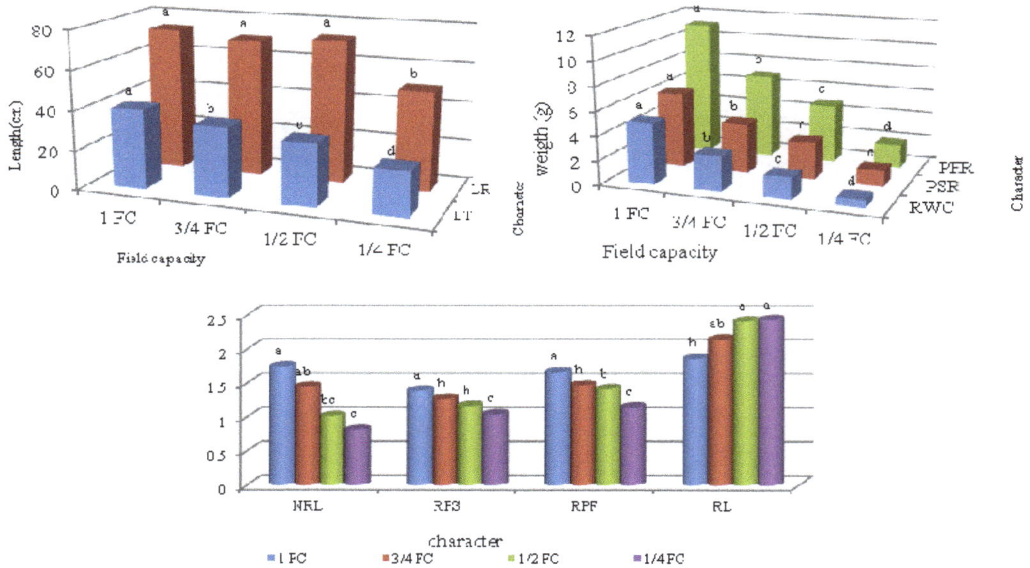

Figure 6. Stem and root length, fresh, dry mass and root water content and their ratios in Argan seedlings by water regime.

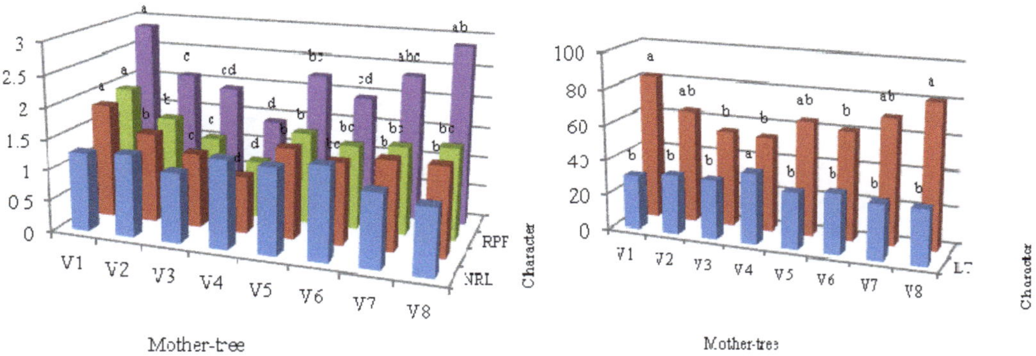

Figure 7. Root and stem length, root and stem ratio of length, fresh, dry mass and number of lateral roots in Argan seedlings by mother-trees.

main stem (NF), fresh, dry mass and leaf water content (Table 2). At field capacity, leaf number (584.6) is three times higher than that obtained at 1/4 FC (201.8) (Figure 8). Water stress induces a significant reduction of leaf biomass, since fresh weight at field capacity (4 g) was eight times greater than that obtained at 1/4 FC (0.5 g). Leaf dry mass does not exceed 0.3 g at 1/4 FC while it reaches 1.6 g at field capacity. Water content at field capacity (2.4 g) is twelve times greater than that obtained at 1/4 FC (0.2 g).

Mother-tree

Water regime × mother-tree × seedling interaction was not significant for leaf characters. Mother-tree and water regime × mother-tree were significant for leaf number but not significant for the three other characters (Table 2). Leaf number varied between 437.1 in seedlings of V4 and 319.3 in seedlings of V5. Fresh and dry biomass varied from 1.7 g, 0.78 g in seedlings of V5 and 2.36 g, 1.01 g in seedlings of V3. Leaf water content varied from 0.91 in seedlings of V5 and 1.34 in seedlings of V3 (Table 3).

Field test

Before transplanting

Seedling, mother-tree × seedling, water regime × seedling and water regime × mother-tree × seedling were

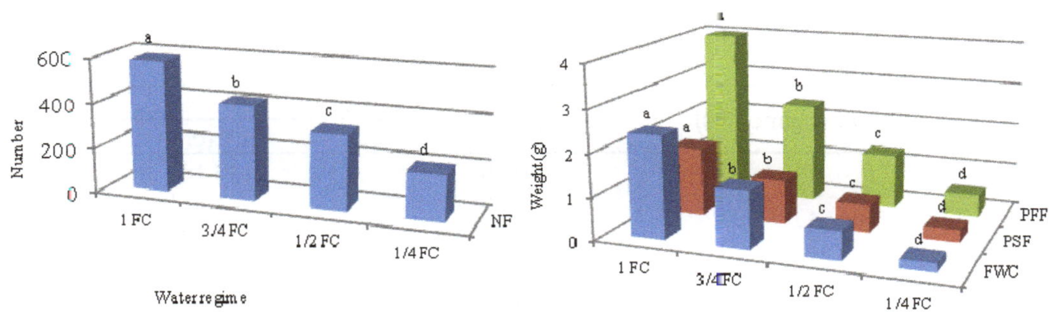

Figure 8. Leaf number on the main stem, fresh, dry mass and water content by water regime.

Table 3. Fresh (PFF), dry mass (PSF), leaf water content (FWC) and leaf number by mother-tree.

Mother-tree	PSF	FWC	PFF	NF
V1	0.82	0.99	1.81	380.4^{bc}
V2	0.95	1.16	2.11	413.1^{ab}
V3	1.01	1.34	2.36	379^{bc}
V4	1.01	1.29	2.3	437.1^{a}
V5	0.78	0.91	1.7	319.3^{d}
V6	0.88	1.26	2.14	388.1^{b}
V7	0.81	1.06	1.87	369.4^{c}
V8	0.85	1.21	2.06	390.4^{b}

Means followed by letters are significant.

Table 4. Variance analysis for seedling height (LT), basal diameter (DT) under five water regimes grown in three containers for seven months.

Source of variation	DF	Average	
		DT	LT
Block	3	0.005^{ns}	2.09^{ns}
Mother-tree	7	0.029*	330.95**
Water regime	4	0.56*	67.77*
Seedling	1	0.007^{ns}	25.34^{ns}
Water regime × mother-tree	28	0.005^{ns}	19.04*
Mother-tree × seedling	7	0.003^{ns}	4.02^{ns}
Water regime × seedling	4	0.001^{ns}	2.43^{ns}
Water regime × mother-tree × seedling	28	0.003^{ns}	7.11^{ns}
Error	236	0.004	10.79

ns: not significant; *: significant at 5%; **: significant at 1%.

not significant for stem diameter (DT) and stem length (LT). Water regime × mother-tree is significant for LT but not significant for DT. Mother-tree and water regime were significant for the two characters of seedlings grown for seven months under five water regimes in three containers (Table 4). Water stress causes a decrease in DT from 0.45 cm at field capacity to 0.21 at 0 FC. Stem length varied from 16.15 cm at 1 FC to 13.44 cm at ¼ FC (Table 5). Stem length and diameter varied in large proportions between mother-trees. Values ranged respectively between 0.29, 10.92 cm in seedlings of V8 and 0.37, 20.22 cm in seedlings of V4 (Table 6). Great variability in leaf area was observed between mother-trees and between water regimes for the same mother-tree. Values varied between 0.43 cm^2 in seedlings of V7 at 1 FC and [illegible].37 cm^2 at 0 FC in seedlings of V3 (Table 6).

After transplanting

Water regime was significant for stem diameter (DT), but

Table 5. Seedling height (LT) and stem diameter (DT) in Argan seedlings grown for seven months under five water regimes.

Water regime (FC)	Average	
	DT (cm)	LT (cm)
1	0.45[a]	16.15[a]
¾	0.38[b]	14.44[b]
½	0.33[c]	14.24[b]
¼	0.26[d]	13.44[b]
0	0.21[e]	13.90[b]

Means followed by letters are significant.

Table 6. Seedling height (LT) and stem diameter (DT) in Argan seedlings grown for seven months under five water regimes by mother-tree.

Mother-tree	Average	
	DT	LT
V1	0.31[cd]	12.49[cd]
V2	0.3[cd]	14.17[c]
V3	0.36[ab]	16.71[c]
V4	0.37[a]	20.22[a]
V5	0.31[cd]	14.22[c]
V6	0.33[bc]	13.97[c]
V7	0.32[cd]	12.76[cd]
V8	0.29[d]	10.92[d]

Means followed by letters are significant.

Table 7. Variance analysis for stem diameter (DT), stem length (LT) in Argan seedlings transplanted in two study plots (Ouled Teima and Essaouira).

Source of variation	DF	Mean square	
		LT (cm)	DT (cm)
Block	3	35.96[ns]	1.1[ns]
Mother-tree	7	2071.5**	1.31[ns]
Water regime	4	1125.7**	4 96**
Observation date	4	3769**	1.23[ns]
Mother-tree × water regime	28	178**	1.08[ns]
Mother-tree × observation date	28	3.8[ns]	1.03[ns]
Water regime × observation date	16	455.73**	1.06[ns]
Mother-tree × water regime × observation date	112	7.73[ns]	1.02[ns]
Error	1387	30.5	1.04

Ns: Not significant, **: significant at 1%.

mother-tree, observation date, mother-tree × water regime, mother-tree × observation date and mother-tree × observation date × water regime interactions were not significant for this character. Water regime, mother-tree, observation date, mother-tree × water regime, water regime × observation date interactions were significant for stem length (LT). Mother-tree × observation date and mother-tree × water regime × observation date interactions were not significant for this character (Table 7). The reducing of water availability causes a reduction of stem length since values are from 28.03 cm at field capacity to 14.08 cm at 0 FC. Stem diameter varied from 0.52 cm at ¾ FC and 0.21 cm at 0 FC (Table 8). Stem diameter and length varied in large proportions between families. Values ranged respectively between 0.3 cm, 17.44 cm in seedlings of V8 and 0.38 cm, 27.27 cm in seedlings of V4 (Table 9). Values reached in seedlings in field test remained lower than pot test even when they

Table 8. Stem length (LT), stem diameter (DT) in Argan seedling of eight mother-trees grown in field and subjected to five water regime.

Water regime (FC)	Average	
	LT (cm)	DT (cm)
1	28.03[a]	0.45[ab]
¾	24.17[b]	0.52[a]
2/4	20.8[c]	0.35[bc]
¼	15.09[d]	0.28[c]
0	14.08[e]	0.21[c]

Values followed by letters are significantly different.

Table 9. Stem length (cm) and stem diameter (cm) by mother-tree having undergone five water regimes.

Mother-tree	Average	
	LT (cm)	DT (cm)
V1	17.55[d]	0.32[c]
V2	20.29[c]	0.31[c]
V3	22.44[b]	0.37[a]
V4	27.27[a]	0.38[a]
V5	20.04[c]	0.33[bc]
V6	19.9[c]	0.35[b]
V7	18.43[d]	0.33[bc]
V8	17.44[d]	0.3[c]

Values followed by letters are significantly different.

were grown for the same period about 14 months (Table 10). Seedling, water regime × mother-tree, mother-tree × seedling, water regime × seedling and water regime × mother-tree × seedling were not significant for leaf area (SF). Mother-tree and water regime were significant for leaf area in Argan seedlings grown for seven months and subjected to five water regime (Table 11).

Leaf area decrease with water deficit, it varied between 0.71 cm^2 at 1 FC and 0.66 cm^2 except at 0 FC (0.76 cm^2) (Table 12). Some genotypes as V3 and V4 have greater leaf areas than seedlings from V7, V8 and V1.

DISCUSSION

The mechanisms adopted by Argan seedlings to withstand drought are still largely unknown. Here, we report data of experiments aimed at investigating the effect of water stress on Argan seedlings growth issued from eigth mother-tree originating from three populations in south west Morocco. In our study, the growth reduction in both seedling height and basal diameter of the stem was observed under moderate (3/4 FC) and severe stress (1/4 or 0 FC). The same result is obtained by Kaabouss (1992) since the stem length of Argan seedlings subjected to water stress did not exceed 22 cm after 160 days of transplantation. In seedlings of *Adenanthera pavonina*, Paliwal and Kannan (1999) showed that water deficit causes a reduction of the stem height. This result suggests that even at reduced soil water availability, the Argan seedlings are able to grow. Our results agree with the findings of Achten et al. (2010) in *J. curcas* since under medium stress (40% plant available water) the plants maintained a similar stem shape, although they grew at lower rate (stem length: 0.28 cm / day; dry biomass production: 0.64 g / day). Seedlings under extreme drought stress (no irrigation) stopped growing, started shedding leaves and showed shrinking stem diameter from the 12th day after the start of the drought treatment. This result showed that water withhold would arrest growth but maintaining plants at low soil water availability (40%) would allow them to continue growing, although at a slower rate than fully irrigated plants (Achten et al., 2010; Sapeta et al., 2013).

Decrease in irrigation level leads to a low level of shoot (secondary and tertiary shoots) and spines production which agree results obtained in adult trees since branching is very low in dry season than in wet season. In addition, the production of spines is not a response to drought (Zahidi, 1997). Drought stress leads also to the decrease of stem fresh mass, dry mass and stem water content in argan seedlings at the end of experiment.

Table 10. Stem length (cm) and stem diameter (cm) in transplanted Argan seedlings grown under five water regimes.

Observation date	Average	
	LT (cm)	DT (cm)
July 19	16.3^{e}	0.27
August 19	18.27^{d}	0.43
September 19	20.01^{c}	0.32
October 19	22.67^{b}	0.37
November 19	24.92^{a}	0.4

Values followed by letters are significantly different.

Table 11. Variance analysis for leaf area (SF) in Argan seedlings transplanted in two study plots (Ouled Teima and Essaouira).

Source of variation	DF	SF
Block	3	0.34^{ns}
Mother-tree	7	6.43**
Water regime	4	0.65*
Seedling	1	0.04^{ns}
Water regime × mother-tree	28	0.53^{ns}
Mother-tree × seedling	7	0.37^{ns}
Water regime × seedling	4	0.11^{ns}
Water regime × mother-tree × seedling	28	0.33**
Water regime × mother-tree × seedling × leaf (error)	1506	0.19

Ns: Not significant, *: significant at 5%, **: significant at 1%.

Table 12. Variation of leaf area by mother-tree and water regime.

Mother-tree	1 FC	3/4 FC	2/4 FC	1/4 FC	0 FC	Average
V1	0.63^{ab}	0.67^{a}	0.64^{ab}	0.56^{bc}	0.48^{c}	0.59^{cd}
V2	0.71^{a}	0.54^{b}	0.51^{b}	0.64^{a}	0.54^{b}	0.58^{cd}
V3	1.07^{ab}	0.82^{b}	0.77^{b}	0.99^{ab}	**1.35**^{a}	1.03^{a}
V4	0.93^{a}	0.88^{b}	0.85^{c}	0.87^{b}	0.96^{a}	0.89^{ab}
V5	0.79^{a}	0.64^{c}	0.77^{ab}	0.66^{bc}	0.8^{a}	0.73^{b}
V6	0.61^{cd}	0.64^{c}	0.91^{a}	0.51^{d}	0.84^{b}	0.7^{bc}
V7	**0.43**^{d}	0.48^{cd}	0.61^{a}	0.51^{bc}	0.57^{ab}	0.52^{d}
V8	0.55^{a}	0.46^{b}	0.45^{b}	0.57^{a}	0.51^{ab}	0.5^{d}
Average	0.71^{ab}	0.64c	0.68b^{c}	0.66b^{c}	0.76^{a}	0.69

Values followed by letters are significantly different.

Drought stress obviously diminished the biomass and their components, and reduced shoot percentages in *S. davidii* seedlings (Wu et al., 2008). Additionally, photosynthesis and growth (biomass production) are the primary processes to be affected by drought (Chaves and Oliveira, 2004; Sapeta et al., 2013). Drought not only changed plant growth and structure [shoot height (Ht), total biomass (Tb)] in *P. davidiana*, but also affected plant physiological properties and constitutes a very important limiting factor at the initial phase of seedling growth and establishment (Zhang et al., 2004). Drought changed also root growth, root to shoot length which is twice higher under sever stress (at 1/4 FC). Thus, water stress induces the development of root to allow the seedling in search of water in depth.

The decrease in the amount of irrigation water leads to a reduction in lateral roots number, fresh, dry mass, root water content, root to shoot fresh and dry mass ratio. Our results agree with the finding in other species, Wu et al. (2008) showed that drought stress diminished the root biomass and their components, but increased below-ground percentages. Drought stress significantly decreased plant total dry mass, but the proportion of changes differed among root and stem. The change in root to shoot dry mass ratio has been considered as one of the mechanisms involved in the adaptation of plants to drought stress (Turner, 1997). So, in order to diminish consumption and increase absorption of water, plants in dry conditions often decrease their growth rate and biomass production, and contribute more biomass to roots (Martin and Stephens, 2006; Wu et al., 2008).

Additionally, the importance of root systems in acquiring water has long been recognized as crucial to cope with drought conditions in acquiring water. A prolific root system can confer the advantage to support accelerated plant growth during the early growth stage after transplantation and extract water from shallow soil layers that is otherwise easily lost by evaporation especially in arid environments (Johansen et al., 1994). Water stress induces a significant reduction of leaf number, leaf biomass (fresh, dry mass) and leaf water content in Argan seedlings. In general, under drought stress (3/4, ½ and ¼ FC), all genotypes had lower values of leaf area than the well-watered regime (1 FC) except at 0 FC. However, drought resulted in a reduction of total leaf area in seedlings of several species. Reduced leaf number and area was a drought avoidance strategy for the seedlings by reducing transpiration. These results confirm the conclusion that drought constrains leaf growth to a much greater extent, indicating that this abiotic factor led to a more conservative balance between losing and obtaining water by different organs of plant (James and William, 1998; Liu and Stützel, 2004; Zhang et al., 2004; Villagra and Cavagnaro, 2006; Wu et al., 2008). Some genotypes as V4 and V3 unlike V1 and V8 are able to grow in length and diameter, able to produce more shoots, spines and biomass (fresh mass, dry mass) and with much water content in the stem even under water stress. These results suggest the existence of mechanisms of drought tolerance in these genotypes.

As proposed by Liu and Stützel (2004), conservative shoot growth during drought could be advantageous, especially if root growth is promoted. Genotypes that sustain shoot growth during drought may have greater marketability, which is particularly important for leafy vegetable crops. In Argan, under drought, genotype V4 and V3 have taproots exceeding 50 cm, which are accompanied by a large number of lateral roots, more leaves, a large biomass (fresh, dry mass and water content) and great leaf areas even under severe stress (1/4 and 0 FC). Those genotypes are promising for reforestation programs in an environment where drought is a limiting factor for the natural regeneration. Therefore, the most difficult in vegetation growth in this area appears in summer for newly planted seedlings, their survival or death will depend on adaptation to the habitat through this crucial period as reported for four shrub species in arid valley of Minjiang River of China (Li et al., 2007). Genotypes (mother-tree of seedlings) V1 and V8 have produced longer roots, higher root to shoot ratio and root biomass production (fresh, dry weight ratios and water content), lower leaves number, a relatively lower biomass (fresh, dry mass and water content) and smaller leaf areas than seedlings from genotypes V4 and V3. In many species,root system in acquiring water has long been recognized as crucial to cope with drought conditions (Kashiwagi et al., 2006).

The increased root (root length, root biomass production) and leaf characters (biomass and area) under drought stress observed in genotypes V1 and V8 is in line with the theory of the functional balance, which predicts that plants will react to a limited water availability with a relative increase in the flow of assimilates to the root leading to an increased root biomass. So, in order to diminish consumption and increase absorption of water, plants in dry conditions often decrease their growth rate and biomass production, and contribute more biomass to roots (Villagra and Cavagnaro, 2006). Partitioning more biomass to below-ground and maintaining higher root to shoot ratio may be beneficial to enhanced capacity of water uptake, by maintaining the shoot in a well-hydrated condition (Blum, 1996; Liu and Stützel, 2004; Zhang et al., 2004; Kashiwagi et al., 2006; Wu et al., 2008). At this stage of seedlings development, this case will be an adaptative response of the plant morphology which may be a primary mechanism by which this species can cope with the environmental characteristics of south west Morocco as reported for other species (Patterson et al., 1997; Achten et al., 2010; Sapeta et al., 2013).

Conclusion

The Argan seedlings grown in pot and field showed that water stress was a very important limiting factor at the initial phase of growth. Drought stress significantly decreased seedling height, root growth, leaf area and biomass production, but the proportion of changes differed among root, stem and leaf. Great genetic variability between seedlings to water stress. Even under severe stress some genotypes increased root to shoot fresh and dry mass ratio and maintained higher investment in the taproot and lateral roots. This adaptive response in morphology may be a primary mechanism by which the Argan seedlings can cope with the environmental characteristics. For newly planted seedlings, the most difficult period for plants growth in this area appears from May to October since rainfall not exceeds 40 mm. So, in order to enhance the survival rate and conserving rate of seedlings planting, appropriate human intervention is required to reduce the damage to Argan seedlings resulting from drought. These results will be useful towards transplantation, and may serve as a guide to initiate effective measures in tree planting to enhance survival rate and conserving rate for Argan forest restoration.

ACKNOWLEDGEMENTS

We gratefully acknowledge anonymous reviewers and office journal which provided helpful comments that greatly improved the manuscript. We thank the Morocco-Germany Co-operative Project 'Conservation Project and Development the Argan forest' (PCDA-GTZ) and the project Pars-Agro 128 of the Morocan Ministry of Scientific Research for financial support.

REFERENCES

Achten WMJ, Maes WH, Reubens B, Mathijs E, Singh VP, Verchot L, Muys B (2010). Biomass production and allocation in *Jatropha curcas* L. seedlings under different levels of drought stress. Biomass Bioenergy 34:667-676.

Bani-Aameur F (1997). Diversity, phenology, regeneration and conservation of the argan tree. Annals of Forestry Research in Morocco. Acts on Improving, Conservation and Use of Moroccan Forest Genetic Resour. 08-10 May: pp. 169-177.

Bani-Aameur F, Alouani M (1999). Viability and dormancy of seeds of argan tree (*Argania spinosa* L.) Skeels). Ecologia mediditerranea. 25(1):75-86.

Bani-Aameur F (2001). flowring phenology in Argania spinosa (L.) Skeels. Genetic Ressour. Crop Evolution 49:75-86.

Bani-Aameur F, Zahidi A (2005). Variability of leaf stomatal density of adult trees of *Argania spinosa* (L.) Skeels in the field. Acta Botanica Gallica 152:182-188.

Bani-Aameur F (2007). Dendromatique description of natural populations of the argan tree in Morocco. Ann. For. Rech. Moroc. 38:79-91.

Blum A (1983). Genetic and physiological relationships in plant breeding for drought resistance. Developments in Agricultural and Manag. Forest Ecol. 12:195-205.

Blum A (1996). Crop response to drought and the interpretation of adaptation. Plant Growth Regul. 20:135-148.

Boudy P (1950). Forest Economy in North Africa. monograph and water regime of forest species. Ed. Larose, Paris (V):382-416.

Boudy P (1952). Forest North Africa Guide. Rustique home, Paris, pp. 184-195.

Chaves MM, Oliveira MM (2004). Mechanisms underlying plant resilience to water deficits: prospects for water-saving agriculture. J. Exp. Bot. 55:2365-2384.

Dagnelie P (1984). Theories and statistical methods. Agricultural Application, T. I, 2nd Ed, The Agricultural presses ASBL Gembloux (Belgium), P. 378.

El Aboudi A, Carlier G, Peltier JP (1991). Water regime of the argan tree (*Argania spinosa* (L.) Skeels) in Souss (Morocco). Physiology of trees and shrubs in arid and semi-arid. Study Group of the tree - Paris, France, pp. 389-403.

Emberger L (1924). About the geographical distribution of the argan tree. Bull. Science Nat. and phys. of Morocco. 4(7):151-153.

Emberger L (1925). The domaine of the argan tree. Bulletin of the French Botanical Society. 72: 770-747.

Emberger L (1955). Biogeographical classification of climates. Fac. Sci. Montp., Bot. Ser. 7:3-45.

Ferradous A, Bani-Aameur F, Dupuis P (1996). Stationnel climate, phenology and fruiting of the argan tree (*Argania Spinosa* (L.) Skeels). Act. Inst. Agron. Vet. Hassan II. 17(1):51-60.

Ferradous M, ElAboudi A, Peltier JP (2005). Biodiversity and biogeography of Moroccan argan tree. Cahiers Agric. 14(4):1-17.

French RJ, Turner NC (1991). Water deficit change dry matter partitioning and seed yield in narrow-leafed lupins (*Lupinus angustifolius* L.). Aust. J. Agric. Res. 42:471-484.

Harrouni MC, Zahri S, Elhemaid A (1995). Transplanting of the argan seedlings. Combined effects of cultural techniques and water stress. Acts of International Symposium: Forest Facing Desertification: The case of argan. Fac. Sci. Agadir. pp. 115-126.

James AZ, William RG (1998). Leaf water relations and plant development of three freeman maple cultivars subjected to drought. J. Am. Soc. Hort. Sci. 123:371-375.

Johansen C, Krishnamurthy L, Saxena NP, Sethi SC (1994). Genetic variation in moisture stress of chickpea grown under line-source sprinklers in a semi-arid tropical environment. Field Crop Res. 37:103-112.

Kaabouss M (1992). Effect of soil moisture on the growth of argan seedlings (*Argania spinosa* L.) Skeels). Dissertation studies for obtaining the horticultural engineering degree option horticulture, IAV Hassan II, Agadir. P. 50.

Kashiwagi J, Krishnamurthy L, Crouch JH, Serraj R (2006). Variability of root length density and its contributions to seed yield in chickpea (*Cicer arietinum* L.) under terminal drought stress. Field Crops Res. 95:171-181.

Li Y, Liu X, Zheng S, Chen H, Yue Y, Mu C, Liu J (2007). Drought resistant physiological characteristics of four shrub species in arid valley of Minjiang River, China. Acta Ecol. Sin. 27(3):870-878.

Liu F, Stützel H (2004). Biomass partitioning, specific leaf area, and water use efficiency of vegetable amaranth (*Amaranthus spp.*) in response to drought stress. Sci. Hortic. 102:15-27.

Loss SP, Siddique KHM (1994). Morphological and physiological traits associated with wheat yield increases in Mediterranean Environments. Adv. Agron. 52:229-276.

Martin PJ, Stephens W (2006). Willow growth in response to nutrients moisture on a clay landfill cap soil. I. Growth and biomass production. Bioresour. Technol. 97:437-448.

M'hirit O, Benzyane M, Benchekroun F, Elyousfi SM, Bendaanoun M (1998). Argan, a forest, fruit species, with multiple uses. Sprimont, Belgium: Margada Edition. P. 150.

Mosseddaq F (1988). Comparison of some methods of measuring leaf area of wheat (*Triticum aestivum* L.). Acts Agronomic Vet. Inst. 8:29-34.

Paliwal K, Kannan D (1999). Growth and nutritional characteristics of four woody species under nursery conditions and growth after transplantation in semi arid field conditions at Madurai, India. J. Arid Environ. 43:133-141.

Sapeta H, Costa JM, Lourenço T, Marocod J, Lindee PV, Oliveiraa MM (2013). Drought stress response in *Jatropha curcas*: growth and physiology. Environ. Exp. Bot. 85:76-84.

Tucker CJ, Goward SN (1987). Satellite remote sensing of drought conditions. In: Whilhite, D.A., Easterling, W.E. (Eds.), planning for drought: toward a reduction of societal vulnerability. Westview Press, Boulder, CO.

Turner NC (1997). Further progress in crop water relations. Advan. Agron. 58:293-338.

Villagra PE, Cavagnaro JB (2006). Water stress effects on the seedling growth of *Prosopis argentina* and *Prosopis alpataco*. J. Arid Environ. 64:390-400.

Vessereau P (1988). Statistical Methods in Biology and in Agronomy. Tec. and Doc-Lavoisier (Paris). J. b. Baillere.

Wu F, Baoa W, Lia F, Wua N (2008). Effects of drought stress and N supply on the growth, biomass partitioning and water-use efficiency of *Sophora davidii* seedlings. Environ. Exp. Bot. 63:248-255.

Yin CY, Wu X, Duan BL, Luo JX, Li CY (2005). Early growth, dry matter allocation and water use efficiency of two sympatric Populus species as affected by water stress. Environ. Exp. Bot. 53:315-322.

Younis ME, El-Shahaby OA, Hasaneen MNA, Gaber AM (1993). Plant growth, metabolism and adaptation in relation to stress conditions: XVII. Influence of different water treatments on stomatal apparatus, pigments and photosynthetic capacity in *Vicia faba*. J. Arid Environ. 25:221-232.

Zahidi A, Bani-Aameur F (1997a). Genetic variability of branching characters in argan seedlings. The sixth Scientific Meeting of the Network of Biotechnology - Genetic engineering of plants AUPELF.UREF. Orsay University Center, Paris Sud XI University. June 30 to July 3.

Zahidi A, Bani-Aameur F (1997b). Argan genetic diversity of root and shoot characters. Botanical Society of America Annual Meetings. 3 August. Palais des Congrés. Montreal. Quebec.

Zahidi A (1997). Phenology, typology and genetic variability of branching and foliation traits *Argania spinosa* (L.) Skeels. Third cycle thesis, Faculty of Sciences, Ibn Zohr University. Agadir appendices. P. 177.

Zahidi A, Bani-Aameur F (1998). Argan seedling damping-off under nursery conditions: effect of mother-tree genotype, Kernel origin and seedling age. Ecol. Mediterr. 24(1):27-32.

Zahidi A, Bani-Aameur F, El Mousadik A (2013). Seasonal change effects on phenology of *Argania spinosa* (L.) in the fields. J. Ecol. Nat. Environ. 5:189-205.

Zhang X, Zang R, Lia C (2004). Population differences in physiological and morphological adaptations of *Populus davidiana* seedlings in response to progressive drought stress. Plant Sci. 166:791-797.

34

Alleviation of salt stress on *Moringa peregrina* using foliar application of nanofertilizers

Amira Sh. Soliman[1], Souad A. El-feky[2] and Essam Darwish[3]

[1]Natural Resources Department, Institute of African Research and Studies, Cairo University, Giza, Egypt.
[2]Department of Laser Applications in Metrology, Photochemistry and Agriculture, National Institute of Laser Enhanced Science (NILES), Cairo University, Egypt.
[3]Plant Physiology Section, Faculty of Agriculture, Cairo University, Egypt.

***Moringa peregrina* plants were grown under four levels of saline water (0, 3000, 6000, 9000 ppm), and treated with sprayed Hoagland solution containing different concentrations of ZnO and Fe_3O_4 Nano-Particles (NP) (30, 60 and 90 mg/L); the normal Hoagland solution was used as a control. Results show that salinity levels significantly reduced growth parameters (plant height, root length, number of leaves, number of branches, shoot and root fresh and dry weights). Also, chlorophyll, carotenoids and crude protein levels decreased meanwhile proline and total carbohydrate levels, antioxidant non-enzymes (vitamins A and C) and enzymes (POD and SOD) increased. *Moringa* plants sprayed with Hoagland-containing ZnO and Fe_3O_4 NP showed an enhancement in growth parameters either under normal or saline conditions when compared to control. Also, spraying plants with Hoagland-containing ZnO and Fe_3O_4 NP resulted in significant reduction in Na^+ and Cl^- and an increase in N, P, K^+, Mg^{2+}, Mn^{2+}, Fe, Zn; total chlorophyll, carotenoids, proline, carbohydrates, crude protein levels, antioxidant non-enzymes and enzymes when compared to control, normal Hoagland sprayed-plants. Generally, this enhancement of salt tolerance was considerable in plants sprayed with 60 mg/L ZnO and Fe_3O_4 NP and grown either in saline and non-saline conditions.**

Key words: *Moringa peregrine*, nanofertilizers, salt stress, growth parameters, chemical composition.

INTRODUCTION

Moringa peregrina (Forssk.) Fiori (Moringaceae) is a tree (4 to 15 m) (Boulos, 1999). Its seeds have different economic and medical importance. Due to its unique composition, the extracted oil is highly valued for preparing cosmetics, cooking, and lubricating purposes (Somali et al., 1984). *Moringa* plants are considered a valuable source for many useful components such as vitamins A, B and C, and provide humans with minerals, protein and amino acids (Price, 2000). As a result of uncontrolled and indiscriminate use of this plant in many activities, the tree has decreased in numbers and become rare in Egypt (Zaghloul et al., 2008). New lands are considered as promising areas to cultivate this crop. The notable problems facing the plants cultivation in the reclaimed lands are drought, salt and heat stresses (abiotic stresses) which adversely affect the growth and productivity of the plants. Salt stress is one of the most devastating problems that limits the crop's production worldwide by imposing its effect through osmotic stress, Na^+ and Cl^- toxicity and ions uptake imbalance leading to deficiency in N, P, K^+, Ca^{2+} and micronutrients (Munns, 2005). Salt stress was reported to decrease the growth and yield of the plants as it affects the organic, ion contents and metabolic activity in the stressed plants. Accumulation of organic solutes is another mechanism that enables the plant to tolerate salt stress. Osmoprotectants (sugars, glycine betaine, proline, mannitol, etc) generally found in cytosol, plays an important role in osmotic adjustment as well as protection of enzymes and proteins (Munns and Tester, 2008). It was suggested that this osmprotecants work as scavengers of ROS (reactive oxygen species) which are induced by salt stress and negatively affect the lipid membrane and enzyme activity. Attempts have been made to increase plants' tolerance against salt stress. These efforts include classical breeding, gene transfer, seed priming, foliar application of osmoprotectants and inorganic compounds (Chen et al., 2007).

Nanoparticles (nano-scale particles; NSPs) are atomic or molecular aggregates with at least one dimension between 1 and 100 nm (Ball, 2002). Nanofertilizers have been developed and have provided a new efficient alternative to normal regular fertilizers. The properties of nano-particles (more surface area) may help in increasing the reactive points of these particles and hence increase the reactivity of these nanoparticles. This leads to changes in the physio-chemical properties of these nanoparticles which help in the absorption of fertilizers in plants (Anonymous, 2009). The promoting effect of nano-particles on seedling growth and development were reported by Zhu et al. (2008). Also, nano-iron oxide compared to other treatments such as organic materials and iron citrate facilitated photosynthesis and iron transfer in peanut leaves (Liu et al., 2005). Nanoparticles can be divided into groups; metal based materials such as nanogold, nanozinc, nanoaluminum; and nanoscale metal oxides like TiO_2, ZnO and Al_2O (Ruffini and Roberto, 2009).

Foliar application of macro and micronutrients has been reported as an effective method to increase salt tolerance in plants and have been suggested to ameliorate the adverse effect of salt stress (Hamayun et al., 2011). This promoting effect can be attributed to the increased and enhanced nutrient uptake of micronutrients through the leaf or root as a result of root improvement. Foliar application may also offer a solution to overcome root restriction caused by salt stress (El-Fouly et al., 2004). Fe is critical for chlorophyll formation and photosynthesis and is important in enzyme systems and plant respiration (Malakouti and Tehrani, 2005). For most plants, zinc is an essential component of enzymes and participates in the synthesis of chlorophyll and other proteins (Vallee and Auld, 1990). The effect of nano fertilizers on plant growth in general and specifically under salt stress by investigating the effect of Hoagland solution containing ZnO and Fe_3O_4 NPs on *M. peregrina* plants grown under different levels of salinity is therefore the aim of this study.

MATERIALS AND METHODS

This study was conducted at the Experimental Laboratories of the Natural Resources Department, Institute of African Research and National Institute of Laser Enhanced Sciences (NILES) Cairo University, Giza. Egypt during the two seasons of 2013 and 2014.

One year old seedlings of *M. bergrina* were obtained on the 1st of May in the first and second seasons, respectively, from Orman Botanical Garden, Cairo, Egypt. Then the seedlings were transplanted into 25 cm diameter-plastic bags filled with 6 kg sandy soil, and watered every 3 days with Hoagland's nutrient solution (Hoagland and Arnon, 1950) for plant maintenance.

Soil analysis

The soil texture was sandy having the following characteristics: 30.82% coarse sand, 62.61% fine sand, 1.22% silt, 5.35% clay, pH 7.75, EC 1.15 dS/m, organic matter 0.08%, available N 6.9 ppm, available P 6.2 ppm, available K 64 ppm, $CaCO_3$ 0.26%, and water holding capacity 14.5%.

Salinity treatments

Two weeks after transplanting (in both seasons), the salinity treatments were initiated after 10 days. Four levels of salinity (Control, 3000, 6000, and 9000 ppm) were used for testing salt stress. The different saline water concentrations were prepared using a mixture of synthetic seawater salt obtained from Sigma Company. At each irrigation, the plants were watered till 100% of soil field capacity (F.C.). To maintain the required soil medium salt levels, the soil EC was measured periodically by portable EC meter.

Nano treatments

***Synthesis of ZnO and Fe_3O_4 magnetic nanoparticles (NPs)*:** Zinc acetate [Zn (H_3COO) $2H_2O$.], NaOH and isopropyl alcohol (2-propanol) with 99.5% reagent grade were obtained from Sigma Aldrich (99.9%). 0.073 mmol Zn (OAc). $2H_2O$ was dissolved at 60°C in 50 ml 2-propanol under stirring. In a second flask, 1.5 mmol NaOH was dissolved under vigorous stirring in 25 ml^2 propanol at 60°C. NaOH solution was added drop wise under stirring to the acetate solution The product was stirred for an hour at 60°C and then cooled to room temperature. The precipitate was washed twice with 2-propanol and centrifuged at 4500 rpm for 30 min (Bardhan et al., 2007). In addition, the Fe_3O_4 magnetic nanoparticles were prepared by coprecipitation of Fe^{3+} and Fe^{2+} at a molar ratio of 2:1 with aqueous ammonia (0.3 mol/L) as precipitating agent (Laurent et al., 2008).

Characterization of ZnO and Fe_3O_4 NPs: The size and shape of

ZnO and Fe_3O_4 nanoparticles were observed directly by transmission electron microscope (TEM) (FEI, Netherland) The TEM samples were prepared by placing a few drops of the solution on a carbon-coated copper grid (Okenshoji Co., Ltd.).

Seedlings were sprayed monthly with Hoagland solution which replaced Zn and Fe with mixed ZnO and Fe_3O_4 NPs (30 (T1), 60 (T2), and 90 (T3) mg/L) after 10 days of adding salinity. Also, the normal Hoagland solution was used as a control (T0). Spraying was carried out between 09:00 and 11:00 AM.

Experimental design

The experiment was based on a Randomized Complete Block Design (RCBD) with two factors, including 16 treatments and three replicates. The first factor was control (without NPs application) and 3 levels of mixed ZnO and Fe_3O_4 NPs applications; the second factor had four irrigation water salinity treatments with each block consisting of 80 plants (five plants/ treatment). The seedlings were harvested at 90 days (in the two seasons, respectively) in order to determine the growth parameters and carry out chemical analysis.

Growth parameters

Plant height (cm), root length (cm), number of branches/plant, numbers of leaves/plant, stem diameter (cm), fresh and dry weight of shoots (leaves and stems) and roots (g/ plant) were also recorded.

Chemical analysis

Leaf pigments and total carbohydrates: Total chlorophyll and carotenoid contents were extracted using the method described by Nornai (1982). Total carbohydrates (%) in the dried leaves were also determined as described by Dubois et al. (1956).

Determination of macro and micronutrients and crude protein: Dried leaves samples were digested and the extract analyzed to determine nitrogen (N%) using the modified micro-Kjeldahl method, phosphorus (%) by Jackson (1967); K and Na% using a flame spectrophotometer (Jameel and Kahayri, 2002); while Ca, Fe, and Zn were determined by atomic absorption (Allen et al., 1984). The proline content in fresh leaves was also determined according to Bates et al. (1973). Also, protein % was determined as described by James (1995).

Antioxidant non-enzymes and enzymes determination: Antioxidant non-enzymes (Vitamins A and C) were measured according to AOAC (1999) using dried leaves. Meanwhile, enzymes extraction was carried out using fresh leaf tissues at 40°C in buffer solution (3: 1 buffer:fresh weight v/v) in a pastel. It was mortared with 100 mM potassium phosphate buffer (at pH 7.5) containing 1 mM EDTA, 3 mM DL-dithiothreitol and 5% (w/v) insoluble polyvinyl pyrolidone. The homogenates were centrifuged at 10000 g for 30 min and then the supernatants were stored in separate aliquots at 8°C (Vitoria et al., 2001). Antioxidant enzymes were assayed as follows; peroxidase (POD) by spectrophotochemically according to Amako et al. (1994) and superoxide dismutase (SOD) by photochemical method as described by Giannopolitis and Ries (1977). Enzymes activities were expressed as units/min/mg protein.

Statistical analysis

The data were subjected to statistical analysis of variance and the means were compared using the least significant difference (LSD) test at the 5% level, as described by Little and Hills (1978).

RESULTS AND DISCUSSION

Characterization of photo-catalysts by TEM

The shape and diameter of the nanoparticles used were observed with TEM. TEM image emphasized that ZnO presents in spherical nanoparticle form, with a diameter range of about 10 to 15 nm (Figure 1a) while Fe_3O_4 nanoparticles diameter ranges from approximately 10 to 12 nm with an almost spherical shape (Figure 1b).

Effect of salt stress on salt stressed-plants

Table 1 shows that growth parameters (plant height, root length, stem diameter, number of leaves, number of branches) decreased in response to different concentrations of salinity and this reduction was significant in plants treated with the two levels of salinity (6000 and 9000 ppm). Furthermore, the shoot and root fresh and dry mass of the *Moringa* plants decreased significantly under salinity conditions compared to those of control plant.

Soil salinity adversely affects plant growth through several physiological and biochemical means like ion toxicity, osmotic stress, nutritional imbalance, biochemical and physiological disorders (Kao et al., 2003). Salt stress resulted in the reduction in the number of leaves and branches and stunted shoot growth in *Acacia saligna* (Soliman et al., 2012). Moreover, Bello and Igbokwe (2013) reported that salt stress reduced height of both *Acacia senegal* and *Parkia biglobosa*. The first reduction in plant growth may be attributed to the initial sudden increase in osmotic pressure as stated by Hajibagheri et al., (1989) thus suggesting that high salinity might inhibit root and shoot elongation due to slowing down of water uptake by the plant. Over time, Na^+ and Cl^- will accumulate to toxic concentrations in the shoot resulting in premature leaf senescence and death due to the ionic component of salt (Munns and Tester, 2008; Hairmansis et al., 2014). The accumulated amounts of ions enter the plant through the transpiration stream thereby causing cells injury in the transpiring leaves which may cause further reductions in photosynthesis processes thereby leading to growth reduction (El-Fouly et al., 2002; Munns et al., 2006).

Total chlorophyll (Chl a and b) and carotenoides contents were significantly lower in plants grown under salt stress conditions than those recorded in control plants (Table, 2). A reduction of 48% in total chlorophyll contents of *Moringa* leaves was recorded at the third salinity concentration (9000 ppm). An inhibition in chlorophyll biosynthesis, activation in the chlorophyllase and/or destruction of chloroplast structure (Gunes et al., 1996) could have contributed to lowering the pigment content under saline conditions.

The results also showed an increase in proline and carbohydrate concentrations in leaves of *Moriga* plants in response to different levels of salt stress (Table 2). In this

(a)

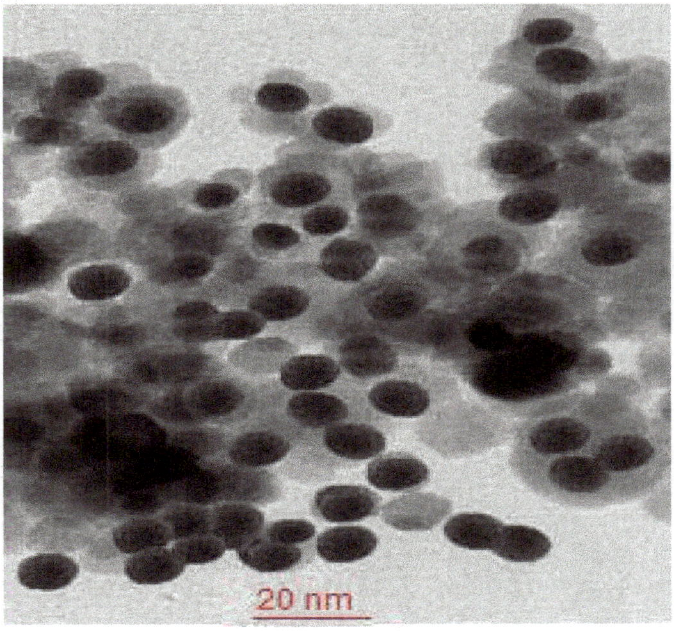

(b)

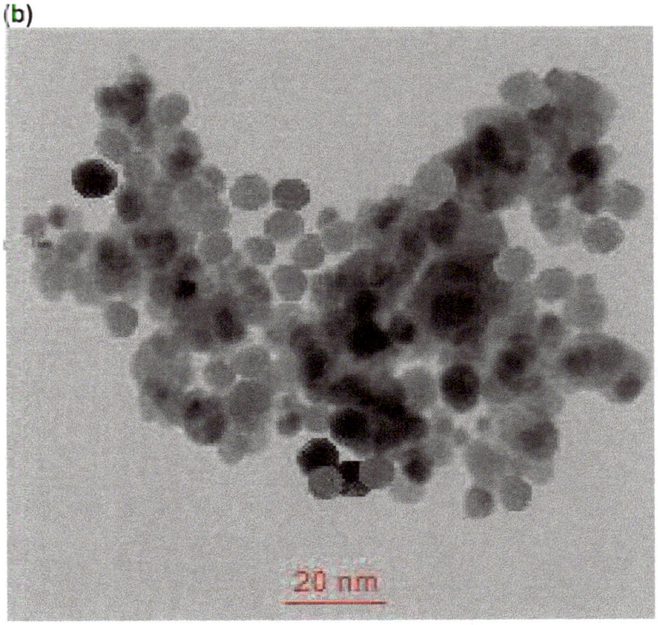

Figure 1. TEM image of the prepared nanoparticles. (a) ZnO, (b) Fe_3O_4.

this regard, increased free amino acids and proline in response to salt stress has been investigated by many researchers in many plants (Gunes et al., 1996; Sharma et al., 2010). One of the strategies that plants use to cope with salt stress is osmoprotectant synthesis of low molecular weight molecules such as sugars, proline and glycine betaine which play an important role in osmotic adjustments and protection of protein and lipids from (ROS). These further results in the protection of plasma membrane integrity and enzyme function. Also, it plays an important role as a scavenger for free radicals which protects cells from ROS actions. Proline serves as a storage sink for carbon and nitrogen and it is a free-radical. It also stabilizes subcellular structures (membranes and proteins), and buffers cellular redox potential. Hence, these organic osmolytes are known as osmoprotectants. These organic solutes may contribute to osmotic adjustment, protecting cell structure and function, and/or may serve as a metabolic or an energetic reserve (Chen and Murata, 2000).

Crude protein was found to decrease in response to salt stress. Protein synthesis has been considered as a possible primary target of salt toxicity because in vitro protein synthesis systems are dependent on physiological potassium and are inhibited by sodium and chloride (Morant-Avice et al., 1998). Considering the evidences on plant soluble protein response to salinity, there is a marked difference between the species and varieties. Thus, proteins may play a role in osmotic adjustment. According to Pareek et al. (1997), proteins may be synthesized *de novo* in response to salt stress, or may be present constitutively in low concentrations and increased when plants are exposed to salt stress.

Raising the salt concentration significantly increased antioxidant non-enzymes (vitamins A and C) and enzymes (POD, and SOD) in tissues of *Moringa* leaves (Figure 2) in both seasons. Accordingly, the lowest values of the non-enzymatic and enzymatic antioxidants were found in control plants irrigated with tap water, whereas the highest values were found in plants irrigated with water containing the highest salt concentration (9000 ppm). Such results are in harmony with Foyer and Noctor (2009), Cazzonelli and Pogson (2010) and Boguszewska and Zagdańska (2012). They found that many plants produce significant amount of a potential source of compounds such as non-enzymatic (vitamins A, and C) and enzymatic antioxidants (POD and SOD) to prevent oxidative stress caused by oxygen and photons. Piotr and Klobus (2005) and Wu et al., (2007) reported that ascorbic acid is an important antioxidant which reacts not only with H_2O_2 but also with O_2, OH and lipid hydroperoxidases. In addition, Shao et al. (2006) and

Abogadallah (2010) indicated that ascorbic acid concentration significantly increases in turf grass during water deficiency. Mittler (2002) and Akram et al. (2012) reported that the enzymatic antioxidants SOD and POD are considered to be the first line of defense against ROS thus the simultaneous increase in the activity of these enzymes contributes to a decrease in the deleterious effects of H_2O_2 under stress. Also, POD activity increased in eggplant plants under saline conditions (Shaheen et al., 2013). Other studies also reported that salt stress-induced enhanced POD and SOD activities were observed in sunflower (Akram et al., 2012) and pistachio plants (Abbaspour, 2012). Thus, it becomes clearly evident that

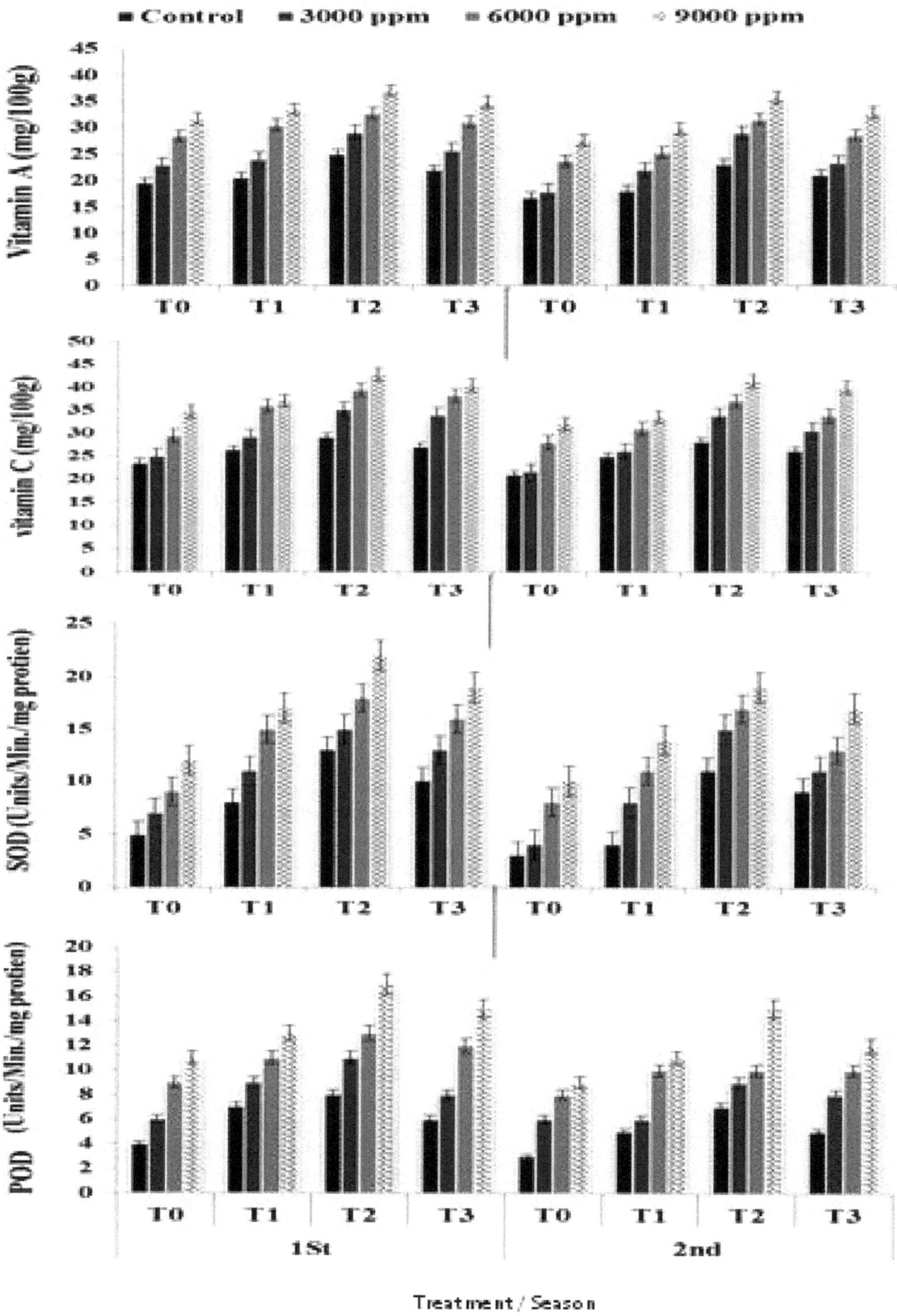

Figure 2. Effect of salt stress and nanofertilization on non-enzymatic (vitamins A, and C) and enzymatic antioxidants (POD, and SOD) in *M. pregrina* during 2013 and 2014.
T0= normal Hoagland solution (control), T1 = 30 mg/L ZnO and Fe_3O_4 NPs, T2= 60 mg/L ZnO and Fe_3O_4 NPs and T3= 90 mg/L ZnO and Fe_3O_4 NPs.

Table 1. Effect of salt stress and nanofertilization on growth parameters of *M. pregrina* during 2013 and 2014.

Treatment			Plant height (cm)	Root length (cm)	Stem diameter (cm)	No. of leaves/ plants	No. of branches/ plant	Fresh weight of leaves and stems (g/ plant)	Dry weight of leaves and stems (g/ plant)	Fresh weight of roots (g/ plant)	Dry weight of roots (g/ plant)
Control	1st	T0	45.00	24.00	2.00	29.00	11.33	21.13	8.90	95.40	46.70
		T1	59.33	29.33	2.37	39.00	23.00	24.80	11.07	128.50	63.25
		T2	70.00	38.33	3.00	52.00	27.67	32.10	16.05	149.43	74.05
		T3	67.00	33.00	2.70	48.33	26.00	28.30	13.15	142.60	70.63
	2nd	T0	41.33	19.33	0.90	24.00	9.00	18.57	7.95	91.70	45.18
		T1	50.00	24.00	1.60	35.00	16.00	21.90	9.28	125.90	61.95
		T2	67.67	32.00	2.00	47.00	22.00	29.20	14.60	140.30	69.65
		T3	60.33	29.67	1.90	43.00	20.00	28.00	12.00	136.80	67.73
3000 ppm	1st	T0	39.33	22.00	1.80	25.00	9.00	18.80	8.73	92.70	45.35
		T1	46.33	24.00	2.10	34.67	18.00	23.43	10.41	121.90	60.28
		T2	66.67	35.00	2.80	49.67	24.00	30.70	14.68	142.07	70.68
		T3	59.67	29.33	2.50	43.00	22.00	26.63	12.32	134.80	66.73
	2nd	T0	34.00	17.33	0.70	21.00	8.00	17.20	7.72	87.90	42.95
		T1	40.00	21.00	1.20	30.00	12.00	19.50	8.75	118.50	58.25
		T2	61.67	30.00	1.90	43.00	19.00	26.67	11.67	136.80	67.73
		T3	53.00	27.00	1.70	41.00	16.33	21.90	9.62	130.40	64.53
6000 ppm	1st	T0	31.33	18.67	1.20	19.67	7.00	15.30	6.95	84.10	41.38
		T1	40.00	20.00	1.73	26.00	11.33	20.09	9.38	117.60	58.80
		T2	52.67	30.33	2.10	37.00	18.33	27.37	12.96	136.70	68.02
		T3	47.33	25.00	2.00	32.00	15.67	24.47	11.91	128.90	63.73
	2nd	T0	25.00	13.00	0.53	16.67	5.00	12.79	5.59	80.40	39.53
		T1	38.33	17.33	1.17	22.33	8.00	17.90	8.02	114.9	56.45
		T2	47.33	25.00	1.60	33.00	15.00	24.44	10.89	131.60	65.30
		T3	41.67	21.00	1.30	30.00	11.00	20.13	8.73	123.70	61.35
9000 ppm	1st	T0	28.33	13.33	0.90	15.00	4.67	13.44	6.31	71.10	35.12
		T1	31.33	18.00	1.30	20.00	8.33	15.97	7.56	113.50	56.08
		T2	43.33	28.67	1.60	30.00	13.00	21.57	10.35	128.90	64.08
		T3	38.33	21.33	1.45	24.00	10.33	18.85	8.94	120.40	59.50
	2nd	T0	21.00	11.00	0.43	12.33	3.33	11.80	4.70	67.40	33.03
		T1	27.67	14.33	0.80	16.33	5.33	13.65	6.42	109.80	54.23
		T2	40.00	20.00	1.17	28.67	11.00	18.87	8.83	123.23	61.12
		T3	32.00	18.33	1.00	22.33	9.00	15.90	7.18	118.70	58.85
LSD (0.05)											
S	1st		1.69	4.07	0.30	8.01	1.53	2.92	1.13	5.92	2.36
N			5.73	6.53	0.66	5.04	2.03	5.06	2.38	3.10	1.59
N× S			11.46	13.06	1.33	10.08	4.07	10.12	4.76	6.20	3.18
S	2nd		2.44	2.64	0.26	8.33	1.24	5.59	2.38	10.88	5.26
N			5.93	5.99	0.28	6.18	2.23	3.69	1.43	6.81	3.34
N× S			11.58	11.98	0.56	12.36	4.47	7.38	2.86	13.61	6.69

T0= normal Hoagland solution (control), T1 = 30 mg/L ZnO and Fe_3O_4 NPs, T2= 60 mg/L ZnO and Fe_3O_4 NPs and T3= 90 mg/L ZnO and Fe_3O_4 NPs. S= salinity treatments, N= nano treatments. 1st = First season, 2nd = Second season.

Table 2. Effect of salt stress and nanofertilization on chemical composition of *M. pregrina* during 2013 and 2014 seasons.

Treatment			Total chlorophylls content (mg/g fresh weight)	Carotenoids content (mg/g fresh weight)	Total carbohydrate (% of dry weight)	Proline content (μ moles/g fresh weight)	Crude Protein (%)
Control	T0	1st	1.66	0.75	19.00	13.00	18.94
	T1		1.70	0.84	23.00	15.00	20.88
	T2		2.29	1.14	29.33	19.33	22.63
	T3		1.96	0.92	27.67	18.00	22.13
	T0	2nd	1.47	0.68	15.67	17.33	18.06
	T1		1.53	0.72	19.33	18.67	19.94
	T2		2.15	1.01	25.67	23.67	21.50
	T3		1.69	0.84	23.33	21.33	20.63
3000 ppm	T0	1st	1.63	0.72	22.33	16.00	17.81
	T1		1.59	0.78	28.33	19.33	18.69
	T2		2.24	1.11	33.67	22.67	20.06
	T3		1.81	0.89	30.33	20.33	19.31
	T0	2nd	1.22	0.59	17.67	19.67	16.69
	T1		1.43	0.69	24.67	21.33	17.69
	T2		2.18	1.06	29.33	28.33	19.38
	T3		1.67	0.84	27.33	25.00	18.13
6000 ppm	T0	1st	1.09	0.55	26.00	20.33	12.81
	T1		1.32	0.68	33.00	25.00	14.31
	T2		2.18	1.08	39.33	29.33	17.19
	T3		1.59	0.81	37.33	27.33	15.25
	T0	2nd	0.99	0.43	23.33	23.00	12.31
	T1		1.25	0.58	29.67	30.33	13.50
	T2		2.02	0.98	35.33	33.00	15.13
	T3		1.51	0.74	31.33	30.33	14.31
9000 ppm	T0	1st	0.86	0.41	32.33	25.67	11.00
	T1		1.05	0.50	37.67	27.67	11.88
	T2		1.49	0.79	43.33	32.33	14.19
	T3		1.13	0.57	40.00	30.67	12.63
	T0	2nd	0.75	0.32	28.00	28.00	9.69
	T1		1.01	0.48	33.00	31.33	11.69
	T2		1.30	0.66	39.33	38.33	13.63
	T3		1.02	0.51	37.67	35.33	12.38
LSD (0.05)							
S		1st	0.03	0.03	0.94	1.76	0.09
N			0.08	0.05	3.89	4.02	0.47
N×S			0.15	0.09	7.78	8.04	0.94
S		2nd	0.06	0.06	1.03	1.54	0.33
N			0.07	0.05	3.05	3.27	0.50
N×S			0.13	0.11	6.10	6.53	1.00

T0= normal Hoagland solution (control), T1 = 30 mg/L ZnO and Fe_3O_4 NPs, T2= 60 mg/L ZnO and Fe_3O_4 NPs and T3= 90 mg/L ZnO and Fe_3O_4 NPs. S= salinity treatments, N= nano treatments. 1st = First season, 2nd = Second season.

non-enzymatic and enzymatic antioxidant status of plants for ROS scavenging is an important salt tolerant trait.

Salinity stress significantly increased percentage Na, Cl and Ca and reduced percentage K, Mg and P in the leaves of *Moringa* plants (Table 3). Salinity may result in the disturbance of uptake and utilization of essential nutrients due to competition and interactions of soluble salts with mineral nutrients (Gouia et al., 1994). Ionic imbalance occurs in the cells due to over accumulation of Na^+ and Cl^- and reduced uptake of other mineral nutrients, such as K^+, Ca^{2+}, Mg^{+2} and No^- and Mn^{2+} thus leading to growth suppression (Karimi et al., 2005).

Effect of ZnO and Fe_3O_4 NPs- containing Hoagland solution on salt stressed-plants

The foliar application of nano-iron and -zinc containing-Hoagland solution caused a significant increase in previously mentioned growth parameters in comparison to control plants (Table 1). This promoting effect of these nano-applications was not only noticed in the growth of salt-stressed plants, but also did promote the growth in plants grown under normal conditions. The most interesting result is that under the highest salinity level 9000 ppm, the increment in shoots fresh weight, number of leaves and plant height reaches up to 60, 100 and 53 % respectively in plants treated with the T2 (6 mg/l ZnO and Fe_3O_4 NPs-containing Hoagland solution) over control plants (Hoagland-sprayed plants). This increment was recorded in the first and second seasons. It means that the T2 treatment has a strong promoting effect either in stress or non-stress conditions. It also noticed that T1 (3 mg/l ZnO and Fe_3O_4 NPs-containing Hoagland solution) and T3 (9 mg/l ZnO and Fe_3O_4 NPs-containing Hoagland solution) also has promoting effect on growth parameters of plants grown under both stress and non-stress conditions in comparison to Hoagland-sprayed plants but is however less than those found in the T2 treatment. The aforementioned data are in trustworthiness with Aslam et al. (1993) who mentioned that growth parameters have been used as an indicator of salt tolerance in plants e.g. shoot weight. Meanwhile, significant increase in biomass, with respect to length or diameter of stem, leaves and dry weight (DW) of plants was observed by spraying *Moringa* plants with the combination of zinc and iron nano fertilizers. This indicates that proper concentration of zinc is required for dry matter accumulation and plant growth (Dimkpa et al., 2013).

Improved salt tolerance by addition of nutrients has been reported in many plants (Zhu et al., 2004 on cucumber; Al-Aghabary et al., 2005 on tomato). In addition, application of micronutrients is reported to enhance photosynthetic activities which lead to an increase in cell division and elongation thereby increasing vegetative biomass. It was also found that foliar spray of zinc sulfate (Yildirim et al., 2008) and treatment of seedlings with zinc sulfate before transplanting (Tzortzakis, 2010) leads to relieve symptoms of salt stress.

Nano-technology can offer opportunities to enhance yield and counter environmental stress. By using nano-particles, we aim to delay releasing fertilizers. Nano-particles have high reactivity because of the larger specific surface area and increased reactivity of these areas on the particle surface. These features simplify the absorption of fertilizers and pesticides that are produced in nano scale (Anonymous, 2009). The application of nano-particles to plants can be beneficial (seedling growth and development) or non-beneficial (prevent root growth) (Zhu et al., 2008). These results are in agreement with the findings of Liu et al., (2005) who concluded that nano-iron oxide facilitated photosynthesis and iron transfer to the leaves of peanut when compared to organic materials and iron citrate. In addition, Sheykhbaglou et al. (2010) found that the nano-iron oxide had significant effects on the dry pod weight; leaf with dry pod, and yield of soybean compared to other treatments. In pumpkin, iron oxide NPs increased root elongation which was attributed to Fe dissolution (Wang et al., 2011). Thus, the positive effects of appropriate zinc and Fe concentrations on fresh and dry weight, plant height, number of leaves and branches under NaCl stress could be explained by the replacement of Fe and Zn with nano forms.

Foliar applications with nano-iron and nano-zinc containing-Hoagland solutions at different concentrations lead to increased total chlorophyll, carotenoids, proline content, total carbohydrates and crude protein percentage more than those recorded in Hoagland-sprayed plants either in non-stress or stress conditions (Table 2). At the highest level of salinity (9000 ppm), increased percentage values in chlorophyll content resulted from the application of the nano form of Fe and Zn Hoagland solution. This increase reached 73% in both seasons when treated with T2 and was noticed in increased leaf numbers. In addition, iron plays an important role in the photosynthetic reactions as it is a component of ferrodoxin, an electron transport protein associated with chloroplast (Hazra et al., 1987). Iron also activates several enzymes and contributes in RNA synthesis and improves the performance of photosystems (Malakouti and Tehrani, 2005). Moreover, iron oxide NPs have been reported as facilitators for iron and photosynthate transfer to the leaves of peanut (Liu et al., 2005). Meanwhile, Zn plays an important role in many biochemical reactions within the plants like chlorophyll and carbohydrate formation (Corredor et al., 2009), increased photochemical reduction rates (Kumar et al. 1988), chloroplast structure, photosynthetic electron transfer as well as photosynthesis (Romheld and Marschner, 1991); in enzyme structure involved in amino acid biosynthesis (Cakmak et al., 1989). These results agree with those of El-Kereti et al. (2013) and El-Feky et al. (2013). The seasons

Table 3. Effect of salt stress and nanofertilization on macro and micro nutrients in *M. pregrina* during 2013 and 2014 seasons.

Treatment			N (%)	P (%)	K (%)	Na (%)	Cl (%)	Ca (%)	Mg (%)	Fe ppm	Zn ppm
Control	T0	1St	3.03	0.30	2.19	0.35	0.18	0.55	0.55	87.46	95.40
	T1		3.34	0.43	2.39	0.31	0.17	0.74	0.90	95.61	103.73
	T2		3.62	0.59	2.62	0.23	0.11	1.02	1.32	121.25	140.50
	T3		3.54	0.47	2.45	0.29	0.13	0.88	1.18	110.50	128.90
	T0	2nd	2.89	0.26	1.89	0.41	0.23	0.47	0.48	82.72	88.50
	T1		3.19	0.37	2.14	0.37	0.21	0.60	0.78	90.85	99.87
	T2		3.44	0.48	2.33	0.29	0.16	0.93	1.19	117.54	123.80
	T3		3.30	0.40	2.20	0.35	0.19	0.79	0.93	100.25	110.60
3000 ppm	T0	1St	2.85	0.27	1.77	0.30	0.27	0.76	0.41	80.26	86.77
	T1		2.99	0.32	2.00	0.35	0.25	1.10	0.82	87.92	95.63
	T2		3.21	0.48	2.29	0.29	0.18	1.25	1.02	100.53	119.48
	T3		3.09	0.37	2.07	0.31	0.23	1.16	0.96	93.51	105.56
	T0	2nd	2.67	0.24	1.63	0.42	0.31	0.63	0.40	76.77	70.29
	T1		2.83	0.28	1.83	0.44	0.28	0.99	0.68	80.39	89.80
	T2		3.10	0.39	2.00	0.34	0.21	1.17	0.95	93.34	106.58
	T3		2.90	0.32	1.90	0.41	0.25	1.10	0.79	88.36	97.28
6000 ppm	T0	1St	2.05	0.22	1.59	0.47	0.40	1.01	0.32	72.70	69.34
	T1		2.29	0.27	1.79	0.41	0.33	1.30	0.64	80.26	80.48
	T2		2.75	0.36	1.93	0.33	0.26	1.46	0.95	98.42	102.80
	T3		2.44	0.30	1.88	0.37	0.29	1.39	0.77	90.53	88.77
	T0	2nd	1.97	0.19	1.44	0.54	0.47	0.85	0.28	64.36	57.14
	T1		2.16	0.22	1.63	0.49	0.39	1.19	0.57	74.61	77.83
	T2		2.42	0.31	1.77	0.38	0.29	1.35	0.90	95.43	93.32
	T3		2.29	0.26	1.70	0.45	0.34	1.26	0.73	87.33	80.87
9000 ppm	T0	1St	1.76	0.17	1.37	0.64	0.45	1.15	0.24	39.30	44.81
	T1		1.90	0.23	1.45	0.48	0.40	1.54	0.48	55.21	76.92
	T2		2.27	0.29	1.67	0.40	0.30	1.73	0.87	67.85	90.41
	T3		2.02	0.25	1.58	0.45	0.36	1.60	0.59	60.44	87.33
	T0	2nd	1.55	0.12	1.30	0.70	0.50	1.10	0.21	32.45	30.20
	T1		1.87	0.16	1.37	0.55	0.46	1.30	0.35	49.47	69.47
	T2		2.18	0.25	1.59	0.47	0.35	1.50	0.81	63.32	87.83
	T3		1.98	0.21	1.46	0.51	0.41	1.39	0.53	57.48	80.30
LSD (0.05)											
	S	1St	0.001	0.12	0.22	0.001	0.05	0.08	0.20	3.98	5.57
	N		0.08	0.07	0.23	0.04	0.03	0.08	0.06	6.78	5.87
	N× S		0.15	0.14	0.47	0.08	0.05	0.17	0.12	13.57	11.73
	S	2nd	0.05	0.03	0.33	0.03	0.05	0.08	0.07	4.99	6.28
	N		0.08	0.06	0.22	0.04	0.04	0.07	0.10	4.98	9.23
	N×S		0.16	0.12	0.44	0.08	0.08	0.14	0.19	9.97	18.46

T0= normal Hoagland solution (control), T1 = 30 mg/L ZnO and Fe_3O_4 NPs, T2= 60 mg/L ZnO and Fe_3O_4 NPs and T3= 90 mg/L ZnO and Fe_3O_4 NPs.

S= salinity treatments, N= nano treatments. 1st = First season, 2nd = Second season.

seasons, the foliar application of a combination of ZnO and Fe_3O_4 NPs in Hoagland solution significantly increased non-enzymatic (vitamins A, and C) and enzymatic antioxidants (POD and SOD) in *Moringa* seedlings in comparison with control plants. The elevated amount in non-enzymatic and enzymatic antioxidants may be attributed to the beneficial effects of Fe and Zn represented in the increasing liberation of more nutrients from the unavailable reserves through correcting iron and zinc deficiency thus resulting in photosynthesis efficiency, increasing amino acids and vitamins to be absorbed by plant roots. This may be attributed to the importance of iron as a cofactor for many enzymes that catalyze unique biochemical reactions that are essential plant development such as chlorophyll and thylakoid syntheses and chloroplast development (Miller et al., 1995). Meanwhile, zinc is an essential element for plants that act as a metal component of various enzymes or as a functional structure or regulatory cofactor for protein synthesis and photosynthesis (Marschner, 1995). Also, Chang and Sung (1998) concluded that priming with antioxidant compounds such as ascorbic acid could increase free radical scavenging enzymes such as superoxide dismutase (SOD), and peroxidase in seeds.

Salt-stressed *Moringa* plants accumulated lower amounts of Na^+, Cl^- and higher amount of N, K^+, P, Ca^{2+}, Mg^{2+}, Fe and Zn upon foliar application of ZnO and Fe_3O_4 NPs-containing Hoagland solution when compared to those of the salt-stressed plants that received only foliar application of Hoagland solution (Table 3). The accumulation of less Na^+ is an important indicator of salt tolerance in plants as those subjected to foliar applications with ZnO and Fe_3O_4 NPs-containing Hoagland solution showed less accumulation of Na^+ in their shoots either in stress or non-stress conditions. The reduction of Na^+ in shoots of *Moringa* plants grown under the highest salinity level and sprayed with T2 reached 37 and 32 % in first and second seasons, respectively, in comparison to plants that received only Hoagland solution and grown under the same salinity level (9000 pm). At highest salinity level, the increase of K^+ in T2- sprayed plants reached 21 and 22% in both the first and second seasons, respectively, over Hoagland-sprayed plants. The importance of determining percentage Na^+ and K^+ in the plants is because they reflect salt tolerance in plants (Tunçtürk et al., 2011). Foliar feeding with micronutrients could partially alleviate the adverse effect of NaCl on nutrients uptake through improving root growth and preventing nutritional disorders and consequently resulting in an increase in nutrients uptake by the roots (El-Fouly et al., 2002). Also, zinc may help nutrient translocation from aged cells to newborn cells (Rockenfeller and Madeo, 2008). Zinc may, therefore, play an important role in membrane permeability, phospholipids (P) accumulation, and free oxygen radical scavenging. These results correlate with the findings of Qu et al. (2009) who reported that zinc application could alleviate possible Na^+ and Cl^- injury in plants.

Our results reveal that salt toxicity in *Moringa* plants can be alleviated by foliar spray of nano- zinc and iron. The results are consistent with Cakmak and Marschner (1988) who reported that zinc could play an important role in the maintenance of the structural integrity of the plasma membrane and thus control Na and other toxic ions uptake. Similarly, Saleh and Maftoun (2008) observed that zinc application reduced Na^+ concentration in rice shoot. Cakmak and Marschner (1988) reported that under zinc application, the activity of membrane-bound nicotinamide adenine dinucleotide phosphate (NADPH) oxidase producing reactive oxygen species (ROS), decreased.

In this study, alleviation of salt stress can be attributed to two reasons: first, promoting effect of spraying nutrients in Hoagland solution on *Moringa* plants grown under salt stress conditions and control conditions; second, the properties of ZnO and Fe_3O_4 NPs (larger specific surface area and moew reactive areas) that help in enhanced enzyme activity related to salt tolerance. Thus the Fe_3O_4 NPs were found to induce oxidative stress and higher antioxidative enzyme activity than the bulk Fe_3O_4 particles.

Conclusion

In this study, we succeeded in showing that salt stress can be alleviated in *Moringa* plants using foliar applications of ZnO and Fe_3O_4 NPs-containing Hoagland solution in comparison to spraying only with normal solution. Growth parameters and chemical composition related to salt tolerance were enhanced when nano-forms of Fe and Zn were used in Hoagland solution (60 mg/L).

Conflict of Interest

The authors have not declared any conflict of interest.

REFERENCES

Abbaspour H (2012). Effect of salt stress on lipid peroxidation, antioxidative enzymes, and proline accumulation in pistachio plants. J. Med. Plants Res. 6:526-529.

Abogadallah GM (2010). Antioxidative defense under salt stress. Plant Signal. Behav. 5(4):369-374.

Akram NA, Ashraf M, Al-Qurainy F (2012). Aminolevulinic acid-induced changes in some key physiological attributes and activities of antioxidant enzymes in sunflower (*Helianthus annuus* L.) plants under saline regimes. Sci. Hort. 142:143- 148.

Al-Aghabary K, Zhu ZJ, Shi QH (2005). Influence of Silicon Supply on Chlorophyll Content, Chlorophyll Fluorescence and Antioxidative Enzyme Activities in Tomato Plants under Salt Stress. J. Plant Nutr. 27:2101–2115.

Allen SF, Grimshaw HF, Rowl AB (1984). Chemical Analysis. In: Moor PD, SB Chapman (Eds.), Methods in plant Ecology. Blackwell, Oxford. pp. 185-344.

Amako A, Chen K, Asada K (1994). Separate assays specific for ascorbate peroxidase and for chloroplastic and cytosolic isoenzymes of ascorbate peroxidase in plants. Plant Cell Physiol. 35:497-504.

Anonymous (2009). Nano technology in agriculture. J. Agric. Technol.

114:54-65.
Association of Official Analytical Chemists (AOAC) (1999). Official Methods of Analysis (21st edition). Washington DC, USA. pp. 600-792.
Aslam M, Qureshi RH, Ahmed N (1993). A rapid screening technique for salt tolerance in rice *(Oryza sativa* L.*)*. Plant Soil 150(1):99–107.
Ball P (2002). Natural strategies for the molecular engineer. Nanotechnology 13:15-28.
Bardhan R, Wang H, Tam F, Halas NJ (2007). Facile chemical approach to ZnO submicrometer particles with controllable morphologies. Langmuir 23:5843-5847.
Bates LE, Waldern RP, Teare ID (1973). Rapid determination of free proline for water stress studies. Plant Soil 39:205-207.
Bello AG, Igbokwe GO (2013). Germination and seedling growth of *Acacia senegal* (l.) Willd and *Parkia biglobosa* (jacq.) as influenced by salinity in Sokoto state Nigeria. Int. J. Forest Soil Erosion 3(1):32-34.
Boguszewska D, Zagdańska B (2012). ROS as Signaling Molecules and Enzymes of Plant Response to Unfavorable Environmental Conditions, Oxidative Stress - Molecular Mechanisms and Biological Effects.
Boulos L (1999). Flora of Egypt, Vol. I (Azollaceae - Oxalidaceae). Al-Hadara publ., Cairo, Egypt. 417P
Cakmak I, Marschner H (1988). Enhanced superoxide radical production in roots of zinc deficient plants. J. Exp. Botany 39:1449-1460.
Cakmak I, Marschner H, Bangert F (1989). Effect of zinc nutritional status on growth, protein metabolism and levels of indole-3-acetic acid and other phytohormones in bean (*Phaseolus vulgaris* L.). J. Exp. Bot. 40:404-412.
Cazzonelli CI, Pogson BJ (2010). Source to sink: regulation of carotenoid biosynthesis in plants. Trends Plant Sci. 15:266-274.
Chang SM, Sung JM (1998). Deteriorative changes in primed shunken-2 sweet corn seeds during storage. Seed Sci. Technol. 26:613-626.
Chen THH, Murata N (2000). Enhancement of tolerance of abiotic stress by metabolic engineering of betaines and other compatible solutes. Plant Biol. 5:250-257.
Chen Z, Cuin TA, Zhou M, Twomey A, Naidu BP, Shabala S (2007). Compatible solute accumulation and stress-mitigating effects in barley genotypes contrasting in their salt tolerance. J. Exp. Bot. 58:4245-4255.
Corredor E, Testillano PS, Coronado MJ (2009). Nanoparticle penetration and transport in living pumpkin plants: in situ subcellular identification. BMC Plant Biol. 9:45-54.
Dimkpa CO, Latta DE, McLean JE, Britt DW, Boyanov MI, Anderson AJ (2013). Fate of CuO and ZnO Nano- and Microparticles in the Plant Environment, Environ. Sci. Technol. 47(9):4734–4742.
Dubois M, Smith F, Gilles KA, Hamilton JK, Rebers PA (1956). Colorimetric method for determination of sugars and related substances. Anal. Chem. 28(3):350-356.
El-Feky SA, Mohammed MA, Khater MS, Osman YAH, Elsherbini E (2013). Effect of magnetite Nano-Fertilizer on Growth and yield of *Ocimum basilicum* L. Int. J. Indigenous Med. Plants 46(3):1286-1293.
El-Fouly MM, Mobarak ZM, Salama ZA (2002). Micronutrient foliar application increases salt tolerance of tomato seedlings. Proceedings of the Symposium on Techiques to Control Salination for Horticultural Productivity. Acta Hort. 573:377-385.
El-Fouly MM, Abou El-Nour EAA, Abdel-Maguid AA (2004). Counteracting effect of foliar application of macronutrients on spinach beet (*Beta vulgaris* var. cycla) grown under NaCl-salinity stress. Agric. Cairo Univ. 55:587-602.
El-Kereti MA, El-Feky SA, Khater MS, Osman YA, El-sherbini EA (2013). ZnO Nanofertilizer and He Ne Laser Irradiation for Promoting Growth and Yield of Sweet Basil Plant. Recent Patents on Food Nutr. Agric. 5(3):1-13.
Foyer CH, Noctor G (2009). Redox regulation in photosynthetic organisms: signaling, acclimation, and practical implications. Antioxid. Redox Signal 11:861- 905.
Ginnopolitis NC, Ries SK (1977). Superoxide dismutases: I. Occurrence in higher plants. Plant Physiol. 68:548-552.
Gouia H, Ghorbal MH, Touraine B (1994). Effects of NaCl on flows of N and mineral ions and NO_3^- reductase rate within whole plants of salt-sensitive bean and salt-tolerant cotton. Plant Physiol. 105:1407-1418.
Gunes A, Inal A, Alpaslan M (1996). Effect of salinity on stomatal resistance, proline and mineral composition of pepper. J. Plant Nutr. 19(2):389-396.
Hairmansis A, Berger B, Tester M, Roy JS (2014). Image-based phenotyping for non-destructive screening of different salinity tolerance traits in rice. Rice 6:17.
Hajibagheri MA, Yeo AR, Flowers TJ, Collins JC (1989). Salinity resistance in *Zea mays* fluxes of potassium, sodium and chloride, cytoplasmic concentrations and microsomal membrane lipids. Plant Cell Environ. 12:753-757.
Hamayun M, Khan SA, Khan AL, Shinwari ZK, Ahmad N, Kim Y, Lee I (2011). Effect of foliar and soil application of nitrogen, phosphorus and potassium on yield components of Lentil. Pak. J. Bot. 43(1):391-396.
Hazra GC, Mandal B, Mandal LN (1987). Distribution of zinc fractions and their transformation in submerged rice soils. Plant Soil 104:175-181.
Hoagland DR, Arnon DI (1950). The water culture method for growing plant without soil. University of California Agriculure Experiment Station. Circular No. 347: 39 p.
Jackson ML (1967). Soil chemical analysis advanced course. University of Wisconsin, Madison, USA. 930p.
Jameel A, Kahayri M (2002). Growth, proline accumulation and ion content in sodium chloride-stressed callus of date palm. In vitro Cell. Dev. Biol. 38:79-82.
James CS (1995). Analytical chemistry of foods. Blackie Academic & Professional, London. pp: 91-105.
Kao WY, Tsai TT, Shih CN (2003). Photosynthetic gas exchange and chlorophyll a fluorescence of three wild soybean species in response to NaCl treatments. Photosynthetica 41:415-419.
Karimi M, De Meyer B, Hilson P (2005). Modular cloning in plant cells. Trends Plant Sci. 10:103-105.
Kumar K, Arvind K, Vidyasagar R, Rao K (1988). Studies on growth and activity of photosynthetic enzymes on *Sorghum bicolor* L. as influenced by micronutrients. Proc. Natl. Sci. Acad. Part B Biol. Sci. 54:75-80.
Laurent S, Forge D, Port M, Roch A, Robic C, Elst LV, Muller RN (2008). Magnetic Iron Oxide Nanoparticles: Synthesis, Stabilization, Vectorization, Physicochemical Characterizations, and Biological Applications. Chem. Rev. 108:2064-2110.
Little TM, Hills FJ (1978). Agricultural Experimentation Design and Analysis. John Wiley & Sons, Inc. New York, USA. pp. 53-63.
Liu XM, Zhang FD, Zhang SQ, He XS, Fang R, Feng Z, Wang Y (2005). Effects of nano-ferric oxide on the growth and nutrients absorption of peanut. Plant Nutr. Fert. Sci. 11:14-18.
Malakouti M, Tehrani M (2005). Effects of micronutrient on the yield and quality of agricultural products: Micronutrient with macro-effects. Tarbiat Modarres University Press, Tehran, Iran. P 230.
Marschner H (1995). Mineral nutrient of higher plants. Second Ed., Academic Press Limited. Harcourt Brace and Company Publishers, London. pp. 347-364.
Morant-Avice A, Pradier E, Houchi R (1998). Osmotic adjustment in triticales grown in presence of NaCl. Biol. Plant 41:227-234.
Miller GW, Huang J, Welkie GW, Pushmik JC (1995). Function of iron in plants with special emphasis on chloroplasts and photosynthetic activity. In: Abadia J (Ed.), Iron nutrition in soils and Plants. Kluwer Academic Publishers, Dordecht. pp. 19-28.
Mittler R (2002). Oxidative stress, antioxidants and stress tolerance-Review. Trends Plant Sci. 7 (9): 405-410.
Munns R (2005). Genes and salt tolerance: bringing them together. New Phytol. 67:645-663.
Munns R, James RA, Läuchli A (2006). Approaches to increasing the salt tolerance of wheat and other cereals. Exp. Bot. 57(5):1025-43.
Munns R, Tester M (2008). Mechanisms of salinity tolerance. Annu. Rev. Plant Biol. 59:651-681.
Nornai R (1982). Formula for determination of chlorophyll pigments extracted with N.N. dimethyl formamide. Plant Physiol. 69:1371-1381.
Pareek A, Singla SL, Grover A (1997). Salt responsive proteins/genes in crop plants. In: Jaiwal PK, Singh RP, Gulati A (eds.), Strategies For Improving Salt tolerance in Higher Plants. Oxford and IBH Publ. Co., New Delhi, India. pp. 365-391.
Price ML (2000). The Moringa Tree. Echo Technical Note. ECHO, North

Ft. Myers, Florida, USA. Available at: http://chenetwork.org/files_pdf/Moringa.pdf

Piotr S, Klobus G (2005). Antioxidant defense in the leaves of C3 and C4 plants under salinity stress. Physiol. Plant. 125:31-40.

Qu YN, Zhou Q, Yu BJ (2009). Effects of Zn^{2+} and Niflumic Acid on Photosynthesis in *Glycine soja* and *Glycine max* Seedlings under NaCl Stress, Environ. Exp. Bot. 65:304-309.

Rockenfeller P, Madeo F (2008). Apoptotic death of ageing yeast. Exp. Gerontol. 43:876-881.

Romheld V, Marschner H (1991). Function of micronutrients in plants. In: "Micronutrients in Agriculture." Soil Sci. Society of America Inc., Madison Wisconsin, USA. pp. 297-299.

Ruffini CM, Roberto C (2009). Nanoparticles and higher plants. Caryologia 62(2):161165.

Saleh J, Maftoun M (2008) Interactive effects of NaCl levels and zinc sources and levels on the growth and mineral composition of rice. J. Agric. Sci. Technol. 10:325-336.

Shaheen S, Naseer S, Ashraf M, Akram NA (2013). Salt stress affects water relations, photosynthesis and oxidative defense mechanisms in *Solanum melongena* L. J. Plant Interact. 8:85-96.

Shao HB, Chen XY, Chu LY (2006). Investigation on the relationship of Proline with wheat antidrought under soil water deficits. Biointerfaces 53:113-119.

Sharma P, Jha AB, Dubey RS (2010). Oxidative stress and antioxidative defense system in plants growing under abiotic stresses. In: Pessarakli M (ed.), Handbook of Plant and Crop Stress, 3rd edititon. CRC Press; Florida, USA. pp. 89–138.

Sheykhbaglou R, Sedghi M, Shishvan MT, Sharifi SR (2010). Effect of nano iron particles on agronomic traits of soybean. Notulae Sci. Biol. 2:112-113.

Soliman ASH, Shanan NT, Massoud ON, Swelim DM (2012). Improving salinity tolerance of *Acacia saligna* (Labill.) plant by arbuscular mycorrhizal fungi and Rhizobium inoculation. Afr. J. Biotechnol. 11(5):1259-1266.

Somali MA, Bajnedi MA, Al-Fhaimani SS (1984). Chemical composition and characteristics of *Moringa peregrine* seeds and seed oil. J. Am. Oil Chem. Soc. 61:85-86.

Tunçtürk M, Tunçtürk R, Yildirim B, Çiftçi V (2011). Effect of salinity stress on plant fresh weight and nutrient composition of some Canola (*Brassica napus* L.) cultivars. Afr. J. Biotechnol. 10(10):1827-1832.

Tzortzakis NG (2010). Potassium and Calcium Enrichment Alleviate Salinity- Induced Stress in Hydroponically Grown Endives, J. Hortic. Sci. 37:155- 162.

Vallee BL, Auld DS (1990). Zinc Coordination, Function, and Structure of Zinc Enzymes and Other Proteins. Biochemistry 29: 5647-5659.

Vitoria AP, Lea PJ, Azevado RA (2001). Antioxidant enzymes responses to cadmium in radish tissues. Phytochemistry 57:701-710.

Wang H, Kou X, Pei Z, Xiao JQ, Shan X, Xing B (2011). Physiological effects of magnetite (Fe_3O_4) nanoparticles on perennial ryegrass (*Lolium perenne* L.) and pumpkin (*Cucurbita mixta*) plants, Nanotoxicology 5(1):30-42.

Wu G, Wei ZK, Shao HB (2007). The mutual responses of higher plants to environment: physiological and microbiological aspects. Biointerfaces 59:113-119.

Yildirim E, Turan M, Guvenc I (2008). Effect of Foliar Salicylic Acid Applications on Growth, Chlorophyll, and Mineral Content of Cucumber Grown under Salt Stress. J. Plant Nutr. 31: 593-612.

Zaghloul MS, Abdel-Wahab RH and Moustafa AA (2008). Conservation of *Acacia tortilis* subsp. *raddiana* populations in Southern Sinai, Egypt. III- Population Structure and Dynamics. Bull. Assuit Univ. 37(1):85-113.

Zhu H, Han J, Xiao JQ, Jin Y (2008). Uptake, translocation and accumulation of manufactured iron oxide nanoparticles by pumpkin plants, J. Environ. Monitor. 10:713-717.

Zhu ZJ, Wei GQ, Li J, Qian QQ, Yu JQ (2004). Silicon alleviates salt stress and increases antioxidant enzymes activity in leaves of Salt-stressed Cucumber (*Cucumis sativus* L.). Plant Sci. 167:527–533.

Phylogenetic relationship among Mango (*Mangifera indica* L.) Landraces of Saurashtra based on DNA fingerprinting

R. S. Tomar[1]*, H. P. Gajera[1], R. R. Viradiya[2], S. V. Patel[1] and B. A. Golakiya[1]

[1]Department of Biotechnology, Junagadh Agricultural University, Junagadh – 362 001, Gujarat, India.
[2]Department of Horticulture, Junagadh Agricultural University, Junagadh – 362 001, Gujarat, India.

Twenty mango cultivars (*Mangifera indica* Linn) collected from Gir region of Saurashtra were examined by ISSR markers. Of the 50 primers screened, 21 primers gave reproducible, polymorphic DNA amplification patterns, and were selected to construct a DNA fingerprinting map to distinguish the genotypes of mango. According to the banding patterns obtained with 21 selected primers, all cultivars tested in this study except Jamadar and Kesar were distinguished from each other and showed ample genetic diversity, indicating that ISSR-PCR was an effective method for cultivar identification of mangoes. Based on 125 selected bands, all Gir mango landraces tested were clustered into a three big groups with 'Kaju' and 'Khodi' in first group, Dudh Pendo, Sopari, Jamadar, Kesar and Ashadhiya in second group, while the third cluster was composed of Agargato, Amir Pasand, Pethal, Gajariyo, Chhappaniyo, Alphanso, Neelum, Jamrukhiyo, Kavasji Patel, Giriraj, Amrutiyo, Dasheri and Deshi based on UPGMA analysis, indicating that some Gir landraces had a close relationship with each other, while some were drastically dissimilar from other landraces.

Key words: *Mangifera indica* L., Genetic diversity, ISSR markers, Mango landraces.

INTRODUCTION

Mango (*Mangifera indica* L.) a diploid fruit tree with 2n=40 chromosomes (Mukherjee, 1950) originated in the Indo-Burma region during the earlier period of the Cretaceous era (Yonemori et al., 2002) and gradually spread to the tropical and subtropical regions of the world. India is thought to be the primary centre of diversity along with its status as the centre of origin for mango. Presently, India harbours more than 1000 mango cultivars and represents the biggest mango germ pool in the world. Australia, China, USA, Israel and Thailand are the other regions that maintain such healthy mango germ pool.

In the last decade molecular methods have been implemented in mango and these included isozymes (Degani et al., 1990), variable number tandem repeats (VNTR) (Adato et al. 1995), random amplified polymorphic DNA (RAPDs) (Schnell et al. 1995; Bally et al.1996; Lopez-Valenzuela et al., 1997; Ravishankar et al.2000; Karihaloo et al. 2003); inter-simple-sequence-repeats (ISSRs) (Eiadthong et al., 1999; Gonzalez et al., 2002) and amplified fragment length polymorphism (AFLPs) (Fang et al., 1999; Eiadthong et al., 2000). RAPDs and AFLPs have also been applied in discrimination of mango cultivars in China (Xu et al., 1998; Fang et al., 1999, 2000, 2001), while ISSRs in mango have not been reported in India. Among these, Inter Simple Sequence Repeat, ISSR (Zietkiewicz et al., 1994) is a reproducible semi-arbitrary primed PCR method that uses simple sequence repeats as primers, combining most of the advantages of microsatellites and Amplified Fragment Length Polymorphism (AFLP), to the universality of Randomly Amplified Polymorphic DNA. ISSRs offer greater probability than any other PCR marker system in the repeat regions of the genome, which are the most potent regions for producing cultivar-specific markers. This is also the attribute of ISSRs, which renders them useful asa supplementary system to any of the random dominant marker systems. Automated PCR base makes ISSRs the markers of choice for screening genotypes.

In Gujarat, during the Mugal empire lots of mango cultivars were planted in the surrounding of Gir region. Among these

*Corresponding author. E-mail: rukam@jau.in.

Table 1. Polymorphism obtained with different ISSR primers generated from 20 *M. indica* genotypes.

S/No.	Name of Primer	Primer sequence 5'-3'	Tm (^{0}C)	% GC
1	UBC-807	$(AG)_8T$	42.4	47.1
2	UBC-808	$(AG)_8C$	43.7	52.9
3	UBC-809	$(AG)_8G$	43.5	52.9
4	UBC-811	$(GA)_8C$	43.2	52.9
5	UBC-812	$(GA)_8A$	44.3	47.1
6	UBC-817	$(CA)_8A$	52.7	47.1
7	UBC-825	$(AC)_8T$	49.2	47.1
8	UBC-834	$(AG)_8YT$	45.3	44.4
9	UBC-835	$(AG)_8YC$	45.5	50.0
10	UBC-836	$(AG)_8YA$	43.3	44.4
11	UBC-840	$(GA)_8YT$	45.7	44.4
12	UBC-844	$(CT)_8RC$	46.5	50.0
13	UBC-845	$(CT)_8RG$	47.7	50.0
14	UBC-848	$(CA)_8RG$	55.5	50.0
15	UBC-855	$(AC)_8YT$	51.9	44.4
16	UBC-856	$(AC)_8YA$	49.8	44.4
17	UBC-857	$(AC)_8YG$	43.7	50.0
18	UBC-864	$(ATG)_6$	51.2	33.3
19	UBC-884	HBH $(AG)_7$	41.9	41.2
20	UBC-889	DBD $(AC)_7$	47.0	41.2
21	UBC-891	HVH $(TG)_7$	51.7	41.2

cultivars Kesar variety is famous throughout the world for its taste and flavor. Along with Kesar there are many other landraces which has the equal potential. Precise information on the genetic relationships within such germplasm diversity is needed for carrying out efficient breeding programmes. In view of this, the present experiment was conducted during December 2009 with 20 mango landraces of Gir region. These 20 mango landraces were analyzed for phylogenetic relationship with ISSR's in order to assess the genetic diversity.

MATERIALS AND METHODS

Plant material

For the present experiment, 20 landraces of Gir region were selected on the basis of their consistency in behavior for the last 30 years at their growing region. Their fruit size, sweetness and yielding behavior were also considered for selection.

DNA isolation

DNA was extracted from mango leaves using the hexadecyltrimethylammonium bromide (CTAB) method described by Doyle and Doyle (1990) with a little modification. Sample of 0.5 g leaves were ground to fine power in liquid nitrogen and added into a 10 ml tube with 4 ml preheated (65°C) extraction buffer consisting of 2% CTAB, 1.4 mol L^{-1} NaCl, 0.1% (v/v) β-mercaptoethanol, 20 mmol L^{-1} EDTA, 100 mmol L^{-1} Tris-HCl (pH 8.0), and 1% (w/v) PVP-40. The homogenates were incubated at 65°C for 1 h and extracted with 4 ml chloroform: isoamyl alcohol (24:1) solution. The tubes were mixed for about 5 min, and spun at 12000 rpm for 10 min. The supernatant was transferred to a new tube containing twice volume of 100% ethanol and 1/10 volume 3 mol L^{-1} NaAc, mixed gently, left at -20°C for 1 h and spun at 12000 rpm for 15 min. The supernatant was then discarded and the pellet washed twice with 70% ethanol. The pellet was dried at room temperature, and resuspended into 200 µl 0.1 × TE with RNase. The DNA was precipitated with 400 µl 100% ethanol at 20°C for 1 h and spun at 12000 rpm for 15 min. The pellet was resuspended into 100 µl 0.1 × TE. DNA was quantified by picodrop.

ISSR markers

In this study, initially 50 ISSR primers were used for screening of mango landraces. But only the polymorphic ones were used for further analyses and estimating of genetic diversity among mango landraces. ISSR primers synthesized by Bangalore Genei were used for PCR. The polymorphic ISSR markers are given in Table 1 along with Tm (°C) and %GC content. An amplification reaction was performed as follows:

25 ng of template DNA, 200 µM of each dNTPs (Bangalore Genei), 20 ng of primer (Bangalore Genei), 0.5 U of Taq DNA polymerase (Bangalore Genei, India) and 1× reaction buffer (Banglore Genei, India) in a total volume of 25 µl. Amplification was carried out for 45 cycles of 1 min at 94°C, 1 min at 45 (± 5)°C and 2 min at 72°C in a Bio-Rad thermal cycler.

Agarose gel electrophoresis

Amplified products were separated in 1.5% agarose gel containing ethidium bromide using 1× TAE buffer. A constant voltage of 55 was provided for 3 to 4 h. DNA fragments were visualized under

Table 2. Polymorphism obtained with different ISSR primers generated from 20 *M. indica* genotypes.

S/No.	Name of primer	Polymorphic bands			Monomorphic bands	Total bands	Polymorphic (%)	Average No. of loci produced in 20 genotypes	Resolution power
		S	U	T					
1	UBC-807	2	2	4	1	5	80.0	1.40	2.80
2	UBC-808	4	2	6	0	6	100.0	1.55	3.10
3	UBC-809	4	0	4	1	5	80.0	3.40	6.80
4	UBC-811	2	1	3	2	5	60.0	3.55	7.10
5	UBC-812	5	1	6	0	6	100.0	1.65	3.30
6	UBC-817	4	0	4	0	4	100.0	1.10	2.20
7	UBC-825	4	0	4	0	4	100.0	1.60	3.20
8	UBC-834	4	2	6	0	6	100.0	2.15	4.30
9	UBC-835	7	1	8	0	8	100.0	4.40	8.80
10	UBC-836	5	0	5	1	6	83.3	3.05	6.10
11	UBC-840	5	6	11	0	11	100.0	2.75	5.50
12	UBC-844	2	2	4	0	4	100.0	0.40	0.80
13	UBC-845	3	2	5	0	5	100.0	0.60	1.20
14	UBC-848	4	4	8	0	8	100.0	0.90	1.80
15	UBC-855	4	5	9	0	9	100.0	1.50	3.00
16	UBC-856	5	2	7	0	7	100.0	0.70	1.40
17	UBC-857	5	2	7	0	7	100.0	1.05	2.10
18	UBC-864	6	0	6	0	6	100.0	1.20	2.40
19	UBC-884	3	1	4	0	4	100.0	1.70	3.40
20	UBC-889	3	1	4	1	5	80.0	2.25	4.50
21	UBC-891	4	0	4	0	4	100.0	0.70	1.40
Total		85	34	119	6	125	94.4	1.80	3.58

S = Shared; U = Unique; T = Total polymorphic bands.

UV light. The patterns were photographed using Geldoc system(Bio-Rad) and stored as digital pictures. The reproducibility of the the amplification was confirmed by repeating each experiment three times.

Data collection and analysis

Since ISSR primers are dominant markers, amplified bands were scored for presence (1) or absence (0) of bands. Statistical analysis of data was performed by means of NTSYS-pc software (Rohlf, 2000). Cluster analysis demonstrating genetic relationships of accessions were generated using unweighted pair-group method using arithmetic averages (UPGMA) and simple matching coefficient. Resolution power was calculated according to Prevost and Wilkinson (Prevost and Wilkinson, 1999), which states that the Resolving power (Rp) of a primer is:

$$Rp = \sum IB$$

where IB (band informativeness) takes the value of: $1 - [2 \times (0.5 - p)]$, where p being the proportion of the 20 landraces (Mango) containing the band.

RESULTS AND DICUSSION

A set of 50 ISSR primers were used for initial screening of mango landraces, of which 21 primers (Table 1) were selected for further analysis of all the landraces depending on the amplification of the DNA. The information obtained on banding pattern by the analysis of twenty landraces of Gir region with 21 ISSR primers is summarized in Table 2. Twenty one ISSR primers used in this study yielded a total 125 bands and 119 scorable polymorphic

Table 3. Jaccard's similarity matrix for twenty mango landrace of Gir region.

	1	2	3	4	5	6	7	8	9	10	11	12	13	14	15	16	17	18	19	20
1	1.00																			
2	0.45	1.00																		
3	0.40	0.38	1.00																	
4	0.35	0.27	0.27	1.00																
5	0.42	0.38	0.44	0.27	1.00															
6	0.47	0.42	0.40	0.27	0.54	1.00														
7	0.37	0.42	0.37	0.24	0.45	0.47	1.00													
8	0.38	0.36	0.24	0.30	0.46	0.45	0.45	1.00												
9	0.38	0.37	0.34	0.29	0.47	0.47	0.46	0.42	1.00											
10	0.34	0.29	0.33	0.24	0.45	0.38	0.44	0.33	0.54	1.00										
11	0.40	0.42	0.31	0.33	0.40	0.44	0.36	0.36	0.46	0.38	1.00									
12	0.31	0.36	0.35	0.41	0.40	0.31	0.33	0.33	0.35	0.36	0.36	1.00								
13	0.38	0.39	0.38	0.29	0.47	0.37	0.43	0.37	0.42	0.40	0.41	0.37	1.00							
14	0.31	0.26	0.22	0.21	0.43	0.33	0.38	0.56	0.47	0.31	0.33	0.23	0.39	1.00						
15	0.34	0.41	0.40	0.20	0.39	0.46	0.36	0.33	0.32	0.28	0.33	0.26	0.34	0.27	1.00					
16	0.48	0.40	0.33	0.26	0.43	0.49	0.48	0.40	0.45	0.38	0.48	0.33	0.34	0.39	0.33	1.00				
17	0.50	0.41	0.37	0.36	0.50	0.53	0.46	0.48	0.55	0.44	0.52	0.41	0.46	0.50	0.33	0.55	1.00			
18	0.40	0.36	0.35	0.39	0.44	0.47	0.40	0.41	0.55	0.47	0.46	0.33	0.41	0.35	0.33	0.41	0.56	1.00		
19	0.36	0.43	0.37	0.21	0.45	0.47	0.35	0.36	0.47	0.37	0.42	0.31	0.38	0.31	0.59	0.38	0.39	0.39	1.00	
20	0.32	0.36	0.37	0.28	0.41	0.45	0.33	0.31	0.47	0.43	0.42	0.38	0.38	0.27	0.37	0.33	0.47	0.42	0.49	1.00

1. Deshi; 2. Ashadhiya; 3. DudhPendo; 4. Khodi; 5. Giriraj; 6. KavasjiPatel; 7. Gajariyo; 8. AmirPasand; 9. Jamrukhiyo; 10. Chhappniyo; 11. Amrutiyo; 12. Kaju; 13. Pethal; 14. Agargato; 15. Kesar; 16. Dasheri; 17. Neelum; 18. Alphanso; 19. Jamadar; 20. Sopari

markers, accounting for 95.2% of total reproducible amplification products in the range 42 to 2522 bp after amplification of total genomic DNA of twenty landraces of mango. Each primer could amplify 4 to 11 DNA bands, of which primer UBC-840 generated the highest number of bands followed by UBC 855, UBC 835, and UBC 848. While primer UBC 817, UBC 825, UBC 844, UBC 884 and UBC 891 showed lowest number of DNA bands.

Out of the total 119 polymorphic DNA bands, 34 DNA bands were unique indicating of its presence only in one of the landraces. While 85 polymorphic DNA bands were shared among few landraces. Except UBC 807, UBC 809, UBC 811, UBC 836 and UBC 889 all other primers showed 100% polymorphism. The average number of loci produced in twenty genotypes by the primers ranged from 0.40 (UBC 844) to 4.40 (UBC 835) with an average of 1.80. The resolution power (Rp) of 21 ISSR primers ranged from 0.80 to 8.80 with an average of 3.58 (Table 2). The clearest 125 amplified bands were treated as dominant genetic markers. For each sample ISSR bands were scored as 1 (present) or 0 (absent) and these binary data were used to assemble a rectangular matrix. The use of UPGMA algorithm permitted to cluster the data and to draw the relationships between the tested landraces. Jaccard's similarity between landraces ranged from 0.21 to 0.59 (Table 3), with maximum genetic similarity between 'Jamadar' and 'Kesar' and lowest genetic similarity between Khodi and Agargato. The landraces having more than 50% similarity are as follow:

Giriraj and KavasjiPatel; KavasjiPatel and Neelum; AmirPasand and Agargato; Jamrukhiyo and Chhappaniyo; Jamrukhiyo and Neelum; Jamrukhiyo and Kesar; Amrutiyo and Neelum; Neelum and Alphanso.

In addition to this, relationships among the different landraces was distributed among three

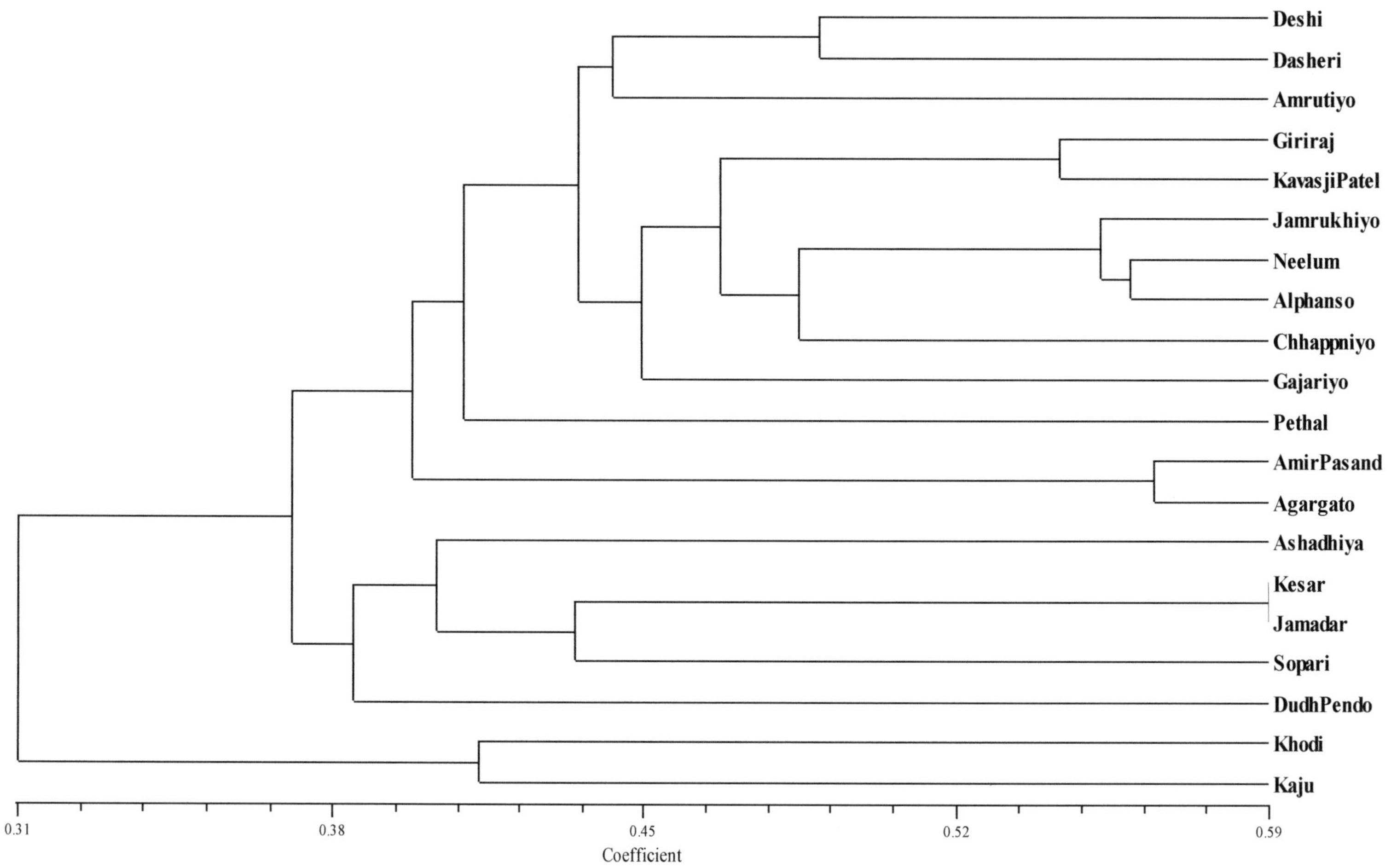

Figure 1. Dendrogram of phylogenetic relationship among 20 mango landraces by ISSR markers based on UPGMA analysis.

main divergent clusters significantly (Figure 1). The first cluster consists of 'Kaju' and 'Khodi'. The second cluster was composed of two subclusters (A and B): the subcluster A consists only of'Dudh Pendo', while the subcluster B consists of 'Sopari', 'Jamadar', 'Kesar' and 'Ashadhiya'. Again the third cluster was composed of two subclusters (C and D). Landraces Agargato and Amir Pasand were in subcluster C, while Pethal, Gajariyo, Chhappaniyo, Alphanso, Neelum, Jamrukhiyo, Kavasji Patel, Giriraj, Amrutiyo, Dasheri and Deshi were in cluster D.

In Figure 2 first two eigenvectors were plotted indicating the separation of population in four clear groups. These groups indicate the genotypic similarity among the groups. The first group consists of only Khodi, indicating genotypic diversity present from other genotypes. The second group consists of Agargato, Kaju, DudhPendo and Kesar, while group three consist of maximum genotypes indicating similarity among these

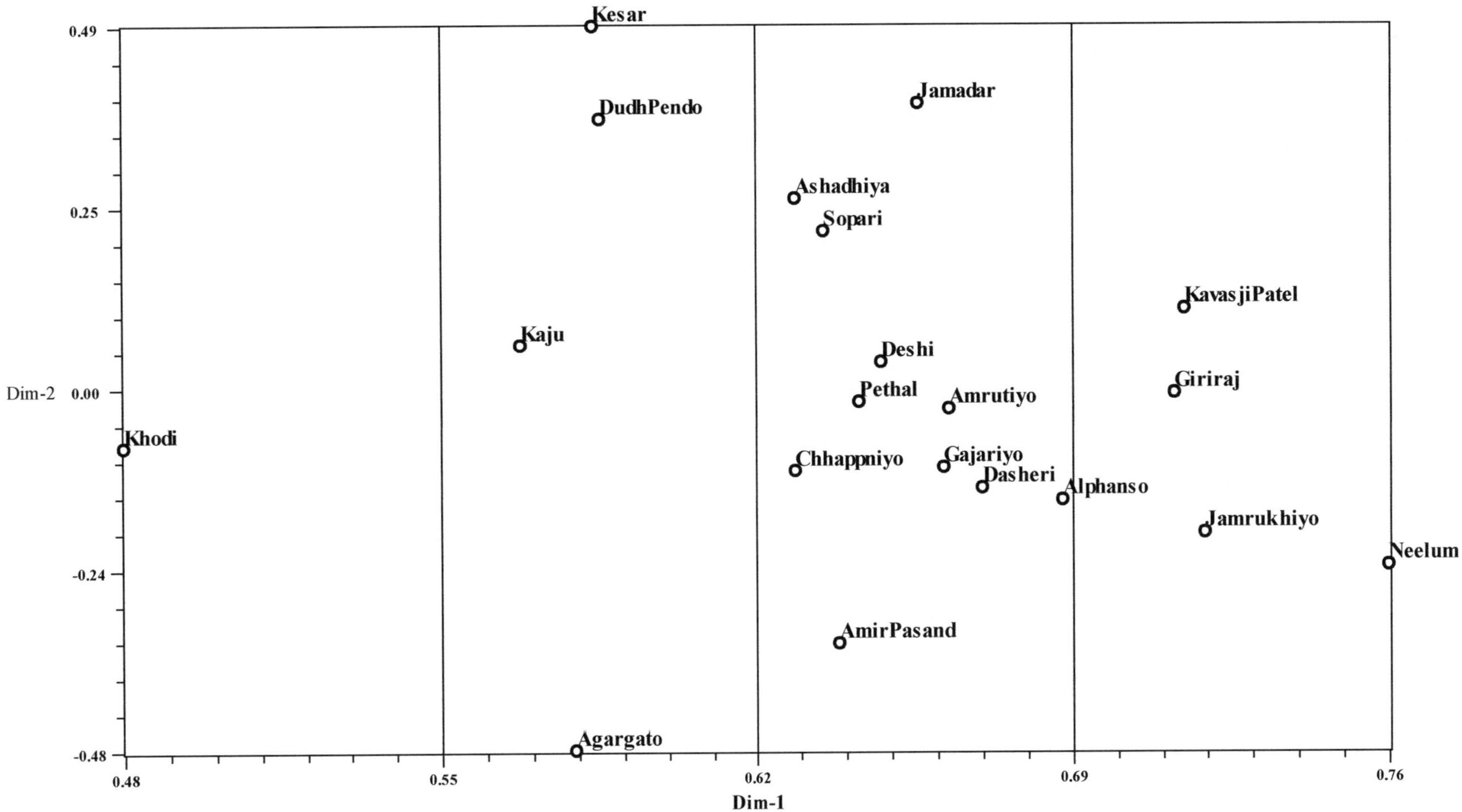

Figure 2. Principal coordinate analysis of 20 mango landraces using ISSR markers.

genotypes. The fourth group consists of Jamrukhiyo, Giriraj, KavasjiPatel and Neelum.

In this study, ISSR profiling technology was used in order to enlarge the number of molecular markers that are suitable in the molecular characterization and examination of the genetic relationships between the landraces of mango. The present work provides evidence that the ISSRs appear to be effective to explore the molecular polymorphism and to assess the genetic relationships in the mango.

Using ISSR assessment, it was demonstrated that most of the cultivars can be easily distinguished. Moreover, some fragments were uniquely amplified or absent in some of the landraces. These fragments are of great interest in optical management and genetic identification of *M. indica* accessions in the germplasm collection. Over all these data extends the knowledge of ISSR application as a molecular tool in mango as reported by Yonemori et al. (2002), Adato et al. (1995), Schnell et al. (1995), Lopez-Valenzuela et al. (1997), Karihaloo et al. (2003), Eiadthong et al. (1999), Eiadthong et al. (2000), Gonzalez et al.

(2002), Fang et al. (1999), Fang et al. (2000); Fang et al. (2001) and Xu et al. (1998), who have used ISSR and other markers for molecular characterization of mango.

REFERENCES

Adato A, Sharon D, Lavi U, Hillel J, Gazit S (1995). Application of DNA fingerprints for identification and genetic analyses of mango (*Mangifera indica*) genotypes. J. Am. Soc. Hortic. Sci., 120: 259-264.

Bally ISE, Graham GC, Henry RJ (1996). Genetic diversity of Kensington mango in Australia. Aus. J. Exp. Agri., 36: 243-247.

Degani C, El-Batsri R, Gazit S (1990). Enzyme polymorphism in mango. J. Am. Soc. Hortic. Sci., 115: 844-847.

Eiadthong W, Yonemori K, Kansaki S, Sugiura A, Utsunomiya N, Subhadrabandhu S (2000). Amplified fragments length polymorphism analysis for studying genetic relationships among *Mangifera* species in Thailand. J. Am. Soc. Hortic. Sci., 125: 160-164.

Eiadthong W, Yonemori K, Sugiura A, Utsunomiya N, Subhadrabandhu S (1999). Identification of mango cultivars of Thailand and evaluation of their genetic variation using the amplified fragments by simple sequence repeat- (SSR-) anchored primers. Sci Hortic., 82: 57-66.

Fang JG, Liu DJ, Zhang Z, Hillel J, Lavi U (1999). The construction of the fingerprinting of two mango cultivars using AFLP. J. Nanjing Agril. Univ., 22(2): 25-27.

Fang JG, Qiao YS, Zhang Z (2001). Application of AFLP in the cultivar identification of mango. Guihaia, 21(3): 281-283.

Fang JG, Zhang Z, Ma ZQ, Liu DJ, Wang SH, Lavi U (2000). The polymorphism and segregation patterns of AFLP markers in the F1 progenies from the cross of the two mango cultivars. Sci. Agric. Sin., 33(3): 19-24.

Gonzalez A, Coulson M, Brettell R (2002). Development of DNA Markers (ISSRs) in Mango. Acta Hortic., 575: 139-143.

Karihaloo JL, Dwivedi YK, Archak S, Gaikwad AB (2003). Analysis of genetic diversity of Indian mango cultivars using RAPD markers J. Hort. Sci. Biotechnol., 78: 285-289.

Lopez VJA, Martinez O, Paredes LO (1997). Geographic differentiation and embryo type identification in *Mangifera indica* L. cultivars using RAPD markers. Hortscience. 32(6): 1105-1108.

Mukherjee SK (1950). Mango: its allopolyploid nature. Nature, 166: 196-197.

Prevost A, Wilkinson MJ (1999). A new system of comparing PCR primers applied to ISSR fingerprinting of potato cultivars. TAG, 98: 107-112.

Ravishankar KV, Anand L, Dinesh MR (2000). Assessment of genetic relatedness among mango cultivars of Indian using RAPD Markers. J. Hortic. Sci. Biotechnol., 75: 198-201.

Rohlf FJ (2000). NTSYS Pc Numerical Taxonomy and Multivariate Analysis System User Guide, New York University Press, New York, USA, 38p.

Schnell RJ, Ronning CM, Knight RJ Jr (1995). Identification of cultivars and validation of genetic relationships in Mangifera indica L. using RAPD markers. TAG. 90: 269-274.

Xu BY, Jin ZQ, Peng SQ, Xu SP, Zhang XQ (1998). RAPD analysis of genomic DNA in mango cultivars in Hainan Island. Chin. J. Trop. Crop. 19(3): 33-37.

Yonemori K, Honsho C, Kanzaki S, Eiadthong W, Sugiura A (2002). Phylogenetic relationships of Mangifera species revealed by ITS sequences of nuclear ribosomal DNA and a possibility of their hybrid origin. Plant System. Evol., 231: 59–75.

Zietkiewicz E, Rafalski A, Labuda D (1994). Genome fingerprinting by simple sequence repeat (SSR)-anchored polymerase chain reaction amplification. Genomics, 20: 176–183.

Evaluation of garlic (*Allium sativum*) genotypes for yield and susceptibility to purple blotch

Ankur Agarwal[1] and R. S. Tiwari[2]

[1]Defence Institute of Bio-Energy Research, Defence R&D Organization, Goraparao, Haldwani, Uttarakhand, 263139 India.
[2]GBPUA & T, Pantnagar, Uttarakhand, 263145 India.

Twenty one genotypes of garlic (*Allium sativum* L.) were screened for yield and purple blotch susceptibility under field conditions for two years. Reaction of genotypes to purple blotch disease differed significantly during both years but the pooled analysis revealed that most of the variations were due to genotype × environment interaction. Nine genotypes viz., G-1, G-50, G-323, PGS-4, PGS-14, G-313, PGS-99-1, PGS-98 and PGS-99-2 fell into the category of resistant, five showed moderately resistant reaction to disease (PGS-13, PGS-17, KGS-1, HG-17 and PGS-99-3), four were mild susceptible (G-41, DG-2, DARL-52 and DARL-53) whereas three (G-282, DG-1 and RAUC-5) were susceptible. Genotype PGS-313 exhibited maximum yield (199.6 quintal/ha) followed by PGS-99-2 (154.9 quintal/ha). A strong and negative correlation was observed between purple blotch incidence and bulb weight at genotypic level (-0.99).

Key words: Garlic, purple blotch, correlation, yield, genotype × environment interaction.

INTRODUCTION

Purple blotch (*Alternaria porri* L.) has been identified as an important, yield limiting disease in garlic (*Allium sativum* L.) (Miller, 1983; Bisht and Thomas, 1992). In India, early symptoms can be seen in the last week of January which gets severe as the temperature rises (Figure 1). Under acute disease pressure, plants are almost devoid of leaves due to drying of diseased leaves leading to heavy losses. A foliar infection of up to 90% has been reported in susceptible cultivars of garlic (Bisht and Agarwal, 1993) with a significant reduction in bulb yield ranging between 25 to 60% due to varying degrees of defoliation (Bisht et al., 1993). Considering the poor economic condition of farmers, high cost of disease control and environmental concerns, it is necessary to select germplasm with resistance or tolerance with purple blotch disease as a pre-requisite to carry out a successful breeding and crop improvement programme. Therefore, the present study is aimed at to screen the garlic germplasm for yield and susceptibility to purple blotch disease and correlation among the traits.

MATERIALS AND METHODS

The experiment was conducted at the Vegetable Research Centre, Patharchatta of Govind Ballabh Pant University of Agriculture and Technology, Pantnagar, Uttarakhand, India during the winter season of 2000 to 2001 and 2001 to 2002. The material comprised 21 genotypes of garlic (Table 1) grown in randomized block design r season of 2000 to 2001 and 2001 to 2002. The material comprised 21 genotypes of garlic (Table 1) grown in randomized block design with 3 replications in each year. The crop was planted

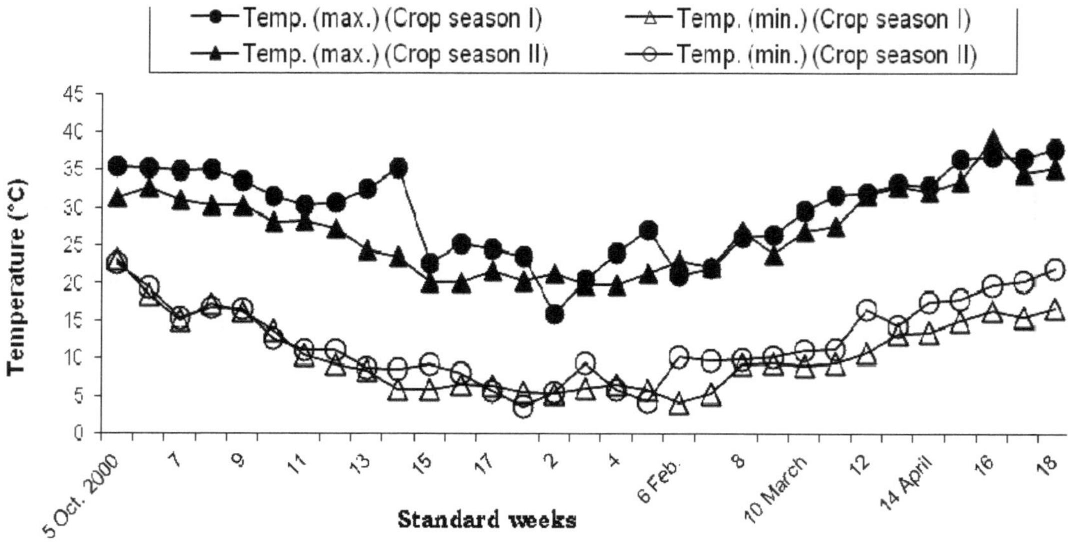

Figure 1. Weekly temperature variation during crop season I and II.

Table 1. Names and sources of genotypes under study.

S/N	Genotypes	Name after release	Source of collection
1	DARL-52	-	Agricultural Research Unit of Defence Research Laboratory, Almora
2	DARL-53	-	Agricultural Research Unit of Defence Research Laboratory, Almora
3	G-1	Yamuna Safed	National Horticultural Research and Development Foundation, Nashik
4	G-41	Agrifound White	National Horticultural Research and Development Foundation, Nashik
5	G-50	Yamuna Safed-1	National Horticultural Research and Development Foundation, Nashik
6	G-323	G-323	National Horticultural Research and Development Foundation, Nashik
7	G-313	Agrifound Parvati	National Horticultural Research and Development Foundation, Nashik
8	G-282	Yamuna Safed-3	National Horticultural Research and Development Foundation, Nashik
9	HG-17	-	C.C.S. Haryana Agricultural University, Hisar
10	KGS-1	-	C.S. Azad University of Agriculture and Technology, Kanpur
11	PGS-4	-	G.B. Pant Univ. of Agric. and Tech., Pantnagar
12	PGS-13	-	G.B. Pant Univ. of Agric. and Tech., Pantnagar
13	PGS-14	Pant Lohit	G.B. Pant Univ. of Agric. and Tech., Pantnagar
14	PGS-17	-	G.B. Pant Univ. of Agric. and Tech., Pantnagar
15	PGS-98	-	G.B. Pant Univ. of Agric. and Tech., Pantnagar
16	PGS-99-1	-	G.B. Pant Univ. of Agric. and Tech., Pantnagar
17	PGS-99-2	-	G.B. Pant Univ. of Agric. and Tech., Pantnagar
18	PGS-99-3	-	G.B. Pant Univ. of Agric. and Tech., Pantnagar
19	DG-1	-	Agricultural Research Station, Durgapur, Rajasthan
20	DG-2	-	Agricultural Research Station, Durgapur, Rajasthan
21	RAUC-5	-	Rajendra Agriculture University, Sabour

in the last week of October at a spacing of 15 (row to row) × 10 cm (plant to plant). Fertilizer application of 100:50:50 kg NPK/ha in form of urea, diammonium phosphate and muriate of potash, respectively was applied. Total phosphorus and potassium and half of the nitrogen was applied before planting and rest amount of nitrogen was top dressed in two equal splits 30 and 45 days after planting when hand weeding was carried out in the experimental plots. Five flood irrigations were given to the crop except fungicide or insecticide spray to ensure a good crop. Plot size was 2.0 × 1.5 m. Disease severity was assessed 120 days after planting during both seasons.

Ten plants from each plot were randomly tagged for disease assessment on the following 0 to 5 rating scale: 0 (Immune) = No infection; 1 (Resistant) = 1 to 10% leaf area infected (LAI); 2 (Moderately resistant) = 11 to 20% LAI; 3 (Mild susceptible) = 21 to 50% LAI; 4 (Susceptible) = 51 to 75% LAI; and 5 (Highly susceptible) = 76 to 100 LAI (Anonymous, 1987). Observation on bulb weight and bulb yield per plot was also taken at the time of

Table 2. Performance of garlic genotypes for purple blotch susceptibility (0-5 scale) bulb weight and yield.

S/N	Genotypes	Mean purple blotch severity index		Average Bulb weight (g)		Bulb yield (quintal/ha)	
		2000-2001	2001-2002	2000-2001	2001-2002	2000-2001	2001-2002
1	G-1	0.2	0.6	25.3	20.6	110.2	93.6
2	G-323	0.2	0.6	28.9	24.0	150.0	125.9
3	PGS-14	0.2	0.6	25.2	22.3	130.5	122.0
4	PGS-4	0.2	0.7	24.7	21.7	128.2	120.5
5	G-313	0.1	0.8	36.5	33.3	213.1	186.0
6	PGS-98	0.2	0.9	29.4	24.7	150.2	126.4
7	PGS-99-1	0.3	0.9	23.1	19.1	112.7	97.7
8	G-50	0.3	0.9	23.3	20.4	116.6	102.6
9	PGS-99-2	0.3	1.0	31.8	28.1	169.8	140.2
10	HG-17	0.2	1.3	24.9	21.3	119.7	107.6
11	PGS-13	0.3	1.4	23.1	20.0	197.6	98.5
12	PGS-17	0.2	1.5	22.9	20.9	112.1	110.3
13	PGS-99-3	0.4	1.5	21.5	17.2	105.7	85.7
14	KGS-1	0.3	1.7	21.2	16.9	95.8	86.0
15	DG-2	0.3	2.3	19.3	15.6	96.7	68.7
16	DARL-52	0.3	2.5	21.7	17.7	112.6	90.8
17	DARL-53	0.4	2.5	20.8	17.8	109.1	94.7
18	G-41	0.5	2.5	23.0	17.0	96.6	78.2
19	DG-1	0.5	3.4	18.7	15.3	92.3	65.9
20	G-282	0.3	3.5	27.0	20.3	134.1	102.5
21	RAUC-5	0.6	3.6	19.8	14.4	97.9	65.8
General mean		0.3	1.6	24.4	20.5	121.5	103.3
SEm±		0.06	0.22	0.72	1.38	6.39	7.50
LSD		0.18	0.63	2.06	3.89	18.27	21.45

harvesting. Average bulb weight was recorded on ten randomly selected plants. Plot yield was recorded for all the lines in all replications and bulb yield for a hectare was extrapolated in quintals. Metrological data was procured from the Agro-meteorological Department of the university.

RESULTS AND DISCUSSION

In the first year (2000 to 2001) trial, disease was very mild and all genotypes showed a disease index value <1 (range 0.1 to 0.5) or as environment was not conducive for disease development (Table 2). In the second year 2001 to 2002 trial environment was more conducive for disease development and genotypes expressed their potential to tolerate the disease. Nine genotypes viz., G-1, G-50, G-323, PGS-4, PGS-14, G-313, PGS-99-1, PGS-98 and PGS-99-2 fell into the category of resistant, five showed moderately resistant reaction to disease (PGS-13, PGS-17, KGS-1, HG-17 and PGS-99-3), four were mildly susceptible (G-41, DG-2, DARL-52 and DARL-53) whereas three (G-282, DG-1 and RAUC-5) were susceptible (Table 2). Bulb weight and yield were negatively correlated with disease severity and variations were attributed to genotypic differences (Table 3). Bisht and Thomas (1992) found only nine accessions among 900 as resistant against purple blotch and stemphyllium blight. Bisht and Agarwal (1993) also categorized G-41 as highly susceptible genotype. Pandey et al. (2000) screened 35 genotypes of garlic and reported resistance in G-1, G-323 and G-282 with percent disease index (PDI) value from 5.4 to 10.0 % while four lines (G-41, RAUC-5, IC-53282 and IC-71983) showing PDI values from 84.3 to 96.8 % were highly susceptible. Poor bulb weight and yield during the second year can be attributed to the severe infection of purple blotch disease at full grown stage of plant which gradually killed the leaves and reduced the leaf area index of plants. The yield reduction from a given percentage of foliage removal has been found maximal when defoliation of foliage occurs during the early stages of bulb development (Baker and Wilcox, 1961). Brewster et al. (1986) also reported that bulb yield depends critically on leaf area index established during bulb development. Higher bulb yield in genotypes with higher leaf area index might be attributed to their ability for higher net assimilation rate resulting into higher production of photosynthates. During 2011 to 2002, the trial of environmental conditions also favored the disease by regular rainfall (Figure 2) and cloudy weather,

Table 3. Analysis of variance for purple blotch and bulb yield in garlic.

Traits	Source of variation					
	First year			Second year		
	Rep	Genotype	Error	Rep	Genotype	Error
df	2	20	40	2	20	40
Purple blotch	0.002	0.048**	0.012	0.621	3.00**	0.149
Bulb weight	18.80	57.44**	1.55	16.44	62.97**	5.75
Bulb yield	539.0	2600**	122.5	225.8	2336**	168.9

** and *, Significance at 1 and 5% level.

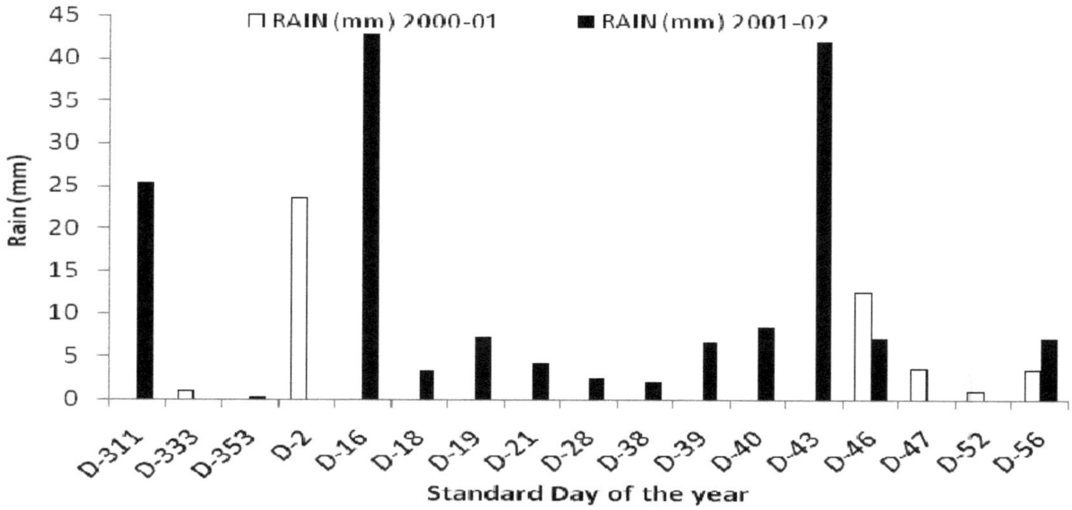

Figure 2. Comparison of daily rainfall during crop season I and II.

Table 4. Correlation among the purple blotch and yield traits.

Traits	Season	Bulb weight		Bulb yield	
		Genotypic	Phenotypic	Genotypic	Phenotypic
	2000-2001	-0.695**	-0.487*	-0.619**	-0.454*
Purple blotch	2001-2002	-0.692**	-0.552**	-0.691**	-0.569**
	Pooled	-0.990**	-0.507*	-0.998**	-0.465*

** and * Significance at 1 and 5% level.

because sporulation of conidia of *A. porri* on affected leaves has been reported to increase after rain (Meredith, 1966) and its subsequent dispersal by winds (Skiles, 1953). Maude (1990) reported that pathogen requires temperature of 21 to 30°C for development and hot humid climate for its severity. Poor weather conditions might also indirectly affect the photosynthetic efficiency of plants because the development of bulbs is mediated mainly by day/night temperature differences and to lesser extent by photoperiod (Abdalla, 1967). Mean performance of genotypes indicated superiority of G-313, PGS-99-2, PGS-98, G-323 and PGS-14 for bulb yield as well as bulb weight also. A strong and negative correlation coefficient was observed between the purple blotch incidence and yield attributing traits (Table 4). Therefore, it is advisable to grow these genotypes in the purple blotch disease prone areas due to their inherent ability of better disease tolerance and high yield potential.

REFERENCES

Abdalla AA (1967). Effect of temperature and photoperiod on bulbing of the common onion under arid tropical conditions of Sudan. Exp. Agric. 3:137.

Anonymous (1987). Directorate of Vegetable Research (ICAR). Proceedings of IX workshop, Jan. 13-16, 1987, NDUA&T, Faizabad, India. P. 95.

Baker RS, Wilcox GE (1961). Effect of foliage damage and stand reduction on onion yield. Proc. Am. Soc. Hort. Sci. 78:400.

Bisht IS, Agarwal RC (1993). Susceptibility to purple blotch (*Alternaria porri*) in garlic. Ann. Appl. Biol. 122:31-38.

Bisht IS, Thomas TA (1992). Field screening of garlic germplasm against purple blotch and stemphylium blight. Indian Phytopathol. 45:244-245.

Bisht IS, Agarwal RC, Venkateshwaran K (1993). Yield loss in different genotypes of garlic due to purple blotch. Indian Phytopathol. 46:89-91.

Brewster JL, Mondal MF, Morris GEL (1986). Bulb development in onion (Allium cepa) IV. Influence on yield of radiation interception its efficiency of conversion, the duration of growth and dry matter partitioning. Ann. Bot. (London) 58:221.

Maude RB (1990). Leaf diseases of onions. In: H.D. Rabinowitch, and J.L., Brewster (eds.), Onion. Allied. Crops .3:173-189. CRC Press Inc. Boca Raton, Florida.

Meredith DS (1966). Spore dispersal in Alternaria porri on onion in Nebraska. Ann. Appl. Biol. 57:67.

Miller ME (1983). Relationship between onion leaf age and susceptibility to *Alternaria porri.* Plant Dis. 67:284.

Pandey KK, Kumar S, Pandey PK (2000). Screening of garlic genotypes against purple blotch and stemphylium complex. Proceedings of National Symposium on Onion-Garlic Production and Post-harvest Management, Challenges and Strategies, Nov. 19-21, 2000. Nashik. pp. 157-159.

Skiles RL (1953). Purple and brown blotch of onions. Phytopathology 43:409.

Performance of soybean plants originated from seeds of high and low vigor submitted to water deficit

Lizandro Ciciliano Tavares[1]*, Cassyo de Araújo Rufino[1], Lilian Madruga de Tunes[2] and Antonio Carlos Souza Albuquerque Barros[1]

[1]Department of Plant Science, Federal University of Pelotas, UFPel/FAEM, Brazil.
[2]Department of Plant Science, Federal University of Santa Maria, UFSM, Brazil.

This study had the objective of evaluating the effect of water deficit on the growth of soybean plants from seeds with high and low vigor. It was conducted at the Faculdade de Agronomia "Eliseu Maciel" (FAEM-UFPel) using the cultivar M-SOY 8008 RR. The experimental design was a completely randomized in factorial AXB (Factor A: Variables 2, 3, 4 and 5 water deficits; Factor B: two levels of seed vigor, high and low) and a total of ten treatments with four replications. Evaluations of fresh mater, plant height, leaf area and dry weight, at the 10, 20, 30 and 40 DAE. Soybean plants from high vigor seeds have higher fresh weight, plant height, leaf area and dry weight until 40 DAE, and provide higher rates of growth up to 40 DAE. High vigor seeds have better performance than plants of low vigor, when submitted to water deficit until 40 DAE. Water deficit reduces the performance of both plants grown from seeds of high and low vigor.

Key words: *Glycine max* (L.) Merrill, initial growth, physiological quality, growth rate.

INTRODUCTION

The frequent water deficit in southern Brazil causes major crop damage, as rainfall is the main source of water for crops. The higher frequency of water deficit occurs in summer, affecting crops of great importance concerning the acreage and production, such as soybeans and corn. The water deficit is characterized by sweating that exceeds the water absorption rate and thus, acts directly on water relations of plants (Costa et al., 2008) in which damage to the plant depend on the intensity and exposure time. With a period of 20 to 30 years, data from 40 meteorological stations in Southern Brazil were observed and it was detected that water availability is a variable that limits the expression of yield potential of soybeans, regardless of the growing cycle, sowing date and location (Cunha et al., 1998). The reduction in available soil water for the plant, affects negatively the growth and development (Sinclair and Ludlow, 1986).

Levit (1980) stresses the importance of analyzing the responses of plants and their defense mechanisms to soil water deficit.

The success of a crop is influenced, directly, by a high-quality seed to be sown and contributes, significantly, to high yields, while low-quality seeds, undertake to obtain an adequate plant stand, directly influencing the productivity of a crop (Krzyzanowski and France Neto, 2003). Early in the development, more vigorous seeds are more resistant to conditions of less water availability, favoring the establishment of field plant population (Tekrony and Egli, 1991).

Despite the complexity involved in growing of plant species, growth analysis is an efficient way to monitor progress and measure the contribution of different physiological processes on the plant behavior (Benincasa, 2003). It can also use the analysis of growth in the observations of physiological parameters indicative of safe methods to increase productivity (Castro, 1974). Campos et al. (2008), consider the analysis of growth as an important measure to explain the growth of plants, because about 90% of dry matter accumulation by plants throughout their development results of photosynthetic activity, allowing to evaluate the growth of the plant as a

*Corresponding author. E-mail: lizandro_cicilianotavares@yahoo.com.br.

final whole and the contribution of different organs in overall development.

This study was done to evaluate the effect of water deficit on the growth of soybean plants grown from seeds with high and low vigor.

MATERIALS AND METHODS

This study was conducted in the crop year 2009/2010, at the Laboratório Didático de Análise de Sementes and in the greenhouse of the Faculdade de Agronomia "Eliseu Maciel", Universidade Federal de Pelotas, city of Capão do Leão - RS.

It were filled with soil, pots with capacity of 15 L and the soil was collected from the A1 horizon of a Typic Hapludalf Eutrophic solodic (Embrapa, 2006) belonging to the mapping unit Pelotas. The fertilization was performed according to CFQS RS (Rio Grande do Sul, state)/SC (Santa Catarina, state) (Committee of Fertility and Soil Chemistry - RS / SC, 2004), incorporating the nutrients to the soil, five days before sowing and liming was performed sixty days before sowing.

The cultivar used was M-SOY 8008 RR. Prior to sowing, seeds were treated with inoculant Gelfix 5®, 200 ml/100 kg of seed and fungicide (chemical group Fenilpirrol Acilalaninato + and + fludioxinil active ingredient Metalaxyl-M (25 ± 10 gL^{-1})), Maxim-XL at a dosage of 100 ml/100 kg seed.

For the selection of seeds that are to be used, it was evaluated that germination (G) was performed with four replicates of 50 seeds for each vigor level, placed on a substrate of germination paper ('germitest"), previously soaked in water, using 2.5 times the mass of dry paper, and kept at 25°C. The evaluations were done according to the RAS -Rules for Testing Seeds (Brazil, 2009) and the evaluation carried out on the eighth day after sowing and results were expressed as percentage of normal seedlings. The first count of germination (FCG) was conducted together with the germination test and evaluated on the fifth day after sowing. The field emergence test (FE) was carried out in plots containing soil, and the manual seeding depth from 0.02 to 0.03 m, using 400 seeds for each level of vigor (four replicates of 100 seeds). The emergence percentage was obtained by counting emerged seedlings in the twenty-first day after sowing.

The physiological quality of seeds was obtained from seed lots produced in the growing season of 2008/2009. The high vigor seed showed 88% germination (G), 82% in the first count of germination (FCG) and 87% field emergence (FE), while the low vigor lot showed 75% G, 61% FCG and 74% FE. It was sown 10 seeds per pot to allow further thinning, removing seedlings earlier in the lot of low vigor and later in the high vigor, with the intention of using the seedlings emerged in the days of major emergency frequency for each level of vigor, leaving at the end, four plants per pot.

The different water restrictions used, which correspond to the treatments, were applied when 51% of the plants were in the stage VE, according to the scale of Fehr and Caviness (1977). Considering the viable seeds used, it was reached, approximately, five to seven days after sowing, for both, high and low vigor seeds.

The treatments consisted of five periods of water deficit and two levels of seed vigor. The periods of water deficits were: control, with irrigation, water deficit from 1 to 10 days after emergence (DAE), water deficit from 11 to 20 DAE, water deficit from 21 to 30 DAE and water deficit from 31 to 40 DAE. Vigor levels studied were high and low, with a total of ten treatments with four replications. Therefore, before and after periods of water deficit, the experimental units were irrigated daily.

Assessments of the effect of physiological quality combined with the water deficit, the initial growth was performed on isolated plants. The following determinations were done: fresh matter (FM), plant height (PH), leaf area (LA) and dry matter (DM). For these determinations, one plant was collected, cut at ground level at 10, 20, 30 and 40 days after emergence (DAE), remaining three plants, two plants and one plant per pot, respectively. For measuring the fresh matter, it used an analytical balance with precision composition. The leaf area of the aerial part of plants was determined using photoelectric determiner (Area Meter, model LI-3100 Li-color Ltda.) giving direct reading in cm^2. To determine the plant height, the measurement was carried out with the aid of rule and the results were expressed in centimeters. The dry matter of shoots was evaluated by a stove at 60ºC, in which the seedlings were kept for 72 h in the oven and after, weighed on an analytical balance with precision composition.

By the results of leaf area and dry weight, the following were determined: growth rate of culture - GRC (mg pl^{-1} day^{-1}), relative growth rate - RGR (mg g^{-1} day^{-1}) and net assimilation rate - NAR (mg cm^{-2} day^{-1}). These determinations were based on the methodology described in Gardner et al. (1985), in which: GRC = $(DM_2 - DM_1) / (T_2 - T_1)$, RGR = $(\ln DM_2 - \ln DM_1) / (T_2 - T_1)$, NAR = $(DM_2 - DM_1) / (T_2 - T_1) * (LA_2 \ln - \ln LA_1) / (LA_2 - LA_1)$, where: DM: dry mass, T: time, LA: leaf area.

The statistical analysis used the Statistical Analysis System for Windows - WinStat - Version 2.0 (Machado and Conceição, 2003). The experimental design was randomized, and the data subjected to analysis of variance and in the presence of significant interaction, proceeded to the developments needed. Means were compared using the Duncan test at 5% probability, in factorial AXB (variables 2, 3, 4 and 5 periods of water deficit and two levels of vigor), with four replications.

RESULTS AND DISCUSSION

The results presented in Table 1 shows that plants from high vigor seeds presented higher values of MF, PH, LA, DM, than plants from low vigor, in both water regimes. Positive effects of physiological seed quality on the height were also found in other studies; however, they were evaluated in different periods of plant development. In soybean, Vanzolini and Carvalho (2002) observed that seed lots with low physiological quality resulted in plants with lower heights at 18 and 38 days after sowing, compared to lots of medium and high quality. According to the authors, these probably reflect the rate of emergence of seedlings originating from seeds of low physiological quality, significantly lower compared to other lots. Kolchinski et al. (2006), said that soybean plants originated from seeds of higher physiological quality, showed the highest growth rates, as a result of plants with greater dry weight and leaf area at 30 days after emergence and similar results were observed by Machado and Schuch (2004) in oat, in rice (Hofs et al., 2004), and in oats (Schuch et al., 2000). It can also be seen that plants from seeds with high and low vigor when submitted to water deficit, have underperformed plants without water deficit (control) in the variables FM, PH and

Table 1. Fresh mass (MF), plant height (PH), leaf area (LA) and dry mass (DM) of soybean plants originated from seeds with high and low vigor, submitted to drought, 10 DAE. Capão do Leão – RS (2010).

Variable	Deficit/Stage***	Vigor		Means
		High	Low	
MF($g.pl^{-1}$)	Control/VE-V2	3.57	2.63	3.10 a
	1 - 10DAE*/VE-V1	1.45	1.03	1.24 b
	Means	2.51 A**	1.83 B	
	CV (%)	11.62		
PH(cm)	Control/VE-V2	14.0	10.6	12.32 a
	1 - 10DAE*/VE-V1	10.6	7.5	9.06 b
	Means	12.3 A	9.0 B	
	CV (%)	8.03		
LA(cm^2 pl^{-1})	Control/VE-V2	93.65 A a	63.34 B a	78.49
	1 - 10DAE*/VE-V1	49.41 A b	36.74 B b	43.07
	Means	71.53	50.04	
	CV (%)	11.05		
DM($g.pl^{-1}$)	Control/VE-V2	0.22	0.17	0.201 a
	1 to 10DAE*/VE-V1	0.07	0.05	0.069 b
	Means	0.151 A	0.119 B	
	CV (%)	8.22		

*DAE: days after emergence, ** means followed by the same letter in the column and capital on the line do not differ by Duncan test at 5% of probability; *** phenological Stage, according to the scale of Fehr and Caviness (1977).

DM, but no significant effect was observed for LA. The occurrence of severe water deficit in the vegetative stage may compromise the yield due to less development of plants (Mundstock and Thomas, 2005).

The data presented in Table 2 shows that plants grown from seeds with high physiological quality have higher FM, PH, LA and DM in those three periods. Seeds are more vigorous for processing capacity of reserves in storage tissues and there is a greater incorporation on the embryonic axis (Dan et al., 1987). This can result in more rapid and uniform emergence and seedlings with larger initial size (Vanzolini and Carvalho, 2002) thereby, influencing the leaf area and dry matter accumulation. Similar results were found in oats by Schuch et al. (1999) when seedlings from seeds with high vigor, had higher initial size, which, consequently, led to higher rates of crop growth, dry matter production and leaf area, during the initial period of growth. Also in Table 2, it can be realized that plants do not suffer from water deficit (Control), have better performance than that of plants with water deficit from 1 to 10 and 11 to 20 days after emergence in both physiological quality seeds, in the four variables. This result is similar to those reported by Hoogenboom et al. (1987) and Thomas and Costa (1994), that water deficit decreases the height of soybean plants. Observe that plants suffering water deficit in the period from 1 to 10 DAE, despite being watered after that period, they cannot restore normal growth in those variables, when compared to treatments without water deficit. The reduction in relative water content of leaves is caused by deficiency of water in the soil, as occurs during photosynthesis. Loss of water throughout the stomatal mechanism and the rate of assimilation of water is adversely affected during water stress (Versluis et al. 2006).

Regarding the assessment at 30 DAE, in Table 3, it can be considered that plants originated from seeds with high vigor continued with higher FM, PH, LA and DM, in the four water deficit periods studied. Seed with low vigor can be led to reductions in field emergence, speed of emergence and initial plant size (Schuch, 2006). Assessing the behavior of individual plants in communities of soybean, Kolchinski et al. (2005) found that the effect of seed physiological quality on seedling development, determined higher seed yields. According to the authors, the use of seeds with high physiological quality provided greater leaf area and dry matter production and, consequently, initial competitive advantage in the use of environmental resources, which resulted in the later stages of development until the maturation phase, resulting in higher seed yield.

Water deficits modify plant metabolism in different ways.

Table 2. Fresh mass (MF), plant height (PH), leaf area (LA) and dry mass (DM) of soybean plants originated from seeds with high and low vigor, submitted to drought, 20 DAE. Capão do Leão – RS (2010).

Variable	Deficit/Stage ***	Vigor		Means
		High	Low	
	Control/VE-V5	14.33 A a**	9.36 B a	11.84
	1 a 10DAE*/VE-V1	8.15 A b	5.65 B b	6.90
MF ($g.pl^{-1}$)	11 a 20DAE/V1 e V3 e V4	5.54 A c	3.75 B c	4.64
	Means	9.34	6.25	
	CV (%)	3.33		
	Control/VE-V5	35.8	25.5	30.63 a
	1 a 10DAE*/VE-V1	28.7	20.8	24.73 b
PH (cm)	11 a 20DAE/V1 e V3 e V4	24.9	17.9	21.38 c
	Means	29.7 A	21.4 B	
	CV (%)	5.74		
	Control/VE-V5	498.00 A a	323.77 B a	410.88
	1 a 10DAE*/VE-V1	268.63 A b	178.94 B b	223.78
LA (cm^2 pl^{-1})	11 a 20DAE/V1 e V3 e V4	234.57 A b	155.77 B b	195.17
	Means	333.73	219.49	
	CV (%)	11.47		
	Control/VE-V5	2.76 A a	1.53 B a	2.15
	1 a 10DAE*/VE-V1	1.31 A b	0.82 B b	1.07
DM ($g.pl^{-1}$)	11 a 20DAE/V1 e V3 e V4	1.40 A b	0.85 B b	1.13
	Means	1.829	1.07	
	CV (%)	11.96		

*DAE: days after emergence, ** means followed by the same letter in the column and capital on the line, do not differ by Duncan test at 5% of probability; *** phenological stage, according to the scale of Fehr and Caviness (1977).

It is observed in water deficit condition that there is increased synthesis of abscisic acid (ABA) in roots, which is subsequently transported to the shoot via the xylem, after rainfall or irrigation (Hartung et al., 2002). For Thomas and Costa (1994), the photosynthetically active leaf area of plants is the most sensitive to water deficit and one of the factors that affect crop yields. It is inferred that, because of stomatal closure caused by water deficit, the treatments without water deficit, along with treatment with water deficit from 1 to 10 DAE, in both levels of vigor, got the best performance for the variables analyzed (Table 3). The superior performance of the control treatment suggests an occurrence in this treatment that remained with irrigation, during the whole period of conducting the experiment, causing the plants an expression of the full potential of growth, showing the negative effects of water deficits on those parameters evaluated. The level of proline, according to Kishor et al. (1995), increased significantly only after 4 days the plants suffered from water deficit and this accumulation is a characteristic response of plants under abiotic stress, which acts as osmotic regulator, in order to reduce the negative effects in plants under adverse conditions, in addition, it promotes increased resistance of cells in these conditions (Xiong and Zhu, 2002). Similar results on the proline accumulation in plants under water deficit was observed by Sarker et al. (1999) working with cultivars of *Triticum aestivum* and Costa (1999) studying *Vigna unguiculata*.

The assessment at 40 DAE in Table 4, showed that plants from high vigor seeds were superior to the plants originated from seeds of low vigor and the control treatment in four periods with water deficit studied in the four variables. This behavior suggests a direct effect of seed vigor on the ability of the tissues of soybean plants to convert solar radiation into dry matter during this period of growth. However, for Tekrony and Egli (1991), the direct effects of seed vigor on the further development of the plants are unlikely to occur.

According to those authors, the structures present in the seeds are important for growth, only during a short period, immediately after emergence. It states that most of the plant tissues involved in the production of dry matter and yield are formed after the emergency.

Table 3. Fresh mass (MF), plant height (PH), leaf area (LA) and dry mass (DM) of soybean plants originated from seeds with high and low vigor, submitted to drought, 30 DAE. Capão do Leão – RS (2010).

Variable	Deficit/Stage ***	Vigor		Means
		High	Low	
MF ($g.pl^{-1}$)	Control/VE-V7 e V8	24.61	17.92	21.27 a
	1 a 10DAE*/VE e V1	21.14	15.68	18.41 b
	11 a 20DAE/V1 e V3 e V4	13.45	9.81	11.63 d
	21 a 30DAE/V3 e V4-V6 e V7	16.78	11.79	14.28 c
	Means	18.99 A**	13.80 B	
	CV (%)	13.41		
PH (cm)	Control/VE-V7 e V8	45.2	35.1	40.10 a
	1 a 10DAE*/VE e V1	47.7	38.4	43.05 a
	11 a 20DAE/V1 e V3 e V4	34.4	26.9	30.65 c
	21 a 30DAE/V3 e V4-V6 e V7	41.5	30.9	36.18 b
	Means	42.1 A	32.8 B	
	CV (%)	8.75		
LA ($cm^2\ pl^{-1}$)	Control/VE-V7 e V8	917.57 A a	606.06 B a	761.81
	1 a 10DAE*/VE e V1	925.96 A a	657.29 B a	791.62
	11 a 20DAE/V1 e V3 e V4	486.37 A c	346.75 B b	416.56
	21 a 30DAE/V3 e V4-V6 e V7	594.73 A b	392.05 B b	493.39
	Means	731.15	500.53	
	CV (%)			
DM ($g.pl^{-1}$)	Control/VE-V7 e V8	5.60	3.95	4.78 a
	1 a 10DAE*/VE e V1	5.17	3.70	4.44 a
	11 a 20DAE/V1 e V3 e V4	2.69	1.92	2.30 c
	21 a 30DAE/V3 e V4-V6 e V7	4.28	2.98	3.63 b
	Means	4.44 A	3.14 B	
	CV (%)	12.42		

*DAE: days after emergence, ** means followed by the same letter in the column and capital on the line, do not differ by Duncan's test at 5% of probability; *** phenological Stage, according to the scale of Fehr and Caviness (1977).

Severe water deficit in the vegetative phase, according to Bonato (2000), can reduce plant growth and reduce leaf area and yield. Also, it was emphasized that the water deficit plants of smaller stature, stunted with small leaves and short nodes, were determined. Results obtained by Petry (2000) confirm that water deficit reduces applied at vegetative growth of soybean due to the decrease in the number of nodes and the length between NODES. It also can be seen in Table 4 that the treatments performed significantly better than the other treatments were the water deficit periods 1 to 10, 11 to 20, and 31 to 40 DAE, the variables FM, PH and DM, respectively. In variable LA, it is observed that treatments with water deficit from 1 to 10 and 11 to 20 DAE, together with the control treatment differed significantly from other treatments. As a result, it appears that the growth of roots, is a strategy used by plants to capture the substrate water in water deficit conditions (Lobato et al., 2008), in which growth and development of plants are dependent on cell turgor, as the water fills the cell and promotes a positive pressure through this mechanism of tissue extension (Kerbauy, 2004).

It is considered that the analysis of growth is still the most accessible and accurate, for evaluating growth and to infer the contribution of different physiological processes on the plant behavior (Trindade and Oliveira, 1999). By observation of Table 5, it appears that GRC plants originated from seeds of high vigor had improved performance over existing plants from low in the four periods. However in the period from 11 to 20 DAE, there was no statistical difference between high and low seed vigor in the control treatment and treatment with a water deficit of 11 to 20 DAE. Oat seeds of high vigor, (Schuch, 1999) produced plants with greater dry matter production, leaf area and growth rates in the early period of culture. The control treatment differed significantly from other

Table 4. Fresh mass (MF), plant height (PH), leaf area (LA) and dry mass (DM) of soybean plants originated from seeds with high and low vigor, submitted to drought, 40 DAE. Capão do Leão – RS (2010).

Variable	Deficit/Stage***	Vigor		Means
		High	Low	
	Control/VE-V9	39.57	30.03	34.80 b
	1 a 10DAE*/VE-V1	37.15	28.91	33.03 b
	11 a 20DAE/V1 e V3 e V4	47.00	38.00	42.05 a
MF ($g.pl^{-1}$)	21 a 30DAE/V3 e V4 –V6 e V7	31.16	23.67	27.41 c
	31 a 40DAE/V6 e V7 e V8	22.83	17.02	19.92 d
	Means	35.54 A**	27.53 B	
	CV (%)	9.03		
	Control/VE-V9	48.5	40.2	44.32 b
	1 a 10DAE*/VE-V1	54.0	45.1	49.55 a
	11 a 20DAE/V1 e V3 e V4	49.2	42.1	45.61 b
PH (cm)	21 a 30DAE/V3 e V4 –V6 e V7	46.5	38.0	42.26 b
	31 a 40DAE/V6 e V7 e V8	47.5	37.8	42.63 b
	Means	49.12 A	40.63 B	
	CV (%)	7.51		
	Control/VE-V9	957.81	719.3	838.56 a
	1 a 10DAE*/VE-V1	926.95	696.32	811.63 a
	11 a 20DAE/V1 e V3 e V4	914.34	715.87	815.11 a
LA (cm^2 pl^{-1})	21 a 30DAE/V3 e V4 –V6 e V7	796.61	596.8	696.70 b
	31 a 40DAE/V6 e V7 e V8	794.72	565.55	680.13 b
	Means	878.08 A	658.77 B	
	CV (%)	5.57		
	Control/VE-V9	7.79	6.11	6.94 b
	1 a 10DAE*/VE-V1	6.65	5.19	5.91 c
	11 a 20DAE/V1 e V3 e V4	6.88	5.50	6.18 c
DM ($g.pl^{-1}$)	21 a 30DAE/V3 e V4 –V6 e V7	6.39	4.97	5.67 c
	31 a 40DAE/V6 e V7 e V8	8.69	6.42	7.55 a
	Means	7.27 A	5.63 B	
	CV (%)	7.81		

*DAE: days after emergence, ** means followed by the same letter in the column and capital on the line, do not differ by Duncan's test at 5% of probability; *** phenological Stage, according to the scale of Fehr and Caviness (1977).

treatments during periods of GRC 1 to 10 and 11 to 20 DAE. CRG in the period from 21 to 30 DAE result in higher water deficit treatment with 10 to 10 DAE, as the period from 31 to 40 DAE GRC superior performances in dealing with water deficit from 11 to 20 DAE.

It is found RGR that plants from seeds of high vigor have improved performance over existing plants from low in the period from 11 to 20 DAE, but the periods 21 to 30 and 31 to 40 DAE observed that plants grown from seeds of low vigor differed significantly from plants grown from seeds of low vigor. It can also be observed that the water deficit treatment with 1 to 10 DAE was higher than control and also to the other treatments during periods of RGR 11 to 20 and 21 to 30 DAE. However, RGR in the period from 31 to 40 DAE had superior performance in dealing with water deficit from 21 to 30 DAE.

The NAR was not different in plants that originated from seeds with high and low vigor, during the periods from 11 to 20 and 21 to 30 DAE, but the period from 31 to 40 DAE plants originated from seeds of low vigor obtained better results than plants that originated from seeds of high vigor. The NAR, in the period from 11 to 20 DAE showed no significant difference between control and water deficits treatments, from 1 to 10 DAE, as the period from 21 to 30 DAE RAN was superior in performance, when dealing with water deficit from 10 to 10 DAE.

Considering the NAR, in the period from 31 to 40 DAE, it was observed that the water deficit period from 11 to 20

Table 5. Growth rate of culture (CRG), relative growth rate (RGR) and net assimilation rate (RAN) of soybean plants from seeds of high and low vigor, submitted to drought. Capão do Leão – RS (2010).

Variable	Deficit	Vigor		Means
		High	Low	
CRG (1 - 10 DAE)	Control	22.32	17.90	22.11 a
	1 - 10DAE*	7.94	5.89	6.91 b
	Means	15.13 A**	11.90 B	
	CV (%)	8.23		
CRG (11 - 20 DAE)	Control	254.6 A a	135.6 B a	195.1
	1 a 10DAE	123.4 A b	77.0 B b	100.2
	Means	189.0	106.3	
	CV (%)	12.41		
CRG (21 - 30 DAE)	Control	283.47	242.12	262.8 b
	1 a 10DAE	386.50	287.50	337.0 a
	11 a 20DAE	128.92	106.52	117.7 c
	Means	266.4 A	212.0 B	
	CV (%)	23.28		
CRG (31 - 40 DAE)	Control	2189.75	2149.00	2169.3 b
	1 a 10DAE	1473.75	1481.50	1477.6 c
	11 a 20DAE	4181.75	3582.25	3882.0 a
	21 a 30DAE	2103.50	1978.00	2040.7 bc
	Means	2487.1 A	2297.6 A	
	CV (%)	23.56		
RGR (11 - 20 DAE)	Control/V5	0.025	0.021	0.023 b
	1 a 10DAE/V5	0.028	0.026	0.027 a
	Means	0.026 A	0.023 B	
	CV (%)	4.75		
RGR (21 - 30 DAE)	Control	0.069	0.094	0.082 b
	1 a 10DAE	0.137	0.150	0.144 a
	11 a 20DAE	0.064	0.081	0.073 b
	Means	0.090 B	0.108 A	
	CV (%)	19.66		
RGR (31 - 40 DAE)	Control	0.034	0.043	0.039 bc
	1 a 10DAE	0.025	0.033	0.029 c
	11 a 20DAE	0.095	0.105	0.099 a
	21 a 30DAE	0.039	0.051	0.045 b
	Means	0.048 B	0.058 A	
	CV (%)	23.40		
RAN (11 - 20 DAE)	Control	1.052	0.853	0.952 a
	1 a 10DAE	0.963	0.866	0.914 a
	Means	1.008 A	0.859 A	
	CV (%)	14.4		
RAN (21 - 30 DAE)	Control	0.416	0.537	0.476 b
	1 a 10DAE	0.735	0.784	0.760 a
	11 a 20DAE	0.373	0.448	0.411 b
	Means	0.590 A	0.508 A	
	CV (%)	22.91		

Table 5. Contd.

RAN (31 - 40 DAE)	Control	0.233	0.325	0.279 b
	1 a 10DAE	0.158	0.219	0.189 c
	11 a 20DAE	0.614	0.703	0.659 a
	21 a 30DAE	0.308	0.411	0.360 b
	Means	0.328 B	0.415 A	
	CV (%)	22.56		

*DAE: days after emergence, ** means followed by the same letter in the column and capital on the line, do not differ by Duncan's test at 5% probability.

DAE was bigger than other treatments.

CONCLUSIONS

Soybean seeds of high vigor generate plants with higher growth rate of the culture until 30 DAE.

Soybean plants from seeds of higher physiological quality submitted to water deficit, up to 40 DAE, have higher performance than the plants from seeds with low vigor.

Water deficits reduce performance at different stages of growth of soybean plants derived from high and low vigor seeds.

ACKNOWLEDGEMENTS

To CAPES (Coordenação de Aperfeiçoamento de Pessoal de Nível Superior) for the scholarship granted to the first author, during the Master's Degree period in Seed Science and Technology Program / UFPel/Brazil.

REFERENCES

Benincasa MMP (2003). Analysis of plant growth: the basics. 2nd ed. Funep, Jaboticabal, Brazil, pp. 41.

Bonato ER. Stress in soybean (2000). Passo Fundo, Embrapa Trigo, pp. 254.

Brazil (2009). Ministry of Agriculture, Livestock and Supply. Rules for Seed Testing. Ministry of Agriculture, Livestock and Supply. Agriculture Defense Department. Brasília, DF: Mapa / ACS, p. 395.

Campos MF, Ono, EO, Boaro CFS, Roberts JD (2008). Growth analysis of soybean plants treated with regulating substances. Biotemas, 21(3): 53-63.

Castro PRC (1974). Growth analysis of peanut (Arachis hipogaea L.) in relation to pest infestation. Anais da Escola Superior de Agricultura Luiz de Queiroz, 31: 207-215.

Commission of Chemistry and Soil Fertility - RS/SC Manual Fertilization and Liming for the states of Rio Grande do Sul and Santa Catarina (2004). 10th ed. Porto Alegre: NRS/SBCS, p. 400.

Costa RCL (1999). Nitrogen assimilation and osmotic adjustment in nodulated bean plants of *Vigna unguiculata* (L.) Walp under water stress, Ph.D. Thesis. Universidade Federal do Ceára, Brazil.

RCL Costa, AKS Lobato, CF Oliveira Neto, PSP Maia, GAR Alves, Laughinghouse HD (2008). Biochemical and physiological responses in two *Vigna unguiculata* (L.) Walp. Cultivars under water stress. J. Agron., 7: 98-101.

Cunha GR, Haas JC, Dalmago GA, Pasinato A. (1998). Potential revenue loss in soybean in Rio Grande do Sul by water deficiency. Journal of Agrometeorology, Santa Maria, 6(1): 111-119.

Dan EL, Mello VDC, CT Wetzel, Popinigis F, Zonta EP (1987). Transfer of dry matter as a method for evaluating the effect of soybean. Journal of Seeds, Brasília, 9(3): 45-55.

Empresa Brasileira de Pesquisa Agropecuária - Embrapa (2006). National Research Center of Soil. Brazilian system of soil classification. 2.ed. Rio de Janeiro, p. 306.

Fehr WR, Caviness CE (1977). Stages of soybean development. Ames: State University of Science and Technology, p. 11 (Special report, 80).

Gardner FP, Pearce RB, Mitchell RL (1985). Physiology of crop plants Ames: Iowa State University Press, p. 321.

Hartung, W, Sauter A, Hose E (2002). Abscisic acid in the xylem: where does it come from, where does it go to? J. Exp. Bot., 53(366): 27-32.

Höfs A, Schuch LOB Peske ST, Barros ACSA (2004). Emergence and growth of rice seedlings in response to physiological quality of seeds. J. Seeds, Brasilia, 26 (1): 92-97.

Hoogenboom G Peterson CM, Huck MG (1987). Shoot growth rate of soybean Affected by the drought stress. Agron. J., Madison, 79: 598-607.

Kerbauy GB (2004). Plant Physiology. Guanabara Koogan S. A., Rio de Janeiro. 2004 p. 452.

Kishor PBK, Hong Z, Miao G, Hu CAA, Verma DPS (1995) Overexpression of _1-pyrroline-5- carboxylate synthetase increases proline overproduction and confers osmotolerance in transgenic plants. Plant Physiol., 108: 1387-1394.

Kolchinski EM, Schuch LOB, Peske ST (2005). Vigor and intraspecific competition in soybeans. Ciência Rural, Santa Maria, 35(6): 1248-1256.

Kolchinski EM, Schuch LOB, Peske ST (2006). Early growth of soybeans in relation to seed vigor. J. Agrociência, Pelotas, 12: 163-166.

Krzyzanowski FC, France Neto JB (2003). Adding value to soybean seed. Seed News, Pelotas, 7 (5), Available at: <http://www.seednews.inf.br/portugues/seed75/artigocapa75.shtml>. Accessed 18 August 2010.

Levit J (1980). Responses of plants to Environmental stress. II. Water, radiation, salt and get the other stress. New York, Academic Press, p. 606.

AKS Lobato, RCL Costa, CF Oliveira Neto, BG Santos Filho, FJR Cruz, JMN Freitas, Lamb HR (2008). Morphological changes in soybean under progressive water stress. Int. J. Bot., 4: 231-235.

Machado AA, Conceição AR (2003). Statistical analysis system for Windows. WinStat. Version 2.0. UFPel.

Machado RF, Schuch LOB (2004). Forage production and seed of oat in relation to seed vigor and plant populations. Sci. J. Rural, Santa Maria, 9(1): 126-136.

Mundstock CM, Thomas LA (2005). Soybeans: factors affecting the growth and yield. Porto Alegre: Universidade Federal do Rio Grande do Sul, p. 31.

Petry MT (2000). Interaction soil-plant and soil water availability to plants of sorghum and soybeans. 127f. Dissertation (Master in Agronomy) – Universidade Federal de Santa Maria.

Sarker AM, Rahman MS, Paul NK (1999). Effect of soil moisture on relative leaf water content, chlorophyll, proline and sugar Accumulation in wheat. J. Agron. Crop Sci., 183: 225-229.

Schuch LOB, Nedel JL, Maia DM, Assisi FN (1999). Vigor and nitrogen fertilization on oats (*Avena strigosa* Schreb.). J. Seeds, Brasilia, 21(2): 127-134.

Schuch LOB, Nedel JL, Assis FN, Maia DM (2000). Field emergence and early growth of oat in response to seed vigor. J. Agrociência, Pelotas, 6(2): 97-101.

Schuch, LOB, Finatto, JA (2006). Behavior of isolated soybean plants depending on the physiological quality of seeds In: "Congress of scientific initiation, 14., and against the graduate", 7. Pelotas. Proceedings .... Pelotas: Editora e Gráfica Universitária UFPel. 1 CD-ROOM.

Sinclair TR, Ludlow MM (1986). Influence of soil water supply on the plant water balance of four tropical grain legumes. Aust. J. Plant Physiol., 13: 319-340.

Tekrony DM, Egli DB (1991). Relationship of seed vigor to crop yield: a review. Crop Sci., Madison, 31: 816-822.

Thomas AL, Costa JA (1994). Influence of drought on the growth and yield of soybeans. Brazilian Agric. Res., Brasília, 29: 1389-1396.

Trindade HS, Oliveira JRP (1999). Propagation and planting. In: Sanchez, NF, Dantas JLL (Coords.). The cultivation of papaya. Cruz das Almas, Embrapa Cassava and Fruits, pp. 17-26.

Vanzolini S, Chapman NM (2002). Effect of soybean seed vigor on their performance on the field. J. Seeds, Brasilia, 24(1): 33-41.

Verslues PE, Agarwal M, Katiyar-Agarwal S, Zhu J, Zhu JK (2006). Methods and concepts in quantifying resistance to drought, salt and freezing, abiotic stresses that affect plant water status. Plant J., 45: 523-539.

Zhu JK, Xiong L (2002). Molecular and genetic aspects of plant responses to osmotic stress. Plant Cell Environ., 25: 131-139.

Permissions

All chapters in this book were first published in JHF, by Academic Journals; hereby published with permission under the Creative Commons Attribution License or equivalent. Every chapter published in this book has been scrutinized by our experts. Their significance has been extensively debated. The topics covered herein carry significant findings which will fuel the growth of the discipline. They may even be implemented as practical applications or may be referred to as a beginning point for another development.

The contributors of this book come from diverse backgrounds, making this book a truly international effort. This book will bring forth new frontiers with its revolutionizing research information and detailed analysis of the nascent developments around the world.

We would like to thank all the contributing authors for lending their expertise to make the book truly unique. They have played a crucial role in the development of this book. Without their invaluable contributions this book wouldn't have been possible. They have made vital efforts to compile up to date information on the varied aspects of this subject to make this book a valuable addition to the collection of many professionals and students.

This book was conceptualized with the vision of imparting up-to-date information and advanced data in this field. To ensure the same, a matchless editorial board was set up. Every individual on the board went through rigorous rounds of assessment to prove their worth. After which they invested a large part of their time researching and compiling the most relevant data for our readers.

The editorial board has been involved in producing this book since its inception. They have spent rigorous hours researching and exploring the diverse topics which have resulted in the successful publishing of this book. They have passed on their knowledge of decades through this book. To expedite this challenging task, the publisher supported the team at every step. A small team of assistant editors was also appointed to further simplify the editing procedure and attain best results for the readers.

Apart from the editorial board, the designing team has also invested a significant amount of their time in understanding the subject and creating the most relevant covers. They scrutinized every image to scout for the most suitable representation of the subject and create an appropriate cover for the book.

The publishing team has been an ardent support to the editorial, designing and production team. Their endless efforts to recruit the best for this project, has resulted in the accomplishment of this book. They are a veteran in the field of academics and their pool of knowledge is as vast as their experience in printing. Their expertise and guidance has proved useful at every step. Their uncompromising quality standards have made this book an exceptional effort. Their encouragement from time to time has been an inspiration for everyone.

The publisher and the editorial board hope that this book will prove to be a valuable piece of knowledge for researchers, students, practitioners and scholars across the globe.

List of Contributors

Jiangjie Lu
Key Laboratory for Systems Biology of Medicinal and Ornamental Plants, Hangzhou Normal University, Hangzhou 310018, China
Institute of Soybean, Nanjing Agricultural University, Nanjing 210095, China

Xu Hu
Key Laboratory for Systems Biology of Medicinal and Ornamental Plants, Hangzhou Normal University, Hangzhou 310018, China

Junjun Liu
Pacific Forestry Centre, Canadian Forest Service, Natural Resources Canada, Victoria V8Z 1M5, Canada

Huizhong Wang
Key Laboratory for Systems Biology of Medicinal and Ornamental Plants, Hangzhou Normal University, Hangzhou 310018, China

S. Khelifa
Sylvopastoral Institute of Tabarka, B.P 345 - 8110 Tabarka, Tunisia

M. M'Hamdi
Department of Agronomic and Economic Sciences, Higher School of Agriculture of Kef, 7119, Kef Tunisia

H. Rejeb
Department of Biological Sciences and Plant Protection, Higher Institute of Agronomy of Chott Mariem, 4042, Sousse, Tunisia

L. Belbahri
Agronomy Department, HEPIA, School of Engineering of Lullier, 150 Route de Presinge, 1254 Jussy, Switzerland

N. Souayeh
University of Gabes, Cité Erriadh 6072, Zrig, Gabes, Tunisia

Cliff S. Dlamini
Department of Forest and Wood Science, Faculty of Agrisciences, University of Stellenbosch, South Africa

V. C. Adaigbe
Entomology Division, NIFOR, P.M.B 1030, Benin City, Edo State, Nigeria

J. A. Odebiyi
Crop Protection and Environmental Biology, University of Ibadan, Ibadan, Nigeria

A. A. Omoloye
Crop Protection and Environmental Biology, University of Ibadan, Ibadan, Nigeria

C. I. Aisagbonhi
Entomology Division, NIFOR, P.M.B 1030, Benin City, Edo State, Nigeria

O. Iyare
Entomology Division, NIFOR, P.M.B 1030, Benin City, Edo State, Nigeria

Nuchhungi Khawlhring
Department of Horticulture, Aromatic and Medicinal Plants, Mizoram University, Tanhril-796009, Aizawl, Mizoram, India

James L. T. Thanga
Department of Economics, Mizoram University, Tanhril-796009, Aizawl, Mizoram, India

F. Lalnunmawia
Department of Forestry, Mizoram University, Tanhril-796009, Aizawl, Mizoram, India

Simon A. Mng'omba
World Agroforestry Centre (ICRAF), P. O. Box 30798, Lilongwe 3, Malawi

Gudeta W. Sileshi
World Agroforestry Centre (ICRAF), P. O. Box 30798, Lilongwe 3, Malawi

Ramni Jamnadass
World Agroforestry Centre (ICRAF), P. O. Box 30677-00100, Nairobi, Kenya

Festus K. Akinnifesi
World Agroforestry Centre (ICRAF), P. O. Box 30798, Lilongwe 3, Malawi

Jarret Mhango
Mzuzu University, Private Bag 201, Luwinga, Mzuzu 2, Malawi

Mekou Y. Bele
Center for International Forestry Research (CIFOR), Central Africa Regional Office P.O. Box 2008, Yaounde-Cameroon

Derek A. Focho
Faculty of Science, University of Dschang, P. O. Box 222 Dschang, Cameroon
Faculty of Science, University of Buea, P. O. Box 63 Buea, SW Cameroon

Enow A. Egbe
Faculty of Science, University of Buea, P. O. Box 63 Buea, SW Cameroon

Bindeh G. Chuyong
Faculty of Science, University of Buea, P. O. Box 63 Buea, SW Cameroon

P. B. Meshram
Forest Entomology Division, Tropical Forest Research Institute (ICFRE), P. O. RFRC, Mandla Road, Jabalpur-482021, Madhya Pradesh, India

K. K. Soni
Forest Pathology Division, Tropical Forest Research Institute (ICFRE), P. O. RFRC, Mandla Road, Jabalpur-482021, Madhya Pradesh, India

Adugna D. Bote
Jimma University College of Agriculture and Veterinary Medicine, P. O. Box 307, Jimma, Ethiopia
Centre for Crop Systems Analysis, Wageningen University and Research Centre, Droevendaalsesteeg 1, 6708 PB Wageningen, The Netherlands

Paul C. Struik
Centre for Crop Systems Analysis, Wageningen University and Research Centre, Droevendaalsesteeg 1, 6708 PB Wageningen, The Netherlands

V. Sivakumar
Agricultural research Station, Tamil Nadu Agricultural University, Bhavanisagar - 638451, Erode District, Tamil Nadu, India

V. Ponnusami
Horticultural College and Research Institute, Tamil Nadu Agricultural University, Periyakulam - 625 604, Theni District, Tamil Nadu, India

Sayed M. Zain Hasan
Faculty of Agrotechnology and Food Science, University Malaysia Terengganu (UMT), 21030 Kuala Terengganu, Malaysia

Isam Al-Madhagi
Faculty of Agrotechnology and Food Science, University Malaysia Terengganu (UMT), 21030 Kuala Terengganu, Malaysia

Aziz Ahmad
Faculty of Science and Technology, (UMT) University Malaysia Terengganu (UMT), 21030 Kuala Terengganu, Malaysia

Wan Abdullah Bin Yusoff
Malaysian Agricultural Research and Development Institute (MARDI), Cameron Highlands, Pahang, Malaysia

Radji Raoufou
Laboratory of Plant Biology and Ecology, BP 1515 Lomé - Togo

Kokou Kouami
Laboratory of Plant Biology and Ecology, BP 1515 Lomé - Togo

Akpagana Koffi
Laboratory of Plant Biology and Ecology, BP 1515 Lomé - Togo

Manjit Singh
Department of Horticulture, Chauras Campus, HNB Garhwal University, Srinagar (Garhwal) 246174, Uttarakhand, India

D. K. Rana
Department of Horticulture, Chauras Campus, HNB Garhwal University, Srinagar (Garhwal) 246174, Uttarakhand, India

J. M. S. Rawat
Department of Horticulture, Chauras Campus, HNB Garhwal University, Srinagar (Garhwal) 246174, Uttarakhand, India

S. S. Rawat
Department of Horticulture, Chauras Campus, HNB Garhwal University, Srinagar (Garhwal) 246174, Uttarakhand, India

Iulian Costache
University of Craiova, Faculty of Horticulture, Department of Biology, 15 Libertă_ii Street, 200583 Craiova, Romania

Bhupendra Singh
Department of Forestry, Post Box-59, H.N.B. Garhwal University, Srinagar Garhwal, 246 174, Uttarakhand, India

A. K. Uniyal
Department of Forestry, Post Box-59, H.N.B. Garhwal University, Srinagar Garhwal, 246 174, Uttarakhand, India

J. S. M. Rawat
Department of Horticulture, Post Box-59, H.N.B. Garhwal University, Srinagar Garhwal, 246 174, Uttarakhand, India

D. K. Rana
Department of Horticulture, Post Box-59, H.N.B. Garhwal University, Srinagar Garhwal, 246 174, Uttarakhand, India

Yuichi Matsumoto
United Graduate School of Agricultural Science, Tokyo University of Agriculture and Technology, Fuchu, Tokyo 183-8509, Japan

Makoto Miyagi
Plant Biotechnology Institute, Ibaraki Agricultural Center, Kasama, Ibaraki 319-0292, Japan

O. A. Falusi
Department of Biological Sciences, Federal University of Technology, Minna, Nigeria

M. C. Dangana
Department of Biological Sciences, Federal University of Technology, Minna, Nigeria

O. Y. Daudu
Department of Biological Sciences, Federal University of Technology, Minna, Nigeria

Jaime A. Teixeira da Silva
Faculty of Agriculture and Graduate School of Agriculture, Kagawa University, Miki-cho, Ikenobe, Kagawa-ken, Japan

Sharmila Piya
Nepal Agricultural Research Council, Agriculture Research Station, Pakhribas, P. O. Box 29, Sunsari, Nepal

Ajaya Shree Ratna Bajracharya
Nepal Agricultural Research Council, Agriculture Research Station, Pakhribas, P. O. Box 29, Sunsari, Nepal

Jawahar Lal Mandal
Nepal Agricultural Research Council, Agriculture Research Station, Pakhribas, P. O. Box 29, Sunsari, Nepal

Banhu Prasad Choudhary
Nepal Agricultural Research Council, Agriculture Research Station, Pakhribas, P. O. Box 29, Sunsari, Nepal

James Munga Kimondo
Kenya Forestry Research Institute, P. O. Box 20412-00200, Nairobi, Kenya

Jacob Godfrey Agea
Department of Community Forestry and Extension, Makerere University, P. O. Box 7062, Kampala Uganda

Clement Akais Okia
Department of Community Forestry and Extension, Makerere University, P. O. Box 7062, Kampala Uganda

Refaat Atalla Ahmed Abohassan
Faculty of Meteorology, Environment and Arid Land Agriculture, King Abdulaziz University, P. O. Box 80208 Jeddah, 21589 Saudi Arabia

Elizabeth T. N. Nghitoolwa Ndeunyema
Faculty of Agriculture and Natural Resources, The University of Namibia - Ogongo Campus, Private Bag 5520, Oshakati, Namibia

Dino Andrew Woiso
Department of Biological Science, Sokoine University of Agriculture, P. O. Box 3038, Morogoro, Tanzania

Zewge Teklehaimanot
University of Bangor, Bangor Gwynedd, LL57 2UW, UK

Jackson Mulatya
Kenya Forestry Research Institute, P. O. Box 20412-00200, Nairobi, Kenya

Muhammad Shahnawaz
Department of Food Technology, Karakoram International University, Gilgit Pakistan

Saghir Ahmed Sheikh
Institute of Food Science and Technology, Sindh Agriculture University Tandojam, Pakistan

Rasool khatibi
Department of Rangeland and Watershed Management, Faculty of Natural Resources, University of Zabol, 98615, Zabol, Iran

Jahanbakhsh Teymorri
Department of Rangeland and Watershed Management, Faculty of Natural Resources, University of Zabol, 98615, Zabol, Iran

A. O. ADETAYO
Institute of Agricultural Research and Training, Obafemi Awolowo University, Ibadan, Oyo, Nigeria

V. O. ADURAMIGBA-MODUPE
Institute of Agricultural Research and Training, Obafemi Awolowo University, Ibadan, Oyo, Nigeria

H. Ibrahim
Department of Crop Production, School of Agriculture and Agricultural Technology, Federal University of Technology, Minna, Niger State, Nigeria

J. A. Oladiran
Department of Crop Production, School of Agriculture and Agricultural Technology, Federal University of Technology, Minna, Niger State, Nigeria

R. A. Adeyemi
Department of Crop Production, School of Agriculture and Agricultural Technology, Federal University of Technology, Minna, Niger State, Nigeria

M. M. Danladi
Department of Crop Production, School of Agriculture and Agricultural Technology, Federal University of Technology, Minna, Niger State, Nigeria

Rasool Khatibi
Department of Natural Resource, University of Zabol, Iran

Mohammadreza DahmardehGhaleno
Department of Raneland and Watershed Managemennt, Facultyof Natural Resources Uuniversity of Zabol, 98615, Zabol

Akbar Fakhireh
Department of Raneland and Watershed Managemennt, Facultyof Natural Resources Uuniversity of Zabol, 98615, Zabol, Iran

G. M. Anjah
Department of Plant Biology, University of Dschang, P. O. Box 67, Dschang, Cameroon

V. F. Nguetsop
Department of Plant Biology, University of Dschang, P. O. Box 67, Dschang, Cameroon

F. M. Tsombou
Department of Plant Biology, University of Dschang, P. O. Box 67, Dschang, Cameroon

M. T. M. Njoya
Department of Biological Sciences, the University of Bamenda, P. O. Box 39- Cameroon

C.S. Bao
Hangzhou Botanical Garden, Hangzhou 310013, P. R. China

P.C. Zhang
Hangzhou Botanical Garden, Hangzhou 310013, P. R. China

C. Chen
Hangzhou Botanical Garden, Hangzhou 310013, P. R. China

H. Zhou
Hangzhou Botanical Garden, Hangzhou 310013, P. R. China

Sisay Feleke
Ethiopian Institute of Agricultural Research, Forest Products Utilization Research Coordination, P. O. Box 2322, Addis Ababa, Ethiopia

R. Chitra
Horticultural College and Research Institute, Tamil Nadu Agricultural University, Coimbatore-641 003, Coimbatore, India

P. Muthulakshmi
Horticultural College and Research Institute, Tamil Nadu Agricultural University, Coimbatore-641 003, Coimbatore, India

K. Rajamani
Horticultural College and Research Institute, Tamil Nadu Agricultural University, Coimbatore-641 003, Coimbatore, India

K. S. Navaneethakrishnan
Department of Fruit Science, Punjab Agricultural University, Ludhiana-141004, Punjab, India

M. I. S. Gill
Department of Fruit Science, Punjab Agricultural University, Ludhiana-141004, Punjab, India

S. Ramesh Kumar
Department of Horticulture, Vanavarayar Institute of Agriculture, Manakkadavu, Pollachi-642 103, Tamil Nadu Agricultural University, Tamil Nadu, India

D. V. Pathak
Chaudhary Charan Singh Harayana Agricultura University (CCSHAU) Regional Research Station, Bawal (Haryana), India

Surender Singh
Chaudhary Charan Singh Harayana Agricultura University (CCSHAU) Regional Research Station, Bawal (Haryana), India

R. S. Saini
Chaudhary Charan Singh Harayana Agricultura University (CCSHAU) Regional Research Station, Bawal (Haryana), India

A. A. H. El-Shaieny
Faculty of Agriculture, South Valley University, Qena, 83523 Egypt

Y. Y. Abdel-Ati
Faculty of Agriculture, Minia University, Egypt

A. M. El-Damarany
Faculty of Agriculture, Sohag University, Egypt

A. M. Rashwan
Faculty of Agriculture, South Valley University, Qena, 83523 Egypt

F. BANI-AAMEUR
Laboratory of Biotechnologies and Valorization of Naturals Ressources, Faculty of Sciences, Ibn Zohr University, BP 8106 Agadir 80000 Morocco

A. EL MOUSADIK
Laboratory of Biotechnologies and Valorization of Naturals Ressources, Faculty of Sciences, Ibn Zohr University, BP 8106 Agadir 80000 Morocco

A. ZAHIDI
Laboratory of Biotechnologies and Valorization of Naturals Ressources, Faculty of Sciences, Ibn Zohr University, BP 8106 Agadir 80000 Morocco

F. BANI-AAMEUR
Laboratory of Biotechnologies and Valorization of Naturals Ressources, Faculty of Sciences, Ibn Zohr University, BP 8106 Agadir 80000 Morocco

A. EL MOUSADIK
Laboratory of Biotechnologies and Valorization of Naturals Ressources, Faculty of Sciences, Ibn Zohr University, BP 8106 Agadir 80000 Morocco

F. BANI-AAMEUR
Laboratory of Biotechnologies and Valorization of Naturals Ressources, Faculty of Sciences, Ibn Zohr University, BP 8106 Agadir 80000 Morocco

A. EL MOUSADIK
Laboratory of Biotechnologies and Valorization of Naturals Ressources, Faculty of Sciences, Ibn Zohr University, BP 8106 Agadir 80000 Morocco

Amira Sh. Soliman
Natural Resources Department, Institute of African Research and Studies, Cairo University, Giza, Egypt

Souad A. El-feky
Department of Laser Applications in Metrology, Photochemistry and Agriculture, National Institute of Laser Enhanced Science (NILES), Cairo University, Egypt

Essam Darwish
Plant Physiology Section, Faculty of Agriculture, Cairo University, Egypt

R. S. Tomar
Department of Biotechnology, Junagadh Agricultural University, Junagadh - 362 001, Gujarat, India

H. P. Gajera
Department of Biotechnology, Junagadh Agricultural University, Junagadh - 362 001, Gujarat, India

R. R. Viradiya
Department of Horticulture, Junagadh Agricultural University, Junagadh - 362 001, Gujarat, India

S. V. Patel
Department of Biotechnology, Junagadh Agricultural University, Junagadh - 362 001, Gujarat, India

B. A. Golakiya
Department of Biotechnology, Junagadh Agricultural University, Junagadh - 362 001, Gujarat, India

Ankur Agarwal
Defence Institute of Bio-Energy Research, Defence R&D Organization, Goraparao, Haldwani, Uttarakhand, 263139 India

R. S. Tiwari
GBPUA & T, Pantnagar, Uttarakhand, 263145 India

Lizandro Ciciliano Tavares
Department of Plant Science, Federal University of Pelotas, UFPel/FAEM, Brazil

Cassyo de Araújo Rufino
Department of Plant Science, Federal University of Pelotas, UFPel/FAEM, Brazil

Lilian Madruga de Tunes
Department of Plant Science, Federal University of Santa Maria, UFSM, Brazil

Antonio Carlos Souza Albuquerque Barros
Department of Plant Science, Federal University of Pelotas, UFPel/FAEM, Brazil